# THE SOLAR SYSTEM
## Observations
## and Interpretations

W. W. Rubey (1898–1974)

MARGARET G. KIVELSON

*Editor*

# THE SOLAR SYSTEM
## Observations and Interpretations

Rubey Volume IV

Prentice-Hall, Englewood Cliffs, New Jersey 07632

*Library of Congress Cataloging-in-Publication Data*
Main entry under title:

The Solar system.

   (Rubey volume; 4)
   Includes bibliographies and index.
   1. Solar system—Congresses.  I. Kivelson,
M. G. (Margaret Galland), (date).  II. Series.
QB500.5.S65  1986     523.2     85-12446
ISBN 0-13-821927-3

Editorial/production supervision:
  *Karen J. Clemments and Joan McCulley*
Manufacturing buyer: *John B. Hall*

© 1986 by Prentice-Hall, Inc.
A Division of Simon & Schuster
Englewood Cliffs, New Jersey 07632

All rights reserved. No part of this book may be
reproduced, in any form or by any means,
without permission in writing from the publisher.

Printed in the United States of America

10  9  8  7  6  5  4  3  2

ISBN 0-13-821927-3 01

Prentice-Hall International (UK) Limited, *London*
Prentice-Hall of Australia Pty. Limited, *Sydney*
Prentice-Hall of Canada Inc., *Toronto*
Prentice-Hall Hispanoamericana, S.A., *Mexico*
Prentice-Hall of India Private Limited, *New Delhi*
Prentice-Hall of Japan, Inc., *Tokyo*
Prentice-Hall of Southeast Asia Pte. Ltd., *Singapore*
Editora Prentice-Hall do Brasil, Ltda., *Rio de Janeiro*

# CONTENTS

PREFACE   *ix*
   Margaret G. Kivelson

1   INTRODUCTION: THE SUN, THE SOLAR NEBULA, AND THE PLANETARY SYSTEM   *1*
   John T. Wasson and Margaret G. Kivelson

2   THE SUN, INSIDE AND OUT   *15*
   H. Zirin

3   THE NEBULAR ORIGIN OF THE SOLAR SYSTEM   *28*
   D. N. C. Lin

4 THE INTERIORS OF THE TERRESTRIAL PLANETS:
   THEIR STRUCTURE AND EVOLUTION   78
      William M. Kaula

5 ISOTOPIC CONSTRAINTS ON PLANETARY
   EVOLUTION   94
      Donald J. DePaolo

6 ATMOSPHERES OF THE TERRESTRIAL PLANETS   116
      Margaret G. Kivelson and Gerald Schubert

7 PLANETARY VOLATILE HISTORY: PRINCIPLES
   AND PRACTICE   135
      Fraser P. Fanale

8 TECTONICS OF THE TERRESTRIAL PLANETS   176
      Peter Bird

9 PLANETARY SURFACE WEATHERING   207
      James L. Gooding

10 JUPITER AND SATURN   230
      Andrew P. Ingersoll

11 ORIGIN, EVOLUTION, AND STRUCTURE
   OF THE GIANT PLANETS   254
      David J. Stevenson

12 CONSEQUENCES OF TIDAL EVOLUTION   275
      S. J. Peale

13 THE GENERATION OF MAGNETIC FIELDS
   IN PLANETS   289
      Eugene H. Levy

14 SOLAR AND PLANETARY MAGNETIC FIELDS   311
      C. T. Russell

15 COMETS   359
      Marcia Neugebauer

16  FROM BIG BANG TO BING BANG
    (FROM THE ORIGIN OF THE UNIVERSE
    TO THE ORIGIN OF THE SOLAR SYSTEM)   *374*
      Typhoon Lee

17  FUTURE PLANETARY EXPLORATION   *396*
      Margaret G. Kivelson

    APPENDIX TABLES   *411*

    INDEX   *415*

# PREFACE

The Department of Earth and Space Sciences of the University of California, Los Angeles, conducts an annual lecture series entitled "The Rubey Colloquium." Specialists from research institutions around the country are invited to present lectures on a central theme of broad interest to earth and/or space scientists. Written summaries or elaborations of these presentations have been gathered together as a series of books known as the Rubey Volumes.

Both colloquia and book series are named in honor of the late W. W. Rubey (1898–1974), career geologist with the U.S. Geological Survey and Professor of Geology and Geophysics at UCLA. A brief sketch of Rubey's geologic accomplishments is contained in the preface to Rubey Volume I, *The Geotectonic Development of California* (W. G. Ernst, ed., 1981).

Rubey Volume IV, *The Solar System: Observations and Interpretations*, is based on the material of a Rubey Colloquium presented as an upper-division undergraduate course in January–March 1982. The students, mostly science majors, were not required to demonstrate preparation in each of the subjects (physics, chemistry, geology, astronomy) pertinent to the study of the solar system. Speakers sought to communicate the important conceptual framework of the field and to describe the principal recent developments to their audience without assuming a common background. Their diverse approaches to the challenge, evident in the papers collected in this volume, reflect not only the individual styles but also the characteristics of presentation intrinsic to the specific subjects discussed. The mixture of some unabashedly pedagogical material and some "state-of-the-science" material at the level of research presentations appealed to the students and is consequently retained in this volume.

The scientists who participated in the Rubey Colloquium are among the most distinguished in their field. Despite many pressing demands on their time, most of them made their presentations available to a wider audience through their contributions to this volume. Collectively, the papers provide a good view of solar-system studies, despite a few gaps where desired contributions were not available. The volume illuminates grand questions of origin and destiny and puts forward the evidence on which current consensus is based. Perhaps more significantly, the lack of complete agreement on many matters is acknowledged, thus providing the reader some insight into the disputes that precede satisfactory understanding.

As editor of this volume, I should like to thank Gary Ernst for encouraging me to undertake a colloquium on the solar system; Don Browne, William I. Newman, and Lee Bargatze for helping me plan and organize it; Lisa Clayton for her efforts in editing the manuscript; and Julie Knaack for invaluable assistance in every aspect of the endeavor.

*Los Angeles*                                                                   Margaret Galland Kivelson

# THE SOLAR SYSTEM
## Observations
## and Interpretations

John T. Wasson and Margaret G. Kivelson
Department of Earth and Space Sciences, and
Institute of Geophysics and Planetary Physics
University of California
Los Angeles, California 90024

# 1

# INTRODUCTION: THE SUN, THE SOLAR NEBULA, AND THE PLANETARY SYSTEM

## ABSTRACT

Here is an overview of the solar system, an introduction designed to help place the subsequent chapters in context. Think of it as a view through a zoom lens with a rapidly changing field of view. Subsequent chapters dwell on elements of the system, give descriptions of formation and evolution of the solar system and its components, and enlarge upon some of the ideas first mentioned here.

## 1-1 INTRODUCTION

The Sun is a common type of star in the outer part of a common type of star assembly called a *galaxy*. That view of our place in the universe would doubtless have surprised the ancients, who believed that the entire universe revolved around the Earth.

The Sun, then, is part of the Milky Way Galaxy, a collection of gas, dust, and approximately $10^{11}$ stars having a mass roughly estimated to be $2 \times 10^{11}$ $M_\odot$ ($M_\odot$ is a *solar* mass, the mass of the Sun). The Sun and the planets were formed by the gravitational collapse of a fragment of an interstellar cloud about 4.5 Gy ago (see De Paolo's discussion of some dating techniques in Chapter 5 and also Lee's discussion in Chapter 16). In the unit Gy, $G$ stands for giga ($10^9$), and $y$ stands for years. The age of the galaxy is estimated at about 13 Gy. The age of the solar system is thus about one-third that of the galaxy.

The Sun is not only at the center of the solar system, but it also contains virtually all (99.87%) of the mass, as is evident from Table A-1 at the end of the book. The planets and minor bodies such as asteroids and comets contain most of the remaining mass; the mass of interplanetary gases and dust is negligible.

The planets and all other solar-system objects follow elliptical orbits around the Sun. As shown in Fig. 1-1, the Sun occupies one focus of the ellipse; the different types of elliptical orbits are characterized by parameters that specify the dimensions and shape of the ellipse (see text in box on page 4) and its orientation in space. For objects in the solar system, orientation in space is described in terms of the inclination of the orbital plane of the object relative to the plane of Earth's orbit. The inclination is thus the angle between the vectors perpendicular to the respective orbital planes and pointing northward, as illustrated in Fig. 1-2.

In Table A-1 of the Appendix, the planets are listed in order of their distance from the Sun. The four inner planets (Mercury, Venus, Earth, and Mars) consist mainly of metal and silicates, and their densities are high. Compared with the Sun, these planets have low abundances of the most volatile elements: hydrogen (H), carbon (C), nitrogen (N), oxygen (O), and the rate gases helium (He), neon (Ne), argon (Ar), krypton (Kr), and xenon (Xe).

The outer planets, Jupiter, Saturn, Uranus, and Neptune, are less dense than the inner planets. Jupiter and Saturn have compositions that are nearly the same as the Sun's, although there is evidence of small enhancements of less volatile substances relative to solar abundances. Uranus and Neptune are strongly depleted in H and the rare gases compared with solar abundances. They consist mainly of compounds containing H and C, N, and O.

Between the inner and the outer planets lies the asteroid belt, a region occupied by large numbers of small silicate-metal objects called *asteroids*, as well as still smaller

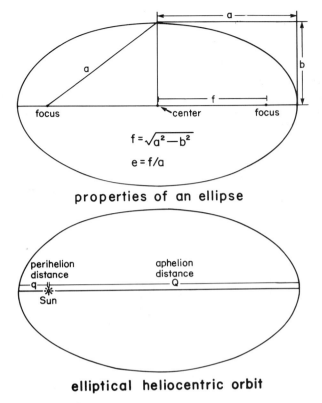

Fig. 1-1 (Top) The properties of an ellipse showing the definition of the *semimajor axis*, *a*, *the semiminor axis*, *b*, and the location of the foci. (Bottom) The definition of the *perihelion*, *q* (the distance of closest approach to the Sun), and the *aphelion*, *Q*, (the distance of greatest recession from the Sun). The Sun occupies one focus (shown here on the left) of elliptical heliocentric orbits.

asteroid fragments. The largest of these asteroids, Ceres, has a radius of 500 km, and is small not only in comparison with the planets but even in comparison with the Moon, whose radius is more than 1700 km. It was once believed that the asteroids were fragments of a shattered planet, but the present consensus is that the largest asteroids have remained unchanged in size since they formed.

Pluto, the ninth and most distant "planet," probably consists mainly of compounds of C, N, and O, like Uranus and Neptune. The recent discovery of a satellite of Pluto led to a determination of Pluto's mass. (The mass is inferred from the period of rotation of the satellite and its distance from the planet using Newton's laws of motion in a gravitational field.) Pluto's mass is only one-fifth that of the Moon. As a result, there are some, including the authors, who doubt that Pluto deserves the status of a planet. It seems more sensible to categorize it as a stray asteroidal-cometary object. Chiron, another such object having a radius of approximately 500 km, was recently discovered in an orbit between those of Saturn and Uranus. It seems probable that many objects having radii

## Elliptical Orbits

Ellipses are described mathematically in terms of parameters illustrated in Fig. 1-1. The long axis is called the *major axis*. Its length is $2a$, so $a$ is the length of the semimajor axis; $b$ is the length of the semiminor axis. An ellipse has two *foci*. The combined distance from the foci to any point on the ellipse is equal to $2a$. The point on the ellipse closest to a focus is called the *pericenter*, and the point farthest from the focus is called the *apocenter*. If the Sun (Earth, Moon, and so forth) is at the focus, these names are replaced by *perihelion* and *aphelion*, respectively (*perigee* and *apogee* for the Earth, *perilune* and *apolune* for the Moon, and so forth). The degree of departure of the shape of an ellipse from circular is described by the parameter $e$, called *eccentricity*, with

$$e = \left(\frac{1-b^2}{a^2}\right)^{1/2} = \left(\frac{1-q}{a}\right)$$

where $q$ is the distance between the focus and the pericenter. A circle has eccentricity equal to 0 and highly elongated ellipses have eccentricities approaching 1. For orbital calculations, it is useful to express the distance between the focus (i.e., the position of the central body) and an object on the ellipse in terms of $\theta$, the angle pericenter-focus-object, as

$$r = \frac{a(1-e^2)}{(1+e\cos\theta)}$$

$\geqslant 100$ km are present in the outer solar system and will be found in the future with improved observational techniques, for example, infrared telescopes mounted on artificial satellites.

From Table A-1, it is clear that the semimajor axes of the planets increase in a regular fashion that can be represented by a simple equation known as the *Titius-Bode rule* or, more commonly, as *Bode's law*:

$$a_n = 0.4 + (0.3 \times 2^n)$$

where $a_n$ is the semimajor axis in astronomical units (AU) and $n$ is an integer assigned sequentially to each planet starting with Venus = 0 and continuing with Earth = 1, Mars = 2, Ceres = 3, Jupiter = 4, and so forth. Mercury fits the sequence if it is assigned $n = -\infty$, an anomaly that casts doubt on the scientific significance of the equation. The unit AU is a convenient unit of distance equal to the radius of the Earth's orbit.

More reasonable from a scientific viewpoint is a "modified Bode's law":

$$a_n = 0.36 \times 1.73^n$$

where $n$ is 0 for Mercury, 1 for Venus, 2 for Earth, and so forth. This equation can be understood by a simple model in which, during its formation, each planet swept out all

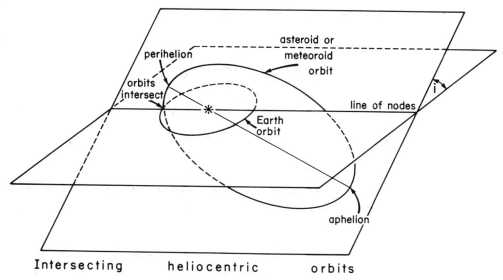

Fig. 1-2  Relative orientation of two helocentric orbits. The angle between the planes of the two orbits is called the *inclination*, $i$, the line formed by the intersection of the two planes is called the *line of nodes*. The case shown is that of the Earth. Such an intersection can occur only if the orbit passes 1 AU from the Sun in the direction of the lines of nodes.

the mass in a region extending from approximately 0.8 to 1.3 times its orbital radius. As shown in Table 1-1, both equations give quite good approximations to the observed semimajor axes. The original Bode's law has one distinct advantage: It is much easier to remember than the modified law and can be a useful aid in recalling the values of the planetary semimajor axes.

Small bodies of the solar system include comets (see Chapter 15 by Neugebauer) in addition to asteroids mentioned above. Comets are icy objects that develop comas, transient atmospheres of gas and dust, when heated by sunlight in the inner solar system.

TABLE 1-1  A comparison of the orbital semimajor axes of the planets (in astronomical units) with those predicted by Bode's law and by a power law called the *modified Bode's law*.

| Planet | Bode's | Actual | Mod. Bode's |
|---|---|---|---|
| Mercury | 0.40 | 0.387 | 0.360 |
| Venus | 0.70 | 0.723 | 0.623 |
| Earth | 1.00 | 1.000 | 1.077 |
| Mars | 1.60 | 1.52 | 1.86 |
| Asteroids | 2.80 | ~2.8 | 3.22 |
| Jupiter | 5.20 | 5.20 | 5.58 |
| Saturn | 10.0 | 9.58 | 9.65 |
| Uranus | 19.6 | 19.1 | 16.7 |
| Neptune | 38.8 | 30.2 | 28.9 |

When viewed through a telescope, comets appear fuzzy because of these extended atmospheres. Earth-based viewers may occasionally observe a grand show if the Earth and a comet are on the same side of the Sun when the coma forms. The solid nuclei of most comets are small, the typical radius being approximately 1 km and that of the largest known comet about 50 km. There is little quantitative evidence regarding the composition or structure of comets (a prime reason to send a spacecraft to rendezvous with one), but there is general agreement that they are composed of roughly equal parts of solid metal oxides (mainly silicates) and ice (mainly $H_2O$).

The orbits of comets that pass through the inner solar system can be strongly perturbed by close encounters with the planets, particularly Jupiter. As a result, about half are ejected from the solar system within about $10^3$ to $10^5$ passages inside 5 AU. The lifetime of periodic comets have orbital periods of less than 300 y is thus much less than the age of the solar system, and the fraction remaining continuously in such orbits since the formation of the solar system is less than $10^{-100}$, a vanishingly small number. This number is far too small to allow any historically observed periodic comets to be part of the original population. As a result, a source is required from which new comets are continuously released into the inner solar system. Neugebauer discusses the source of comets in the outer solar system in Chapter 15.

## 1-2 THE ORIGIN OF THE PLANETS

Other chapters of this volume describe the formation and evolution of the solar system (see Chapters 3 and 16 by Lin and Lee, respectively) and the planets. Our purpose in this chapter is to introduce some of the important ideas in order to provide a context for the more complete discussions. The formation of the solar system is linked to the formation of its central star, the Sun, so some features of star formation need to be considered. Then some properties of the matter bound to the Sun but not incorporated within it are summarized. In particular, we discuss the importance of certain meteorites in the study of the solar system.

### 1-2-1 Stellar Processes: The Hertzsprung-Russell Diagram

Many aspects of stellar evolution can be illustrated in a plot of stellar luminosity (the total output of light) versus the surface temperature of the visible stars. Such a diagram is called a *Hertzsprung-Russell (HR) diagram* and is shown in Fig. 1-3, in which both axes are logarithmic and, by tradition, temperature decreases to the right. Stars are assigned to classes, designated with letters, on the basis of their temperatures. The nominal ranges corresponding to these classes are shown at the top of the diagram, from the hottest (O) to the coldest (M). The Sun is a G star. Luminosity is given in terms of $L_\odot$, the solar luminosity. Most visible stars plot along a band crossing the HR diagram diagonally from the upper left to the lower right. Several additional fields in which visible stars are found are also indicated and labeled.

Two features of the HR diagram are worth stressing. First, the reason for the clustering of stars in distinct fields is that there are several different modes of energy production in stars, related both to their masses and their ages. Stars falling along the main sequence

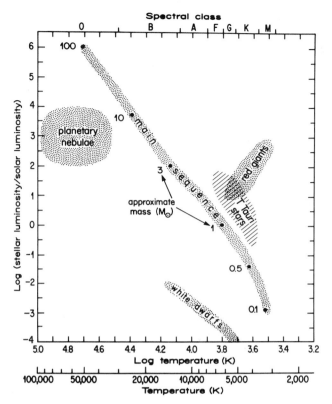

Fig. 1-3 In this Hertzsprung-Russell diagram, log stellar luminosity is plotted against log temperature. Stars that are burning hydrogen fall along the main sequence; the red giants, planetary nebulae, and white dwarfs are more evolved. The T Tauri stars are young stars that have not yet evolved into the main sequence. The spectral (temperature) classes are shown along the top of the diagram.

are burning hydrogen; those in the second most populous field, the red giants, have exhausted their hydrogen and are burning helium.

The second point to note is that the information on the HR diagram can tell us something about the ages of the different classes of stars. We must recognize that the luminosity measures the rate of energy release. The total energy available for release is proportional to the mass of a star. One can then estimate the relative lifetimes of the stars on the main sequence as follows. Relative to the lifetime of the Sun ($\tau_\odot$), the lifetime ($\tau$) of another star on the main sequence is given by

$$\frac{\tau}{\tau_\odot} = \frac{ML^{-1}}{M_\odot L_\odot^{-1}}$$

Thus an M-type star with $M \simeq 0.1 \text{ M}_\odot$ and $L \simeq 10^{-3} \text{ L}_\odot$ will have a lifetime 100 times longer than the Sun, while a bright O-type star with $M \simeq 100 \text{ M}_\odot$ and $L \simeq 10^6 \text{ L}_\odot$ will use up its fuel very rapidly and will burn up its hydrogen in a time $\tau_\odot/10^4$. It is estimated that the hydrogen-burning lifetime of the Sun will be about 10 Gy, meaning the Sun is now a middle-aged G star. It follows that the most massive ($m = 10^6$ $M_\odot$) and brightest O-type stars can last only about 1 My (million years). This implies that all the O stars visible in the sky are very young compared with the 13-Gy age of the galaxy. Thus star formation must be occurring now or must have occurred recently (on galactic time scales) in any region of the galaxy that contains O-type stars.

What happens to an O-type star when it has finished burning its hydrogen fuel? Such massive stars begin to collapse, become unstable, and explode, producing intense radiation for a short period of time. The exploding star is called a *supernova*. Material from the stellar interior is ejected, a process that introduces elements more massive than helium into the interstellar medium. This partially accounts for the great range of elements available for incorporation into new stars (and present in solar-system matter). The explosion also generates a shock wave that can compress the interstellar medium and initiate the formation of new stars.

### 1-2-2 Star Formation

Star formation is observed to occur in interstellar clouds, particularly those with densities of approximately $10^3$ H atoms $cm^{-3}$ or more. One such region in our part of the galaxy is the Orion Nebula, located near the center of the "sword" in the constellation of Orion, a familiar sight in the winter night sky seen from the Northern Hemisphere. The Orion Nebula is only approximately $10^8$ AU away, a relatively near neighbor by galactic standards. The youngest stars in Orion belong to classes O and B and are thus no older than about 1 My.

The actual process of star formation is difficult to study by telescopic techniques, in part because time scales are short ($\leqslant 1$ My) and thus only a few stellar objects occupy this evolutionary stage. As well, there is much obscuring dust in the regions where star formation is occurring. In any one region, the period of star formation probably lasts no more than 100 My. The obscuring dust is gradually dissipated, pushed out by the "pressure" of light from the new stars. Some stars escape from their placental star clusters, and the mean distances among remaining stars increase. The Pleiades, a constellation that is recognizable with the naked eye, is a young, local cluster of stars having ages approximately equal to 100 My.

It seems probable that the Sun formed under circumstances similar to those prevailing today in Orion. It was probably a member of a cluster, but, if so, it escaped so long ago that its siblings can no longer be identified. But whether or not the Sun formed within a cluster, it is widely believed that during or soon after its formation, it was surrounded by a gas and dust cloud called the *solar nebula*. The planets formed by gradual accumulation of materials from this nebula.

### 1-2-3 Planet Formation

The description of the process leading from interstellar cloud to solar system rests only partially on observation, which must be supplemented by arguments based on models. Recent models vary widely, but typical theories include some variants of the following sequence of steps leading from the precursor interstellar cloud to the planets:

1. An interstellar cloud with dimensions of the order of $10^5$ AU was sufficiently compressed by a shock wave or a gravity wave, so that it became unstable with respect to collapse.
2. Collapse occurred; as the density increased, subregions of the cloud began independent collapse. The final fragment that formed the solar system had a mass of 1.1 to 2.0 $M_\odot$.
3. As the collapse proceeded, gravitational potential energy was converted to kinetic energy, resulting in significant heating. The heat vaporized a fraction of the original solid matter present in the interstellar cloud. The fraction vaporized is uncertain, but many models suggest that it was more than 90% in the inner solar system.
4. As the interstellar cloud became depleted, the rate of collapse decreased, heat energy was radiated into space more rapidly than it was released by collapse, and the nebula began to cool.
5. As the nebula cooled, the most *refractory* of the evaporated elements began to recondense. (The most refractory elements are those that can be found in solid form at the highest temperature, in particular, calcium-rich and aluminum-rich minerals.)
6. Cooling continued, and major elements such as iron (Fe), magnesium (Mg), and silicon (Si) condensed as mineral grains.
7. Because small particles and mineral grains were in motion relative to one another, collisions were frequent. Collisions at low relative velocity are more likely to lead to adhesion than destruction, so the particles grew in size. Gravitational attraction led to settling in the nebular median plane, creating a disk-shaped particle cloud.
8. The particle density in the midplane increased to a degree that resulted in gravitational instability. The resulting collapse produced low-density, kilometer-sized objects called *planetesimals*. Some models emphasize the important consequences of nebular angular momentum in describing this stage of evolution.
9. Random differences of planetesimal orbits resulted in collisions. Collisions among planetesimals whose orbits were similar (i.e., within a narrow range of semimajor axes) merged them into small, asteroidal-sized planets.
10. The gravitational fields of the largest of these small planets perturbed the orbits of smaller objects over a larger range of semimajor axes, thus sweeping in the objects over increasingly large distances. The scale of this accumulation process increased in size and distance until most of the solid matter had been swept up and the present set of planets was produced.

The processes described above seem adequate to account for the formation of the asteroids and the inner or terrestrial planets. It seems clear that the asteroidal-sized bodies also formed in the regions now occupied by the inner planets because of the presence of such bodies in the evolutionary path to planets described in steps 9 and 10. The asteroids of today are believed to have remained separate, unincorporated into a planet, because any incipient concentrations of matter were disrupted by the perturbing influence of Jupiter.

The outer planets differ both in size and composition from the terrestrial planets, so the account of their formation must be a modified version of the description above. The decrease of temperature in the solar nebula with distance from the Sun explains some aspects of the differences between inner and outer planets. When the precursor planetesimals of these outermost planets were forming, the temperature of the outer solar system (heliocentric distances greater than 5 AU) was sufficiently low that $H_2O$ condensed to the solid phase. In contrast, temperatures in the inner solar system were never low enough to permit $H_2O$ to condense. Earth, typifying the inner planets, has a bulk $H_2O$ content of only ~0.1% by weight. The solid grains and planetesimals from which it formed contained very little water bound to silicate minerals or as ice. The outer planets, on the other hand, formed by accretion of cometlike, icy planetesimals.

Other compositional differences are important as well. Jupiter and Saturn, for example, consist largely of H and He, both of which are too volatile ever to have existed as solids in the solar nebula. Possibly these planets also grew by the accumulation of condensible solids until they had masses somewhat greater than the mass of the Earth while H and He were still present in the nebula. A subsequent stage, still not fully described, would allow them to accumulate gases and retain them through gravitational attraction until the total planetary mass exceeded the mass of the original solid planet by large factors (300 times for Jupiter).

The densities of Uranus and Neptune are considerably greater than Saturn's density, and these bodies appear to have cometlike compositions, that is, to consist largely of compounds formed from C, N, and O, elements that can condense from the solar nebula at low temperatures. The higher density of Uranus and Neptune corresponds to a reduced concentration of H and He relative to Saturn and Jupiter, and is thought to indicate that the nebular gases had disappeared before Uranus and Neptune grew large enough to gravitationally capture $H_2$ and He.

## 1-3 COMPOSITION OF THE INNER PLANETS

### 1-3-1 Meteorites as Rosetta Stones

Our discussion of planetary evolution has referred repeatedly to relative abundances of various elements. Arguments based on abundances assume that we know the average composition of the solar nebula (see Chapter 16 by Lee). In part, our information rests on studies of the solar spectrum, from which we infer relative abundances that are thought to be those of the solar nebula. However, planets of the inner solar system were formed from condensed solids, and we seek clues to the composition of those condensed solids by studying the composition of meteorites.

What are meteorites? They are rocks from space that have survived their encounter with the Earth's atmosphere and reached the surface of the Earth. Their source is the stony and metallic debris of interplanetary space, fragments of material resulting from collisions between asteroids. It is possible that a few are from comets. When in orbit around the Sun, these fragments are called *meteoroids*. They are gravitationally attracted to Earth or other large bodies if their orbits bring them close by, and they may fall toward the surface. Passing through the atmosphere, they are heated, and we see them with glowing tails as "shooting stars." Small or weak meteoroids vaporize completely or fragment to dust before they reach the surface, but larger, stronger ones survive. It is the survivors that we dub *meteorites*. In the laboratory we seek to decode the information that these rocks contain regarding the early epochs of solar-system formation.

Although it seems probable that most meteorites come from asteroids, we are not yet able to associate groups of meteorites with observed asteroids. This can probably be accomplished only by recovering samples of candidate asteroids by spacecraft missions so that compositions can be compared. We do, however, recognize some of the processes that bring asteroidal fragments into Earth's neighborhood.

Most asteroids are in the Asteroid Belt, the region 2.2 to 3.4 AU from the Sun, intermediate between the orbits of Mars and Jupiter. A few asteroids are in eccentric orbits of Mars and Jupiter. A few asteroids are in eccentric orbits that have aphelia in the Asteroid Belt but perihelia inside the orbits of Mars or Earth. There is no doubt that many meteorites are fragments of impact ejecta from Earth-crossing asteroids, but the lifetime of objects in such orbits is only approximately 10 My, far less than the 4.5-Gy age of the solar system. As a result, the present population of Earth-crossing asteroids must have been injected into such orbits within the past 100 My. A key, and incompletely understood, question is how this injection occurred. The distribution of belt-asteroid semimajor axes is not continuous; there are gaps at semimajor axes corresponding to orbital periods that are whole-number fractions of Jupiter's period. In Fig. 1-4, such gaps are especially prominent at periods corresponding to $3/5$, $1/2$, $3/7$, $2/5$, and $1/3$ of Jupiter's period. At these semimajor axes, resonance gravitational interactions with Jupiter increase the eccentricities of the asteroids; this causes some to become Mars-crossing and also increases the probability that disruptive collisions will occur. Gravitational perturbations by Mars reduce the perihelia of some of the objects below 1 AU, and in this way replenish the supply of Earth-crossing asteroids or meteoroids. Short-period comets from which ices have evaporated are also sources of Earth-crossing asteroids.

Meteorites can be grouped in classes that differ in appearance and composition. Most meteorites show signs of *differentiation* (i.e., separation of constituents following melting) that may have occurred in their parent bodies, probably asteroids. There is, however, one category of meteorites, the chondritic meteorites, that appears to have formed directly in the solar nebula; some of these have experienced minimal subsequent alteration. Chondritic meteorites, for example, share rather similar bulk compositions. To compare their compositions with those of other bodies, it is convenient to compare the number of atoms of each element per atom of silicon (Si), a ratio referred to as the *abundance*. Abundances of nonvolatile elements in chondritic meteorites (also called *chondrites*) never differ from solar abundances by more than a factor or two. Most chondrites contain *chondrules*, spheroidal silicate grains that originally formed in the nebula as solidified droplets. Spheroidal grains are not found in terrestrial igneous rocks in such large amounts.

The properties noted (chondrules, solar composition), as well as isotopic evidence based on decay of short-lived radioactive isotopes (not discussed here), constitute strong evidence that chondrites formed in the nebula and were almost certainly in the inner portion of the solar system at the time planetary formation began. It follows that the chondrites are the "building blocks" of the inner planets and that a reasonable working hypothesis is that each of the inner planets should have a "chondritic" bulk composition, that is, relative abundances of condensable elements should be in the range found in chondrites.

Testing the hypothesis described is not straightforward. The planets have experienced differentiation, so the composition of surface material is not representative of bulk composition. The only well-determined planetary property that can be used to test the hypothesis is the average density, that is, the mass of the planet divided by its volume.

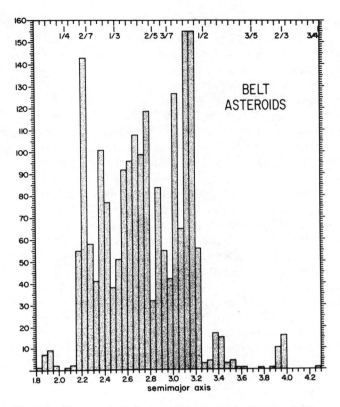

Fig. 1-4 Most asteroids have semimajor axes in the range 2.15 to 3.45 AU; this region (or a larger one extending from 1.8 to 4.3 AU) is known as the *Asteroid Belt*. Closer to the Sun, asteroids have been removed as a result of gravitational interactions with Mars; farther from the Sun, they have been removed by gravitational interactions with Jupiter. Along the top of the diagram are marks indicating the positions of semimajor axes yielding periods related to Jupiter's period as the ratio of small integers. Inside 3.9 AU, minima called *Kirkwood gaps* appear near these resonance locations, whereas near 4.0 and 4.3 AU, maxima are associated with these resonances.

Even this average density needs to be corrected in order to compare one planet with another. The correction accounts for the compression of the matter in the planetary interior. Increase of pressure is usually accompanied by an increase of temperature, which also affects the density. Consequently, in making comparisons it is important to correct the measured densities so that they all refer to the same pressure and temperature.

Corrected densities (see reduced densities listed in Table 4-1 in Chapter 4) for Venus, Earth, and Mars range from 3.7 to 4.0 g · cm$^{-3}$, values that are "chondritic" to within the relatively large uncertainties ($\pm 0.3$ cm$^{-3}$) associated with the correction procedures. The corrected density of Mercury is about 5.3 g · cm$^{-3}$, a value so high that it requires a severalfold enrichment in Fe (mostly as metallic Fe-Ni) compared with values observed in the chondrites. Does this observation mean that the hypothesis of chondritic building blocks needs to be abandoned? Not necessarily. Perhaps, for example, one needs to consider the way in which the Sun's gravitational field perturbed the mechanical agglomeration processes in Mercury's vicinity. Metal particles may have settled more efficiently to the nebular midplane in regions near the Sun, whereas silicate particles tended to remain suspended in the nebular gas. This argument could account for the higher density of Mercury, but it does not rest on direct evidence.

## 1-3-2 Melting and Fractionation

The processes of melting and fractionation have been alluded to repeatedly; these processes account for the existence of dense cores (see Chapters 4 and 5 by Kaula and DePaolo, respectively) and less dense outer layers. As we have described, chondritic meteorites with large amounts of metallic Fe-Ni and the iron sulfide (FeS) are most closely related to the inner solar system. If such chondritic materials are melted, immiscible silicate and metal sulfide melts are produced. The denser (metal sulfide) material would be expected to migrate toward the center in the gravitational field of a planet. This process is believed to account for the formation of the metal-sulfide cores of the inner planets during the later stages of the formational history of the planets.

What was the source of heat that melted the planets? The melting probably occurred in the late stages of planetary formation as a direct consequence of bombardment of the surface by the accumulating planetesimals. The models that yield this result are complex, but the source of heat is readily understood. As a body in a gravitational field falls towards the center, it accelerates (i.e., gravitational potential energy is converted into kinetic energy of motion). The minimum velocity attained at the planetary surface is the same in magnitude (but opposite in direction) as the velocity needed to allow an object to escape from the planetary surface to infinity, that is, the "escape velocity." A body of mass $m$ moving with velocity $v$ has kinetic energy $\frac{1}{2}mv^2$. If the energy of a typical silicate rock moving at 1 km · s$^{-1}$ is completely converted to heat, the rock temperature will rise high enough for it to melt. The escape velocity at Earth is 11 km · s$^{-1}$. Late in its history, then, each object accreted delivered enough kinetic energy to melt a mass approximately 120 times its own mass. (Melting as a consequence of impacts is discussed by Fanale in Section 7-2-6.) It now appears that more than 1% of the kinetic energy of impact was retained as heat inside the Earth, and that Earth was totally molten during the late portion of its formational history. Thus the conditions for differentiation of the chondritic materials were present.

The kinetic energy of impact accounts for melting of the terrestrial planets, but earlier we remarked that many meteorites appear to be the debris of asteroids in which melting and fractionation had occurred. Asteroids are small bodies with small escape velocities; they cannot be heated significantly by bombardment during accretion. What was the source of heat for them? We cannot answer with confidence. A possible heat source was from radioactive decay of a short-lived isotope of aluminum (see the discussion of $^{26}$Al in Chapter 5 by DePaolo and also in Chapter 16 by Lee) that could have heated the asteroids in the first few million years of the existence of the solar nebula. There is considerable doubt, though, that, when the asteroids formed, enough of this isotope remained to cause melting. Another theory suggests that the asteroids were melted by electrical currents induced by the dynamo action of the solar-wind magnetic field (see Chapters 13 and 14 by Levy and Russell, respectively), but details are not worked out.

One might also speculate on how the Moon, whose corrected density is much lower than that of Earth, fits into our picture. One plausible model suggests that the Moon formed in Earth's orbit from low-density asteroid fragments. The argument is based on the idea that collisions among asteroids in heliocentric orbits (i.e., orbits around the Sun) fragmented the low-density outer layers, scattering them widely and allowing them to be captured into Earth's orbit where they agglomerated to form the Moon. Because of their high strengths, the cores of the asteroids did not fragment but remained in their original orbits until they made a direct hit on the Earth or Venus or were removed from the inner solar system by drastic orbital perturbation. Thus, according to this model, the low-metal density of the Moon reflects the fact that it mainly formed from the mantles and crusts of differentiated asteroids.

In another model, the Moon is proposed to have resulted from the splashing off of some of the Earth's mantle as the result of an impact of a Mars-sized object onto the Earth. The absence of metal in the Moon results because the impact was unable to excavate core materials.

## 1-4 A CAVEAT AND AN INVITATION

It should be apparent to the reader that there is much speculative material in this chapter. Less obvious immediately is the fact that new data and improved calculations have considerably tightened the constraints on possible models during the past decade. Speculation is a necessary stage in scientific progress. Properly approached, it brings coherence to established facts and suggests pertinent new experiments or observations. The "dynamic tension" (for those who remember the old Charles Atlas advertisements) between the generation of new facts and the development of plausible scenarios can be recognized in most of the remaining chapters of this book.

At any stage of the study of a complex scientific problem, different experts advocate different interpretations, as will become obvious from reading on. The book gives you, the reader, an opportunity to eavesdrop on the ongoing dialogue among the experts, the scholarly dispute from which consensus emerges. Part of the excitement of studying the solar system is that each one of us can become a member of the jury, entitled to weigh the evidence and to support the ideas we find convincing. You are invited to learn enough to become a participant.

## 1-5 SUGGESTED ADDITIONAL READING

Beatty, J. K., B. O'Leary and A. Chaikin (eds.), 1981, *The New Solar System*, 2nd edition, Sky Publishing Corp., Cambridge, Mass., 221 pp.

Wasson, J. T., 1985, *Meteorites–Their Record of Early Solar System History*, Freeman, New York, N.Y., 267 pp.

Wood, J. A., 1979, *The Solar System*, Prentice-Hall, Inc., Englewood Cliffs, N.J., 196 pp.

H. Zirin
Big Bear Solar Observatory
California Institute of Technology
Pasadena, California 91125

# 2

# THE SUN, INSIDE AND OUT

# ABSTRACT

The energy of the Sun is produced by conversion of hydrogen into helium deep in its interior and carried to the surface by radiation. So far as we can tell, its emission is constant for billions of years. Unknown magnetic processes deep in the interior produce the 22-year magnetic cycle in which the sunspots peak in number every 11 years and reverse their magnetic polarity each half cycle. The strong magnetic fields control the structure of the solar atmosphere and the heating of the corona, the million-degree envelope of the Sun. Spots emerge in a regular pattern and a specific polarity structure. Interaction between moving magnetic fields gives rise to great eruptions called solar flares, the effects of which reach out to the Earth. Beside the flare effects, the Earth is affected by the solar wind, which is produced by evaporation of hot coronal material.

## 2-1 INTRODUCTION

The Sun is not only the source of our life and energy but a touchstone to understanding the rest of the universe. It is the only star that we may study in any detail, and as advances in technology enable us to learn more details about other stars, we are able to detect various phenomena which we are familiar with and understand in the Sun.

The Sun is a rather ordinary star of somewhat average size. However, if we consider that most of the stars in the universe are actually small dwarfs barely detectable by our large telescopes, the Sun is, in fact, larger than average. It only looks average when compared with those stars readily visible to the eye or with a small telescope. It is quite an old star since it radiates energy slowly enough that it should not change over several billions of years. Yet if we study its chemical composition by analysis of the lines in its spectrum, or by analysis of the material in the solar wind, we find that its composition is remarkably similar to most of the rest of the universe. This means that the heavy elements had already been formed by some kind of processing of the material of the original solar nebula before the Sun formed (see Chapter 16 by Lee).

When a cloud of gas or a nebula collapses to form a star, it cannot get rid of angular momentum; in fact, that is one of the difficulties in understanding how condensation might occur (see Chapter 3 by Lin). We actually observe that the youngest stars in the universe, hot O stars that will probably live only a few hundred thousand years, are rotating rapidly. Other stars that may be interpreted as being young because they are associated with young clusters are also found to be rapid rotators. Thus, almost all stars younger than the Sun rotate more rapidly.

How did the Sun get rid of its angular momentum? It is conceivable that it did not, that the center of the Sun is, in fact, rapidly rotating. This was the conjecture of Robert Dicke, who tried to develop a new, non-Einstein cosmology by explaining the advance in the perihelion of Mercury as produced by the oblateness of the Sun. He postulated that most of the mass of the Sun, deep in its interior, is rotating rather rapidly and that the Sun therefore has a greater diameter at the equator than at the pole. The effect of this is a quadrupole term in the solar attraction that would advance the perihelion of Mercury. It would also make the apparent solar disk ever so slightly larger (about 1 part in 200,000) than the polar diameter. Dicke had thought he measured this solar oblateness, but the new observations establish that there is no excess oblateness.

## 2-2 ENERGY PRODUCTION AND TRANSPORT IN THE SUN

Life on Earth is made possible by the endless supply of energy emitted by the Sun's surface. The Sun radiates $2 \times 10^{23}$ kW. (A big power plant produces $10^6$ kW. Thus, the Sun's energy production corresponds to that of nearly a billion billion power plants.) For every gram of mass in the Sun, 2 ergs are radiated every second (1 kW = $10^{10}$ erg · s$^{-1}$). Since the fossil record shows that the Sun's luminosity has been essentially unchanged for at least 1 by ($3 \times 10^{16}$ s), each gram must be able to give at least $6 \times 10^{16}$ ergs, if the Sun is to continue in its present state. The only possible source of so much energy is nuclear reactions; since the Sun is mostly hydrogen, the conversion of hydrogen to helium will do it. Each hydrogen nucleus weighs 1.0078 atomic units. When four are converted into a helium nucleus by the proton-proton reaction, they weight only 4.003 units. The remaining 0.028 units, 0.7 percent of the original mass, are converted into energy, and, by the Einstein formula $E = mc^2$, we get $0.007 \times 9 \times 10^{20} = 6 \times 10^{18}$ erg/g. Thus, total conversion of hydrogen into helium would run the Sun for 100 by.

To understand if this process actually takes place, we must deduce the exact physical conditions inside the Sun from its exterior properties. We know the mass of the Sun ($2 \times 10^{23}$ g) from its gravitational attraction for the planets, its surface temperature (6000°K) from the color and intensity of its light, and its distance and hence its size from planetary motions. The Sun is so hot that no solid or even liquid molecules can exist in its atmosphere or interior. In fact, virtually all the atoms are ionized (broken up into nuclei and electrons). Since the gas has no rigidity, the surface must be supported by its pressure alone all the way down to the center, where the pressure is enormous. The density there is nearly 100 times that of water, and the temperature nearly 100 million degrees Kelvin. At this high temperature, the protons that are the nuclei of hydrogen atoms collide frequently, with enough energy to overcome their mutual electrostatic repulsion and to combine to form deuterons. The deuterons collide with more protons to produce $^3$He and then $^4$He; in each of these steps a sizable amount of energy is released in the form of gamma rays and fast electrons, which after collisions and absorptions end up providing the luminosity of the Sun.

The slowest, and controlling, process is the proton-proton reaction, the rate of which we cannot measure in the laboratory. The devices used for controlled fusion use a similar process but must employ tritium because they cannot duplicate the great temperatures and pressure inside the Sun. One byproduct of the proton-proton chain in the Sun that might be detected at the Earth is the *neutrino*, produced each time a pair of protons combine. Since the neutrinos have no charge, they pass freely through the Sun and everything else. So many protons are combining that the flux at the Earth is about a trillion ($1 \times 10^{12}$) neutrinos per square centimeter per second. Hold out your hand (or keep it in your pocket), and 10 trillion neutrinos pass through it every second. Unfortunately, since these neutrinos penetrate everything, there is at the moment no feasible way to measure them (methods under consideration involve tons of rare gallium or vast quantities of sea water). There are other neutrinos produced by boron isotopes as a byproduct of a side reaction in the proton-proton chain that have a high enough energy to be detected, but only by heroic efforts. A huge tank of cleaning fluid—400,000 ℓ—is buried in a deep gold mine in South Dakota. The neutrinos combine with the chlorine atoms in the fluid to produce a few argon atoms each month. The numbers detected have been very few. The small numbers of neutrinos reaching Earth allow us to conclude that the Sun's energy is not produced by alternative nuclear reactions, which would produce many more neutrinos than the proton-proton chain described above. Even the proton-proton chain described above just barely fits the data, and a fit requires that we make substantial changes in present solar models. Lately questions have been raised about the rate of the boron side reaction. One can conclude that the proton-proton reaction is most likely the source of solar energy, but conclusive proof and a believable model are not quite in hand.

Is there any other way beside the elusive neutrino to probe the hidden depths of the Sun? What we should like to know is the mass distribution (which determines the central temperature and nuclear reaction rates) and the distribution of rotation. The mass distribution would affect the motion of a Sun-grazing spacecraft, and there are plans to send a spacecraft very near to the Sun to determine the mass distribution. As noted earlier, attempts to detect a bulge have been made with no clear-cut results.

What happens to the energy produced at the center of the Sun? The fast particles coming from the nuclear reactions collide with other atoms, creating photons or simply sharing their energy with them; the gamma rays, which are just very energetic photons, are scattered or absorbed and reemitted. None of the energy gets very far: It heats the local gas, which radiates new photons, and the energy slowly works its way out by successive absorption and reradiation of photons. If the nuclear furnace at the center were turned off, it would still take a very long time, perhaps 100,000 y, for any effects to show up at the surface.

The transfer of radiation through the Sun is quite efficient in the solar interior because the atoms are all highly ionized and do not strongly absorb. Near the surface, however, a small fraction of the hydrogen atoms, which make up 90 percent of the population, and the helium atoms, which are about 9 percent, can hang onto their electrons and absorb radiation voraciously. The photons cannot get through, a steep temperature gradient develops, the hot gas rises, and convection occurs, with streams of hot gas going up and cooled gas falling back. Because the hydrogen atoms are now partly neutral, they are ionized at the bases of the hot columns and give up the energy of ionization to the cooler layers into which they rise.

A similar process occurs in the eye of a hurricane: Air is heated at the surface of the sea, and water is vaporized. The warm air rises, and the heat of vaporization is deposited much higher up when the water condenses out. On the Sun, the convective process carries the energy up past the blanket of absorbing atoms to the surface. But what is the surface? Clearly, since the matter is all gaseous, there is no hard surface, and the gas simply gets thinner and thinner until it is transparent and the radiation can freely escape. This point is called the *photosphere* and is the surface of the Sun that we see. The photosphere displays the underlying convection in a cellular pattern called *granulation*. Doppler measurements show the bright granules are in fact rising. The individual granules last about 8 min, and the cool gas sinks around their edges.

When we study the photosphere, we find a whole range of phenomena that hint at the large-scale processes below the surface. There is a strong oscillation up and down with a 5-min period, which shows up also as a weak fluctuation in brightness; the velocity amplitude is about 0.3 km/s (about the speed of a jet plane). When the 5-min oscillation is filtered out, weaker oscillations are found, with periods depending on their scale. The relation of period to size, known as a *dispersion relation*, depends on a large-scale underlying structure. A whole new field of "solar seismology" has sprung up, devoted to using the measurement of these pulsation patterns to infer properties of the Sun's interior.

The total radiative output of the Sun is called the *solar constant*, a term that instantly conveys the attitude of scientists toward its variability. We do not expect it to change much because of the obvious stability of the energy source: If the center of the Sun cools a bit, the pressure is brought up again by shrinking of the outer layers that depend on that pressure for support. The increased central pressure in turn leads to an increased rate of nuclear reactions. Even if changes do occur, the bulk of the Sun is so great that 100,000 y must pass before effects show up at the surface. The fossil record indicates that the temperature at the Earth's surface has been substantially unchanged for 1 by. In fact this constancy is embarrassing, since one would expect that as the Sun uses up its hydrogen it should shrink, the temperature at the center should rise, and the brightness should increase. Models predict a 10 percent increase in brightness in 1 by; this would have produced substantial changes in the Earth's temperature. Although it is possible that changes in the Earth's atmosphere have compensated for the increase in solar brightness, it is more likely that our picture of the solar interior, while accurate in general, has errors in detail.

## 2-3 THE SURFACE OF THE SUN: SUNSPOTS

The interior of the Sun, while fascinating for its long-term significance to terrestrial life and for the intellectual challenge of unraveling its mysteries, does not present the aesthetic interest of the changing surface phenomena, which we may observe every clear day, and the effects of these rapidly evolving phenomena, which are felt rather rapidly on Earth.

Anyone who observes the Sun with a simple telescope (or even an image projected by a pinhole) will see sunspots (see Fig. 2-1) and will find that they slowly move from east to west on succeeding days.[1] A simple explanation, confirmed by Doppler measurements, is that the Sun rotates once in 27 days. More careful observations reveal that the sunspots at the equator rotate faster (27 days) than those at higher latitudes near the poles (29 days). Again, Doppler measurements confirm this. Even better measurements of sunspots and other features that may be identified in H$\alpha$ photos show that some of the high-latitude spots (and other features) move faster than others; in fact, the rate of rotation seems to change a bit from day to day, although the average remains constant. Since it is hard to believe that the Sun speeds up and slows down from day to day, we conclude either that there are giant currents on the surface of the Sun or that the coupling to a more rapidly rotating interior changes from time to time. How this coupling and the currents are connected to the sunspot activity is not clear.

[1] To avoid injuring one's vision, observations should not be made by looking directly at the Sun through the telescope, but by observing the image cast onto a sheet of paper or other surface.

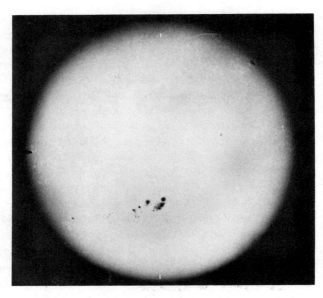

Fig. 2-1 The Sun photographed in white light, March 31, 1979, showing the photosphere. This picture, taken near sunspot maximum, shows several sunspot groups (Big Bear Solar Observatory).

*Sunspots* are dark, cool spots that are the locus of strong magnetic fields; the fields appear to suppress the convection of energy from below. Sunspots are the main manifestation of the solar magnetic cycle. In white light, the magnetic effects are only seen in the sunspots, but in the atmosphere the pressure is much lower, and the magnetic fields spreading from the spots dominate everything.

How can we see the solar atmosphere? Although the gases are transparent, there are certain wavelengths that are absorbed and emitted by electrons jumping between the shells of the atoms and ions; by looking at these wavelengths, we see the atoms, hence the atmosphere. With special narrow-band filters we isolate those wavelengths. The red spectrum line, Hα, corresponding to electrons jumping from the third to the second energy level in hydrogen, is most often employed. The pictures made at this wavelength reveal a wealth of complex detail that shows how the atmosphere is controlled by the magnetic fields of sunspots. Thin, elegant dark *fibrils* go from one sunspot to another, bright patches called *plages* mark magnetic fields too weak to show up in white light. These features can be seen in Figs. 2-2 and 2-3. The region just above the photosphere seen in Hα is called the *chromosphere* (because it emits the red Hα light). It is filled with jets and loops and in a movie it sloshes endlessly back and forth. Wherever the magnetic field is vertical, we see bright plages. These can be strong near sunspots or distributed in a coarse network in the "quiet" areas away from sunspots; we call this the *chromospheric network*. Each element of the network is marked by a weak plage but also by a cluster of thin jets squirting up and down, called *spicules*. The network pattern probably marks a convective pattern deeper than the granulation.

Just as a vertical magnetic field is marked by a bright plage, the horizontal magnetic field is marked by dark fibrils, and two sunspots of opposite polarity are connected by a

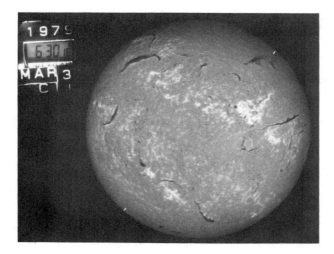

Fig. 2-2 The same, but photographed on the same day through the Hα line of hydrogen, which enables us to see the solar atmosphere. We now see bright areas (plages) near the sunspots, marking additional areas of strong magnetic fields (Big Bear Solar Observatory).

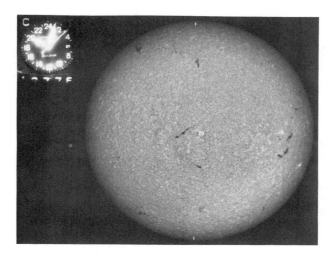

Fig. 2-3 The same, but near sunspot minimum, in 1976, when very little magnetic structure appears (Big Bear Solar Observatory).

nest of curving dark fibrils. This is illustrated in Fig. 2-4. The magnetic fields spread out from the sunspots and from extensive regions of one polarity. Where two such regions of opposite polarity meet, they are connected by a long, sheared, neutral line occupied by a dark cloud called a *filament*, evident in Fig. 2-5. When we see the cloud at the edge of the Sun, it looks bright against the sky and is called a *prominence*. The material in the prominence is supported against gravity by the horizontal magnetic field. If we take a

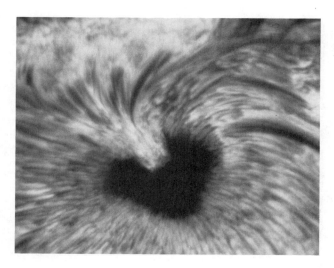

Fig. 2-4  A sunspot photographed in Hα light, greatly enlarged. The spot diameter is four times that of the Earth. Dark fibrils mark the magnetic connection between the spot and the bright plage, where much of the sunspot field reenters the surface. But the other curving lines show that sunspot connects to fields far away (Big Bear Solar Observatory).

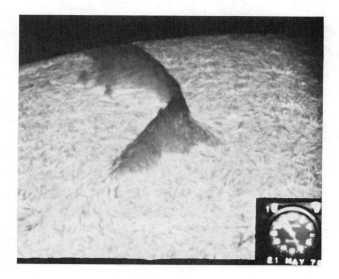

Fig. 2-5  A filament seen on the disk in Hα. These clouds, suspended above the surface by magnetic fields, appear bright against the sky but dark against the solar disk. They divide great areas of one magnetic polarity from another. They may live for months, but when the magnetic boundary becomes unstable, they erupt upward in a great spectacular bright arch (Big Bear Solar Observatory).

good Hα picture, we can size up the three-dimensional distribution of the field by following all these features, which always bear the same relationship to the form of the field.

Higher in the atmosphere the gas takes a remarkable form: It is extremely hot and transparent but can be seen as a bright halo around the Sun at a total eclipse. For this reason it is called the *corona*. The corona is full of arches and loops marking the magnetic field. These are even better seen in X-ray pictures from satellites. The temperature of the corona is more than 1 million degrees Kelvin, and it radiates intensely in X-rays, whereas the cool surface is not even visible in a map of X-ray intensity.

## 2-4 SOLAR MAGNETIC ACTIVITY

The magnetic phenomena of the Sun are not just random but occur in a remarkable cycle of solar activity which goes on over the centuries (see Chapter 14 by Russell). At the beginning of the cycle, we see just a few sunspots at higher latitudes, 20° to 30°N and S, with a few as far above the equator as 40°N. The spots appear in pairs of opposite polarity lined up east-west. We call those leading in the direction of rotation *preceding* (p) and the trailing ones, *following* (f). The rotation is westward from the point of view of an observer fixed on Earth. In a given hemisphere, the same sign of magnetic field always leads [say, for example, the north pole (N)], and the leading polarity in the other hemisphere is always the opposite in sign (S). The field of the "preceding" sunspots is the same as the general solar field at the pole in that hemisphere. The first spot groups are small and last only a few days. The level of activity gradually increases, and the groups occur closer to the equator. Early in a cycle, we may still see a few old-cycle sunspots at low latitudes, recognized by their opposite magnetic polarity. After a year or so of growth in fits and starts, the first really big new-cycle spots occur, and activity, as measured by the number and area of sunspots, rises to a peak about 3 y later. The biggest and most active spot groups, however, occur with equal frequency for 5 y or 6 y. Individual spot groups move very little, but the remnant fields spread outward from the group centers. Shortly after the activity peaks, the remnant magnetic fields from "following" sunspots spread poleward and replace the old polar fields; the Sun becomes a dipole of polarity opposite to that at the beginning of the cycle. The remnant magnetic fields of "preceding" sunspots penetrate the field-free equatorial zone and intermingle there. After a long decline punctuated by sporadic outbreaks of activity, the sunspots die away, and a period of low activity ensues; the minimum is reached after 11 y. When the cycle revives, the magnetic polarities in both hemispheres are reversed: Where N spots once led, S spots lead and N spots all follow. So the sunspot number has an 11-y period, but the magnetic field, the fundamental driver, follows a 22-y cycle.

During the period of 140 y that sunspots have been closely observed, the magnetic cycle has continued faithfully with only a little irregularity. More significantly, there have been periods where all the activity was concentrated in one hemisphere. For example, the dominant activity in the 1957 and 1969 cycles was in the northern hemisphere, but in the present cycle, the southern hemisphere dominates. At other times, only one range of longitudes is active, and sometimes flux emerges all over the Sun at once.

We really cannot explain the magnetic cycle. Various models have been advanced.

Some involve regeneration of existing fields or leftover fields from the previous cycle, others involve oscillations of a deep-seated solar field. The former can reproduce many of the observations but cannot explain the recovery of the cycle after dormant periods like the *Maunder Minimum* (See Chapter 14 by Russell) or the dominance of one hemisphere. The global models are somewhat artificial, because so little is known about the subsurface fields. The author personally favors a deep-seated source which spins off the eleven-year cycles.

The birth and growth of sunspots are fascinating. A bright cloud crossed by dark arches appear in the chromosphere, and a few little sunspots of opposite magnetic polarities appear in the photosphere at either end. The appearance is always the same: We call it an emerging flux region (EFR). As the spots grow, they separate rapidly, the p (preceding) spots moving westward and the f (following) spots fixed. The arches mark lines of flux pushing up from below, and the spots spread as the arch rises. The f spots die out, in about three out of four cases, leaving a plage. Little p spots flow together (we find over and over in solar phenomena that like polarities attract, just the opposite of normal experience). When the growth stops, we typically see a big p spot followed by a plage of following magnetic polarity. In 9 out of 10 spot groups, the growth goes on for a few days, the region matures, lasts a week or two, dies away, and leaves puddles of p and f polarities behind. Almost all spot groups merge with the magnetic axis (the line connecting the p and f polarities) parallel to the equator, and the proper magnetic polarity leading. About 1 in 10 will emerge with the axis tilted 90° to the equator, or even totally reversed. A strong, unknown force opposes this: Usually the group dies out prematurely. Sometimes the p spot twists around and takes the lead as it should, and sometimes the inverted group grows to a huge, complex center of flare activity.

Sunspots are dark probably because the strong magnetic fields suppress the outward transport of energy by convection. The fields are apparently strong enough to produce this effect below the surface. At the surface, the field acts to lower the temperature as well as the pressure, and the magnetic region is squeezed together by the pressure of the surrounding photosphere. The central dark part of the sunspot, the *umbra*, is thus made up of a tight clump of vertically oriented magnetic fields; surrounding is the *penumbra*, less dark and less compressed, with magnetic field lines radiating outward. Although the sunspot is cool and the gas motions supposedly suppressed by the strong magnetic field, we in fact see in the umbra strong oscillations with a period of about 150 sec and waves running out through the penumbra with about twice that period.

Although most sunspot groups are relatively stable, about 10% become quite active. The activity occurs when the normal expansion or realignment of new flux pushes one strong magnetic field against another. For example, in inverted-polarity emerging flux regions, where the flux loop must be twisted below the surface, the p flux, even though behind, forms a spot and runs ahead. As it pushes into other flux, it produces steep field gradients and complexities, leading to solar flares. In other cases, a mature spot is the site of a new resurgence of flux eruption, the new flux pushing the old aside and producing strongly twisted fields. The really big spot groups come up in complexes of dipoles, dozens of spots pushing up from below and producing complex, steep field gradients as they jostle for room.

Wherever the magnetic fields are complex, twisted, and have high gradients, solar flares occur. (See Fig. 2-6 for an exceptional example.) They are the means by which the

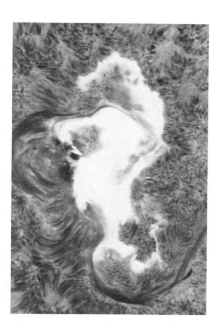

Fig. 2-6 The great "sea-horse" flare of August 7, 1972, which produced shock waves and storms of particles which reached the Earth, causing bright auroras, radio blackouts, and power line surges. The area of this flare is about that of Jupiter, i.e., 100 times the size of Earth (Big Bear Solar Observatory).

magnetic field simplifies. The field cannot do this gradually; normal diffusion processes would take thousands of years. So the moving spots push and push until cataclysmic instabilities occur (like those that plague experimenters in controlled fusion), and the tremendous turbulent release of magnetic energy that we call a *solar flare* takes place. When that energy is suddenly released, the plasma is heated to over 30 million degrees Kelvin, great shock waves and clouds of material are ejected from the Sun, and copious quantities of high-energy electrons and protons are produced. Since the flares take place in the tenuous, transparent atmosphere, only the greatest are visible in integrated light, but many are beautifully visible in Hα.

Flares are dynamic and exciting and produce considerable terrestrial effects. Except in the biggest flares, the energy is released in less than 1 min, sometimes in 10 sec or less. A burst of hard X-rays is observed in the rise phase, along with a pulse of microwave emission; both are produced by energetic electrons. These emissions are observed at the Earth simultaneously with the Hα brightening, since they too are electromagnetic and travel with the speed of light. If the interplanetary conditions are right, we observe a cloud of energetic protons and electrons—the electrons after 20 min and the nucleons, heavier and slower, hours later. The big interplanetary shock wave produced by larger flares travels about 1000 km/sec and takes two days to reach the Earth. Trapped in the shock wave are many energetic particles. The arrival of the shock may trigger the Aurora Borealis and geomagnetic storms on the Earth; the early nucleons can also produce blackouts of radio transmission in the polar regions. In some cases, the terrestrial magnetic fluctuations have produced large voltage pulses in long-distance transmission lines, oil-well logging lines, and telephone cables. The shock wave is often seen traveling across the surface of the Sun.

After the initial fireworks, the flare enters a stable phase. A hot cloud of gas 30 million degrees Kelvin appears above the surface, probably evaporated out by the tremendous flare energy. This cloud is a strong source of soft X-rays. As it cools, the gas condenses into elegant loop prominences that flow down to the surface; thermal energy then flows down the loops and produces emissions below. The mature great flare has a characteristic two-strand form because every line of force has two ends.

Almost every flare starts at a filament marking a magnetic neutral line (where the magnetic fields of two opposite orientations are separated by a region of zero field strength). As magnetic fields of opposite sign are pushed together, it appears that a filament always forms at the boundary. This is because the fields are not connected (otherwise they would not get pushed together), and their field lines connecting elsewhere get pushed together to form a sheer layer at their boundary. This layer then fills with neutral prominence material, which stabilizes it. But there is a lower-energy state where these fields are directly connected. After a flare results, we often see such a direct connection. The material in the prominence is heated or expelled from the Sun. All the energy came from the motions pushing the stuff together.

Curiously, old prominences never fade away, nor do they dribble down to the surface and dissipate. Instead, they erupt outward (again a measure of the stresses they absorb) in spectacular flares. Satellite cameras have photographed the great ejections and shock waves from the filament eruptions at great distances from the Sun. The total amount of matter ejected is comparable to the solar wind. (The solar wind is composed of ionized gases that expand from the corona and flow radially out from the Sun through the solar system at a million miles an hour.) The filaments have many analogies to earthquake faults on the Earth.

## 2-5 THE CORONA AND THE SOLAR WIND

The solar corona, high above the surface, is very hot, at least a million degrees. Because it is so tenuous, it does not radiate efficiently. The coronal atoms are highly ionized and do not absorb visible light well, but, because all the ions have so much energy, they radiate most strongly in the ultraviolet and X-ray range. Because these wavelengths are absorbed by the Earth's atmosphere, their contribution must, of course, be studied from above the atmosphere. An X-ray picture of the Sun does not show the disk at all, only the hot corona filled with elegant loops connecting the different poles on the surface. Each individual sunspot group or active region has a bright cloud of coronal material above it. Often weaker loops connect very distant spot groups. They emerge individually, unconnected; the interaction of the surface magnetic fields connects them in processes that play an important role in the diffusion of solar fields.

The most remarkable features seen in the corona are *coronal holes*, great regions of the surface where no corona at all is seen. Coronal holes can cover a quarter of the Sun and last for many rotations. They are almost always present at the polar caps so long as one magnetic polarity is dominant there. In fact, they occur any place on the Sun where the magnetic field lines do not close back on the surface but connect to distant places—open field lines leading to interplanetary space. The coronal holes have also been found to be sources of high-velocity streams in the solar wind; however, to understand this and why they exist at all, we must understand the solar wind.

Because the corona is so hot, the atomic velocities there are high, but not quite high enough to escape the Sun's gravity; the Sun retains its hold up to about 4 million degrees Kelvin. The conductivity of this gas, because of the great velocities, is very high, and at 5 solar radii the temperature is still nearly 1 million degrees Kelvin. There the Sun's gravity is down by a factor of 25 and can no longer hold onto the gas. It flows outward, and the gas that diffuses from lower down does too. So everything flows out, and most of the internal energy of the coronal gas is converted into the flow energy of the solar wind, flying outward at a speed of 300 to 500 km/sec. The weaker magnetic fields looping up from the surface are dragged outward by the solar wind and end up in an Archimedean spiral outward. Where the magnetic fields are strong and closed, the ionized gas cannot escape: The flow is limited to the coronal holes. In the coronal holes, the field lines reach out far enough to be pulled outward by the wind flow, and an open avenue into interstellar space results. The coronal holes at the poles result from the fact that the field lines from one pole must go all the way to the other pole and, of course, are dragged outward by the solar-wind flow. Because the coronal holes last a long time, a big coronal hole will cause geomagnetic disturbances affecting the upper atmosphere on Earth every time it comes around to face us. In this case, at least, we can predict such events will occur 27 days after the last one, and about 3 days after the hole itself is seen to cross the solar meridian.

## 2-6 CLOSING THOUGHTS

And so the nuclear furnace of the Sun goes on. The nuclei at the center pack closer and closer together, giving up their energy. This energy flows outward, generating convective currents and magnetic fields and producing the interesting phenomena we have described. Eventually, a few atoms sail out into space, but it is not those atoms that keep us warm and well on Earth, but the great photon stream from the surface.

## 2-7 SUGGESTED ADDITIONAL READING

Eddy, J. A., 1981, The Sun, pp. 11–22 in *The New Solar System*, (2nd edition, Betty, J. K., B. O'Leary, and A. Chaikin, eds.), Sky Publishing, Cambrige, Mass.
Noyes, R., 1983, *The Sun, Our Star*, Harvard, Cambridge, Mass.
Parker, E. N., 1975, The Sun, *Sci. Am., 233*, 42–50.
Zirin, H., 1966. *The Solar Atmosphere*, Blaisdell Publishing, Waltham, Mass., 502 pp.

D. N. C. Lin
Lick Observatory
University of California
Santa Cruz, California 95064

# 3

# THE NEBULAR ORIGIN OF THE SOLAR SYSTEM

# ABSTRACT

It is generally believed that the Sun and its planetary system were formed about 4.6 billion years ago from a rotating cloud of interstellar gas and dust. The detailed processes of solar-system formation, however, remain a topic of great controversy and uncertainty. Modern theories of cosmogony suggest that as the cloud collapsed due to its own gravity, gas condensing into the central region of the cloud formed a young stellar object, the protosun. Revolving around the protosun was a gaseous disk out of which the planetary system was formed. This gaseous disk is commonly referred to as the *primordial solar nebula*. In general, the basic physical properties of the primordial solar nebula are very similar to those of an accretion disk. The structure and evolution of an accretion disk are primarily determined by the efficiency of the angular momentum and heat transport of the disk. The dominant transport mechanisms may be turbulence and convection. Irrespective of the initial distribution of the nebular material, under the action of viscous stress arising from convectively-driven turbulence, the central regions of the nebula always evolve into a quasi-stationary state such that the nebular material viscously diffuses toward the Sun with a mass flux that is independent of the distance from the Sun. The radial distribution of other physical quantities such as temperature, density, pressure, and thickness of the nebula can also be determined at any given epoch. These quantities change with time as the nebular material is gradually depleted.

Observational data, such as the spatial and mass distribution of the planets, as well as meteorite data, are utilized to provide constraints on these models. The most favorable model is the one in which the protogiant planets as well as the prototerrestrial planets formed through several stages of coagulation, accumulation, and accretion. Owing to constraints on (1) the grains' condensation temperature in the nebula, (2) the minimum nebular mass required to allow accretion of the protoplanets, and (3) the magnitude of nebular viscosity in the vicinity of protogiant planets' orbits around the protosun, the final epoch, when the protoplanets acquired most of their masses, must have lasted less than a million years. Furthermore, during this epoch the solar nebula was considerably less massive than the present Sun.

While the model for the formation of the solar system described above is supported by both theoretical calculation and currently available observational data, it may be verified and tested further by additional data. These data can be provided not only by meteorite analysis but also by observation of young stars, such as those classified as T Tauri stars, which are in the process of formation. For example, both the rate of energy release and the spectral characteristics, calculated for the solar nebula, may be applied to these systems and checked for consistency. In addition, the theoretical methods used to determine the mass limit and mass distribution of the planets may be utilized for other stars to provide the best candidate for future observational detection of extrasolar planetary systems.

## 3-1 INTRODUCTION

The first well-posed conceptual model of the origin of the solar system is attributed to René Descartes, the French philosopher. In his *Principia Philosophia*, published in 1644, Descartes speculated that the universe was filled with matter and ether and that vortices were set up as soon as motion was imparted. These vortices were broken up into spherical eddies that eventually evolved into various planets. Descartes' theory, however, lacked a dynamical basis. When Isaac Newton, the English mathematician and natural philosopher, established in 1687 the revolutionary concepts of the laws of motion and the principles of universal gravitation in his greatly acclaimed, monumental work, *Principia*, Descartes' speculation was immediately invalidated since it did not provide the basis to describe the origin of planetary motions. For more than a century, philosophers and mathematicians searched vigorously for alternative theories. In 1745, Georges Louis Leclerc, Comte de Buffon, the French philosopher, suggested that the planets were formed by material that was torn out of the Sun during a close encounter between the Sun and another massive body. (Buffon suggested a comet.) In the two and a half centuries following, almost all the theories that were proposed tended to elaborate on either Descartes' monistic hypothesis or Buffon's dualistic approach. In 1755, Immanuel Kant, the German philosopher, postulated that the solar system condensed out of a rotating gas cloud. As the cloud contracted under its own gravity to form the Sun, its rotation speed increased with diminishing radius owing to the conservation of angular momentum. Eventually the cloud was rotating so fast that it flattened into a disk. While the central region formed the Sun, secondary condensations formed the planets and their satellites. In 1796, Pierre Simon de Laplace, the French mathematician, extended Kant's conjecture by arguing that in order for the central region of the rapidly rotating disk to condense into a protosun, the disk, referred to by Laplace as the *solar nebula*, must have ejected several rings along with the bulk of its angular momentum. Laplace proposed that a planet formed from each of the ejected rings. Although Laplace's nebular hypothesis provided a natural and attractive scenario for the origin of the solar system, one major problem remained to prevent it from being generally accepted until after the Second World War. The argument was that if the Sun formed in this way, it should at present be still rotating on the verge of rotational breakup. Instead, Buffon's theory became very fashionable when it was improved by the assumption that the hypothetical massive object which was supposed to have had a close encounter with the Sun was another star (Jeans, 1916; Jeffreys, 1918; Woolfson, 1978).

Today, Buffon-type catastrophe theories have generally been abandoned despite the lack of rigorous disproof. We now realize that it would be very difficult for the hot tidal debris to condense into a planetary system that has the present observed properties without either falling back into the two closely approaching stars or being dispersed into space (Russell, 1935; Spitzer, 1939; Woolfson, 1978). There is also a philosophical argument for the general abandonment of the catastrophe theory: The likelihood for the occurrences of the required close encounters is too small to produce more than a handful of planetary systems in the Milky Way. In the spirit of the Copernican principle, it may be argued that since we do not expect the solar system to be unique, the solar system may not be formed from a highly improbable accident. However, if the Sun formed as other stars are observed to do, in a cluster, the chance of an encounter is highest during the formation stage of the Sun. During this epoch, the material which eventually formed the Sun and the solar system may be distended in an envelope considerably larger than the present size of the Sun so that a close encounter is most likely to have a strong effect (Kobrick and Kaula, 1979). The probability for such an event appears modest, but is not zero.

In the past few decades, the nebular hypothesis has gained considerable acceptance. The original criticism regarding the angular momentum of the Sun may be discounted by the argument that the Sun has lost a considerable amount of its initial spin angular momentum. In particular, the Sun is continually losing matter through a strong solar wind. Near the Sun the motion of this wind is constrained to follow the path determined by the solar magnetic field. Such a magnetically coupled solar wind provides a strong torque to remove spin angular momentum from the Sun. If the solar magnetic field strength and the solar wind were at least as strong in the past as today, the Sun may indeed have been a rapid rotator when it was formed and then lost most of its angular momentum to evolve into the present state (Mestel, 1968a, 1968b). Data obtained from meteoritic analyses provide some indication that the solar wind may have been considerably more intense and the solar magnetic field considerably stronger in the past, so that the solar spin may have decreased relatively quickly (Black, 1972a, 1972b; Brecher and Arrhenius, 1974). Although these mass- and angular-momentum–loss mechanisms remove a major obstacle for general acceptance of the nebular hypothesis, they have not disposed of all the objections. In this chapter I describe some of the recent developments in the nebular hypothesis, particularly in the area of how the initial angular momentum was distributed and rearranged through a process of dynamical evolution.

## 3-2 MODERN THEORIES OF COSMOGONY

Today the basic concept of solar-system formation has a close resemblance to the Kant-Laplace nebular hypothesis. However, in view of the vast amount of data accumulated over the past three decades from meteorite analyses, ground-based observations, and interplanetary probes, today's theories are confronted with much greater constraints than was the original theory. Consequently, the modern theories necessarily pay more attention to the detailed processes at each stage of solar-system formation. It is hoped that a close compatibility between theory and observation on many minor issues may eventually provide an incontrovertible body of evidence to support a grand unified theory of the origin of the solar system.

### 3-2-1 Star Formation

One of the many wonders provided by modern technology in the form of radio telescopes is the discovery of cold, dense, interstellar molecular clouds in our and other galaxies. These clouds have typical temperatures around $10°K$, densities of several thousand molecules per cubic centimeter, and sizes that are a few thousand times larger than the diameter of the Earth's orbit around the Sun. Young, bright stars are found near the clouds with a frequency higher than chance association would predict, strongly suggesting that these giant molecular clouds are indeed the birthplace of stars. Although the exact processes that cause these clouds to be transformed into young stars are not well understood, it is generally believed that, through either occasional collisions between clouds, compression due to the blast wave from a nearby supernova explosion (Cameron and Truran, 1977; Lada et al., 1978), or compression due to passage through the spiral arms of the Milky Way Galaxy (Shu et al., 1973), the internal density of regions of these clouds can increase by an amount sufficiently large that the self-gravity of the region dominates and produces a collapse. (Further discussion of this topic is outlined in Chapter 16 by Lee.) During the early stage of the collapse, gravitational energy is released and dissipated into heat. However, the clouds are essentially transparent to the radiation released owing to their low temperature and density. Consequently, the regions collapse isothermally and fragment into protostars. Eventually the internal density becomes sufficiently large for the protostars to become optically thick (i.e., photons can no longer emerge from the inteior of the protostars without being absorbed or scattered), so that the gravitational energy released is temporarily trapped. Thereafter the internal temperature of the protostars increases quickly, and a large outwardly directed force owing to the gas-pressure gradient is established. This pressure gradient can slow down the collapse, and eventually the entire protostar reaches a hydrostatic equilibrium in which the inwardly directed gravitational force is balanced by the outwardly directed pressure-gradient effect. At this point the protostar is first observable as a stellar object. There is some observational evidence to suggest that the T Tauri stars are currently evolving through such a phase (Larson, 1969). During the *T Tauri phase*, a 1-solar-mass protostar releases strong and rapidly varying infrared, optical, and ultraviolet radiation. Its spectrum shows strong emission lines and indicates large-scale inflow and outflow from the system. These features gradually disappear after a few hundred thousand to a few million years (Herbig, 1962; Cohen and Kuhi, 1979). As energy is radiated from its surface, a young star slowly contracts, and its central temperature gradually increases. After 40 My or so, for a 1-solar-mass star, the central temperature becomes so high that nuclear burning begins, and the T Tauri phase is terminated. At this point the star has arrived at the main sequence (Iben, 1965; Ezer and Cameron, 1967). Stars remain on the main sequence for most of their lifespans. (This sequence is illustrated in Chapter 1 by Wasson and Kivelson.)

Although the specific angular momenta of the molecular clouds are typically much larger than that of the solar system, rotational effects do not prevent these gas clouds from collapsing into stars. One possible scenario of the loss of angular momentum relies on the presence of a galactic magnetic field. In the early stages of the collapse, the cloud is transparent to background cosmic ray and ultraviolet radiation, and a few particles may be ionized. The motion of these ionized particles may be constrained by the presence of the global galactic magnetic field such that, as they take part in the rotation of the cloud, the field line is twisted. Through collision between the ionized and the nonionized particles that make up the rest of the cloud, the twisted magnetic field indirectly provides a torque to slow down the rotation of the clouds (Mestel and Spitzer, 1956). When the cloud density is increased as a consequence of collapse, the cloud becomes optically thick to cosmic rays and ultraviolet radiation that induce the ionization of the particles. Without the ionized particles, the magnetic field does not exert any influence on the rotation of the cloud, so that the magnetic braking mechanism becomes an inefficient process to transfer away angular momentum in order to allow further collapse. (Mouschovias, 1978; Mouschovias and Paleologou, 1980). Recent theories of star formation, however, indicate that when a cloud is sufficiently flattened by rotation, it can fragment into two or more pieces by rotational instability (Bodenheimer et al., 1980). As a result, the bulk of the angular momentum of the cloud is stored in the relative orbital motion of the fragments that can continue to contract in size by a factor of more than a hundred before their own internal rotation is built up again to a value sufficient to flatten them. Bodenheimer (1978) has shown that successive fragmentations and conversion of spin angular momentum into orbital motion can lead directly from a rotating interstellar cloud to stable multiple systems near the main sequence whose masses and orbital angular momentum are consistent with observations of multiple-stellar systems. The principal question still remaining is how to produce a stable configuration of the protosun plus a nebular that is almost symmetric about this spin axis of the protosun before the onset of nonaxisymmetric instability (i.e., breakup into two stars that are gravitationally bound to each other in very close orbits). If some processes of angular-momentum transport were to redistribute the angular momentum from the inner to the outer regions, the system could be stabilized. Today, despite the many theoretical models that have been proposed, there is still no general consensus on the critical process that determines whether a cloud fragment becomes a binary star or a star with a rotating nebula. Nor do the observational data provide any definite answer to this question. Some of the data indicate that a binary or multiple system is the usual outcome of star formation (Abt and Levy, 1976); other data indicate that most young T Tauri stars are not rapidly rotating (Cohen and Kuhi, 1979). This problem may indeed be one of the central issues concerning the formation of the Sun that need to be addressed in the near future.

## 3-2-2 Phenomenological Models of the Solar Nebula

In modern theories of cosmogony, the protoplanets were formed at the same time as the Sun, about 4.6 Gy ago. Since all the major planets lie very close to a plane, they were presumably formed from a disk—the rotationally flattened solar nebula revolving around the protosun. In order to understand the origin of the planets, the environment in which they were formed must be examined in detail. The simplest approach to determine the structure of the solar nebula is to construct a series of phenomenological models.

Early quantitative investigations of the structure of the solar nebula were based on the hypothesis that the planets and satellites collected all the rocky and ice-rocky material within the solar nebula. By adopting the assumption that the chemical composition of the original solar nebula was similar to that of the Sun today, and that most of the volatile materials in the nebula were not condensed into the planets and satellites, a lower limit of the nebular mass may be determined by augmenting the mass of the planets to account for the missing constituents, namely hydrogen and helium gas (Cameron, 1962). The mass obtained this way is about 2 percent of the present mass of the Sun. If the efficiency of collection of rocky or ice-rocky material by the protoplanets is less than unity, the mass of the nebula must have been larger than this amount. Although this phenomenological model does not provide us with an accurate estimate of the mass of the nebula, it does provide a constraint that the nebular mass must be greater than 2 percent of the solar mass. For the remainder of this chapter, such a model is referred to as the *minimum-mass nebula.*

Once the nebula is formed, the protoplanetary formation processes depend critically on the thermal properties of the nebula. The spatial distribution of various physical parameters such as mass, temperature, density, pressure, and thickness of the nebula must be specified or determined before any viable model of protoplanetary formation may be constructed. The simplest phenomenological model for the mass distribution of the rocky material is based on the assumption that the rocky material in the solar system has preserved its original spatial distribution. Consequently, the mass distribution of all the nebular material may be constructed, provided a limited spatial mixing is allowed (Kusaka et al., 1970; Cameron, 1973a, 1973b). Through careful analyses of the temperature required for various chemical elements to condense into rocky materials, the temperature distribution of the nebula may also be deduced under the same assumption (Lewis, 1972, 1974). These results indicate that the nebular temperature was roughly inversely proportional to the distance to the Sun. Thus, the primary source of thermal energy of the nebula cannot have been due to absorption of the solar radiation; otherwise the temperature would be inversely proportional to the square root of the distance to the Sun. Since the nebula has a tendency to attain thermal equilibrium, the heat gain must be balanced by radiation loss in the nebula. Normally the nebula is in a state of hydrostatic equilibrium such that in the vertical direction, that is, the direction perpendicular to the plane of the nebula, the gravity effect, which pulls the nebular gas toward the midplane, is compensated by a pressure-gradient effect that supports the gas above the midplane. Under these conditions, the thickness of the nebula may be deduced. Utilizing information on the mass and temperature distribution in the nebula deduced earlier, the density distribution is now determinable (Safronov, 1962; Kusaka et al., 1970). Although these early phenomenological models provided useful quantitative results, the basic assumptions they depended on cannot be fully justified, since both the nebular material and the protoplanets may have undergone considerable orbital evolution during the course of protoplanetary formation.

## 3-3 EVOLUTION OF THE SOLAR NEBULA

In the past few years, considerable progress has been made in a variety of related astronomical and astrophysical problems. These advancements are brought about by the rapidly increasing volume of observational data obtained by interplanetary probes, such as the *Voyager* mission, and by large, ground-based telescopes. Important theoretical understanding of the dynamics of rotating gaseous disks has also been gained in the context of interacting binary stars and quasars. The new information and knowledge have been applied to and have considerably substantiated the models of solar-system formation. In the remainder of this chapter, I present the currently acceptable models. Although some of these models are supported by both observational data and theoretical computations, they are, nonetheless, models. The details of these models will no doubt change with increasing knowledge and time.

After the formation of the protosun, there may be a considerable amount of high-angular-momentum material that revolves around the protosun in the form of a disk or a flattened nebula. The nebular material adjusts its rotational velocity to achieve a hydrostatic equilibrium in which the centrifugal force, due to its orbital motion, is balanced against the gravitational force due to the protosun. In this state the rotational angular velocity of the nebular material decreases outward, whereas the specific angular momentum (i.e., angular momentum per unit mass of material) increases outward. Consider the gas flow at a given nebular radius. There is a change in the velocity of the gas, a velocity gradient, as one moves outward from the Sun because the gas in the inner nebula revolves around the incipient Sun faster than does the gas in the outer nebula.

In most instances of gas flow, molecules in different regions of the flow can communicate with one another through molecular collisions. Such collisions provide *friction between parts of the flow with different velocities.* In fluid dynamics, such a frictional phenomenon is normally referred to as *viscous stress.* The general effect of viscous stress on a flow with a velocity gradient is to reduce that gradient. In the case of a differentially rotating disk, such as the solar nebula, viscous stress induces a viscous coupling that acts to reduce the difference in the angular velocity between gases in adjacent regions. In order to speed up the slower-moving gas in the outer region of the nebula, angular momentum must be transferred from the inner to the outer region (see Fig. 3-1). Upon acquiring more angular momentum, however, the gas in the outer region experiences a stronger centrifugal force, and it expands to larger radii until a new force balance can be achieved. Similarly, upon losing its angular momentum, gas in the inner part of the nebula drifts inward. Consequently, the action of a viscous couple results in a spread of the nebular material instead of a reduction in the velocity gradient. The accumulated effect of the viscous couple over the entire nebula is to transport angular momentum continuously to the outer regions of the nebula. Once sufficient time has elapsed, a small fraction of the

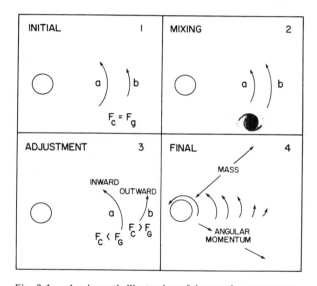

Fig. 3-1  A schematic illustration of the angular-momentum-transfer process in a differentially rotating nebula. (1) Initially gas elements a and b revolve around the protosun in circular orbits and with velocities such that the gravitational force $F_g$ is balanced everywhere by the centrifugal force $F_c$. (2) Through turbulent mixing, viscous friction induces fluid element a to slow down and fluid element b to speed up in order to reduce the local differential rotation around the protosun. (3) In doing so, fluid element a loses angular momentum and spirals inward, whereas element b receives angular momentum and thereby spirals outward. (4) Overall, the viscous stress causes an outward transport of angular momentum and inward diffusion of matter. A small fraction of the nebular material carries most of the angular momentum and moves outwards.

> **Conservation Laws**
>
> In this chapter, arguments often appeal to some conservation law, in particular, conservation of angular momentum and of energy. In other contexts, theoretical arguments frequently refer to conservation of quantities such as mass or linear momentum. To understand the concept of conservation of energy, one needs to recognize that what is conserved is energy in all its forms. For an element of nebular gas, these forms include the thermal energy arising from the random motion of the molecules of the gas, the kinetic energy arising from directed flow of the gas as a whole, and the potential energy arising from the presence of gravitational fields. Angular momentum is also conserved if no external influences (torques) act to change it. Collisions between two bodies within the nebula do not involve external influences, so the sum of the angular momentum of the two bodies must remain the same before and after the collision. Since angular momentum is proportional to velocity and radial distance from the axis of rotation, a decrease in velocity must be compensated by an increase in radial distance. This idea is used, for example, in the discussion of Fig. 3-1.

nebular material at the outer edge absorbs most of the angular momentum and expands considerably. The bulk of the nebular material loses angular momentum and thereby drifts inward. Since the larger fraction of the material drifts toward the central region of the nebula, where the gravitational potential is more negative, there is a continuous release of gravitational energy from the disk. This energy is dissipated into heat by frictional force or equivalently by viscous stress exerted on the differentially rotating gas in the nebula. Consequently, the thermal properties of the nebula may also be determined by the action of viscous stress.

Under these physical conditions, the dynamical properties of the solar nebula are very similar to those of accretion disks, which have been investigated extensively in other astrophysical contexts. Although the foundation of an accretion-disk theory was already established by Lüst in 1952, it remained poorly known until it was rediscovered independently by Lynden-Bell and Pringle (1974). The general concept of an accretion-disk theory is precisely to consider the effect of internal friction on a differentially rotating disk of gas in which the centrifugal force acting on the gas is everywhere balanced by gravitational force. The primary conclusion of these investigations is that the structural evolution of an accretion disk is determined by the efficiency of angular-momentum transport, which itself is determined by the magnitude of the internal frictional force in the flow.

### 3-3-1 Effective Transfer of Angular Momentum in the Solar Nebula

What is the primary source of friction in the solar nebula? In steady laboratory flows, the internal friction is generally induced by collisions between molecules with different velocities, so that the magnitude of the friction is determined by the product of the velocity gradient and a quantity, normally referred to as *viscosity*, which measures the particles' ability to reduce the local velocity gradient by collisions. If molecular collisions provide the only source of friction in the solar nebula, the time scale for mass redistribution is much longer than the present age of the solar system. In this case, the gas collected into the nebula lacks any appreciable motion apart from its circular motion around the central region, and a central condensation is unlikely to develop. In such a nebula, density perturbations cannot grow gradually into protoplanets because they cannot accrete material beyond the immediate vicinity of their orbits. Furthermore, because this type of nebula does not generate appreciable internal energy, it cools off quickly. As the nebula cools off, the pressure effect decreases so that the normal state of hydrostatic equilibrium is temporarily disrupted. Owing to the temporary dominance of gravity effects, the nebular gas contracts toward the midplane to adjust to a new hydrodynamic equilibrium, and thereby the thickness of the nebula is reduced. The time scale for such a contraction is comparable to the cooling time of the nebula. If such a trend continues, eventually the nebula becomes unstable against its own gravity so that it breaks up into several pieces. The actual temperature at which the nebula first becomes unstable can be determined by quantitative calculations (Safronov, 1960; Goldreich and Ward, 1973). The critical temperature is proportional to the square of the surface density (mass per unit area) and the cube of the radius of the nebula, so that it is everywhere relatively low for a low-mass nebula. For a minimum-mass nebula, the critical temperature at a distance from the Sun where Jupiter is now located is actually below the background temperature of the $10°K$ molecular clouds. From quantitative analyses, the mass of the largest fragments may also be determined. In general, the mass and the number of fragments deduced from these analyses do not agree with those of the planetary system today, although the number of fragments may be readjusted through subsequent coagulational, collisional, and dynamical evolution.

Although the arguments above do not directly lead to a workable model by which the physical characteristics of the present solar system may be satisfactorily explained, they do indicate the likelihood of instability in the nebula that can provide a huge degree of chaotic motion (Cameron and Pine, 1973; Cameron, 1978). This situation is somewhat analogous to that of certain laboratory flows in which the velocity gradient of the flow may be too large for the inefficient momentum-transfer process, provided by particle collisions, to smooth them out within a time scale that permits the flow to move through a characteristic dynamical distance scale. In laboratory flows, the efficiency of momentum transfer through viscous processes is normally measured in terms of dimensionless *Reynolds number*, which is the ratio of the product of the characteristic velocity and length scale to the magnitude of viscosity. Flows with a large Reynolds number are relatively inviscid. In most laboratory flows with the Reynolds number larger than an order of a thousand, small perturbations, either intrinsic to the flow or due to the physical conditions at the flow boundaries, can grow vigorously into large-scale turbulence. In turbulent flows, gas or liquid from different regions of the flow are mixed together mechanically with an efficiency considerably higher than that provided by molecular collisions, such that the effective Reynolds number is on the order of a few hundred to a few thousand (Townsend, 1976).

In the case of the solar nebula, if the molecular collisions are the only mechanism that can induce momentum transfer, the molecular Reynolds number may be larger than

$10^{14}$. If the results obtained from laboratory flows may be extrapolated to apply to the solar nebula, the nebular flow is almost certainly prone to become unstable and turbulent (Lynden-Bell and Pringle, 1974). However, this is not a sufficient reason to assume a priori that the nebular flow is unstable. There are several possible processes that may cause the gas flow to become turbulent in the nebula. The first of these mechanisms is associated with rotation. In a nonrotating and perfectly spherically-symmetric star, gravity is isotropically balanced by the pressure-gradient effect, such that the density, pressure, potential gradient, and pressure gradient are all constant on the surfaces of constant gravity. There is no force that acts on the star to change its shape from such a state. In a rotating star, the centrifugal force acts to flatten the star, so that the surfaces of constant gravitational potential are also distorted into surfaces of ellipsoids. In general, the potential gradient and consequently the force exerted on the gas are not constant on the equipotential surfaces. In order to establish a state of hydrodynamic equilibrium on the equipotential surfaces, there must be a pressure gradient along these surfaces that acts to induce a circulation flow pattern (Mestel, 1965). Such a flow pattern is commonly referred to as *meridional circulation*. In the case of the solar nebula, the maximum velocity of the meridional circulation current may be comparable to the speed of sound (Cameron and Pine, 1973). Since meridional circulation provides a mixing mechanism for gas from different regions of the nebula, it generally induces angular-momentum transfer and matter redistribution.

A second source of mixing is associated with the infall of material onto the nebula, which is expected to continue for some time after the formation of the nebula. In general, the infalling material is expected to have velocity and angular momentum substantially different from those in the nebula at the arrival points. The interaction between the infalling and the nebular material may cause turbulence and mixing of material from different regions of the nebula. Furthermore, the mixing of angular momentum between the infalling and nebular material can provide additional stirring (Cameron, 1976).

Third, because of photoionization processes, at least a small fraction of the nebular material may be ionized. These ionized particles are easily trapped into a weak magnetic field in the nebula. Through collisions between the ionized and the neutral particles in the nebula, the ionized particles are forced to move together with the neutral particles. Consequently, the field is dragged along the flow and is stretched by differential rotation of the nebula. (Similar processes are discussed in Chapters 13 and 14 by Levy and Russell, respectively.) The field strength, as a result, grows until field lines break and reconnect. In general, provided there is a sufficient degree of ionization, the nebular magnetic field may be maintained at a level of a few to a few hundred Gauss, and the magnetic energy is comparable to the local thermal energy. If the motion of gas in the solar nebula responds to these fields, they couple gas from different regions of the nebula and lead to transfer of angular momentum between regions (Hoyle, 1960; Hayashi, 1982). Today it is still an open question whether a significant fraction of gas may be ionized to allow this process to operate.

Although these mechanisms can provide much more efficient angular-momentum transport than particle viscosity, the theoretical basis for their effectiveness depends on many weak and ad hoc assumptions. It is often difficult to assess just how much confidence should be attached to each of these assumptions. There is, however, another mechanism for efficient angular-momentum transport. Its origin can be deduced rigorously, and the magnitude of the effective viscosity can be determined from a still approximate but more physical basis. This mechanism is convection.

## 3-3-2 Convective Instability

In most regions of the nebula, the temperature is well below 2000°K, so that micron-sized dust grains can form. If the dust grains are well mixed with the gas, they provide the dominant constituent that absorbs the radiation released from the nebula. The efficiency of absorption in a medium is normally referred to as *opacity*, whose magnitude is generally a function of temperature, density, and chemical composition. For dust grains, the magnitude of the opacity roughly varies as the square of the temperature (Kellman and Gaustad, 1969; DeCampli and Cameron, 1979). Above certain temperatures, species of dust grains may be evaporated, so that the opacity can decrease rapidly over narrow ranges of temperature. For example, due to a first-order phase transition, ice grains evaporate at 160°K. In the absence of ice grains, the magnitude of the opacity at this temperature can decrease by a factor of 30. An important effect provided by such a temperature-dependent opacity law is that a cooler region of the nebula has a lower opacity and therefore can cool off more quickly than the hotter regions in the nebula.

Consider the vertical structure of the nebula, that is, in the direction perpendicular to the plane of the nebula. Normally gas is in a state of hydrostatic equilibrium, such that the gravity effect is balanced by a pressure-gradient effect. The top regions of the nebula, being the most outlying, cool off first. As their temperature decreases, these regions become easier to cool. Eventually the thermal structure of the nebula adjusts to a new equilibrium in which the top surface layers of the nebula are considerably cooler than the midplane. Owing to the large temperature and pressure gradient in the vertical direction, the nebula may become convectively unstable in the following sense. When an element of gas, with size smaller than the vertical scale height (i.e., the characteristic length scale over which density changes in the vertical direction), is displaced upward by a small distance, it expands, without any loss of energy, until the pressure inside the element is balanced by that of the ambient medium (see Fig. 3-2). After such an adjustment, the density inside the element may not be the same as that in the surroundings. If the internal density is greater than that of the ambient medium, the buoyancy effect acts on the element to pull it back to its original position. When the temperature gradient in the ambient medium is sufficiently large, however, the density in the ambient medium is larger than that within the element, so that the buoyancy effect forces the element to continue its upward motion. Consequently, radiative equilibrium is unstable and convection sets in.

The condition for convective instability is well worked out for general fluid flow (see, e.g., Schwarzschild, 1958; Cox and Guili, 1968). The criterion for convective instability can be rigorously applied to the solar nebula. For regions where the temperature is less than 2000°K, the nebula is intrinsically unstable to convection, provided it is *optically thick*, that is, on average every photon released near the midplane is absorbed and re-emitted or scattered more than once before it escapes the surface of the nebula (Lin and Papaloizou, 1980). In the region of the nebula very near the Sun, however, the dominant opacity source is no longer dust grains but molecular transitions and negative

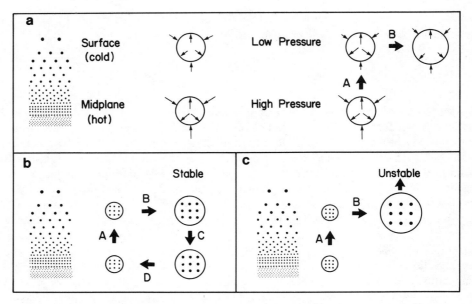

Fig. 3-2 A schematic illustration of convective instability. (a) The solar nebula is stratified in the vertical direction such that temperature, pressure, and density are higher at the midplane than at the surface. When a fluid element is displaced away from the midplane ($A$), the pressure within the element is larger than that in the ambient medium. The element expands to adjust to pressure equilibrium ($B$). (b) If, after the adjustment, the density interior to the element is larger than that of the ambient medium, the element returns to the midplane ($C$) and contracts to its original state ($D$). (c) If the density interior to the element is smaller than that of the ambient medium, the element continues to rise under buoyancy force. The flow is convectively unstable in this case.

hydrogen ions. Nonetheless, the magnitude of opacity still increases rapidly with the nebular temperature, so that this region is also likely to be convective.

### 3-3-3 Convective Accretion-Disk Model for the Solar Nebula

Since the nebula is convectively unstable, buoyancy effects act on a perturbed element that is displaced upward in such a way as to induce a continuation of its upward motion. Similarly, a downwardly displaced element is denser than its surroundings and therefore continues its downward motion. Since the internal temperature of the upward-moving element must exceed that of its surroundings, that element carries an excess of thermal energy. Similarly, the downward-moving element carries a deficit of thermal energy, so that convective motion contributes to a net transfer of thermal energy from the midplane to the surface layers.

After the onset of convection, gas participating in such a vertical circulation may break up into a collection of convective eddies. Through such a circulatory motion of the convective eddies, gas over a region comparable to the size of a typical convective eddy is mixed together. Since the nebula is differentially rotating, such a mixing process results in angular-momentum transfer in the nebula (see Fig. 3-3). As in the case of laboratory turbulent flow, the largest eddies are the most efficient agents of angular-momentum transport simply because they have a relatively large mixing length. While angular momentum is being transferred by the turbulence, kinetic energy is continuously transferred from the main flow, namely, the circular motion around the protosun, into the largest eddies to sustain the turbulence. However, eddies break up on a time scale comparable to that required to travel one mixing length through their random motion. In the process of breaking up, the random kinetic energy carried by the large eddies is transferred down to smaller scales. At the small-scale end, the random velocity between eddies is relatively high and the velocity gradient thus relatively large. Eventually, the eddies decay down to a scale such that the local viscous stress, which is the product of the local velocity gradient and the molecular viscosity, is sufficiently large, so that the random kinetic energy of the small eddies is dissipated into heat before the eddies can break up any further. The heat released from this energy dissipation provides the primary source of thermal energy for the nebula. The drain of kinetic energy from the main flow is eventually compensated by the release of the gravitational energy as material drifts toward the protosun.

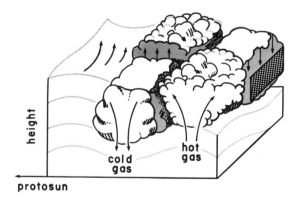

Fig. 3-3 A schematic illustration of convective motion in a differentially rotating nebula. Convective currents circulate up and down in the vertical direction of the nebula. In doing so, convective eddies not only transport heat from the hot nebular midplane to the nebular surface layers, but also induce mixing in the radial direction.

As a result, convectively driven turbulence has three major functions in the evolution of the nebula. First, it provides a mechanism for angular-momentum transfer, so that the nebular material may be redistributed on a time scale considerably shorter than that provided by molecular collisions. Second, it provides a dissipational mechanism for turning kinetic energy into heat and thereby provides a source of thermal energy. Finally, it provides a mechanism for transfer of thermal energy from the nebular interior to the nebular surface layers.

## 3-4 DETAILED MODELS

To model the fine details of turbulent shear flow is not a simple matter. Even relatively simple laboratory turbulent flows are not sufficiently well understood to be described by a simple mathematical model. The gross features of turbulent flow, however, may be studied under considerably simplified assumptions. In normal practice, a simple model called a *mixing length model* is induced to mimic the effect of turbulent mixing. By utilizing such a model, the vertical structure of the nebula may be determined (Lin and Papaloizou, 1980). Although this type of model is still not free from gross assumptions, it neither requires any ad hoc assumption concerning the origin of turbulence in the solar nebula, nor is it based on phenomenological models that cannot be justified.

In order to obtain actual quantitative results, we must solve several important vertical-structure equations, such as the equation that describes the force balance between gravity and the pressure effect and that which describes the transport of radiation from the midplane to the surface layer of the nebula. Furthermore, these equations must be solved simultaneously with the radial-diffusion equations, such as the equation that describes the transport of angular momentum induced by viscous frictional forces and that which describes the transport of nebular material caused by its inward drift. The enormous complexity of the problem is best handled with numerical techniques using a modern computer.

### 3-4-1 Vertical Structure of the Nebula

In order to obtain a set of numerical solutions to describe the physical characteristics of the nebula, appropriate conditions must be specified for its midplane and surface layer. Since thermal energy is generated throughout the nebula by viscous dissipation and is transported from the midplane to the surface layers, the nebula always adjusts to an equilibrium configuration in which the heat flux emerging from any surface, at a given distance above the midplane, is generated between that surface and the midplane. Therefore, an appropriate physical condition for the midplane is that the heat flux is reduced to zero there. At the surface layer, since the nebula is, in general, opaque to the radiation it releases, the emerging radiation obeys the *black-body-radiation law*, which says that an opaque object at temperature $T$ radiates energy $\propto T^4$ per unit area per second. Thus, the radiative heat flux is proportional to the fourth power of the surface temperature, so that once the surface heat flux is specified, the surface temperature is automatically determined.

With these physical conditions at the midplane and the surface layer of the nebula, the numerical solution for a given surface heat flux or equivalently for surface temperature (see Fig. 3-4) can be obtained (Lin and Papaloizou, 1980). These solutions indicate that (1) over substantial regions of the nebula, the gas flow is convective. In general, a convective region is centered on the midplane and there is a transition to a radiative layer near the surface. (2) Convection induces very efficient angular momentum as well as heat transfer. The effective Reynolds number is reduced to values typically on the order of a few thousand. (3) Turbulent motion induced by convection provides an efficient dissipation process that can convert kinetic energy into heat. (4) The energy ultimately comes

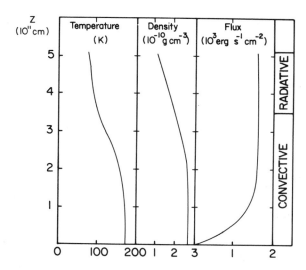

Fig. 3-4 The vertical structure of the nebula. The nebular temperature decreases, whereas the surface flux increases, with distance from the midplane, z. In this case, the convective region constitutes up to two-thirds of the nebula in the vertical direction. Since kinetic energy of the gas is dissipated into heat only in the convective region where the flow is turbulent, the heat flux increases with z throughout the convective zone and then becomes constant in the radiative zone.

from the release of gravitational energy of inwardly drifting nebular material. (5) The gas temperature decreases with distance from the midplane. The temperatures at the midplane and the surface layers normally differ by a factor greater than 1 but less than 10. (6) The gas density also decreases with distance from the midplane, often by a factor of 10 or more. (7) The heat flux increases with distance from the midplane.

Obtaining these results for various values of surface heat flux or surface temperature is a straightforward procedure. Since the thermal energy emerging from the surface layer ultimately arises from the gravitational energy of the inwardly diffusing gas in the nebula, a relatively large heat flux can only result from a relatively high mass-transfer rate. This deduction is in good agreement with the expectation that a nebula with relatively high heat flux is relatively hot and therefore undergoes more vigorous mixing. Consequently, angular-momentum transport is relatively efficient, and therefore the mass flux is relatively high. Numerical studies of models with different surface flux indeed agree with this general expectation (see Fig. 3-5). Thus temperature, density, and pressure at the midplane of the nebula as well as mass flux, surface density (mass per unit area), and thickness of the nebula increase with the surface heat flux.

In the convective accretion-disk models, the thickness of the convection zone is particularly important. The flow becomes turbulent only in this zone, and therefore its thickness effectively provides a natural length scale for the largest eddies, which are the most effective medium of angular-momentum transport. As indicated in Fig. 3-4, the convection zone does not extend over the entire nebular thickness, since the outer layers are radiative and approximately isothermal. Convection only sets in at the point where the

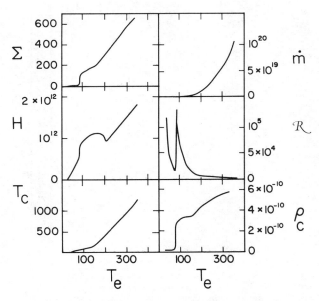

Fig. 3-5 Numerical solutions as functions of the effective temperature $T_c$ at a distance $10^{13}$ cm from a 1-solar-mass protosun. $H$ and $\Sigma$ are thickness and surface density of the nebula, respectively, and are measured in units of cm and g·cm$^{-2}$, respectively. $T_c$ and $\rho$ are the temperature and density, respectively, of the gas near the midplane of the nebula, $\dot{m}$ is the mass flux of the nebula and is measured in gm·s$^{-1}$. The Reynolds number, $R$, the ratio of specific angular momentum and viscosity of the flow, is dimensionless.

nebula becomes opaque, so that a sufficiently large temperature gradient can be established to induce the convective instability. Consequently, the thickness of the convection zone is a function of the magnitude of opacity.

In those regions of the nebula where the temperature is approximately 160°K, the condition for ice-grain condensation is marginal. The presence of ice grains can change the opacity by a factor of 30, so that the nebula undergoes a transition in the ice-grain condensation zone. Numerical results indicate that the convection zone is much thinner if the ice grains cannot condense, and this would greatly reduce the efficiency of angular-momentum transport (see Fig. 3-5). In order to maintain the nebular temperature above 160°K, however, there must be a corresponding heat flux that is ultimately determined by the rate of gravitational energy release by inwardly drifting nebular matter. In order to maintain a relatively high mass flux with a low radial velocity, the surface density must be relatively high. This expectation is again confirmed by the numerical models.

### 3-4-2 Evolution of the Nebula

Based on purely qualitative arguments, we have already shown that the general evolutionary tendency for a nebula is to continuously transport angular momentum outward and therefore to allow the nebular material to fall inward. The evolution of a nebula may be divided into three major stages: the infall stage, the viscous stage, and the clearing stage. In the early phase of the evolution (the infall stage), material from the original collapsing, rotating cloud continuously falls toward the protosun and the solar nebula. The evolution of the nebula is strongly influenced by the influx of external material. In the viscous stage, after the external supply of material is exhausted, the evolution of the nebula is primarily determined by its internal viscous frictional force. In the clearing stage, most of the nebular material is eventually accreted by either the protosun or the protoplanets. Through various physical processes the remaining nebular material is driven away. We now examine each stage in detail.

**The infall stage** In the early stages of nebular formation, matter from the collapsing cloud continuously arrives at the nebula at a rate faster than that of matter redistribution induced by the viscous frictional force, so that the mass of the nebula gradually builds up. Such a mass buildup cannot continue indefinitely because (1) the cloud material is eventually exhausted (Cameron, 1976, 1978a, 1978b) and (2) the viscous frictional force increases with the mass of the nebula until eventually the viscous diffusion rate becomes comparable to the infall rate (Cassen and Moosman, 1982; Lin and Bodenheimer, 1982). Although the duration of the infall stage is somewhat uncertain, it is unlikely to last for as long as 1 My. This constraint is based on the estimate for the time scale required for the original cold and dense interstellar molecular cloud to condense into a protosun. If we assume that the original cloud is as cold as other interstellar molecular clouds, that is, with a temperature around $10°K$, and contains roughly one solar mass of gas, the duration of this infall stage could be as short as 60,000 y (Cameron, 1976, 1978a, 1978b).

Taking into consideration the uncertainty in the mass flux, angular-momentum flux, and the duration of the infall stage, the evolution of the nebular may be investigated quantitatively under two different scenarios. In the first of these, the protosun and the nebula form at the same time, and a considerable amount of material, approximately one solar mass, with relatively large angular momentum, falls rapidly onto the nebula in, say, 60,000 y (Cameron, 1978a, 1978b). With such a large infall rate, the mass of the nebula rapidly increases initially in order to establish an equilibrium in which the mass-diffusion flux is matched by the infall flux. Furthermore, the nebula extends well beyond the present radius of the solar system. In fact, the nebular mass builds up so quickly that the local self-gravitational effect quickly dominates the flow. Consequently, a nebula of this kind is gravitationally unstable and can break up into many large pieces, each having a mass of around 10 percent of the present mass of the Sun. Cameron proposed that these were the "seeds" that produced the planets in the present solar system. We return to the topic of protoplanet formation later.

In the second scenario, the rapid initial infall forms the protosun first. Subsequently, only a fraction of a solar mass of material falls onto the solar nebula during a time interval of 100,000 y. The infall material, furthermore, has a relatively low angular momentum, so that the nebula attains more or less the same size as the present diameter of the solar system. In this case, the frictional force builds up to establish a mass diffusion equilibrium in the nebula (Lin and Bodenheimer, 1982). In this equilibrium, the nebular mass, thickness, temperature, density, surface density, and pressure all increase with the mass infall rate. The time scale for the establishment of such an equilibrium is less than 100,000 y. Because of the strong gravitational influence of the central condensed protosun, the mass of the nebula is never sufficiently large to allow gravitational instabilities to develop in this case.

There are two important physical characteristics, both of which enhance the mixing of the nebular material during the infall stage. First, during this stage the protosun is less massive than the present Sun. Thus, the gravity effect that pulls the nebular material toward the midplane is relatively low, so that the vertical scale height of the nebula and the size of the largest eddies are relatively large. Second, the newly arrived infalling material may have very different angular momentum from that of the nebular material in the regions where it joins the nebula. As the gas mixes, a relatively large region of the nebula may be affected. Since both effects enhance the mixing, evolution in the infall phase may progress on a relatively short time scale.

**The viscous stage**  After the termination of the infall, the solar nebula evolves under the action of its own internal friction. In general, angular momentum is transported outward to allow most of the matter to fall inward, so that the nebular mass decreases and the nebular size increases with time. In the process of gradually exhausting its mass, the nebula continuously adjusts to a new equilibrium by which the vertical force balance is always maintained (see Fig. 3-6). In doing so, the nebular temperature, density, surface density, pressure, and thickness gradually decrease in time. An important question to

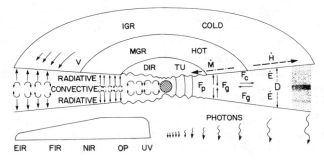

Fig. 3-6  A schematic illustration of the global structure of the solar nebula. In the vertical direction, pressure-gradient effect, $F_p$, balances gravity, $F_g$. Near the midplane, the nebula is convective, and convection induces viscous dissipation, $D$, and angular-momentum transport, $\dot{H}$. Near the surface, the nebula is radiative and the heat generated through viscous dissipation is transported away from the midplane, $\dot{E}$. The size distribution of the grains may be stratified in the vertical direction. In the radial direction, the nebula evolves around the protosun with differential velocities $V$. In general, matter diffuses inward, $\dot{M}$, whereas angular momentum is transported outward, $\dot{H}$. The nebula may be divided into three regions: (1) the ice-grain region, IGR, where it is relatively cold, say below 160°K; (2) the metal-grain region, MGR, where it is relatively hot with temperatures up to 2000°K; (3) the dissociation and ionization region, DIR, where it is very hot, say, up to 10,000 to 20,000°K. The DIR is also thermally unstable, TU. The cooler regions of the nebula release extreme infrared radiation, EIR, and far-infrared radiation, FIR. The MGR region releases near-infrared radiation, NIR. The typical spectrum in the infrared region may be a power law such that the energy flux per frequency may be proportional to the one-third power of the frequency. Optical, OP, and ultraviolet radiation UV, may be released from the DIR and the surface of the protosun.

address is how long a minimum-mass nebula can be sustained in the absence of infall. Utilizing the numerical results obtained for the vertical structure of the nebula, the quantitative evolutionary trend of the nebula may be studied on a computer. Considering the physical condition of the nebula during the viscous stage, the following assumptions are expected to be satisfied: (1) The nebular mass is small compared with the mass of the protosun, (2) the nebula is always in vertical pressure equilibrium, (3) all the energy generated by viscous dissipation is transported to the surface layer and radiated away locally, and (4) convection provides the dominant mechanism for angular-momentum transport. Using these assumptions, the computational procedures are much simplified (Lin and Bodenheimer, 1982). The main results of the numerical computations are: (1) The evolution of a minimum-mass nebula with a physical dimension comparable to that of the present solar system takes place on a time scale of $10^5$ y to $10^6$ y. (2) The nebular mass, surface density, density, temperature, and thickness at a given distance from the Sun all decrease with time (see Fig. 3-7). Since the scale height also decreases with time, angular-momentum transport gradually becomes less efficient and the evolutionary time scale gradually lengthens. (3) The inner regions of the nebula always evolve into a quasi-steady state in which the mass-transfer rate is independent of the distance from the protosun. (4) After an initial transition phase, the evolution of the nebula, regardless of how the nebular mass was distributed at the termination of the infall stage, follows a unique pattern. Thus there is a self-adjusting process in the overall nebular structure. (5) The nebula has a first-order, phase-transition zone for water molecules. Inside this zone the temperature of the nebula is sufficiently hot, so that most of the water molecules are in a gaseous state. At larger radii, most of the water molecules can undergo a first-order phase transition and condense into ice grains. Similar transition zones exist for other molecules. Because iron and silicon grains have a much higher phase-transition (condensation) temperature than the ice grains, their transition zones are at much smaller radii. Consequently, metal grains are probably found throughout the nebula, whereas ice grains are only found at large radii. (6) Inside the ice-grain–transition zone, the nebular surface density is higher, and the viscosity lower, by an order of magnitude than they are outside the zone. (7) The ice-grain–transition zone evolves inward in time as the nebula gradually cools off and exhausts most of its material. Melting and recondensation of ice grains are possible as a consequence of convective circulation in the transition zone. (8) At least in some outer regions of the nebula, the time scale for the nebular temperature to decrease is shorter than that for the inward viscous diffusion of the gas. Gas elements in these regions, consequently, are gradually cooled off as they evolve viscously. As a result, the gas elements may become supersaturated, creating new nucleation sites that enhance grain condensation. In these regions, grain condensation is a continuous process throughout the viscous stage. These continuous condensation regions generally spread inward in time.

The inner region of the nebula, where the distance to the Sun is less than a few times $10^{14}$ cm, is particularly interesting because this is the region where most of the protoplanets may be formed. This region is relatively simple to analyze mathematically because of the general evolutionary tendency for it to reach a quasi-steady state (Lin, 1981). According to the quasi-steady-state solutions (see Fig. 3-8), we can reach the following conclusions: (1) The mass flux, surface density, and the viscosity are roughly independent of radius in both the ice-grain and the metal-grain regions of the nebula.

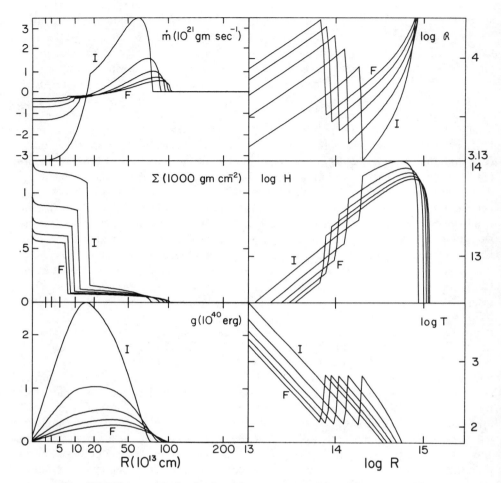

Fig. 3-7 The spatial distribution of various physical quantities at various times. Several quantities are the same as those in Fig. 3-5. The viscous couple, $g$, is a couple that is exerted on the nebular material at a given radius by similar material slightly inside that radius. The five curves in each plot are separated by equal time intervals, 5000 y, starting with $I$, which refers to 5000 y after the beginning of the calculation, and ending with $F$.

The latter quantities undergo a near-discontinuous change at the ice-grain–condensation point. (2) The density near the midplane is inversely proportional to the three-fourths power of the distance to the Sun. (3) The thickness of the nebula is proportional to the three-fourths power of the distance to the Sun, so that the outer regions of the nebula are always shielded by the inner regions from the radiation released by the protosun. (4) The temperature near the midplane of the nebula is inversely proportional to the three-halves power of the distance to the Sun. (5) The surface temperature of the nebula is inversely proportional to the three-fourths power of the distance to the Sun. Typically the temperature at the nebular midplane is hotter than that at the surface by a factor between 1 and 10. (6) Since the mass-transfer rate is independent of the radius, the

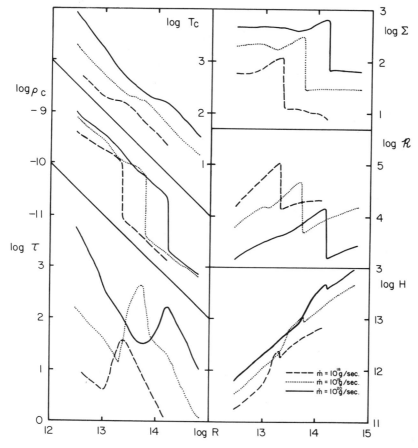

Fig. 3-8 Steady-state solutions obtained for various values of the mass flux $\dot{M}$. Several quantities are identical to those in Fig. 3-5. The optical depth is represented by $\tau$.

observed spectrum of such a nebula, as long as grains dominate the opacity, has a characteristic form in which the intensity-per-unit-frequency interval is inversely proportional to the one-third power of the frequency. Most of the radiation is in the infrared region of the spectrum.

These results may be compared with cosmochemical data (Lewis, 1972, 1974) to provide useful interpretation and implication. A detailed fit between the quasi-steady-state model of the nebula (Lin, 1981) and the cosmochemical data reveals that: (1) The different composition of the satellites of giant planets and that of the terrestrial planets are consistent with the temperature distribution in a quasi-steady-state model with an appropriately chosen mass flux, and (2) the particular model with this appropriate temperature distribution has a mass flux around $10^{-6}$ solar masses per year (see Fig. 3-9). According to the evolutionary calculation (Lin and Bodenheimer, 1982), this mass flux cannot be maintained in the viscous stage for longer than a few hundred thousand years. Of course, a more appropriate comparison between observation and theories is a detailed fit of the

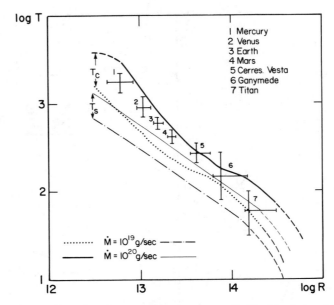

Fig. 3-9 A comparison between the temperature distribution deduced from the steady-state nebular models and that deduced from Lewis's cosmochemical data. In each case, both the nebular surface, $T_S$, and central, $T_C$, temperature distributions are specified. At a large distance, $R$, from the protosun, the quasi-steady-state assumptions must be invalid, and the estimated correction for the temperature distribution is marked with dashed lines. Although the actual condensation location is somewhat uncertain, these results indicate that (1) the condensation temperature of grains in all planets is consistent with the temperature distribution of a single model, and (2) the best-fit model has a mass flux around $10^{-6}$ solar masses per year or equivalently, $10^{20}$ g·s$^{-1}$.

data to a fully evolutionary rather than a quasi-steady-state model. Investigation in this area is still underway.

Some evidence for the time scale of early nebular processes can be obtained from the studies of isotopic composition of meteorites. It has recently been discovered that the Allende meteorite contains an excess of the isotope $^{26}$Mg, a decay product of the short-lived isotope $^{26}$Al. The implication of this important discovery is that $^{26}$Al must have been present at the time of grain formation, as discussed in Chapter 16 by Lee (see also Lee et al., 1976). Chapter 5 by DePaolo explains how the history of the solar nebula may be constructed on the basis of evidence deduced from the studies of isotopes. Since $^{26}$Al can only be produced in a supernova explosion and has a half-life of around 1 My, its presence in the grains is extremely intriguing. If we assume that the grains condensed in the solar nebula, we would deduce from the inclusion of $^{26}$Al that (1) the formation of the solar nebula may have been triggered by a supernova explosion, and (2) the grain-condensation time is less than 1 My (Cameron and Truran, 1977). The grain-condensation time deduced from these meteoritic data is consistent with that deduced above in Fig. 3-9. Fine structure data obtained from the Allende meteorite also indicate that the nebula may have undergone repeated condensation and evaporation (Boynton, 1978). It is interesting that convective eddies may exert a viscous drag on small-sized grains, inducing them to circulate between the relatively cool nebular surface region and the relatively hot nebular interior. For certain types of grains, such a circulation between the hot and cool regions may provide a mechanism for repeated condensation and evaporation (Cameron and Fegley, 1982).

Very near the protosun, the nebula may be too hot for grains to condense, so that the dominant sources of opacity are due to molecular transitions in, for example, $H_2O$ and CO molecules and atomic-hydrogen ionization processes. Since grains cannot condense in this region, protoplanets cannot form here. Nonetheless, most of the gravitational energy of the inwardly drifting material is released in this region, making it an interesting region to study. This region, too, is convectively unstable. However, unlike elsewhere in the solar nebula, most of the energy generated through viscous dissipation is not released locally as heat. Instead it is absorbed by the atoms, the molecules, and the electrons in the dissociation and ionization processes. Consequently, the radiation released from this region lacks the characteristic spectrum we discussed earlier, even if the flow is in the form of a quasi-steady state. This region, furthermore, is likely to be thermally unstable as a consequence of the opacity law. The most likely outcome of such an instability is still somewhat uncertain. It is generally thought that the nebular thickness may vary in time and a nebular wind may be induced. These processes are still under investigation.

**The clearing stage** Today there is very little gas left between the planets. At some stage of the solar nebular evolution, the gas must have been cleared away. There are three basic models to account for the clearing of the nebula. First, sometime after the formation of the protosun, there may have been a very strong stellar wind that swept out the bulk of the nebular material (Elmegreen, 1978; Horedt, 1978). Although there is no satisfactory model to account for the origin of such a wind, mass outflow has been observed in a class of newly forming stars, T Tauri stars (Herbig, 1962; Cohen and Kuhi, 1979). With the typical mass outflux and wind velocity in the T Tauri stars, a minimum-mass nebula may be cleared in a few thousand to a few million years. However, there are observational evidences for anisotropy in the mass outflow from some T Tauri stars (Cohen, 1981, 1982). In these cases, the most likely direction of the T Tauri winds is perpendicular to the plane of the nebula. Consequently, the wind-induced clearing process is unlikely to be efficient in removing gas from most outer regions of the nebula. The second model is based on the general expectation that there may be a hot coronal layer above the surface layer of the nebula that is established by the dissipation of mechanical waves generated by turbulence (Cameron, 1978a, 1978b). Such a model is analogous to one of the two models that account for the solar corona (see Chapter 2 by Zirin). Just as the presence of a hot corona in the Sun causes a hydrodynamic flow of mass away from the Sun, coronal layers high in the atmosphere of the nebula may induce mass flow away from the nebula. The quantitative verification of these qualitative arguments is still to be carried out.

A third model is based on the interaction of protoplanet with nebular gas. Protoplanets with very small mass have little effect on the structure and evolution of the nebula. When they have grown to sufficiently large mass, the protoplanets can induce a tidal transfer of angular momentum between their orbits and the nebular material. (Tidal interactions and angular-momentum transfer are explained in the context of planet-satellite interactions by Peale in Chapter 12.) Since the protoplanets revolve

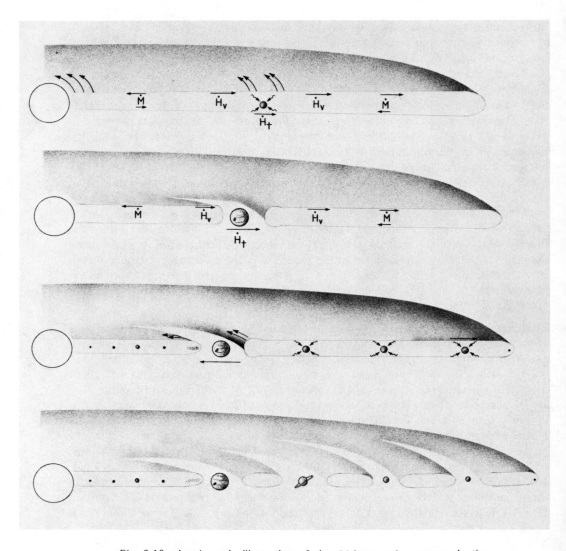

Fig. 3-10 A schematic illustration of the *tidal truncation* process. In the early stages of protoplanetary growth (top), the rate of tidally induced transfer of angular momentum, $\dot{H}_t$, is less efficient than the viscous transfer of angular momentum, $\dot{H}_v$. The overall nebular structure is not strongly influenced by the accretion of nebular material onto the protoplanets, because the nebular material is continually brought to the vicinity of the protoplanets' orbits through viscous diffusion (second). When one major protoplanet—say, Jupiter—has acquired sufficient mass, so that $\dot{H}_t$ is greater than $\dot{H}_v$, its tidal effect opens up a gap in the nebula near its orbit (third). As the nebula evolves viscously, matter inside Jupiter's orbit is gradually depleted, whereas matter outside Jupiter's orbit accumulates. Thereafter Jupiter evolves inward (bottom). All the protogiant planets may eventually tidally truncate the nebula into several rings.

around the Sun faster than does the nebula outside their orbits of the protoplanets, angular momentum is transferred from the protoplanets' orbits to the nebula in order to achieve a lower-energy state, that of uniform rotation. Similarly, angular momentum is transferred from the nebula inside the protoplanets' orbits to the protoplanets. The rate of this tidally induced angular-momentum transfer can be calculated in various ways (Lin and Papaloizou, 1979a, 1978b; Goldreich and Tremaine, 1980). This rate is proportional to the mass of the protoplanets as well as to the nebular surface density. As the nebula evolves, the mass of the protoplanets grows, so that the rate of tidal transfer of angular momentum is increased. At the same time the nebular material is exhausted, and there the rate of the viscous transfer of angular momentum in the nebula is decreased. Eventually the tidal effect can redistribute angular momentum faster than can the viscous effect. When angular momentum is being transported tidally to the protoplanet from the nebular gas that is interior to its orbit, the gas moves inward and away from the protoplanet. Similarly the nebular gas that is exterior to the orbit of the protoplanet moves outward and away from the protoplanet (see Fig. 3-10). Consequently, the tidal effect begins to open up gaps in the nebula in the vicinity of the protoplanets' orbits (Cameron, 1979; Lin and Papaloizou, 1980, 1982).

The actual protoplanetary mass required for this process to take place is given by the condition that the mass ratio of the protoplanet to the protosun is larger than the inverse of the Reynolds number in the nebula. When the protoplanets have grown to such a mass, the tidal effect induces not only a termination of any further protoplanetary mass growth but also expels the remaining nebular material over extended regions in the vicinity of the protoplanets' orbits (Fig. 3-11). The nebular flow between the massive protoplanets subsequently becomes unstable, so that the remaining nebular gas trapped

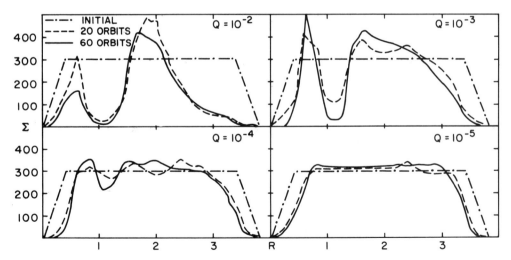

Fig. 3-11 The evolution of the surface density in the nebula as a function of the mass ratio ($Q$) between the protoplanet and the protosun. The Reynolds number of the nebula is $625\ R^{-1}$. When the mass ratio is relatively large, say $Q = 0.01$, that is, when the protoplanet is relatively massive, the nebula is clearly truncated tidally. However, the presence of the protoplanet has very little tidal effect on the nebula when the mass of the protoplanet is relatively small, for example, $Q = 10^{-5}$ case.

between the protoplanets may be ejected from the protoplanetary system by a wind or, perhaps more likely, be driven toward the protosun.

The advantage of this tidal model is that it allows most of the protoplanets to acquire their mass before the remaining nebular gas is cleared away. It also provides a strong dynamical constraint on the mass of the nebula, as we show later.

## 3-5 PROTOPLANET FORMATION

Having constructed a sequence of models for the primordial solar nebula, we are now in the position to examine the physical conditions in the solar nebula at the time the protoplanets acquire most of their mass. These studies can provide information not only on the time scale of protoplanetary formation but also on the mass of the solar nebula during the epoch of protoplanetary formation. Furthermore, they can provide information on the temperature, mass, density, and pressure distribution of the nebula at that epoch, and this information can be verified by the data obtained from meteorite studies. Most importantly, they may provide constraints on and evidence for or against protoplanetary-formation scenarios.

### 3-5-1 Scenarios for Protoplanet Formation

Both the fully time-dependent and the quasi-steady-state solutions of the evolution of the solar nebula indicate that the mass of the nebula reduces to a value below the minimum mass deduced from the phenomenological models within a few hundred thousand to a million years after the termination of infall. Therefore, most of the massive and gaseous giant planets such as Jupiter and Saturn must have been formed within 1 My after the initial formation of the Sun. Although this argument provides a strong limit on the duration of the protoplanetary formation epoch, there is still considerable disagreement on just when and how the protoplanets did form. The diversity of opinion may be divided into three major scenarios: (1) All protoplanets formed within $10^5$ to $10^6$ y during the infall stage by gravitational instability, which occurred on a very massive and extended nebula (Cameron, 1978a, 1978b). (2) The protogiant and terrestrial planets formed at different epochs through different processes. While protogiant planets may be formed from massive gravitational instability, analogous to the first scenario, the prototerrestrial planets formed through a series of gradual accretion processes in an essentially gas-free environment (Safronov, 1969, 1972; Kusaka et al., 1970; Wetherill, 1980). (3) The protogiant as well as prototerrestrial planets formed through a process of gradual accretions, first of dust, then of gases (Hayashi, et al., 1977; Hayashi, 1981; Lin and Papaloizou, 1982).

**The massive gravitational instability scenario** Normally the nebula is in a state of hydrodynamic equilibrium in which the pressure-gradient effect is balanced by gravity in the vertical direction and the centrifugal force is balanced by gravity in the radial direction. When a perturbation is imposed on the nebula, such that the local density is slightly increased, the local pressure is also increased. This increase in the local pressure induces a pressure gradient, which acts to smooth out the perturbation locally. If the perturbation is extended over a large region of the nebula, global shear acts to smooth out the perturbation. However, a density increase can also cause an increase in the local gravity. If the nebula is relatively massive compared with the protosun, the increase in the local gravity may dominate the local-pressure gradient and shearing effects, so that the nebula contracts toward the regions of density enhancement. This phenomenon is commonly known as *gravitational instability* (Safronov, 1960; Toomre, 1964; Goldreich and Lynden-Bell, 1965a, 1965b). According to the first scenario, the protoplanets formed from such a gravitational instability during the infall stage. Quantitative investigations indicate that a nebula may indeed become gravitationally unstable provided it (1) is massive, for instance, 1 solar mass, (2) is extended, for example, 10 times larger than the present diameter of the solar system, and (3) has a protosun in the center with a low mass, for instance, 0.1 solar mass (Cameron, 1978a, 1978b; Lin, 1981). Such a massive and extended nebula, in order to be sustained, must have acquired approximately 1 solar mass of material within $10^5$ y. Since most of the infall material is eventually accreted by the protosun, unless the Sun had a substantial postformation mass loss, this epoch cannot continue for longer than 100,000 y. A massive and extended nebula with a low-mass protosun at its center is unstable because the destabilizing self-gravity effect of the nebula is maximized, whereas the stabilizing pressure-gradient and shearing effects are minimized. Nonetheless, the temperature on the nebula is so high, because of vigorous viscous dissipation, that the mass and orbital radius of the unstable fragments are typically 0.1 solar mass and several dozen astronomical units (Cameron 1976, 1978a, 1978b). Thus, in order for fragments to evolve into the protoplanets of our solar system, the following sequence of events must take place: (1) The protosun must gain the other 90% of its present mass in the postfragmentation period. (2) The massive fragments must lose 99% of their mass to evolve into protogiant planets and 99.99% of their mass to evolve into prototerrestrial planets. (3) Gas must be lost preferentially in order for the fragments to evolve into prototerrestrial planets. (4) The fragments' orbital radii must be reduced by at least an order of magnitude. (5) These evolutionary processes must occur on a time scale of less than a few hundred years in order to prevent orbital instability from randomizing the protoplanets' orbits and inducing the ejection of low-mass fragments and a very large orbital eccentricity in the protoplanetary system, which is not seen today. Cameron (1978b) proposes that the large fragments may indeed have spiraled toward the protosun and lost a lot of mass (see Fig. 3-12). His argument is based on the hypothesis that after the fragmentation of the nebula, 0.9 solar mass of material diffuses toward the protosun. As a result, the protosun's mass and gravity are increased and the fragments' orbital radii reduced. Both the increase in the protosun's mass and the decrease in the orbital radii enhance the tidal influence of the protosun on the envelope of the fragments. Eventually this influence is sufficiently strong to induce the removal of the envelope mass in the fragments (Cameron, 1978a, 1978b).

Although these qualitative arguments describe a possible chain of events that may ultimately lead to the formation of the present solar system, there are several inconsistencies in the gravitational-instability scenario. First, the nebula remains gravitationally unstable unless its gas content is substantially depleted as a consequence of fragmentation. Therefore, there is insufficient gaseous material left, in the postfragmentation era, to feed the protosun in order for it to grow substantially. Second, since the typical mass of the fragments is comparable to that of the protosun, the tidal effect induced by these frag-

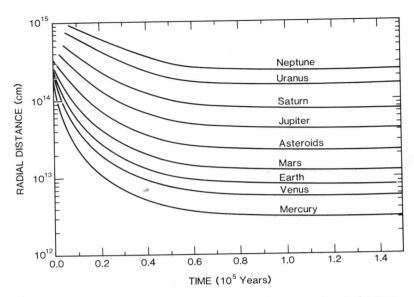

Fig. 3-12 The evolution of the orbital radii of regions of protoplanetary formation, in the gravitational instability scenario. The protoplanetary-formation regions are centered on the specific angular momenta of the present planets. (Cameron, 1978, courtesy T. Gehrels, ed. *Protostars and Planets*, University of Arizona Press.)

ments strongly truncates the nebula, so that it is impossible for any remaining nebular material to stream past the orbit of these fragments. Third, if mass is transferred from the envelope of the fragments to the protosun, the separation between the fragments and the protosun generally increases unless the protosun is less massive than the fragment. Fourth, through gravitational encounters between the massive fragments and the protosun, the orbital eccentricity of the fragments increases rapidly and within a few hundred years may be considerably larger than the present eccentricity of the planets. In fact, the ejection of one or more of the fragments may occur. The four objections noted above are among the many that can be used to argue against the gravitational-instability scenario.

**The gas-free accumulation scenario** In the second scenario, the formation epoch continues for about several dozen million years or longer (Safronov, 1972; Wetherill, 1980). The mass of the nebula, from which the protoplanets formed, is a small fraction—for instance, about 10 to 20 percent—of the present mass of the Sun. Because of the relatively low nebular mass and surface density, the nebula is stable against its own gravity, that is, the pressure-gradient effect is always sufficient to balance the gravity in the vertical direction. In this model, prototerrestrial planets are formed from the accumulation of small planetesimals that are produced from gains (Safronov, 1969). The term *planetesimals* is usually loosely used to refer to solid particles inside the nebula having sizes ranging from a few centimeters to a thousand or so kilometers. These objects are more massive than the grains, and they can be segregated from the nebular gas and settle toward the midplane of the nebula. They are also considerably less massive than the present-day terrestrial planets. The primary motion of the grains is their revolution around the protosun in nearly circular orbits. The small eccentricity among the grains provide a random velocity that acts to balance their self-gravitating effect. The random velocity also causes the grains to collide with one another. During these collisions, the grains sometimes stick to one another, whereas at other times they break one another apart. Through cohesive, inelastic collision, the kinetic energy of the grains' random motion is continually dissipated, so that the grains settle down to the midplane. When a sufficiently dense layer has been formed near the midplane, the grains' random motion is insufficient to overcome the mutual gravity they exert on one another, and they become gravitationally unstable (Safronov, 1972; Goldreich and Ward, 1973; Nakagawa, et al., 1981). In this case, the critical random velocity of the grains is determined by the grains' surface density (i.e., the total mass, in the form of grains, per unit nebular surface area). If the nebula has the same chemical composition as the Sun and if the condensation of the heavy elements into grains is fairly complete, the random velocity of the dust would have to be less than about 10 cm/sec in order for the grains to become gravitationally unstable, on a minimum-mass nebula, in the vicinity of the present orbits of the terrestrial planets. The equivalent thickness of a gravitationally unstable grain layer must be less than about 1000 km. After the onset of the instability, the dense grain layer contracts and fragments until an equilibrium is reached (Cameron, 1975a, 1975b; see Fig. 3-13). Such an equilibrium is determined by a rotational stability requirement. Any further contraction induces rotational fission among the grain fragments. The grain fragments rotate around the protosun with velocities solely determined by the gravitational force in the radial direction. Owing to a pressure gradient in the radial direction, which induces an outward force, the nebular gas rotates around the protosun at lower velocities (Kusaka et al., 1970; Whipple, 1971). Consequently, the grains move relative to the nebular gas. Viscous coupling in the boundary regions between the grains and the nebular gas induces a drag. Such a drag force can act to reduce the random velocities of the grains within the grain fragments and thereby allow the fragments to contract further. The final products are objects roughly 10 km in size that are normally referred to as *planetesimals*.

The discussion above does not take into account the presence of turbulence, which must influence the motion of grains. It has been suggested (Safronov, 1969) that turbulence may inhibit gravitational instability among the grains. As we have already shown, an optically thick nebula is intrinsically unstable against convection. According to the convective-accretion-disk model of the nebula, the convectively driven turbulent eddies may have sizes and random velocities up to the nebular pressure scale-height and a fraction of the nebular sound, speed, respectively. These turbulent eddies exert a viscous drag on the grains in random directions. In particular, a freely falling grain with a linear dimension smaller than about 1 cm achieves its terminal velocity, analogous to falling snow flakes, on time scales that are short compared with the average life span of typical turbulent eddies. The motion of the small grains, consequently, is strongly coupled to that of the turbulent flow. Although larger grains may be decoupled from the turbulent eddies, the viscous drag induced by the eddies can still cause the grains to have random velocities greater than 10 cm/sec, or equivalently induce the grains' vertical scale heights to be greater than 1000 km, unless the grains are larger than several kilometers in diameter. Thus, gravitational instability may not be the process responsible for building kilometer-sized planetesimals. However, the moderately large random velocity of the grains induces frequent collisions among the grains. Through a series of inelastic and cohesive collisions,

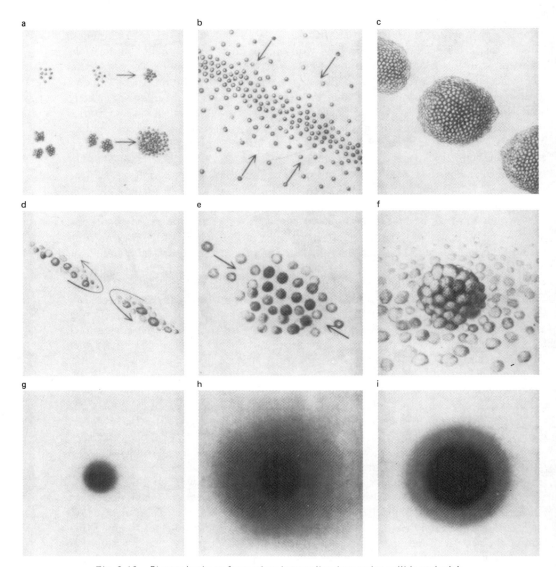

Fig. 3-13 Planets begin to form when interstellar dust grains collide and stick to one another, forming ever larger clumps (a). The clumps fall toward the midplane of the nebula (b) and form a diffuse disk there. Gravitational instabilities collect this material into millions of bodies of asteroid size (c), which collect into gravitating clusters (d). When clusters collide and intermingle (e), their gravitational fields relax, and they coagulate into solid cores, perhaps with some bodies going into orbit around the cores (f). Continued accretion and consolidation may create a planet-sized body (g). If the core gets larger, it may concentrate gas from the nebula gravitationally (h). A large enough core may make the gas collapse into a dense shell that constitutes most of the planet's mass (e). (Adapted from "The Origin and Evolution of the Solar System" by A. G. W. Cameron. Copyright © 1975 by Scientific America, Inc. All rights reserved.)

the grains may also grow into kilometer-sized planetesimals before they settle down to the midplane of the nebula (Weidenschilling, 1980).

By whatever process the kilometer-sized planetesimals may have been formed, they are still considerably less massive than the present-day planets, so that they must grow substantially. For example, a typical planetesimal with a mass around $10^{15}$ to $10^{18}$ g must grow by a factor between $10^9$ and $10^{12}$ before its mass is comparable to that of Earth. However, the exact growth process is somewhat uncertain. The terrestrial planets, which are mostly composed of solid material, may have been formed through a series of dust accretion and cohesive collisions. Since their present gas content is relatively low, the building of the terrestrial planets must have taken place in an environment that was either free of gas or unfavorable for gas accretion. Most investigators of the gas-free scenario focus their attention on the first possibility (Wetherill, 1980). Since all giant planets are gas-rich today, the gas-free scenario is not applicable to their formation process.

Most of the quantitative investigations have addressed various important physical processes such as coagulation, fragmentation, inelastic collision, and gravitational scattering (Safronov, 1969; Wetherill, 1980; Greenberg, 1982). In these investigations, gas is assumed to be removed from the nebula by solar ultraviolet radiation after the condensation of one generation of grains, so that the total mass of the grains available for protoplanetary formation is limited. Under these conditions, the planetesimals grow via collisionally induced coagulation processes. In reality, there are two different types of collisions: gravitational scattering and direct physical collision. *Gravitational scattering* is a process in which the orbits of two nearby planetesimals are influenced by the mutual gravity between the planetesimals. Gravitational scattering is a particularly important process if the random velocity of the planetesimals is relatively small. However, when the random velocity is relatively large, the range over which the planetesimals can effectively influence each other's motion through their mutual gravity is relatively small. But the frequency of direct physical collisions increases with random velocity, so that they may have a dominant effect on the collisional evolution of the planetesimals. The critical random velocity, at which gravitational scatterings and direct physical collisions have comparable effectiveness, is determined by the mass and number-density distribution of the planetesimals. If the density is relatively high and the planetesimals have relatively small mass, as in the early stages of their evolution, direct collisions become important at relatively low random velocities. If, however, planetesimals are sparsely populated and have relatively large masses, as in the final stages of their evolution, gravitational scatterings dominate the dynamics up to relatively large random velocities.

The main purpose of investigation on the collisional evolution of the planetesimals is to evaluate the growth rate of planetesimals due to collisionally induced coagulation processes. But, since gravitational scatterings do not generally involve direct physical contacts between planetesimals, they neither effectively induce coagulation among planetesimals nor lead to significant dissipation of planetesimals' kinetic energy. Direct physical collisions may lead to either coagulation or fragmentation depending on the impact velocity between two colliding planetesimals. A major obstacle for rapid growth of the planetesimals by direct physical collisions resides in the dichotomy between the requirement of relatively large random velocity to enhance the collisional frequency and that of relatively small impact velocity to ensure that collisions result in coagulation rather than fragmentation. In general, if direct physical collisions dominate the dynamics, the planetesimals collide among themselves with impact velocities comparable to their random velocities, so that the rate of growth by coagulation processes is limited.

The planetesimals random velocity may evolve with time. Since the planetesimals are formed in a differentially rotating gaseous nebula, they may amalgamate into a differentially rotating particle disk analogous to planetary rings (Goldreich and Tremaine, 1980). In these particle disks, collisions induce a continuous transfer of energy, from that stored in the systematic, circular, Keplerian motion into that stored in the random motion. When the random velocity is relatively small, this energy transfer is caused by gravitational scattering. Since gravitational scattering is a perfectly elastic process, that is, there is no loss of kinetic energy during each collision, the kinetic energy transferred into the random motion may not be dissipated. Thus the random velocity of the planetesimals increases until it becomes sufficiently large, so that direct physical collisions occur more frequently and begin to dominate the dynamical evolution of the planetesimals. Direct physical collisions are similar to gravitational scatterings in the sense that they also induce energy transfer into random motion. However, they are different from gravitational scattering in the sense that they are partially inelastic so that at least a fraction of the random kinetic energy associated with each collision is dissipated into heat and radiated away. A system of planetesimals may eventually achieve an energy equilibrium when the rate of energy transfer into random motion is balanced by the rate of energy dissipation due to inelastic, direct, physical collisions. As planetesimals coagulate and fragment, their mass distribution must evolve. The evolution of the mass distribution may also have profound influences on the evolution of planetesimals' random velocities (Greenberg et al., 1978a, 1978b; Wetherill, 1980). Under some conditions, the mass distribution may evolve too rapidly for an energy equilibrium to be established by collisions.

Thus, the physical condition under which a collisional equilibrium is achieved may be determined by the poorly known elastic, erosive, and coagulational properties of the direct physical collisions between the planetesimals as well as complicated mutual gravitational interaction among the planetesimals. Due to these complexities and the general lack of experimental data, the growth rate of planetesimals under these conditions has only been studied under grossly simplified and somewhat ad hoc assumptions. Although the results deduced from these investigations may be an inadequate description of the actual physical events that took place in the formation stage of the solar system, they can generally provide useful insights. For example, the critical random velocity, at which the maximum growth rate of the planetesimals is achieved, may be estimated from such an approach. If the planetesimals have roughly the same mass, the maximum growth rate is achieved if the random velocities are maintained at values approximately equal to those of the escape velocities from the surface of the planetesimals. For a more general mass distribution, the critical random velocity for maximum growth rate for planetesimals with certain mass range depends on the assumed mass distribution (Ziglina and Safronov, 1976; Greenberg et al., 1978b; Kaula, 1979; Cox and Lewis, 1980; Wetherill, 1980). Nonetheless, it is related to the surface escape velocity of some characteristic-sized planetesimals within that mass range. With these critical random velocities, the shortest growth time scale for a planetesimal may be estimated.

According to most gas-free scenarios, shortly after the formation of the kilometer-sized planetesimals, their orbits overlap, so that direct or glancing collisions occur frequently. These collisions can lead to both coalescence and disruption of the planetesimals, so that a size distribution among the planetesimals is quickly developed (Zvygina et al., 1973; Greenberg et al., 1978b). After, typically, several thousand years, planetesimals with diameters of the order of 1000 km (comparable to that of the Earth's satellite, the Moon) are produced. Thereafter, the collisional frequency is relatively low unless the planetesimals have considerable random velocities. Random velocities among the planetesimals may be induced by the gravitational influence planetesimals exert on one another when they undergo near but noncollisional scattering. Random velocities may also be damped by dissipative collisions (Safronov, 1969; Kaula, 1979; Wetherill, 1978, 1980; Stewart and Kaula, 1980; Stewart, 1982). Toward the final stages of accumulation, when most of the mass is contained in several massive protoplanetary embryos, it is essential for the embryos to maintain relatively large random velocities until they have coagulated into four embryos corresponding to the progenitors of the four terrestrial planets (Ziglina, 1976; Ziglina and Safronov, 1976; Cox et al., 1978; Cox and Lewis, 1980; Wetherill, 1980). Owing to the low collisional frequency, the typical growth time scale during the final stage is around $10^7$ y to $10^8$ y.

The gas-free scenario provides a natural explanation for the observed similarity of relative chemical abundances of heavy inert gases (e.g., Xe, Kr, $^{36+38}$Ar), with respect to other elements such as $^{16}$O, found on the surface of terrestrial planets and in meteorites (Grossman and Larimer, 1974; also see Chapter 1 by Wasson and Kivelson). Since these gases are inert, they are probably well mixed in the interior of terrestrial planets. Thus the relative abundances found on the surface of the terrestrial planets probably represent the average abundances of these planets. These relative abundances are smaller than those found in the Sun by a factor of $10^{-6}$ to $10^{-9}$. The extremely low relative abundances of these heavy inert gases have been interpreted as evidence that either gas was never accreted onto, or inert gas was blown off, the surface of planetesimals when they had sizes comparable to or larger than the meteorites' parent bodies (Wasson, 1981).

The gas-free scenario provides a more plausible mechanism for prototerrestrial planet formation than does the massive gravitational instability scenario, principally because the mass and the composition of the protoplanets may be deduced without a great number of ad hoc assumptions. However, it is not by any means free of assumptions. Since the random velocities of the planetesimals at each stage depend critically on the balance between the gravitational relaxation process and the poorly understood collisional fragmentation, coalescence, and damping processes, the outcome and the time scale of the accumulation process depend critically on various adopted assumptions.

One of the basic assumptions of the gas-free scenario is that the nebular gas must be removed shortly after the condensation of the grains. Since the first stage of the grain accumulation is very rapid, unless gas is removed within $10^4$ y after the condensation of the grains, planetesimals of all sizes up to that of the Moon may be produced in a gas-rich environment. This argument leads to a crucial constraint on the timing of the clearing stage. The existence of such a precisely timed gas-removal mechanism still needs to be verified. One of the weakest aspects of the scenario is that it does not provide a unified scheme for the formation of all protoplanets. Since the giant planets are gas-rich, they must be formed in a gas-rich environment. However, the nebula is continually heated by viscous dissipation, so that the gaseous environment is relatively hot. If the protogiant planets formed through gradual accumulation processes like the prototerrestrial planets, they would have acquired a solid core with a mass comparable to that of the terrestrial planets before they could have accreted the relatively hot nebula gas efficiently. In this case, it would be rather difficult to avoid the formation of prototerrestrial planets during the formation epoch of the preprotogiant planetary embryos. Thus, according to the basic assumption of the gas-free scenario, the prototerrestrial planets must form not only at a different epoch and at a different distance from the Sun, but also through different processes from those of the protogiant planets (Wetherill, 1980). Around all the giant planets today, however, there are gas-free, solid, or icy-solid satellites which were presumably formed at the terminal stages of giant-planet formation. The chemical compositional gradient found in the Galilean satellites around Jupiter may be interpreted as evidence for the hypothesis that these solid or icy-solid satellites may have formed in a gaseous disk around Jupiter. Because satellite systems around giant planets have much smaller physical dimensions than the planetary system, the gaseous disks out of which these satellites were formed around the protogiant planets evolve much faster than the solar nebula. Thus, in terms of the global evolutionary time scale, these satellites may be considered as being formed at approximately the same time as the protogiant planets. If the giant planets can be coeval with their solid or icy-solid satellites, why cannot they be coeval with the terrestrial planets? To summarize, these arguments imply that, in addition to philosophical preference, there is observational evidence providing support for a grand unified scenario in which the prototerrestrial and giant planets formed in a gas-rich environment at the same time.

**The scenario of accumulation in a gas-rich environment** The third scenario is somewhat similar to the second in which the protoplanets formed through a series of gradual accumulation processes. However, the basic concept of this scenario is that all the protoplanets formed in a gas-rich environment at about the same time (Hayashi et al., 1977; Hayashi, 1981; Lin and Papaloizou, 1985). The starting point of this scenario is identical to that of the gas-free accumulation scenario in the sense that the grains are condensed out of the nebula at an early epoch. Through cohesive collisions, the grains grow to centimeter-sized objects and then segregate from the gas flow. The presence of convectively driven turbulence induces the grains to have sufficiently large velocity dispersion, so that the grain layer is gravitationally stable. Through inelastic and cohesive collisions, grains can quickly build up to produce kilometer-sized planetesimals (Weidenschilling, 1977). It should be pointed out that collisions induce both coalescence and fragmentation, so that there are always ample micron-sized grains to provide the opacity for the nebula to remain optically thick and thereby convective. Furthermore, the grain condensation in the viscous stage is a continuous process, and thus there is a steady supply of newly condensed grains.

In a gas-rich environment, the planetesimals cannot accrete gas until their mass is sufficiently large for their gravity to dominate the random motion of the gas molecules. In a minimum-mass nebula, the planetesimals, near the orbital radius of the Earth, must acquire a solid core comparable to the mass of the Moon before they can accrete gas at all. At larger orbital radii, the nebular gas is cooler so that the critical mass is somewhat smaller. Thus, as in the gas-free scenario, the first stage of the protoplanetary formation proceeds with cohesive collisions among the planetesimals. In the gas-rich scenario, it is postulated that the prototerrestrial planets form in the relatively hot inner region of the nebula so that they are unable to accrete much gas (Lin and Papaloizou, 1985). Later we justify this postulate. The giant planets, composed mostly of hydrogen gas, could have been formed through accretion of gas onto a solid core of planetesimals. The protogiant planets can attain most of their mass during the gas-accretion phase provided that (1) there is at least a minimum of 3 percent of a solar mass of gas left in the nebula to provide the necessary amount of gas that the protogiant planets must accrete, and (2) the effective nebular Reynolds number in the vicinity of the protogiant planets' orbit does not exceed a few thousand so that they can attain their present mass without tidally truncating the nebula. What is the physical implication of these two constraints? In order to satisfy both of these criteria, the inwardly diffusing mass flux in the nebula must be at least a good fraction of one millionth of a solar mass per year. If such a high nebular mass flux is maintained by infall, this stage cannot be continued for several hundred thousand years; otherwise the Sun would have accreted much more mass than it now has. According to the numerical results on the evolution of the nebula in the viscous phase, which are discussed above, such a high nebular mass flux can be maintained at most several hundred thousand years, without infall, before most of the nebular gas is depleted by accretion onto the protosun. Because of these arguments, we are forced to conclude with a very strong constraint that the protoplanets must have acquired most of their mass within 1 My after the formation of the protosun. This time scale is consistent with the constraint for the grain-condensation time scale and the $^{26}$Al-inclusion time scale.

Can protoplanets grow so quickly? The discussion of the gas-free scenario indicates that the formation of prototerrestrial planets, through a gradual accumulation process in a gas-free environment, has a much longer time scale than a 1 My. However, in a gas-rich environment, the accumulation processes may be somewhat speeded up. For example, through the gas-drag effect (Kusaka et al., 1970; Whipple, 1971; Adachi et al., 1976), tidal effects (Goldreich and Tremaine, 1980), and inelastic collisions (Wetherill, 1980), the random motion of the dust grains, planetesimals, and protoplanets may be continually dissipated so that their random velocity is considerably less than their surface escape velocity. In this case, the self-gravity of the planetesimals influences the orbit of neighboring planetesimals over a radius comparable to the radius of the Laplace sphere of influence. The *Laplace sphere of influence* of a planetesimal is normally defined in a frame that rotates around the protosun at the same angular frequency as the planetesimal. It is the average distance from the planetesimal where massless particles would experience equal gravitational force from the protosun and from the planetesimal. Quantitatively, it is proportional to the distance, as well as to the four-tenths power of the mass ratio, between the planetesimal and the protosun, and is generally considerably larger than the linear dimension of the planetesimals. Detailed numerical calculations indicate that if all the planetesimals have very small orbital eccentricities, most planetesimals that come within the Laplace sphere of influence of a planetesimal may either collide with that planetesimal or collide with other deflected planetesimals (Dole, 1962; Lin and Papaloizou, 1982). If there is a significant gaseous envelope within the Laplace sphere of influence, the gas-drag effect may also lead to the captures of intruding planetesimals (Pollack et al., 1979). If a significant fraction of the strongly deflected planetesimals collide cohesively, the capture cross section would increase considerably, so that the growth time scale for Moon-sized planetesimals can be reduced to about $10^4$ y (Lyttleton, 1972). Incidentally, this time scale is the shortest one possible for the growth of solid planetesimals through inelastic collisions. It is also a generally accepted time scale by almost all the scenarios for the information of Moon-sized objects.

The phase of rapid growth leads to the production of Moon-sized planetesimals. Thereafter, most of the planetesimals with overlapping Laplace sphere of influence have coalesced. While there is little disagreement on the relatively short time scale required for the production of Moon-sized planetesimals, there is considerable divergent discussion on the detailed processes under which a few hundred Moon-sized planetesimals coagulate into four terrestrial planets. After most of the mass in heavy elements has been incorporated into Moon-sized planetesimals, the rapid growth rate is terminated unless either (1) the high collisional frequency is retained through a relatively large random motion among the planetesimals, or (2) there is a continuous supply of solid material for the most massive planetesimals to accrete. The first possibility has already been discussed extensively in the gas-free scenario. The typical growth deduced therefrom is $10^7$ to $10^8$ y (Wetherill, 1980), which is much too long to satisfy the deduced growth-time constraint described above. There are two possible mechanisms to provide a continuous supply of solid material for the most massive planetesimals to accrete. First, the gas-drag effect induces low-mass planetesimals to migrate toward the protosun faster than high-mass planetesimals. This effect tends to cause planetesimals with different masses to cross one another's orbit as they evolve. Thus, planetesimals that are orginally located at different radii may be brought together by this differential radial migration, provided their masses are sufficiently different (Hayashi et al., 1977; Nakagawa, 1978). However, if most of the solid materials are already in the form of Moon-sized or larger planetesimals, this process becomes ineffective (Wetherill, 1980). Second, as we have already indicated in our discussion on the evolution of the nebula, grain condensation is a continuous process in the outer region of the nebula during the viscous stage. As the nebula evolves, the continuous-condensation zone spreads inward. The nebular gas-drag effect is always effective in inducing a radial migration of the grains so that the newly condensed grains may be continually brought to the Laplace sphere of influence of the massive planetesimals (Lin and Papaloizou, 1985). Combining these two effects we find that the time scale required for a Moon-sized planetesimal to grow to a mass about that of the Earth, at the Earth's orbital radius, is about $10^5$ to $10^6$ y, provided the nebula is somewhat more massive than the minimum-mass nebula (Lin and Papaloizou, 1985). In fact, such a nebula may be required since a significant fraction of grains may be carried by the gas flow in the nebula and be accreted by the Sun so that the heavy elements in the nebula are not entirely retained by the planetesimals.

Planetesimals with mass larger than that of the Moon can also accrete gas. The gas and dust accretion rates depend on different powers of the nebular temperature and of the planetesimals' mass. There is a critical mass at every orbital radius such that the two accretion rates are equal. For example, in a minimum-mass nebula, this mass is approximately 1 Earth mass at the present oribt of Jupiter. At the present orbit of Earth, this mass is approximately 10 Earth masses. Because the dust accretion rate depends on a lower power of the accretor's mass, for masses much less than 1 Earth mass, dust accretion dominates gas accretion. The gas-accretion rate rapidly exceeds the dust-accretion rate as the accretors grow beyond the critical mass. The gaseous envelope surrounding the solid core mass becomes optically thick, forming a protogiant planet. At the present orbital radius of Jupiter, the fastest time scale for doubling the mass of an object of 1 Earth mass is about $10^5$ y, and that for doubling the Jupiter's mass is about $10^3$ y.

Can this gas-rich scenario provide a satisfactory explanation for the lack of gas in the terrestrial planets today? The numerical solution for the solar nebular indicates that the gas temperature at the midplane is inversely proportional to the three-halves power of the orbital radius, and thus the nebula is considerably hotter than 100 K inside the orbit of Jupiter. At the orbital radius of Earth, the mass of the planetesimals must exceed several times the Earth's present mass before they can accrete gas faster than dust. Thus, even if the prototerrestrial planets collected all the dusty material from the nebula in the vicinity of their present orbits, they still would not have sufficient mass to accrete gas efficiently. Nonetheless, some nebular gas may be accreted. According to one estimate, the protoearth may accrete a primitive atmosphere with a mass approximately 5 percent of the present mass of the Earth (Hayashi et al., 1979). If so, all but a small fraction of the original atmosphere must subsequently be ejected in order for the Earth's atmosphere to evolve into its present state. The protostar ultraviolet radiation and the T Tauri-like wind may induce hydrogen and helium to escape the terrestrial planets after they are accreted (Hunten and Donahue, 1976), while the inert volatile gas could be carried along with the hydrogen and helium if the outflow is sufficiently intense (Sekiya et al., 1980, 1981). It is also possible for the surface of Earth to become relatively hot—say, 4000 K— in the accretion phase, so that some inert gas such as neon can be dissolved into the melted Earth crust (Hayashi et al., 1979). All these processes are very complex and are still the subject of some controversy (Wetherill, 1980). For example, the gas-accretion estimates are based on somewhat ad hoc assumptions made for the structure of the atmosphere. When the environmental effect of the relatively hot nebular gas is appropriately taken into account, the amount of gas that may be accreted would be somewhat smaller than these estimates. We have just begun to address some of these issues. In the near future, the results produced from theoretical investigations may provide some predictions that can be used with experimental data to test the validity of the theory.

What process causes the termination of the growth of protogiant planets? As was already discussed, when the protogiant planets grow to a sufficiently large mass, such that the ratio of their mass to that of the protostar is larger than the inverse of the nebular Reynolds number in the vicinity of their orbit, the nebular-protoplanetary-tidal interaction truncates the nebula and opens up zones of avoidance in the vicinity of the protoplanets, so that their further growth is inhibited (Cameron, 1979; Lin and Papaloizou, 1980, 1982; see Fig. 3-14). The present mass ratio of the giant planets to the Sun is between $10^{-4}$ and $10^{-3}$. If the masses of the planets and the Sun have not changed significantly since their formation, the Reynolds number at their formative radius must be relatively low. This constraint is particularly severe for the most massive planet, Jupiter. Numerical models of the nebula indicate that the regions of the nebula with the lowest Reynolds number are either very near the protosun or just outside the ice-grain condensation zone (see Fig. 3-8). It is unlikely that Jupiter formed very near the protosun because the nebula within its orbital radius would be too hot to allow grain condensation, which is essential for the formation of the terrestrial planets. Just outside the ice-grain condensation zone, however, the Reynolds number may be as low as a few thousand, provided the inwardly drifting mass flux of the nebula is roughly one-millionth of a solar mass per year.

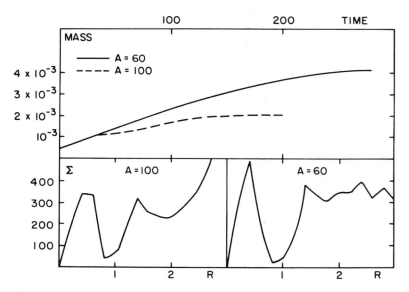

Fig. 3-14  The evolution of the protoplanetary mass as a function of the local Reynolds number. The Reynolds number is defined as being $(0.25A)^2 R^{-1}$, so that it equals $625 R^{-1}$ for $A = 100$ and $225 R^{-1}$ for $A = 60$. Notice the gradual termination of protoplanetary growth as a function of time. Tidal truncation of the nebula occurs when the protoplanet has a relatively low mass in the case $A = 100$ where the Reynolds number is relatively high.

Beyond the ice-grain condensation zone, the effective Reynolds number increases with the square root of distance; that is, if all the protogiant planets were formed at the same epoch, the more distant protogiant planets would terminate their growth at lower masses. If they were not formed at the same epoch, the termination condition must be modified accordingly. Of course the condition outlined above provides upper mass limits to protoplanetary growth. The nebula may very well be depleted of gas before such maxima are reached.

According to the convective nebular models, the prototerrestrial planets' formation region is too warm to permit ice-grain condensation, so that the opacity and the magnitude of the viscosity would be relatively low there. Consequently, the Reynolds number is very large there. Thus, if the prototerrestrial planets were able to collect a sufficient amount of solid material to build up a critical mass for efficient gas accretion to proceed, they would still be unable to accrete very much gas before their tidal effect would induce a truncation of the nebula and thereby force a termination of their growth. It may be of interest to note that although different schools of thought may have different opinions on one or another aspect of the description above, the general physical constraints and the basic physical mechanisms outlined above are generally applicable to all models.

The basic attractiveness of the scenario of planetary formation described above is that it provides a natural explanation for the mass distribution and spatial distribution of the present planetary system. Although some detailed assumptions still need to be justified and some physical processes still need to be examined, so far the scenario presents no difficulties in satisfying several critical constraints on the time scales of protoplanetary formation. Furthermore, this scenario provides a few important constraints on the solar nebular models.

## 3-5-2 The Nebular Structure at the Final Stages of Protoplanetary Formation

**Mass** So far, the discussion has centered on the efficiency and time scale of protoplanetary growth from a relatively low mass nebula. Although the mass of the nebula during this epoch may have been considerably larger than the adopted values, we can justify the approximation noted above by the argument that the evolutionary time scale for a relatively massive nebula is much too short for the protoplanets to grow substantially. A stronger dynamical constraint, which suggests that during this epoch the nebula's mass is indeed between a few and 10 or 20 percent of 1 solar mass, is provided by the nebular-protoplanetary tidal interaction. According to the discussion above, the protoplanets' growth is terminated when the ratio of their masses to the protosun exceeds the inverse of the Reynolds number. The present mass of Jupiter is one-thousandth of a solar mass. If Jupiter's growth was indeed terminated by tidal truncation of the nebula, the Reynolds number close to Jupiter's initial orbital radius must be around 1000. If we adopt Jupiter's initial orbital radius to be comparable to its present value—an assumption that is justified later—the mass of the nebula may be deduced from our numerical models, provided the locations of the inner and outer boundaries of the nebula are specified. The location of the inner boundary is not particularly critical, because the central region of the nebula always tends to evolve into a quasi-steady state that extends all the way to the surface of the protosun. Since the surface density remains more or less independent of distance from the protosun in the inner region of the nebula, the nebular mass there is relatively small. The location of the outer boundary, however, is crucial to the determination of the mass of the nebula. Its value can be deduced from a constraint on how far out the optically thick region of the nebula extends during the epoch of protogiant-planet formation. Numerical models of the solar nebular indicate that the optical depth as well as temperature of the nebula are decreasing functions of the radius. At very large radii, the nebula becomes transparent and convection terminates. Under this situation, the time scale for radiative cooling is short compared with that for viscous dissipation, so that turbulence decays. Unless there is some other mechanism by which efficient angular-momentum transfer may be maintained, the nebular-protoplanetary-tidal interaction becomes dominant. Consequently, protoplanets produced in the optically thin region cannot accrete very much gas from the nebula. Although we have very little observational evidence from which to deduce where the nebula actually became transparent, such a location must have been beyond the initial orbit of Neptune in order for Neptune to have accreted its present mass. The lack of major planets beyond the present orbital radius of Neptune suggests that the optically thick region of the nebula may have indeed terminated there. By applying the constraints on the Reynolds number at the vicinity of Jupiter's orbit and on the outer boundary of the nebula to a numerical solar-nebular model, we deduce that the nebular mass for the epoch of protoplanetary formation was between a few and 20 percent of 1 solar mass. This result is also consistent with the mass deduced for a detailed parameter fit between the cosmochemical data on the condensation temperature of various planets and satellites and the temperature distribution of a quasi-steady-state nebular model (see Fig. 3-9).

**Orbital evolution of the protoplanets** Through the tidal interactions just discussed above, not only can the nebular structure be strongly affected, but the orbits of the protoplanets may evolve. In this case, the time scale for tidally induced orbital evolution would be of the order of a few tens of thousands of years (Goldreich and Tremaine, 1980). Once the nebula is tidally truncated, gas cannot stream from the outer region of the nebula inside the orbit of Jupiter, so that matter accumulates just beyond the orbit. Such an accumulation of nebular material provides favorable conditions for the formation of Saturn. The nebular gas inside Jupiter's orbit, however, can easily transfer its angular momentum to Jupiter so that this gas diffuses toward the protosun. Consequently, the surface density inside Jupiter's orbit gradually decreases, while that outside Jupiter's orbit gradually increases. Eventually, the inner region of the nebula transfers angular momentum to Jupiter at a slower rate than does Jupiter to the outer region, so that Jupiter loses angular momentum and spirals toward the protosun (see Fig. 3-10). Such an inward orbital evolution is terminated when most of the external nebula either is accreted by Saturn or somehow disperses.

What observational evidence is there to support the speculation that Jupiter may have undergone considerable orbital evolution? One possible evidence is the present distribution of asteroids' orbits. Today, in addition to some major gaps where the population of asteroids is relatively small (e.g., the Kirkwood gaps), there are several classes of asteroids (e.g., the Hilda and Trojan groups), which are found to be in orbital resonances with Jupiter (i.e., the orbital period can be expressed as a simple integer multiple of that of Jupiter). The relatively large orbital eccentricities of the resonant asteroids suggest that they cannot have evolved into their current orbital commensurabilities with Jupiter through collisional processes. In fact, collisional processes have a tendency to smear out the gaps and to cause the commensurable asteroids to leave the resonances (Franklin et al., 1980). There is a possible scenario to account for the origin of the orbital distribution of the asteroids. As Jupiter moves inward, its resonances also move inward. As the low-order resonances sweep across the nebula, asteroids may be captured through resonant effects in a way analogous to the model proposed for the origin of the commensurability of Jupiter's inner satellites (Goldreich, 1965). (See a more detailed discussion in Chapter 12 by Peale.) This model of tidally induced resonant capture provides not only a natural explanation for the spatial distribution of the asteroids and the origin of the frequently found commensurability between the orbital periods of Jupiter and asteroids but also a constraint on the extent of orbital evolution of Jupiter. This constraint is based on the scenario that the asteroids trapped in Jupiter's resonances today were once forced by Jupiter's resonant tidal effect to evolve along with the inward migration of Jupiter's orbit. In this scenario, Jupiter's tidal effect induces a progressively increasing eccentricity on the resonant asteroids. Thus, from the magnitude of the eccentricities of the resonant asteroids, it is possible to deduce that the proto-Jupiter's orbit did not migrate by more than a factor of two. The initial orbital radius of Jupiter, therefore, may not have been very different from the present radius of Jupiter's orbit.

We now turn our attention to the evolution of Neptune's orbit. Since it is located near the outermost region of the nebula, Neptune's orbit tends to be driven outward by the nebular-protoplanetary-tidal interaction. It is quite conceivable that during its outward migration, Neptune actually captured Pluto into its presently observed 3:2 resonance. If so, the large magnitude of Pluto's and the small magnitude of Neptune's orbital eccentricities again indicate that the proto-Neptune's orbit did not evolve by more than a factor of two since the resonant capture of Pluto. Combining the results that Jupiter has moved inward and Neptune has moved outward, we deduce that the bulk of the nebular material is located between the orbits of Jupiter and Neptune. The orbits of other protoplanets may also have evolved through their orbital interaction with the nebula. Detailed analysis of this process may provide an explanation for Bode's law (see Chapter 1 by Wasson and Kivelson).

If the final stage of protoplanetary formation coincided with the nebular clearing stage, the gradual depletion of nebular gas could induce a significant orbital evolution of the protoplanetary system (Ward, 1981). In particular, as the gravitational influence of the nebular gas decreases, the gravitational significance of the protoplanetary system begins to emerge and generate orbital interaction. This gradual change causes secular variation in the orbital characteristics of the protoplanetary system and induces temporary secular orbital resonances among the protoplanets. If these resonances persisted for a significant period, say much greater than a few tens of thousands of years, the orbital eccentricities and inclinations of the terrestrial planets would have been forced to increase beyond the presently observed values. Under certain circumstances, the propagation of the secular resonances may induce the orbits of remaining Moon-sized planetesimals between the major planets to become unstable. Such a scenario provides a natural explanation for the absence of Moon-sized objects between the major planets today. The time scale for the propagation of the secular resonance is much shorter than the time scale of the prototerrestrial planets' growth deduced from the gas-free accumulation scenario, but is consistent with the protoplanetary formation time scale deduced from the gas-rich accumulation scenario. But such a phenomenon probably does not occur if the terrestrial planets were formed according to the gas-free scenario, since most of the gas in the nebula would have been depleted before the planetesimals have acquired any appreciable mass.

**The efficiency of heavy-element accumulation** The nebular mass deduced above is considerably smaller than that required for gravitational instability but somewhat larger than that required for the minimum-mass nebula. If the chemical abundances of the nebula were similar to those of the Sun, the total mass of the heavy elements within the orbital radius of Jupiter was only a few times larger than the total mass of the terrestrial planets. Relatively high efficiency in collecting the heavy elements is required for the formation of the terrestrial planets in such an environment. This dilemma implies that the formation of the first-generation planetesimals, both in the ice-grain and the ice-grain-free regions, must have taken place prior to the final stage of Jupiter's growth. At these earlier epochs, the mass of the nebula, both inside and outside Jupiter's orbit, is somewhat larger so that sufficient amounts of heavy elements can be converted into planetesimals even if the conversion efficiency is moderately low. As the nebula evolves, its mass flux decreases. The combined effects of this decrease and Jupiter's growing mass may cause the nebula to be tidally truncated and the gas supply interior to Jupiter's orbit to become sparser. If a sufficient amount of heavy elements has already been converted into planetesimals, however, there is enough material to build up the terrestrial planetary system. This requirement is again in good agreement with the results obtained from the direct comparison between cosmochemical data and quasi-steady-state models (see Fig. 3-9). In particular, both arguments lead to the same conclusions, namely; (1) the condensation of the grains proceeds within a few hundred thousand years, and (2) the prototerrestrial planets form at the same epoch as the protogiant planets.

Throughout our presentation we have analyzed the importance of the nebular-protoplanetary-tidal interaction. This process provides (1) a natural terminal point for protoplanetary growth, (2) a final clearing of the nebula, (3) a mechanism for orbital evolution of the protoplanets, and (4) the basis for several important constraints on the structure of the nebula.

Convection as a mechanism of transporting angular momentum, as well as energy, provides the minimum effective viscosity for the nebula. Indeed, there may be other sources of effective viscosity, such as those suggested by Cameron (1978a, 1978b). However, the effective viscosity cannot be much greater than that provided by convection; otherwise the mass of the nebula would be even smaller than what is deduced above. It is difficult to assess the error in the analysis above. For example, the constraint on Jupiter's mass set by nebular-protoplanetary-tidal interaction is an order-of-magnitude estimate. Some of the constraints imposed above, however, are grossly violated if the mass of the nebula is changed substantially from the values deduced above. Therefore, it seems quite likely that the time span for the formation of the protogiant planets ranged from several hundreds of thousands to a million years.

## 3-6 OBSERVATIONAL TESTS

The models constructed for the solar nebula may be applied to other planetary systems that are now in the process of formation around young stars. The most distinctive features that can be observed in these newly forming systems are the characteristic spectra and the temporal variations in the energy released from the nebula. In particular, the numerical solutions indicate that throughout the viscous phase of the nebular evolution, the inner region of the nebula always evolves toward a steady state. At a temperature below 2000°K, grains provide the dominant opacity source such that the characteristic spectrum of the radiation released from the nebular component is that in which the luminosity per unit frequency is proportional to the one-third power of the frequency. The characteristic wavelength of this radiation is in the far-infrared region, that is, several to several dozen microns.

In the region of the nebula very near the protostar, the temperature may be well above 2000°K, so that molecular dissociation and ionization of hydrogen are very active processes. The energy released from this region is in the near-infrared spectrum. The characteristic intensity in the near-infrared may not increase with frequency as fast as in the far-infrared region, since a considerable amount of energy released through dissipation within this region of the nebula is absorbed in order to provide the energy for hydrogen molecules to dissociate and hydrogen atoms to ionize. Furthermore, this region of the nebula may be thermally unstable. The cause of the thermal instability is the strong dependence of opacity on the nebular temperature. A slight increase in nebular temperature induces a considerable increase in opacity with a resultant drastic reduction in the cooling efficiency. However, a slight increase in the temperature also induces an increase in the convective velocity, so that the rate of energy release due to convectively driven turbulent dissipation is increased at the same time. The rate of heat output consequently exceeds the rate at which heat may be transported, and the nebula continues to heat up.

Without detailed quantitative analysis, the actual outcome of the instability is somewhat uncertain. The unstable region of the nebula, however, may oscillate between a very cold state, that is, less than 2000°K, and a very hot state, that is, greater than 10,000°K, on a time scale ranging from several hours to a few months. It is possible that this thermal instability may induce a fluctuation in the observed luminosity and in the optical and near-infrared spectrum, and may perhaps even induce a wind similar to that observed in T Tauri stars.

In the outer region of the nebula, a steady state cannot be achieved. Since both the energy dissipation rate and the nebular temperature decrease rapidly as a function of the distance from the protosun, less energy is released at lower photon energy, that is, longer wavelength. Consequently, the spectrum in the extreme far-infrared region is expected to increase with frequency faster than that in the far-infrared region (see Fig. 3-6). The overall luminosity from the nebula depends on the total mass of the nebula and the radius of the protostar. If the mass of the nebula is comparable to that deduced for the minimum-mass nebula and if the protostar is a few solar radii in size, the total nebular luminosity may be around a few to a few hundred times that of the present solar luminosity. Most of this energy is radiated in the near-infrared and optical regions. Of course, the protostar may also release intense optical and near-infrared radiation. It may be extremely difficult to separate the contribution between the protostar and the nebula.

Another interesting feature may be observable in the far-infrared if the nebula is in the final stage of the viscous epoch in which the protoplanets are fully grown and are tidally interacting with the nebula. According to a series of computer simulations, when the tidal force of the protoplanets is sufficiently large to truncate the nebula, it also induces a spiral-dissipation pattern analogous to the grand spiral pattern observed in spiral galaxies (Lin and Papaloizou, 1979a, 1979b; see Fig. 3-15). The most probable location for the formation of this spiral pattern is near the ice-grain–transition region. The energy

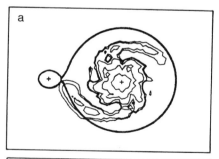

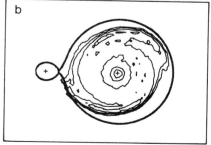

Fig. 3-15 (a) The dissipation pattern of the nebula with a hypothetical protoplanet that has a mass 20 times that of Jupiter. The spiral pattern extends from inside the 2:1 resonance to the edge of the Hill radius. (b) The density contour of a weak spiral pattern is shown.

released from the spiral pattern may have wavelengths ranging from a few to a few tens of microns. The luminosity of the spiral pattern may range from a few percent to an appreciable fraction of the solar luminosity. If the nebula is observed nearly edge on, regular variations in the far-infrared radiation may be observed as the pattern rotates around the protostar. The period of the variation may range from several to several tens of years. Although this period is expected to be regular, it may contain secular variations on the time scale of a few tens of thousands to hundreds of thousands of years. The secular variation is the result of the tidally induced orbital evolution of the protoplanets.

According to the discussion above, aside from the fact that the Sun is not a member of a close, multiple-star system, there are virtually no unique properties of the Sun that are required to induce the formation of the solar system. Planetary systems, therefore, may be abundant in the galaxy. Efforts are being made to detect the existence of these systems (Black, 1980; Black and Scargle, 1982). It may be interesting to utilize the models of solar-nebula and protoplanetary formation described above to predict the existence and characteristics of extrasolar planetary systems. The ultimate discovery of such systems would have a profound philosophical impact. It may constitute the first step in the search for the existence of extraterrestrial intelligent life.

## 3-7 SUMMARY

In the past few years, theoretical investigations on the origin and evolution of the solar nebula have progressed rapidly. Through these investigations, a new grand unified scenario for the formation of the solar nebula has emerged. In this scenario, the nebular phase of the solar system is turbulent, of relatively low mass, and short-lived. Convectively driven turbulence not only provides an energy-transport process in the vertical direction but also induces viscous energy dissipation and mass transfer in the radial direction. Consequently, the presence of turbulence prevents a buildup of the gaseous mass of the nebula. Various dynamical constrainsts suggest that prototerrestrial and protogiant planetary formation proceeds through a series of gradual accumulation processes. During the final epoch of protoplanetary formation, the mass of the nebula was around 0.1 solar mass. Most of the protoplanets probably acquired the bulk of their mass within about 1 My after the formation of the protosun.

## 3-8 ACKNOWLEDGMENTS

I thank Drs. Peter Bodenheimer and John Papaloizou for useful conversations and helpful comments. This work was supported in part by a grant from the National Science Foundation AST 81-00163.

## 3-9 REFERENCES AND ADDITIONAL READING

Abt, H., and S. Levy, 1976, Multiplicity among solar-type stars, *Astrophys. J. Suppl.*, 30, 273-306.

Adachi, I., C. Hayashi, and K. Nakazawa, 1976. The gas drag effect on the elliptic motion of a solid body in the primordial solar nebula, *Prog. Theor. Phys. Japan, 56*, 1756-1771.

Black, D. C., 1972a, On the origins of trapped helium, nitrogen, and argon isotropic variables in meteorites I—gas rich meteorites, lunar soil, and breccia, *Geochim. Cosmochim. Acta, 36*, 347-375.

Black, D. C., 1972b, On the origins of trapped helium, nitrogen, and argon isotropic variables in meteorites II: carbonaceous chondrites, *Geochim. Cosmochim. Acta, 36*, 377-394.

Black, D. C., 1980, The detection of other planetary systems, *Mercury, 9*, 105-111.

Black, D. C., and J. D. Scargle, 1982, On the detection of other planetary systems by astrometric techniques, *Astrophys. J., 263*, 855-869.

Bodenheimer, P., 1978, Evolution of rotating interstellar clouds. III. On the formation of multiple star systems, *Astrophys. J., 224*, 488-496.

Bodenheimer, P., J. Tohline, and D. Black, 1980, Criteria for fragmentation in a collapsing rotating cloud, *Astrophys. J., 242*, 209-218.

Boynton, W. V., 1978, The chaotic solar nebula—evidence for episodic condensation in several distinct zones, in *Protostars and Planets* (T. Gehrels, ed.), Univ. Arizona, Tucson, Ariz., 427-438.

Brecher, A. and G. Arrhenius, 1974, The paleomagnetic record in carbonaceous chondrites—natural remanence and magnetic properties, *J. Geophys. Res., 79*, 2018-2106.

Buffon, G. L. L. Comte de, 1745, *De la Formation des Planetes*, Paris, France.

Cameron, A. G. W., 1962, The formation of the sun and planets, *Icarus, 1*, 13-69.

Cameron, A. G. W., 1973a, Accumulation processes in the primitive solar nebula, *Icarus, 18*, 407-450.

Cameron, A. G. W., 1973b, Abundances of the elements in the solar system, *Space Sci. Rev., 15*, 121-146.

Cameron, A. G. W., 1975a, Clumping of interstellar grains during the formation of the primitive solar nebula, *Icarus, 24*, 128-133.

Cameron, A. G. W., 1975b, The origin and evolution of the solar system, *Sci. Am., 233*, 33-41.

Cameron, A. G. W., 1976, The primitive solar accretion disk and the formation of the planets, in *The Origin of the Solar System* (S. F. Dermott, ed.), John Wiley, New York, 49-74.

Cameron, A. G. W., 1978a, Physics of the primitive solar accretion disk, *Moon and Planets, 18*, 5-40.

Cameron, A. G. W., 1978b, Physics of the primitive solar nebula and the giant gaseous protoplanets, in *Protostars and Planets* (T. Gehrels, ed.), Univ. Arizona, Tucson, Ariz., 453-487.

Cameron, A. G. W., 1979, The interaction between giant gaseous protoplanets and the primitive solar nebula, *Moon and Planets, 21*, 173-183.

Cameron, A. G. W., and M. B. Fegley, 1982, Nucleation and condensation in the primitive solar nebula, *Icarus, 53*, 1-13.

Cameron, A. G. W., and M. R. Pine, 1973, Numerical models of the primitive solar nebula, *Icarus, 18*, 377-406.

Cameron, A. G. W., and J. W. Truran, 1977, The supernova trigger for the formation of the solar system, *Icarus, 30*, 447-461.

Cassen, P., and A. Moosman, 1982, On the formation of protostellar disks, *Icarus* (submitted for publication).

Cohen, M., 1981, Are we beginning to understand T-Tauri stars? *Sky and Telescope, 62,* 300–303.

Cohen, M., and L. V. Kuhi, 1979, Observational studies of pre-main-sequence evolution, *Astrophys. J. Suppl., 41,* 743–843.

Cox, J. P., and R. T. Guili, 1968, *Principles of Stellar Structure,* Gordon and Breach, New York, 281–325.

Cox, L. P., and J. S. Lewis, 1980, Numerical simulation of the final stages of terrestrial planet formation: *Icarus, 44,* 706–721.

Cox, L. P., J. S. Lewis and M. Lecar, 1978, A model for close encounters in the planetary problem, *Icarus, 34,* p. 415–427.

DeCampli, W. M., and A. G. W. Cameron, 1979, Structures and evolution of isolated giant gaseous protoplanets, *Icarus, 38,* 367–391.

Descartes, R., 1644, *Principia Philosophiae* (trans. by V. R. Miller and R. P. Miller), Dordrecht, Holland, 1983.

Dole, S. H., 1962, Gravitational concentration of particles in space near the Earth, *Planet. Space Sci., 9,* 541–553.

Elmegreen, B. G., 1978, On the interaction between a strong stellar wind and a surrounding disk nebula, *Moon and Planets, 19,* 261–277.

Ezer, D., and A. G. W. Cameron, 1967, Early and main sequence evolution of stars in the range 0.5 to 100 solar masses, *Canadian J. Phys., 45,* 3429–3477.

Franklin, F. A., M. Lecar, D. N. C. Lin, and J. Papaloizou, 1980, Tidal torque on infrequently colliding particle disks in binary systems and the truncation of the asteroid belt, *Icarus, 42,* 271–280.

Goldreich, P., 1965, An explanation of the frequent occurrence of commensurable mean motions in the solar system, *Mon. Not. Roy, Astr. Soc., 130,* 159–181.

Goldreich, P., and D. Lynden-Bell, 1965a, Gravitational instability of uniformly rotating disks, *Mon. Not. Roy. Astr. Soc., 130,* 97–124.

Goldreich, P., and D. Lynden-Bell, 1965b, Spiral arms as sheared gravitational instabilities, *Mon. Not. Roy. Astr. Soc., 130,* 125–158.

Goldreich, P., and S. Tremaine, 1980, Disk-satellite interactions, *Astrophys. J., 241,* 425–441.

Goldreich, P., and W. R. Ward, 1973, The formation of the planetesimals, *Astrophys. J., 183,* 1051–1061.

Greenberg, R., 1982, Planetesimals to planets, in *Formation of Planetary System* (A. Brahic, ed.), Centre National d'etudes spatiales, Toulouse, France, 515–569.

Greenberg, R., W. K. Hartmann, C. R. Chapman, and J. F. Wacker, 1978a, The accretion of planets from planetesimals, in *Protostars and Planets* (T. Gehrels, ed.), Univ. Arizona, Tucson, Ariz., 599–624.

Greenberg, R., J. F. Wacker, W. K. Hartmann, and C. R. Chapman, 1978b, Planetesimals to planets: numerical simulation of collisional evolution, *Icarus, 35,* 1–26.

Grossman, L., and J. W. Larimer, 1974, Early history of the solar system, *Rev. Geophys. Space Phys., 12,* 71–101.

Hayashi, C., 1981, Formation of the planets, in *Fundamental Problems in the Theory of Stellar Evolution* (D. Sugimoto, D. Q. Lamb and D. N. Schramm, eds.), Proceeding of IAU Symposium, no. 93, Reidel, Dordrecht, Netherlands, 113–128.

Hayashi, C., 1982. Structure of the solar nebula, growth and decay of magnetic fields and effects of magnetic and turbulent viscosities on the nebular, *Prog. Theor. Phys. Suppl. Japan* (in press).

Hayashi, C., K. Nakazawa, and J. Adachi, 1977, Long-term behavior of planetesimals and the formation of the planets, *Pub. Astr. Soc. Japan, 29,* 163-196.

Hayashi, C., K. Nakazawa, and H. Mizuno, 1979, Earth's melting due to the blanketing effect of the primordial dense atmosphere, *Earth Planet. Sci. Lett., 43,* 22-28.

Herbig, G., 1962, The properties and problems of T Tauri stars and related objects, in *Advances in Astronomy and Astrophysics* (Z. Kopal, ed.), Academic Press, New York, *1,* 47-104.

Horedt, G. P., 1978, Blow-off of the protoplanetary cloud by a T Tauri–like solar wind, *Astr. Astrophys., 64,* 173-178.

Hoyle, F., 1960, On the origin of the solar nebula, *Q. J. Roy. Astr. Soc., 1,* 28-55.

Hunten, D. M., and T. M. Donahue, 1976, Hydrogen loss from terrestrial planets, *Ann. Rev. Earth Planet. Sci., 4,* 265-292.

Iben, I., 1965, Stellar evolution. 1. The approach to the main sequence, *Astrophys. J., 141,* 993-1018.

Jeans, J. H., 1916, The part played by rotation in cosmic evolution, *Mon. Not. Roy. Astr. Soc., 77,* 186-199.

Jeffreys, H., 1918, On the early history of the solar system, *Mon. Not. Roy. Astr. Soc., 78,* 424-441.

Kant, I., 1755, *Allemeine Naturgeschichte und Theories des Himmels* (trans. by W. Hashe), New York, Greenwood Publications, 1968.

Kaula, W. M., 1979, Equilibrium velocities of a planetesimal population: *Icarus, 40,* 262-275.

Kellman, S. A., and J. E. Gaustad, 1969, Rosseland and Planck mean absorption coefficients for particles of ice, graphite, and silicon dioxide, *Astrophys. J., 157,* 1465-1467.

Kobrick, M., and W. M. Kaula, 1979, A tidal theory for the origin of the solar system, *Moon and Planets, 20,* 61-101.

Kusaka, T., T. Nakano, and C. Hayashi, 1970. Growth of solid particles in the primordial solar nebular, *Prog. Theor. Phys. Japan, 44,* 1580-1595.

Lada, C. J., L. Blitz, and B. G. Elmegreen, 1978, Star formation in O-B associations, in *Protostars and Planets* (T. Gehrels, ed.), Univ. Arizona, Tucson, Ariz., 341-367.

Laplace, P. S. de., 1796, *Exposition du System du Monde,* Paris, France.

Larson, R., 1969, Numerical calculation of the dynamics of a collapsing protostar, *Mon. Not. Roy. Astr. Soc., 145,* 271-295.

Lee, T., D. A. Papanastassiou, and G. T. Wasserberg, 1976, Demonstration of $^{26}$Mg excess in Allende and evidence for $^{26}$Al, *Geophys. Res. Lett., 3,* 109-112.

Lewis, J. S., 1972, Low temperature condensation from the solar nebular, *Icarus, 16,* 241-252.

Lewis, J. S., 1974, The temperature gradient in the solar system, *Science, 186,* 440-443.

Lin, D. N. C., 1981, Convective accretion disk model for the primordial solar nebula, *Astrophys. J., 246,* 972-984.

Lin, D. N. C., and P. Bodenheimer, 1982, On the evolution of convective-accretion-disk models of the primordial solar nebula, *Astrophys. J., 262,* 768.

Lin, D. N. C., and J. Papaloizou, 1979a, Tidal torques on accretion disks in binary systems with extreme mass ratios, *Mon. Not. Roy, Astr. Soc., 186,* 799-812.

Lin, D. N. C., and J. Papaloizou, 1979b, On the structure of circumbinary accretion disks at the tidal evolution of commensurable satellites, *Mon. Not. Roy. Astr. Soc., 188,* 191-201.

Lin, D. N. C., and J. Papaloizou, 1980, On the structure and evolution of the primordial solar nebula, *Mon. Not. Roy. Astr. Soc., 191*, 37-48.

Lin, D. N. C., and J. Papaloizou, 1985, On the dynamic origin of the solar system, in *Protostars and Planets II* (D. Black, ed.), Univ. Ariz., Tucson, Ariz. (in press).

Lüst, R., 1952, Die Entwicklung einer um einer zentralkoerper rotierenden Gasmasse, I. Loesungen der hydrodynamischen Gleichungen mit turbulenter Reibung, *Zeit. f. Naturforschung, 7a*, 87-98.

Lynden-Bell, D., and J. E. Pringle, 1974, The evolution of viscous discs and the origin of the nebula variables, *Mon. Not. Roy. Astr. Soc., 168*, 603-637.

Lyttleton, R. A., 1972, On the formation of the planets from the solar nebular, *Mon. Not. Roy. Astr. Soc., 158*, 463-483.

Mestel, L., 1965, Meridian circulation in stars, in *Stellar Structure* (L. H. Aller and D. B. McLaughlin, eds.), Univ. Chicago, Chicago, Ill., 465-498.

Mestel, L., 1968a, Magnetic braking by a stellar wind. I, *Mon. Not. Roy. Astr. Soc., 138*, 359-391.

Mestel, L., 1968b, Magnetic braking by a stellar wind. II, *Mon. Not. Roy. Astr. Soc., 140*, 177-196.

Mestel, L., and L. Spitzer, 1956, Star formation in a magnetic dust cloud, *Mon. Not Roy. Astr. Soc., 116*, 503-514.

Mouschovias, T. Ch., 1978, Formation of stars and planetary systems in magnetic interstellar clouds, in *Protostars and Planets* (T. Gehrels, ed.) Univ. Arizona, Tucson, Ariz. 209-242.

Mouschovias, T. Ch, and E. V. Paleologou, 1980, The angular momentum problem and magnetic braking during star formation—exact solutions for an aligned and a perpendicular rotator, *Moon and Planets, 22*, 31-45.

Nakagawa, Y., 1978, Statistical behavior of planetesimals in the primitive solar system, *Prog. Theor. Phys. Japan, 59*, 1834-1851.

Nakagawa, Y., K. Nakazawa, and C. Hayashi, 1981, Growth and sedimentation of dust grains in the primordial solar nebula, *Icarus, 45*, 517-528.

Newton, I., 1687, *Principia:* London, England.

Pollack, J. B., J. A. Burns, and M. E. Tauber, 1979, Gas drag in primordial circumplanetary envelopes—a mechanism for satellite capture, *Icarus, 37*, 587-611.

Russell, H. N., 1935, *The Solar System and its Origin:* Macmillan, New York.

Safronov, V. S., 1960, On the gravitational instability in flattened systems with axial symmetry and non-uniform rotation, *Annuales d'Astrophysique, 23*, 979-982.

Safronov, V. S., 1962, On the temperature of the dust component of the protoplanetary cloud, *Sov. Astr.–A.J., 6*, 217-225.

Safronov, V. S., 1969, *Evolution of the Protoplanetary Cloud and Formation of the Earth and Planets*, Nauka, Moscow, USSR, (Transl. Israel Program for Scientific Translation, 1972, NASA TTF-677).

Safronov, V. S., 1972, Accumulation of the planets, in *On the Origin of the Solar System* (H. Reeves, ed.), CNRS, Paris, 89-113.

Schwarzschild, M., 1958, *Structure and Evolution of the Stars*, Princeton Univ. Press, Princeton, N.J.

Sekiya, M., K. Nakazawa, and C. Hayashi, 1980, Dissipation of the rare gases contained in the primoridal Earth's atmosphere, *Earth Planet. Sci. Lett., 50*, 197-201.

Sekiya, M., C. Hayashi, and K. Nakazawa, 1981, Dissipation of the primoridal terrestrial atmosphere due to irradiation of the solar far-UV during T Tauri stage, *Prog. Theor. Phys. Japan, 66*, 1301-1316.

Shu, F. H., V. Milione, and W. W. Roberts, Jr., 1973, Nonlinear gaseous density waves and galactic shocks, *Astrophys. J.*, *183*, 819-841.

Spitzer, L., 1939, The dissipation of planetary filaments, *Astrophys. J.*, *90*, 675-688.

Stewart, G. R., 1982, A gravitational kinetic theory for planetesimals, Ph.D. Thesis, Univ. California, Los Angeles, Calif.

Stewart, G. R., and W. M. Kaula, 1980, A gravitational kinetic theory for planetesimals, *Icarus*, *44*, 154-171.

Toomre, A., 1964, On the gravitational stability of a disk of stars, *Astrophys. J.*, *139*, 1217-1238.

Townsend, A. A., 1976, *The Structure of Turbulent Shear Flow*, Cambridge Univ., Cambridge, England.

Ward, W. R., 1981, Solar nebula dispersal and the stability of the planetary system. 1. Scanning secular resonance theory, *Icarus*, *47*, 234-264.

Wasson, J. T., 1974, *Meteorites—Classification and Properties*, Springen-Verlag, New York.

Weidenschilling, S. J., 1976, Accretion of terrestrial planets II, *Icarus*, *27*, 161-170.

Weidenschilling, S. J., 1977, Aerodynamics of solid bodies in the solar nebula, *Mon. Not. Roy. Astr. Soc.*, *180*, 57-70.

Weidenschilling, S. J., 1980, Dust to planetesimals—settling and coagulation in the solar nebula, *Icarus*, *44*, 172-189.

Wetherill, G. W., 1978, Accumulation of the terrestrial planets, in *Protostars and Planets* (T. Gehrels, ed.), Univ. Arizona, Tucson, Ariz. 565-598.

Wetherill, G. W., 1980, Formation of the terrestrial planets, *Ann Rev. Astro. Astrophys.*, *18*, 77-113.

Whipple, F. C., 1971, *From Plasma to Planets*, John Wiley, New York.

Woolfson, M. M., 1978, The capture theory and the origin of the solar system, in *The Origin of the Solar System* (S. F. Dermott, ed.), John Wiley, New York, 179-198.

Ziglina, I. N., 1976, Effect on eccentricity of a planet's orbit of its encounter with bodies of the swarm, *Sov. Astr.—A. J.*, *20*, 730-733.

Ziglina, I. N., and V. S. Safronov, 1976, Averaging of the orbital eccentricity of bodies which are accumulating into a planet, *Sov. Astr.—A. J.*, *20*, 244-248.

Zvygina, Y. V., G. V. Pechernikova, and V. S. Safronov, 1973, Quantitative solution of the coagulation equation with allowance for fragmentation, *Sov. Astr.—A. J.*, *17*, 793-800.

William M. Kaula
Department of Earth and Space Sciences, and
Institute of Geophysics and Planetary Physics
University of California
Los Angeles, California 90024

# 4

# THE INTERIORS OF THE TERRESTRIAL PLANETS: THEIR STRUCTURE AND EVOLUTION

# ABSTRACT

Defined by size and composition, there are seven terrestrial planets: the Earth, the Moon, Mercury, Venus, Mars, Io, and Europa. The Earth is by far the most examined, but the Moon is better understood because it is simpler. The planets in general are declining in activity because of dwindling energy sources and increasingly stable differentiations. The Earth and the Moon indicate strongly that the degree of this decline varies inversely with the size of the body. But there are significant variations in the manifestations of this rule as well as exceptions to it because of circumstances peculiar to each body, such as the high surface temperature of Venus, the apparently low initial heat and volatile content of Mars, the tidal effects of Jupiter on Io, and so forth.

## 4-1 INTRODUCTION

The term *terrestrial planet* is properly applied to the four innermost of the planets orbiting the Sun: Mercury, Venus, Earth, and Mars. However, a planetary interior is extremely insensitive to the orbital circumstances of the planet. Hence a more functional definition is "a body that is composed mainly of rock and iron and more than 1000 km in radius." The specification "mainly rock and iron" is equivalent to having a mean density of more than $2.5 \text{ g} \cdot \text{cm}^{-3}$. The specification "more than 1000 km in radius" essentially means that the body has had significant internally generated evolution since the circumstances attendant on solar-system origin more than 4.5 Gy ago. By this more functional definition we can add three more bodies to our four "proper" planets: the Moon and two of the satellites of Jupiter—Io and Europa. The leading properties of these seven bodies are given in Table 4-1. All these bulk properties are inferred by space techniques: radar distances, spacecraft perturbations, and so forth. The largest omitted body satisfying the mean density specification is the asteroid Vesta, $\frac{1}{200}$ as massive as Europa.

Because the Earth is so much more accessible than the other planets, a plausible discussion of terrestrial planet interiors would first describe the Earth and then the others in terms of differences from the Earth. However, we also know appreciably more about the Moon than about Mars, Venus, or Mercury, thanks to the *Apollo* project. Our knowledge of the Moon is much less than about the Earth, if we measure "knowledge" by the number of bits of information that have been measured. But we understand the Moon significantly better than the Earth because it is a much simpler object.

Our inferences about the interior of the planets other than Earth and Moon are so dependent on assumptions about their evolution—implicitly, if not explicitly—that a discussion organized on the lines of structure or the present state of all the planets before evolution (or how they got the way they are) is awkward. Instead it is more efficient to summarize the structures and the evolutions of both the Earth and the Moon before proceeding to the other planets. It is fortunate that the Earth is the biggest of the set of seven and the Moon is almost the smallest, since interpolation is always easier than extrapolation.

The comparative examination of the interiors of terrestrial planets should be a ripe topic, not only because of the planetary exploration program—capped by the first detailed views of the surfaces of Venus, Io, and Europa in 1980—but also because of improved insights about the Earth's mantle. So far, there are few good works on the subject. An outstanding exception is the Basaltic Volcanism Study Project (1981), an 82-author, 1286-page tome that is much broader in scope than implied by its title and that could almost be used as the sole reference for the present review. References thereto are given in the form (BVSP, 4.5), where 4.5 is the section number. For the Earth there are several good recent collections: McElhinny (1979); Dziewonski and Boschi (1980); Strangway (1980); O'Connell and Fyfe (1981); and an excellent elementary text: Brown and Mussett (1981).

### Moment-Of-Inertia Ratio

The moment of inertia is the integral (or summation) over the volume of a body of the density times the distance squared off an axis through the center of mass. Mathematically,

$$I = \int_{VOL} \rho s^2 dV$$

$$= \int_0^R \int_0^\pi \int_0^{2\pi} \rho (r \sin \theta)^2 d\lambda \sin \theta \, d\theta \, r^2 dr$$

here $R$ is surface radius, $\rho$ is density, and coordinates $r, s, \lambda, \theta$ are defined in Fig. 4-1. If $\rho$ is a function of $r$ only,

$$I = \frac{8\pi}{3} \int_0^R \rho r^4 dr$$

and for a homogeneous body, $\rho$ constant,

$$I = \rho 2\pi 2 \left(1 - \frac{1}{3}\right) \frac{R^5}{5}$$

$$= \frac{8}{15} \pi R^5 \rho = 0.4 MR^2$$

since mass $M = 4\pi R^3 \rho / 3$

The moment of inertia is used as the primary indicator of radial variation in density not only because it is conceptually simple, but also because it is most easily measured, since the response of a planet to a torque is inversely proportionate to the moment-of-inertia ratio (see also footnote d in Table 4-1).

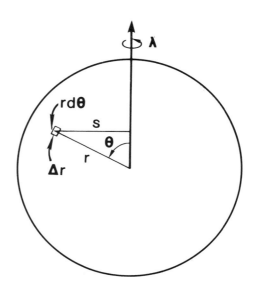

Fig. 4-1 Geometry related to the moment of inertia (see boxed material on p. 80).

TABLE 4-1 Bulk Properties of Terrestrial Planets

| Body | Mass (Earth Masses)[a] | Mean Radius (km) | Mean Density (g·cm$^{-3}$) | Reduced Density (g·cm$^{-3}$)[b] | Moment-of-Inertia Ratio ($I/MR^2$)[c] | Mean Crustal Thickness (km)[d] | Solar Distance (AU)[e] |
|---|---|---|---|---|---|---|---|
| Earth | 1.00000 | 6371.0 | 5.514 | 4.03 | 0.3315 | 14 | 1.00 |
| Moon | 0.01230 | 1737.5 | 3.344 | 3.34 | 0.390 | >60 | 1.00 |
| Mars | 0.10745 | 3390 | 3.934 | 3.71 | 0.365 | >28 | 1.52 |
| Venus | 0.81500 | 6051.4 | 5.245 | 3.95 | ? | ≥21 | 0.72 |
| Mercury | 0.0553 | 2439 | 5.435 | 5.31 | ? | ≥ 7 | 0.39 |
| Io | 0.0149 | 1815 | 3.55 | 3.55 | ? | ? | 5.2 |
| Europa | 0.0080 | 1569 | 3.96 | 2.96 | ? | ? | 5.2 |
| (Vesta | 0.00004 | 269 | 2.9) | | | | |

[a]The mass of the Earth is $5.974 \times 10^{24}$ kg.
[b]Reduced density is the density the planetary material would have if it were all at 10-kb pressure (sufficient to squeeze out voids). These estimates entail the assumption that all metallic iron and nickel are in a core.
[c]See boxed material on p. 80. The expression as a ratio $I/MR^2$ is convenient because this quantity will never differ drastically from the ratio for a homogeneous body, 0.400.
[d]The crust is the layer of rock at the surface which is of lower density (less than 3.0 g·cm$^{-3}$) because it has a higher ratio of calcaluminous to ferromagnesian silicates. The crustal thickness is measured by seismology only on the Earth and in a limited region of the Moon. Elsewhere it can be estimated only by assuming that the support of variations of topographic elevations is the same as for the Earth's continents: variations in crustal thickness. The more data about the planets are acquired, the more this assumption appears dubious.
[e]One AU is 149.6 million km. It is the radius of the Earth's orbit.

## 4-2 STRUCTURE OF THE EARTH

The main parts of the Earth are given in Table 4-2 and are shown schematically in Fig. 4-2. Between the atmosphere, oceans, crust, mantle, and core, there are sharp discontinuities corresponding to changes in composition. The continental and oceanic crusts have significant differences in composition, but the transitions between them are much more gradual than the radial transitions. In reference to solid parts of Earth, the continental-oceanic boundary is the edge of the continental shelf at a depth of about 2 km, so that about 60 percent of the Earth's surface is oceanic and about 40 percent is continental. The density of the mantle corresponds to an Mg/(Mg+Fe) mole (or atomic) ratio of about 0.88, an important parameter for interplanetary comparisons (BVSP, 4.5). A likely compositional difference not appearing in Table 4-2 is that the innermost 5 percent of the core, which is solid, is probably more nearly pure iron and nickel than the fluid outer

TABLE 4-2  Main Compositional Parts of the Earth

|  | Mean Radii | | Mean Density ($g \cdot cm^{-3}$) | Mass ($10^{24}$ kg) | Predominant Composition |
| --- | --- | --- | --- | --- | --- |
|  | Inner (km) | Outer (km) |  |  |  |
| Atmosphere | 6371 | – | – | 0.000005 | Nitrogen and oxygen |
| Oceans | 6367 | 6371 | 1.0 | 0.0014 | Water |
| Oceanic crust | 6360 | 6367 | 2.9 | 0.0062 | Calcaluminous and ferromagnesian silicates |
| Continental crust | 6341 | 6371 | 2.8 | 0.0171 | Calcaluminous silicates[a] |
| Mantle | 3485 | 6352 | 4.5 | 4.02 | Ferromagnesian silicates[b] |
| Core | 0 | 3485 | 10.9 | 1.93 | Iron |

[a]Calcaluminous silicates are combinations of CaO, $Al_2O_3$, and $SiO_2$.
[b]Ferromagnesian silicates are combinations of FeO, MgO, and $SiO_2$.

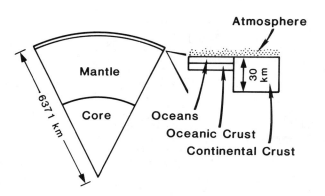

Fig. 4-2  Main compositional parts of the Earth (see also Table 4-2).

core. The density of the outer core requires a minor portion of a lighter constituent such as sulfur or oxygen.

The behavior of material is influenced not only by its composition, but also by its temperature and pressure. Hence another division of the Earth can be made on the basis of mechanical properties, as given in Table 4-3 and Fig. 4-3. Some divisions of the mantle in Table 4-3 are uncertain and rather arbitrary. The lithosphere is a relatively rigid outer layer of high mechanical strength; its thickness varies from 0 km to 150 km. Associated with these variations in lithospheric thickness are tectonic phenomena, as discussed in Chapter 8 by Bird. The asthenosphere is a layer of relative weakness; the location of its lower limit is unsure and a matter of debate. Closely coinciding with the lithosphere and (in some places) the upper asthenosphere, respectively, are layers defined by seismic velocities: the *lid* and the *low velocity channel*. The boundary zone near 5700-km radius (670-km depth) has a pressure corresponding to some silicate phase transitions that may have significant effects on the behavior and evolution of the mantle. The region between these phase transitions and the crust is often called the *upper mantle*. The boundary between the outer and inner core is a transition from liquid to solid.

TABLE 4-3  Main Mechanical Parts of the Earth

| | Mean Radii | | Mean Density (g · cm$^{-3}$) | Mass (10$^{24}$ kg) | State |
|---|---|---|---|---|---|
| | *Inner* (km) | *Outer* (km) | | | |
| Atmosphere | 6371 | — | — | 0.000005 | Gas |
| Oceans | 6367 | 6371 | 1.0 | 0.0014 | Liquid |
| Lithosphere | 6280 | 6368 | 3.4 | 0.121 | Solid |
| Asthenosphere | 6000 | 6280 | 3.4 | 0.45 | Solid |
| Mesosphere | 5700 | 6000 | 3.8 | 0.49 | Solid |
| Lower mantle | 3485 | 5700 | 5.0 | 2.98 | Solid |
| Outer core | 1227 | 3485 | 10.8 | 1.84 | Liquid |
| Inner core | 0 | 1227 | 13.0 | 0.10 | Solid |

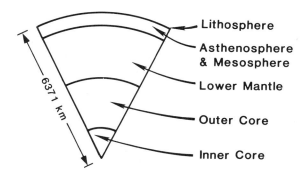

Fig. 4-3  Main mechanical parts of the Earth (see also Table 4-3).

The structure of the Earth as set forth in Tables 4-2 and 4-3 was inferred mainly by seismological techniques, with the help of some constraints from geodesy (see Table 4-1), cosmochemistry (see Chapter 1 by Wasson and Kivelson), petrology (the compositions of crustal rocks and inferences of source regions therefrom), geothermics (temperature gradients inferred from heat flow), and laboratory experiment (changes of iron and silicate properties with temperature, pressure, and composition).

The major problem in understanding Earth structure as a function of radius is inferring the composition which matches the densities at pressures beyond those achieved by static experiments. (Much higher pressures are attained by experiments using explosions, but the applicability of the results is limited and debated.) First, the temperatures are uncertain because of the imperfection of the convective theory for such a complex regime. Second, the most common iron- and magnesium-rich (called *ferromagnesian*) minerals undergo a series of phase transitions at pressures in the range 120 to 350 kb (1 kb = 1000 atm = $10^8$ Pa), corresponding to depths of 380 km to 1100 km. In these phase transitions, the metallic atoms and silica tetrahedra change their arrangement toward closer packing at higher pressures. The density changes associated therewith are uncertain to about 3 percent. Hence, it is still unsure whether the mantle is homogeneous in its major constituents, or whether these are trends in the Mg/(Mg+Fe) ratio (Ringwood, 1975; Anderson, 1977). If such trends exist, their effects on density are smaller by a factor of at least 30 than the effect of the compositional difference between the core and the mantle. However, such slight differences would be enough to have profound effects on mantle convection and thence on long-term evolution (BVSP, 9.5). Isotopic evidences thereon are discussed in Chapter 5 by DePaolo.

Some lateral variations in crustal and lithospheric properties have major implications for the evolution of the Earth. As indicated in Table 4-2, the mean compositions and thicknesses of the oceanic and continental crusts differ appreciably. However, there are also markedly greater variations about these means in the continental crust than in the oceanic. The two types of crusts also have greatly different age distributions. Oceanic crust has roughly the composition that would be expected from an approximately 25 percent partial melt of a mix of the silicates most common in bulk composition. It is largely formed at ocean rises and subducted at island arcs or other compressive features. Hence it is all less than 200 My old. Variations from the mean thickness in oceanic crust are not more than 1 km or so. On the other hand, most continental crust appears to be highly processed material, cycled through a variety of events. Significant differentiations of continental rocks have occurred in a variety of tectonic settings, although subduction zones, where oceanic lithosphere returns to the mantle, appear to be quantitatively most important. The median age of the continental basement (the igneous rocks under the sediments) is about 300 My, but some of it is as old as 3800 My. The continents appear to have grown throughout Earth history.

The thinnest lithosphere is at the ocean rises; the thickest is under the continents. These variations in thickness are inversely correlated with variations in heat flow, which are about 10 times the global mean (0.08 Watts $\cdot$ m$^{-2}$) at ocean rises, but half the global mean in some old continental shields. About 70 percent of the heat flow out of the Earth is by cooling of lithosphere spreading from the rise (Sclater et al., 1980). The associated rate of crustal creation is about 18 km$^3$/y. (Heat flow and crustal thickness are discussed more fully in Chapter 5 by DePaolo.)

The heat generated by radioactivity also varies drastically among rocks. The typical value for oceanic crust is equivalent to a rate of warming of $1° \cdot My^{-1}$ (if the heat were retained); for continental crust, at the surface, about $50° \cdot My^{-1}$; for a few upthrust mantle rocks, as little as $0.01° \cdot My^{-1}$. The mean heat flow of about $0.06$ Watts $\cdot m^{-2}$ associated with cooling oceanic lithosphere corresponds to an average for the mantle of $0.2° \cdot My^{-1}$.

## 4-3 EVOLUTION OF THE EARTH

It is evident that there has been an appreciable upward concentration of heat sources associated with the separation of crustal rocks from the mantle, yet most of the heat is still being brought up from deep in the mantle to be lost by cooling of oceanic lithosphere. This lithosphere is apparently the boundary layer of a mantle convection system, a regime in which heat is transferred by motion of matter, at rates of centimeters per year.

The mechanisms and patterns of contemporary mantle convection are still imperfectly understood, especially whether the convection is throughout the mantle or whether it is divided into upper- and lower-mantle convection. However, it is fairly evident that with time the Earth must be cooling off (due to decline of radioactive heat sources) as well as becoming more heterogeneous (see Chapter 5 by DePaolo). Hence if we look back in time, the Earth's mantle must have been appreciably more active. For example, 3000 My ago the rate of radioactive heat generation was more than twice as great as it is now. In addition, the further we go back in time, the more there remained the effects of the heat associated with the formation of the Earth, about 4600 My ago.

The geologic record remaining from the Archean more than 2500 My ago has many indications that the Earth's interior was indeed much more active then. There was considerably more volcanism, recycling of material, and so forth. Perhaps the most striking evidence is that the oldest rock yet found is only 4100 My old, solidified more than 400 My after the origin of the Earth. Associated with this early activity was most of the outgassing of the ocean and atmosphere, as discussed in Chapter 7 by Fanale. However, the stabilization of most of the continental crust appears to have occurred somewhat later, 2500 to 3000 My ago. Perhaps there had not been sufficient cooling until then to attain lithosphere thick enough to stabilize a crust thick enough to project above sea level (analogous to an iceberg in the ocean).

While the Earth's interior was appreciably more active in the Archean Era, the time of greatest energy availability, and hence activity, was during its formation, some 4600 My ago. As discussed by Wetherill (1981) and also by Lin in Chapter 3, formation is believed to have occurred by collisions among sizable planetesimals. Hence a great amount of energy would have been delivered, with most of it being trapped as heat to drive thermal and tectonic activity. An important byproduct was almost certainly core formation. Concentrations of iron, whether delivered as planetesimals or subsequently collected, would have sunk rapidly through silicate that was nearly molten, while the stresses set up by density inhomogeneities would have broken up any cooler center, if such a cooler center existed (Stevenson, 1980). Ratios of lead isotopes resulting from radioactive decays indicate core formation occurred more than 4300 My ago (see Chapter 5 by DePaolo). Core formation itself would have generated appreciable heat, enough to warm the Earth several hundred degrees.

It thus appears that the overall character of the Earth's evolution has been one of cooling off and differentiation. However, these processes have not varied smoothly but have had some irregularity. As mentioned, core formation was probably rather catastrophic, whereas stabilization of continental crust reached a peak much later. In more recent times, there is evidence of oscillation of the rate of plate tectonics, and hence heat loss, by a factor of almost 2 (Turcotte and Burke, 1978). Although it is important to understand such irregularities and to infer their causes, they do not alter the basic trends. Despite the overall cooling, the Earth is still very active and far from a state of cold quiescence: Apparently it takes a long, long time for a planet of such size to quiet down.

## 4-4 STRUCTURE OF THE MOON

As evidenced by the *moment-of-inertia* ratio 0.390 in Table 4-1 (see boxed material on p. 80), the Moon has very little radial differentiation in its composition. In fact, most of the difference of this ratio from 0.400 arises from the approximately 60 km of lunar crust. This 60-km estimate mainly depends on seismological measurements in a limited region; extrapolation to a global estimate on the basis of apparent isostatic compensation of the topography (see Chapter 8 by Bird for a description of these terms and for the application of these ideas to the analysis of the Moon) is uncertain because of the possibility of mechanisms for compensation besides variations in crustal thickness, such as lateral variations in upper mantle Mg/(Mg+Fe) (Wasson and Warren, 1980).

Other important inferences from seismology on the Moon are the existence of a *lithosphere*—a cold, mechanically strong layer—800 to 1000 km thick, a level of moonquake activity many orders of magnitude smaller than the Earth's seismicity, and the probable absence of a core.

The extreme dryness of the rocks on the Moon indicates that the low mean density of the Moon, 3.34 g/cm$^3$, arises from a lack of iron rather than a surplus of volatiles. The Moon's iron is essentially all incorporated into ferromagnesian silicates and, in proportion to silicon, magnesium, and oxygen, less abundant by a factor of one-third than in the Earth, an important datum for theories of the origin of the Moon. The Moon's density of 3.34 g · cm$^{-3}$ is consistent with a ratio Mg/(Mg+Fe) of 0.8, somewhat lower than the Earth's mantle (BVSP, 4.5), but probably not a significant difference.

Most of the Moon's surface is covered by calcaluminous silicates: This region is called the *highlands*, or *terrae*. About 20 percent of the surface is characterized by darker rocks of higher ferromagnesian content: These regions are the *maria*. Hence the Moon is similar to the Earth in having a correlation of rock composition with topographic elevation. However, the low-density rocks of the *terrae* are explicable by much simpler differentiations than are continental rocks on Earth, whereas the thicknesses of the maria are appreciably less than those of oceanic crust, probably nowhere more than 1 km. In most cases, the basins in which the maria lie have sharp, nearly circular margins, manifesting generation by major impacts rather than by internal activity. The maria are appreciably younger than the highlands, as evidenced by fewer small craters per unit area (see boxed material). But the youngest rock age found at any of the six *Apollo* sites is 3200 My. Extrapolation from this firm data by using crater counts indicates that no region of the Moon's surface is less than 2300 My old (BVSP, 8.4).

> ### Dating By Crater Counts
>
> An object big enough to make a crater at least 4 km across hits the Moon on the average once every million years. Hence there are a lot of big craters on the Moon. If we assume the infall rate to be constant, then for an area large enough to encompass several craters, the number of craters is proportionate to the age of the surface. There are problems with the infall rate not being constant and new craters obliterating old ones, but the technique is good enough to date maria surfaces less than 3900 My old with uncertainties of about 300 My.

The lateral irregularities in topography and gravity also indicate a more quiescent state than the Earth's. The principal constraint on these irregularities is stress, which in turn is proportional to gravity. Gravity on the Moon is one-sixth that on Earth; if stresses were equal, then the irregularities would be six times as great. But they are not; they are only two or three times as great.

## 4-5 EVOLUTION OF THE MOON

The Moon's evolution is like the Earth's in that it is mainly one of cooling and differentiation. However, the time scale is shorter by more than a factor of 10. Lead-isotope data indicate that crustal separation took place more than 4400 My ago, within about 200 My of origin. The energy source for this differentiation must have been the infall energy associated with formation, since 200 My is too short to generate appreciable radiogenic heat, especially because potassium, the shortest lived major radioactive element, is among the volatiles depleted in the Moon. However, the lava flows that created the maria 800 My to 1400 My after formation do appear to depend on radiogenic heat. The sources for the flows are inferred to be more than 200 km deep, where there could conceivably be regions that had not lost their radioactive elements to the crust (see also Chapter 8 by Bird).

An important constraint on long-term evolution of the Moon is the absence of marked crumpling of the surface, as would be expected if the history had been predominantly cooling. There is neither a marked net compression nor a marked net tension in the lunar surface. Hence, to balance the cooling of the outer few 100 km of the Moon, there must have been a heating of the interior to yield about the same integral of temperature change over volume (Solomon and Chaiken, 1976). The inner parts of the Moon must have retained sufficient radioactive elements to provide this heat. The amount required is not extraordinary compared with those allowable by cosmochemical estimates. Corroborative evidence of this heating up is that an asthenosphere extending from the center to between 700- and 900-km radius is inferred from seismological data.

The Moon is thus very quiet compared with the Earth. It lacks sufficient internal energy to generate any tectonics, volcanism, or magnetic field. The main shaper of its surface features has long been the infall of objects from elsewhere in the solar system. The Moon's evolution is no doubt significantly affected by its dearth of volatiles and iron, but the main cause of its quiescent history is undoubtedly its smaller size. It is remarkable that a difference by less than a factor of four in linear dimension can produce such dissimilarity in evolutionary character. If internally generated activity is an essential part of the definition of a terrestrial planet, a 1000-km radius is certainly a generous lower limit on size.

## 4-6 BASES FOR INTERPOLATION

The terrestrial planets are a heterogeneous set, more like a bunch of adoptees than blood siblings formed from a common material pool by the same processes in the origin of the solar system. As previously discussed, the great difference in behavior of the Earth and the Moon appears to be attributable to their difference in size. But the body closest to the Moon in size, Io, appears to have a lot of volcanism even now (Carr et al., 1979; Peale et al., 1979), whereas the body closest to the Earth in size, Venus, appears to be lacking plate tectonics (Kaula and Phillips, 1981).

Hence we should examine other independent variables, the most evident of which is distance from the Sun, which would seem to be pertinent to bulk compositions, if they were affected by temperatures in the primordial solar nebula. There is a trend of the reduced densities in Table 4-1 among the four proper planets Mercury, Venus, Earth, and Mars which plausibly relates to the oxidation level increasing with distance from the Sun. The reduced densities of Venus, Earth, and Mars can be explained as varying degrees of oxidation of the composition of chondritic meteorites, which composition is the same as the Sun's without the Sun's volatiles (see Chapter 1 by Wasson and Kivelson). But the oxidation level does not vary smoothly with distance from the Sun. Furthermore, the high reduced density (see Table 4-1) of Mercury requires that it have an elemental abundance of iron much higher than chondritic. To attribute Mercury's ratio of metallic iron to magnesian silicate solely to temperature effects constrains nebula conditions severely, since the two condensation temperatures differ by less than $100°C$ at pressures plausible for the nebula.

The three satellites listed in Table 4-1 all have bulk compositions much less easily related to distance from the Sun. The deficiency of iron and associated elements (siderophiles) in the Moon is generally held to be evidence of prior differentiation of the Moon's material in a planetary body or bodies. For dynamical reasons (as well as the similar Mg/(Mg+Fe) ratio), the most plausible locus for this differentiation is the Earth, which implies a major impact into the Earth (perhaps a Mars-sized body) to remove the protolunar material. A second possibility is that planetesimals in orbit around the Sun were differentiated and subsequently broken up. These fragments were then captured into a circumterresterial swarm by some sort of selective process (Kaula, 1977; Ringwood, 1979).

Io and Europa do not fit any trend in solar distance. But with the other Galilean satellites, Ganymede and Callisto, they do show an increasing trend in the ratio of ice to rock with Jovian distance, evidently a byproduct of Jupiter's formation.

Acquisition of volatile elements does not appear to be correlated with solar distance, but rather with size among the five terrestrial bodies in the inner solar system. The Earth and Venus could have had initial volatile contents that were quite similar, but Mars, Mercury, and the Moon must have had appreciably less, in the order listed here. This condition, together with the coexistence of iron cores and outer parts containing volatiles, suggests that the volatile contents may have been largely acquired in the later stages of planetary growth, when the capture capability of the larger planets would have been enhanced by their gravitational fields (see also Chapter 7 by Fanale).

The main thing that can be inferred about the terrestrial planets from the circumstances of their origin is that they started out hot, at least in their outer parts. The degree of this heating was positively correlated with size. Otherwise, about all that can be said is that the planets were "born different."

The order of discussion of the remaining terrestrial bodies corresponds to how well they can be understood. This order is also one of increasing size. As already stated about the Moon, the smaller is a planet, the simpler is its history. Exceptions are Io and Europa, which are different not because of size but because of composition and remote location. Hence they are discussed at the end.

## 4-7 MERCURY

Mercury looks very much like the Moon; its surface is characterized mainly by craters. However, a subtle but important difference is that Mercury has lobate scarps and other features that indicate a net shrinkage of the surface of some kilometers. Such a shrinkage would be expected if Mercury's thermal history had been a net cooling of a few 100°C. Cooling of this magnitude implies an early differentiation of the core and an upward differentiation of the radioactive heat sources uranium and thorium. This differentiation took place 4000 My or more ago. The resulting compression would also shut off any volcanism and leave the surface to be modified only by impacts.

Mercury has a perceptible magnetic field, about 0.5 percent of the Earth's in intensity, which is consistent with it having an iron core. A secondary indicator of early core differentiation and subsequent rather passive evolution is the locking of Mercury's rotation in a 3:2 resonance with its orbital motion. This locking requires an equatorial ellipticity in the mass distribution that is not remarkable in magnitude to stabilize it. An internally active Mercury might have its ellipticity pass through zero slowly enough for tidal friction to slow the rotation from the 3:2 commensurability.

Although Mercury is depleted in elemental magnesium, the low oxidation level requires that nearly all the iron is in the core and that the ratio $Mg/(Mg+Fe)$ of the mantle is quite high, as much as 0.98 (BVSP, 4.5). An important unknown is whether Mercury has enough silica ($SiO_2$) (which has a lower condensation temperature than iron) to combine with alumina ($Al_2O_3$) and calcium oxide ($CaO$) (both of higher condensation temperature than iron) to make anorthite, a feldspar necessary for a crustal differentiation like the Moon's. Limited reflection spectrophotometry of Mercury indicates pyroxene, a ferromagnesian silicate, suggesting that Mercury may not have a crust like the Earth's or the Moon's.

## 4-8 MARS

Mars, like Mercury and the Moon, appears to be a one-plate planet. A single coherent lithosphere completely surrounds the planet, so that if there are any internal convective motions they do not reach the surface, and the final heat transfer outward must be by thermal conduction or volcanism through this lithosphere. But Mars differs in that it evidences mainly tensional tectonics, most spectacularly the great Valles Marineris along the equator. This tensional character confirms a thermal history of predominant warming. However, the outer parts have cooled sufficiently to form a considerable lithosphere that is 200 km thick or more. A consequence of this thick lithosphere is that volcanism is confined to a relatively few major features. Hence the Tharsis region with its crowning feature Mons Olympus (27 km high) is the greatest topographic irregularity on any of the terrestrial planets. Associated with this topographic bulge is a great bulge in the gravity field, suggesting that a significant portion of the volcanic flows associated with Tharsis are supported by the mechanical strength of the lithosphere. Mars has the most irregular gravity field of any of the planets, in the sense of the stress-differences in the rocks implied thereby (Phillips and Lambeck, 1980; see also Chapter 8 by Bird).

The moment-of-inertia ratio ($I/MR^2$) of Mars, which is 0.365, together with chondritic Fe:Mg:Si ratios, requires that it have a core that is 15 to 25 percent of the total mass. A consequence of the higher oxidation level is more iron in the mantle, so that the $Mg/(Mg+Fe)$ ratio is about 0.76. Consistent with this higher iron content are the spectral character and gentle slopes (implying low viscosity) of the Martian lava flows.

The inference that the thermal history of Mars is one of predominant warming implies that core formation was relatively late, perhaps less than 2000 My ago. Otherwise, the upward differentiation of radioactive heat sources would have led to a more Moon-like evolution. The ages of major lava flows inferred from crater counts are less than 1000 My each. As discussed in Chapter 7 by Fanale, appreciable outgassing activity continued into these relatively recent times, although Mars must be very depleted in volatiles compared with the Earth.

Mars is thus close to being the geometric mean between the Moon and the Earth in its bulk characteristics and average level of activity. This activity is modest compared with the Earth's, but not nearly as moribund as the Moon's. The main reason Mars isn't a neat interpolation of the Earth and Moon is that Mars probably formed under somewhat cooler circumstances than either the Moon or the Earth, possibly a consequence of dynamical disturbances generated by Jupiter that stunted Mars' growth. This stunting is also a plausible explanation for the dearth of volatiles in Mars. The difference in evolutionary character of Mars and Mercury is probably the most marked example of the effect of bulk composition on evolution.

## 4-9 VENUS

The most evident difference of Venus from the Earth is its massive atmosphere, leading through the greenhouse effect to high surface temperatures and the loss of water, as discussed in Chapters 6 by Kivelson and Schubert and in Chapter 7 by Fanale. Offhand, the consequences of these differences for the solid planet should be minor. One would expect much the same volcanic and tectonic behavior as on Earth, adjusted to run at interior temperatures moderately higher than on Earth, but, because of the temperature dependence of viscosity, not nearly as much as 450°C difference in their surface temperatures. Instead, the *Pioneer Venus* radar altimeter and gravity measurements showed a radically different character.

The frequency distribution of elevations on Venus has one peak, rather than two, like the Earth's continents and oceans. The gravity field on Venus is generally milder, with more variability in its short wavelengths relative to the long, which is suggestive of shallow sources. Furthermore, the gravity field has a striking positive correlation with the topography, in contrast to the Earth. Finally, there is no evidence of a globally interconnected ridge system of relatively uniform elevation like the Earth's ocean rises, the most obvious manifestation of plate tectonics and its underlying process of mantle convection (Kaula and Phillips, 1981; Phillips et al., 1981).

It thus seems clear that the final stage in removal of heat from Venus is thermal conduction or volcanism rather than cooling of a spreading lithosphere. The reason for this difference cannot be a thick lithosphere, as on Mars; on the average, the higher surface temperature entails a thinner lithosphere than on Earth. A feature that could inhibit convective flow from reaching the surface is a low-density layer: that is, a crust. On Earth nearly all oceanic crust is recycled to the interior, but on Venus the higher rock temperatures, thinner lithosphere, and lack of water may greatly inhibit this recycling process. The higher rock temperatures would have led to differentiation of thicker crusts. The thinner lithosphere would have made it more difficult for instabilities that initiate subduction to develop, since there would be a lower ratio of more dense to less dense material in this lithosphere. Finally, the lack of water would have made rocks near the surface stiffer, despite their higher temperature. Thus the surface of Venus is probably all continent: that is, crustal material that permanently floats on top rather than being subducted. The thinner lithosphere also contributes to making this crust global by imposing a shallower depth for stabilization of isostatic compensation, thus inducing accumulations of crustal material to spread out.

The main features most difficult to reconcile with a globally thick crust on Venus are the high plateaus Ishtar, Aphrodite, and Beta. The high ratio of gravity to topography for these features implies compensation depths more than 100 km, seemingly too deep for stabilized compensation such as exists under the Earth's continents. Marked lateral variations in upper-mantle temperatures may exist on Venus. More likely, these plateaus are the surface manifestations of dynamical pheonmena reaching hundreds of kilometers deep. These effects may arise from interaction of convective upstreams with the thick crust. A thick, low-density layer on Venus may have the effect of concentrating surface expressions of interior convection in a relatively few locations, analogous to the concentrating effect of a thick lithosphere on Mars.

The slightly lower reduced density of Venus, compared with that of the Earth, may be the consequence of the higher temperature, leading not only to thermal expansion but also to more calcaluminous material being incorporated in the uppermost layers as basalt rather than deeper as eclogite, a rock that is denser because of a phase transition occurring in calcaluminous silicates at about 10 kb (Anderson, 1981). The lower density could also arise from higher oxidation level or lower iron content.

The absence of a magnetic field from Venus is consistent with its smaller size and higher interior temperatures, since pressures necessary for solidification of the inner core are not attained (Stevenson et al., 1983). This solidification is believed to be the energy source for the geodynamo in the Earth (see Chapter 13 by Levy).

## 4-10  IO AND EUROPA

The density of Io is higher than that of the Moon because it has more iron. Io's composition could plausibly be the same as fully oxidized chondritic meteorites. Io differs from the Moon in that it has an energy source for tectonics and volcanism. The resonant coupling of the orbits of Io, Europa, and Ganymede not only leads to keeping Io close to Jupiter because tidal transfers of angular momentum from Jupiter's spin to Io's orbit get passed on to the other satellites, but it also leads to a forced eccentricity in Io's orbit sufficient to flex Io appreciably every revolution (see Chapter 12 by Peale). The heating of Io by this flexing apparently has been enough to expel all its $H_2O$ and $CO_2$ (thus driving up its density), leaving sulfur and its compounds as the material repeatedly ejected and recondensed by the volcanic activity. The absence of craters on Io indicates that the level of this activity is considerable.

Europa has a lower density than Io or the Moon because it contains a lot of hydrous and carbonaceous material. The smoothness of Europa's surface topography indicates that much of this ice is in a surface layer probably tens of kilometers thick, sufficient to squelch manifestations of internally generated activity even more effectively than the basaltic layer on Venus. But Europa has a higher ratio of area to mass and a lower ratio of heat source to mass than the Moon, and receives tidal flexing nowhere near Io's. Hence, even in the implausible event that it escaped an early energy pulse associated with formation, Europa should be a quiet place.

## 4-11  SUMMARY

The terrestrial planets are indeed a bunch of "odd balls." But, as we have attempted to show, all their leading peculiarities can be plausibly explained as consequences of known laws of physics and chemistry operating in the circumstances of each. However, the variety of applicable laws and circumstances is great enough that terrestrial planetology is not a neat exercise of consensus identification of problems followed by concentration on technical difficulties. In the next few years, while the extraterrestrial data flow is sparse, we can expect further happy insights using old physics and chemistry (from the fundamentalist's view) about such diverse matters as mantle convection, silicate phase transitions, magmatic differentiation, crustal and lithospheric interactions, impact thermodynamics, planetesimal dynamics, and so forth. Additional study in all these areas will advance our understanding of why the Earth and other terrestrial planets are as they are.

## 4-12  ACKNOWLEDGMENTS

This writing relates somewhat to NASA Grant 05-007-002.

## 4-13 REFERENCES

Anderson, D. L., 1977, Composition of the mantle and core, *Ann. Rev. Earth Planet. Sci., 5,* 179-202.

Anderson, D. L., 1981, Plate tectonics on Venus, *Geophys. Res. Lett, 8,* 309-311.

Basaltic Volcanism Study Project, 1981, *Basaltic Volcanism on the Terrestrial Planets,* Pergamon, Elmsford, New York, 1286 pp.

Brown, G. C., and A. E. Mussett, 1981, *The Inaccessible Earth,* George Allen & Unwin, London, 235 pp.

Carr, M. H., H. Masursky, R. Strom, and R. Terrile, 1979, The volcanic features of Io, *Nature, 280,* 729-733.

Dziewonski, A. M., and E. Boschi, eds., 1980, *Physics of the Earth's Interior,* North Holland Publishing New York, 716 pp.

Kaula, W. M., 1977, On the origin of the Moon, with emphasis on bulk composition. *Proc. Lunar Planet. Sci. Conf. 8th,* 321-331.

Kaula, W. M., and R. J. Phillips, 1981, Quantitative tests for plate tectonics on Venus, *Geophys. Res. Lett, 8,* 1187-1190.

McElhinny, M. W., ed., 1979, *The Earth, Its Origin, Evolution and Structure:* Academic Press, London, 597 pp.

O'Connell, R. J. and W. S. Fyfe, 1981, *Evolution of the Earth,* Am. Geophys. Union, Washington, D.C. and Geol. Soc. America, Boulder, Colo., 282 pp.

Peale, S. J., P. Cassen, and R. T. Reynolds, 1979, Melting of Io by tidal dissipation, *Science, 203,* 892-894.

Phillips, R. J., W. M. Kaula, G. E. McGill, and M. C. Malin, 1981, Tectonics and evolution of Venus, *Science, 212,* 879-887.

Phillips, R. J., and K. Lambeck, 1980, Gravity fields of the terrestrial planets—Longer wavelength anomalies and tectonics, *Rev. Geophys. Space Phys., 18,* 27-76.

Ringwood, A. E., 1975, *Composition and Petrology of the Earth's Mantle,* McGraw-Hill, New York, 618 pp.

Ringwood, A. E., 1979, *Origin of the Earth and Moon,* Springer-Verlag, New York, 295 pp.

Sclater, J. G., C. Jaupart, and D. Galston, 1980, The heat flow through oceanic and continental crust and the heat loss of the Earth, *Rev. Geophys. Space Phys., 18,* 269-311.

Solomon, S. C. and J. Chaiken, 1976, Thermal expansion and thermal stress in the Moon and terrestrial planets, *Proc. Lunar Planet. Sci. Conf.* 7th, 3229-3243.

Stevenson, D. J., 1980, Lunar asymmetry and paleomagnetism, *Nature, 287,* 520-521.

Stevenson, D. J., T. Spohn, and G. Schubert, 1983, Magnetism and the thermal evolution of the terrestrial planets, *Icarus, 54,* 466-489.

Strangway, D. W., 1980, The continental crust and its mineral deposits, *Geol. Assoc. Canada Spec. Paper, 20,* 804 pp.

Turcotte, D. L., and K. Burke, 1978, Global sea-level changes and the thermal structure of the Earth, *Earth Planet. Sci. Lett., 41,* 341-346.

Wasson, J. T., and P. H. Warren, 1980, Contribution of the mantle to lunar asymmetry, *Icarus, 44,* 752-771.

Wetherill, G. W., 1981, The formation of the Earth from planetesimals, *Sci. Am., 244,* 162-174.

Donald J. DePaolo
Department of Earth and Space Sciences, and
Institute of Geophysics and Planetary Physics
University of California
Los Angeles, California 90024

# 5

# ISOTOPIC CONSTRAINTS ON PLANETARY EVOLUTION

# ABSTRACT

Measurements of isotope ratios of Pb, Sr, Nd, Hf, Ar, and other elements provide information about the internal evolution of the Earth and Moon since the time of formation 4.5 billion years ago. The existing evidence indicates that the Earth's core formed more than 4.4 billion years (Gy) ago, possibly while the Earth was still accreting. The crust of the Moon is similarly very old, which shows that even small planets could be initially hot enough to melt and differentiate. The Moon, however, ceased to differentiate after about 1 Gy, whereas the Earth's differentiation has continued to the present. The ages of the rocks of the Earth's continental crust have values such as 3.6, 2.8, and 1.8 Gy, suggesting that the continuing internal evolution of the Earth has been punctuated with cataclysmic episodes. The largest happened 2.8 Gy ago, at which time about half of the present continental mass formed. Data relevant to the degassing of the Earth are sketchy but suggest that a major amount of degassing took place before the time of formation of the oldest preserved continental crust 3.8 Gy ago. Subsequently, there has been continued degassing, but it has affected the mantle nonuniformly. In a general sense, the postcore-formation differentiation of the Earth has been highly nonuniform, the upper mantle appears to be more differentiated and more completely degassed than the lower mantle.

## 5-1 INTRODUCTION

This chapter focuses on lunar and terrestrial rocks and the information they have yielded about the internal evolution of the Earth and the Moon since their formation 4.5 Gy ago. As these two bodies represent nearly the extremes of size found in the population of "terrestrial planets" as defined by Kaula in Chapter 4, the observations provide a valuable basis for understanding all of the rocky planets of the solar system. The "tools" that are used to study internal evolution are the naturally occurring radioactive elements and their decay products. The questions for which answers are sought are (1) when did the major structural elements (core, mantle, and crust) come into being, (and for the Earth, the atmosphere and oceans also) and (2) how have they changed through time. The regularities of radioactive decay provide the "clock" necessary to answer "when" (consult Faure, 1977; York and Farquhar, 1972, for general discussions of geochronology). In some cases, theoretical models have been constructed that predict some of the times of interest (for example, the time of core formation).

However, this discussion concentrates exclusively on direct observations made on terrestrial and lunar materials and on the ways in which those observations are interpreted. In certain instances more questions will be raised than answers offered. Continuing improvements in techniques and enlargements of the data base have led to changing perspectives and attempts to answer more difficult questions with greater precision. Nevertheless, many first-order conclusions can be reached.

## 5-2 AVAILABLE ISOTOPIC TRACERS

The relevant radioactive isotopes (see box page 96) and their decay products are summarized in Table 5-1. The long-lived nuclides are those with half-lives equal to or greater than the age of the Earth. These nuclides provide information about planetary-evolution

## Isotopes and Radioactive Decay

*Chemical elements* can be thought of as composed of nuclei, formed of protons (positively charged) and neutrons (uncharged), embedded in a cloud of electrons (negatively charged). The chemical properties of an element are largely determined by the nuclear charge and are quite insensitive to the nuclear mass. Thus a particular element, say, strontium (Sr), must always have the same number of protons in its nucleus (38 for Sr). But, as it is possible to have different numbers of neutrons in the nucleus, Sr can exist with different nuclear masses. A superscript as a prefix is used to identify the number of neutrons plus protons in a particular nucleus, examples being $^{86}$Sr, $^{87}$Sr, and $^{87}$Rb. Some of these chemical forms are stable and will remain unchanged indefinitely. Others, such as $^{87}$Rb (rubidium), are subject to radioactive decay, which changes their nuclear composition. In particular, $^{87}$Rb has a small but finite probability of decay to $^{87}$Sr. If you wait long enough (48.8 Gy), you will find that 50% of the $^{87}$Rb atoms in a sample will have decayed to $^{87}$Sr. Often we refer to the initial and final nuclei in a radioactive decay as *parent* and *daughter* nuclides. We also refer to the time for decay of half of the sample as the *half-life* for radioactive decay.

TABLE 5-1  Geologically Important Radioactive Nuclides and Their Decay Parameters[a]

|  | Radioactive Parent | Decay Product | Half-life (Gy) |
|---|---|---|---|
| Long half-life ($\tau_{1/2} \geq 4.5$ Gy) | $^{238}$U | $^{206}$Pb | 4.468 |
|  | $^{232}$Th | $^{208}$Pb | 14.01 |
|  | $^{176}$Lu[b] | $^{176}$Hf | 35.7 |
|  | $^{147}$Sm[c] | $^{143}$Nd | 106.0 |
|  | $^{87}$Rb | $^{87}$Sr | 48.8 |
| Intermediate half-life ($\tau_{1/2} < 4.5$ Gy) | $^{235}$U | $^{207}$Pb | 0.7038 |
|  | $^{40}$K | $^{40}$Ar, $^{40}$Ca | 1.250 |
| Short half-life ($\tau_{1/2} \ll 4.5$ Gy) | $^{129}$I | $^{129}$Xe | 0.016 |

[a]Consult Faure (1977) for information on decay constants.
[b]Consult Patchett et al. (1981) for information on decay constants of $^{176}$Lu.
[c]Consult Lugmair et al. (1975) for information on decay constants of $^{147}$Sm.

processes that take place gradually over time periods of billions of years, such as the formation of continents on the Earth and the development of heterogeneity in the Earth's mantle. They are also useful for determining the absolute ages of rocks (see material in box entitled Using Isotopes to Date Rocks). However, they do not usually provide definitive information about processes that may have happened in the first few tens of millions of years of planetary history, because the effects of early processes are overprinted by, and indistinguishable from, the effects of processes that occurred over the subsequent 4.5 Gy.

> Using Isotopes to Date Rocks
>
> The chemical constituents of rocks are established at the time of their formation. Subsequent weathering changes some of the elemental ratios, but for heavy elements, the elemental ratios change mainly through radioactive decay. Thus by measuring the decay products present in a rock sample, it is possible to determine the time since formation of the rock. A relevant example is provided by the elements samarium (Sm) and neodymium (Nd) (see Table 5-1). The ratios $^{147}Sm/^{144}Nd$ and $^{143}Nd/^{144}Nd$ are determined in the laboratory on rock samples. The *increase* of the $^{143}Nd/^{144}Nd$ ratio, since the rock formed is equal to $(2^{T/\tau_{1/2}} - 1)\, ^{147}Sm/^{144}Nd$, where $T$ is the amount of time since the rock formed (that is, its age), and $\tau_{1/2}$ is the half-life of decay. In Fig. 5-3 later in the chapter, the slope of the line defined by separate measurements on different samples of the same rock body is equal to $(2^{T/\tau_{1/2}} - 1)$ and thus gives the age of the rocks. The intercept ($^{147}Sm/^{144}Nd = 0$) gives the $^{143}Nd/^{144}Nd$ ratio at the time the rocks formed.

An example of this limitation is given in Fig. 5-1. The isotope ratio $^{87}Sr/^{86}Sr$, which increases with time as $^{87}Rb$ decays to $^{87}Sr$, has been found to vary from 0.702 to about 0.707 in basaltic lava erupted from oceanic volcanoes (Peterman and Hedge, 1971). This

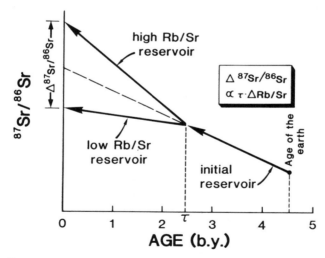

Fig. 5-1 Schematic representation of the growth of the isotopic ratio $^{87}Sr/^{86}Sr$ with time as a result of the decay of $^{87}Rb$ to $^{87}Sr$ (see Table 5-1). Differences of $^{87}Sr/^{86}Sr$ observed today in reservoirs of rock material result from differences of the ratio $^{87}Rb/^{86}Sr$ that have existed for geologically long times. The diagram depicts a situation where prior to time $\tau$, there is only one homogenous reservoir, and at time $\tau$, part of this reservoir is separated into two reservoirs having different, but complementary, Rb/Sr ratios.

indicates that the ratio Rb/Sr has been different in separate parts of the Earth's mantle for a long time. However, the same difference of 0.005 in the ratio $^{87}$Sr/$^{86}$Sr could be generated by small variations of Rb/Sr that formed 4.5 Gy ago, by larger variations that formed, for example, 0.5 Gy ago, or by intermediate-sized variations that formed gradually over all or part of the Earth's lifetime. This is a somewhat overly pessimistic view of the uncertainty in interpretation, as other observations can be brought to bear on the question, but intrinsic uncertainty is nevertheless present. In mathematical terms, the variability of the $^{87}$Sr/$^{86}$Sr ratio ($\Delta^{87}$Sr/$^{86}$Sr) as observed today in different parts of the Earth's mantle, is proportional to the product of the variability of the Rb/Sr ratio ($\Delta$Rb/Sr) and the amount of time ($\tau$), since that variability formed:

$$\Delta^{87}\text{Sr}/^{86}\text{Sr} \propto \Delta\text{Rb/Sr} \times \tau \qquad (5\text{-}1)$$

If $\Delta^{87}$Sr/$^{86}$Sr is known through measurements on rocks, $\tau$ can be calculated if $\Delta$Rb/Sr can be estimated. For the Earth's mantle, $\Delta$Rb/Sr is comparatively small, so clear-cut differences of $^{87}$Sr/$^{86}$Sr take several hundred millions of years to develop, which represents the inherent resolution of this approach.

In contrast, relatively short-lived radioactive elements (half-lives significantly smaller than the age of the Earth) are sensitive to processes that occurred early in Earth history but are insensitive or less sensitive to later processes. For example, large variations in the $^{207}$Pb/$^{204}$Pb ratio in the Earth are more likely to have developed early in Earth history than late because $^{235}$U, the radioactive parent of $^{207}$Pb, has a half-life of only about 700 My. Any variation of the U/Pb ratio that formed 3.5 Gy ago would cause a variation of $^{207}$Pb/$^{204}$Pb today that is almost 20 times larger than the equivalent magnitude variation of U/Pb would produce if it formed 0.5 Gy ago. If we retain the symbol $\tau$ to represent the number of years in the past that the variation of the U/Pb ratio formed, and we used $T$ to represent the age of the Earth,

$$\Delta^{207}\text{Pb}/^{204}\text{Pb} \propto (\Delta\text{U/Pb}) \, (\tfrac{1}{2})^A \qquad (5\text{-}2)$$

where $A = (T - \tau)/\tau^{235}_{1/2}$.

For an element with an even shorter half-life (e.g., $^{129}$I, $\tau_{1/2} = 16$ My), any shift in the parent–daughter ratio that occurred more than approximately $10^8$ y after the Earth formed ($T - \tau > 10^8$ y) will not produce any significant isotopic variations for the daughter species $^{129}$Xe (see Table 5-1). In this case, any observed variations in $^{129}$Xe isotopic abundances must have formed very early in Earth history and thus can be unambiguously traced to processes that occurred then.

Implicit in this discussion is the fact that isotope ratios provide information on planetary differentiation processes only when the *parent*, or radioactive element is at least partially separated from the *daughter* element, the product of decay of the parent in the process being considered. For example, when the Earth's core formed, it is believed to have incorporated a large amount of the Pb isotopes that existed in the Earth at the time, but no uranium. Almost all of the uranium was left in the mantle. Consequently, the core formed with a low ratio of U/Pb, whereas the mantle was left with a relatively high ratio in comparison to the ratio in the Earth as a whole. If the difference in U/Pb ratios between the mantle and core can be estimated, then the ratios $^{207}$Pb/$^{204}$Pb and $^{206}$Pb/$^{204}$Pb in the Earth's mantle can be used to estimate when the core formed [see Eq. (5-2)]. Similarly, when continental crust forms from the mantle, it incorporates relatively more Rb than Sr (see Table 5-1) and therefore has a relatively high ratio of Rb/Sr in comparison with the mantle. The Rb-Sr isotopes thus provide information about when the continents formed [see Eq. (5-1)].

In some cases, the ages of rocks from a planet's surface can be directly measured. (One of the techniques used to determine age is described in section 5.4.) Age determinations provide detailed chronological information on processes that affect the planet's surface geology. For the Earth, where many thousands of rock samples have been dated, there is a detailed chronology available. A sufficient number of rocks from the Moon have been dated, so that lunar chronology is also quite well known. The isotopic evidence for the nature of planetary evolution is thus a combination of (1) inferences based on the relationships between chemical fractionations of parent-daughter ratios (Rb/Sr, U/Pb, Sm/Nd, and so forth) and isotopic variations ($^{87}$Sr/$^{86}$Sr, $^{207}$Pb/$^{204}$Pb, $^{143}$Nd/$^{144}$Nd, and so forth), and (2) age determinations on rock samples from planetary surfaces. In the following sections, applications of isotope techniques to the problems of planetary evolution are discussed.

## 5-3 CORE FORMATION IN THE EARTH

The only isotopic system that may yield useful information on the time of core formation is U-Pb (see Table 5-1), because Pb is the only available parent or daughter element that has the correct chemical properties to have been included in the Earth's metallic core. The age of the oldest rocks on the Earth, dated at close to 3.8 Gy (Moorbath et al., 1972; Goldich and Hedge, 1974; Hamilton et al., 1978) is generally considered to be a lower limit on the age of core formation. The question is whether the time of core formation can be specified to be almost the same as the age of Earth, about 4.55 Gy, as opposed to somewhat later, for example, in the range 4.4 to 4.0 Gy. The difference is important for understanding the early thermal evolution of the Earth (e.g., Kaula, 1979; Solomon, 1979).

The crux of the arguments using U-Pb isotopes is the observation that most solar-system materials, as represented by chondritic meteorites and the Sun, have low values (generally less than 1) of the parameter $\mu$ ($\equiv {}^{238}$U/$^{204}$Pb as it would be measured today) (Mason and Moore, 1982). Based on Pb-isotopic measurements of terrestrial rocks, it has been found that the Earth's mantle has a $\mu$ value of about 8 to 10, and has had such a value since at least about 3.6 Gy ago (Patterson, 1956; Holmes, 1946; Houtermans, 1946; Stacey and Kramers, 1975). If the Earth accreted from materials having a low $\mu$ value, the high mantle $\mu$ may have been caused by removal of Pb into the Earth's core at the time it formed. Based on the initial Pb-isotopic ratios of the oldest rocks, it is clear that the mantle must have had a $\mu$ value much higher than 1 for a substantial time period prior to 3.6 Gy. Gancarz and Wasserburg (1977) summarized the constraints by using a simple, two-stage model and data on initial Pb isotopes from samples of 3.64-Gy-old crustal rocks (see Fig. 5-2). The parameter $T_D$ can be taken as the time of core formation, and $T_A$ is the time of accretion of the Earth. The parameter $\mu_1$ applies to the Earth prior to core formation, and $\mu_2$ applies to the mantle after core formation. If, for ex-

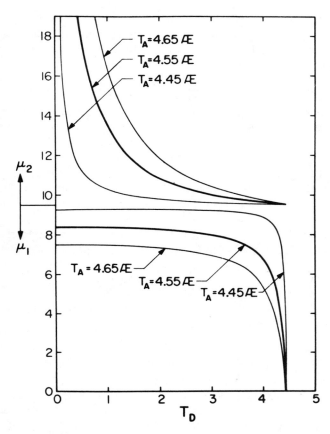

Fig. 5-2 Summary graph of the relationships between the time of the Earth's accretion ($T_A$), the time of core formation ($T_D$), and the $^{238}U/^{204}Pb$ ratios ($\equiv \mu$) in the Earth's mantle before ($\mu_1$) and after ($\mu_2$) the time of core formation. (Æ is equivalent to "billion years".) (Reprinted with permission from *Geochimica et Cosmochimica Acta, 41*, Gancarz, A. J. and G. J. Wasserburg, "Initial Pb of the Amitsoq Gneiss, West Greenland, and its implications for the age of the Earth," Copyright 1977, Pergamon Press.)

ample, $\mu_1$ were equal to 1, and the time of accretion $T_A$, were 4.55 Gy ago ($\equiv$ 4.55 Æ), the lower half of the graph would show that the "time of core formation" must be about 4.4 Gy ago. The limiting value is 4.43 Gy if $\mu_1 = 0$. If the $\mu$ value of the bulk Earth is indeed less than or equal to 2, the core must have formed quite early, although later than 4.55 Gy. On the other hand, if Pb was lost from the Earth by some other process such as volatilization, so that the U/Pb ratio of the Earth is, for example, 4 or 6, then core formation must have taken place later, perhaps 4.1 to 4.3 Gy ago. Ignoring the uncertainties in the model used to construct Fig. 5-1, the strength of the conclusions based on the U-Pb data are dependent on the value of $\mu$ for the bulk Earth as it was accreting. However, if, as proposed recently (Wetherill, 1976; Kaula, 1979; Solomon, 1979) the Earth accreted over a time period of the order $10^8$ y then, for the bulk Earth $\mu$ values of less than or equal to 2, the time of core formation becomes indistinguishable from the time of the later stages of accretion. Insofar as the assumption of low-$\mu$ starting material for the Earth is correct, the U-Pb data strongly support early core formation.

## 5-4 CRUST FORMATION ON THE EARTH AND MOON

Some examples of direct age determinations on lunar and terrestrial rocks are shown in Figs. 5-3 to 5-5. The Sm-Nd age of 3.75 Gy (see Table 5-1) for the metamorphic rocks of Isua, West Greenland, represents the oldest age obtained on terrestrial rocks (Hamilton et al., 1978). One difference between the Earth and the Moon is illustrated in Fig. 5-4 by the Rb-Sr age of 4.51 Gy determined on an igneous rock from the Moon (Papanastassiou and Wasserburg, 1976). All of the rocks from the lunar crust that have been dated thus far have ages in excess of 4.0 Gy, with the exception of the basalt flows filling the maria (Nyquist, 1981; Tera et al., 1977). The Moon's crust clearly formed very early in its history and has been largely preserved since. An example of age determinations on mare

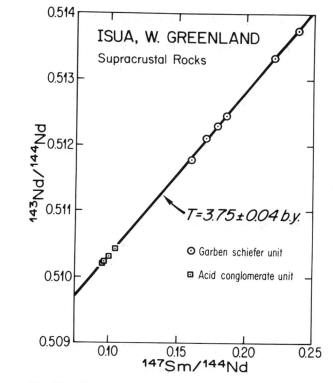

Fig. 5-3 Samarium-neodymium "isochron" determined on metamorphic rocks from the Isua area, near Godthaab, West Greenland. The age 3.75 Gy is the oldest age ever obtained from terrestrial rocks. (Reprinted by permission from Hamilton et al., 1978, Nature, 272, pp. 41-43. Copyright © 1978 Macmillan Journals Limited.)

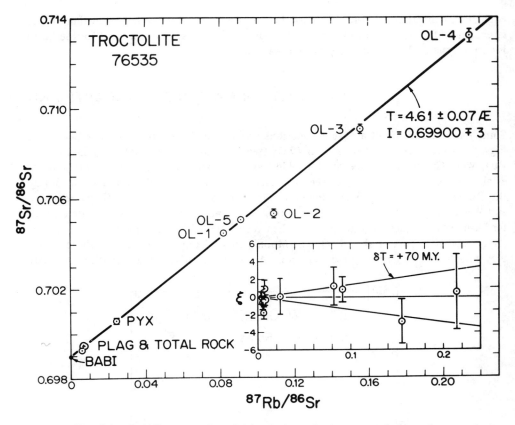

Fig. 5-4  Rubidium–strontium isochron determined on a rock from the ancient lunar-highlands crust. The age of 4.51 Gy (the age of 4.61 Gy shown was calculated using a different value for the half-life of $^{87}$Rb) is nearly equal to the age of the solar system, and it is much older than the oldest terrestrial rocks (see Fig. 5-3). This and other similar age determinations show that the formation of the lunar crust began very soon after the Moon formed. The insert shows the deviation parameter, $\xi$, in parts per 10,000 of the $^{87}$Sr/$^{86}$Sr ratio. (Reprinted with permission from *Proc. Lunar Sci. Conf., 7th*, Papanastassiou, D. A. and G. J. Wasserburg, "Rb-Sr age of troctolite 76535," Copyright 1976, Pergamon Press, Ltd.)

basalts (see Fig. 5-5) shows two samples from the Apollo 11 landing site that have ages substantially less than 4.0 Gy. A clearly resolvable age difference of about 300 My is shown (Papanastassiou et al., 1977). Many more data exist (e.g. Papanastassiou and Wasserburg, 1973; Wasserburg et al., 1977; Stettler et al., 1973; Lugmair et al., 1975), and the general conclusion is that the basaltic volcanism on the Moon did not commence until about 3.9 Gy ago, and lasted at least several hundred million years. Another important observation (see Fig. 5-6) is that the U-Pb data on lunar-highlands rocks show evidence for gross redistribution of U and Pb isotopes at about 3.9 Gy ago (Tera et al., 1974), the approximate time of several large basin-forming impacts. The largest of these impacts produced Mare Imbrium.

The chronology of crust formation on the Moon appears to be relatively simple.

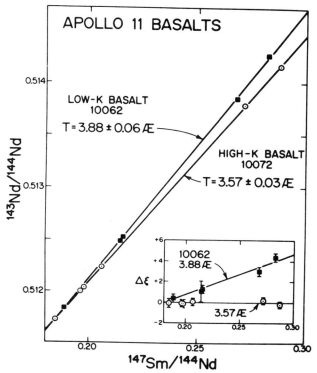

Fig. 5-5  Samarium-neodymium isochrons on two lunar basalt samples from the Sea of Tranquility. Note the resolution in the age difference between the two samples, and that the ages are considerably younger than the highlands sample (see Fig. 5-4). Basalt 10062 is among the oldest mare (basin) basalts. The insert shows the resolution between the two best-fit lines on an expanded vertical scale analogous to that of Fig. 5-4. (Reprinted with permission from *Proc. Lunar Sci. Conf., 8th*, Papanastassiou, D. A., D. J. DePaolo, and G. J. Wasserburg, "Rb-Sr and Sm-Nd chronology and geneology of mare basalts from the Sea of Tranquility," Copyright 1977, Pergamon Press, Ltd.)

Most of the crust was formed during an early planetwide differentiation that may have extended over a period as long as 200 My to 400 My (approximately 4.1 to 4.5 Gy ago). At about 3.9 to 4.0 Gy ago, a series of major impacts caused substantial disruption of the lunar crust. Subsequently, for about 1 Gy, basaltic lava flows poured out over the lunar surface in regions within the boundaries of the major impact basins. The Moon's history is one of early activity followed by a long period of quiescence. In terms of internally generated surface geologic activity, the Moon has been a dead planet for about 3 Gy. The lesson we have learned from the Moon, however, is that the internal evolution of planets starts almost at the time of formation, so the fact that the oldest rocks on the Earth are only 3.8 Gy old is of great significance.

The formation of the crust of the Earth has been a more complex process than on the Moon and has extended over at least 85% of the Earth's history. Today there are two distinct types of crust—oceanic and continental. The oceanic crust forms via the seafloor-spreading process at the globe-encircling oceanic ridge system (for an overview of plate tectonics, consult Cox, 1973). This crust moves away from the ridges at velocities of

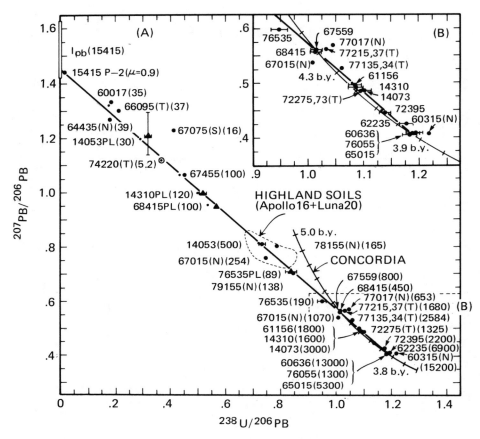

Fig. 5-6 Lead-uranium systematics of lunar-highlands rocks. The insert shows a portion of the curve on an expanded scale. The clustering of measurements near an isochron that intersects the concordia (upper) curve at 4.4 and 3.9 Gy has been interpreted as indicating that the age of the lunar crust is 4.4 Gy. At 3.9 Gy ago the crust was disrupted by several large meteorite impacts that formed the ringed basins (maria), and U-Th-Pb isotopes were grossly redistributed. All of the basalt flows that now fill the basins are 3.9 Gy old or younger. (Reprinted with permission from *Proc. Lunar Sci. Conf., 5th*, Tera, F., and G. J. Wasserburg, "U-Th-Pb systematics on lunar rocks and inferences about lunar evolution and the age of the Moon," Copyright 1974, Pergamon Press, Ltd.)

about 1 cm/y to 10 cm/y as new crust is continually formed at the ridge. The average oceanic crust spends only about 60 My at the surface before it is subducted back into the underlying mantle. The oceanic crust is composed mostly of basalt, with a thin veneer of covering sediment, is relatively thin (about 5 to 8 km), and represents the uppermost portion of a thermal boundary layer separating the hot, convecting mantle from the oceans and atmosphere. No part of the oceanic crust is more than about 200 My old.

The continental crust differs from oceanic crust in that it is richer in alkalies and silica, of lower density, thicker (averaging 40 km), and generally much older (up to 3.8 Gy) (see Moorbath and Windley, 1981). Continental crust is apparently less likely to be subducted and hence has accumulated persistently since 3.8 Gy ago. It is particularly valuable in that it records the history of the Earth over this long span, providing clues about how a planet slowly evolves through time. Some parts of the continents have been remarkably well preserved, having remained almost unchanged for 3 Gy. Other parts have been disturbed by more recent geological events, and the effects of these processes must be understood in order to interpret what the crust was like when it originally formed.

The formation of continental crust is related to subduction of oceanic crust in a fairly direct way (Taylor, 1977). Because of this, it has been implicitly assumed that the rate of formation of continental crust is directly related to the rate of subduction (mass/time) of oceanic crust. Because the subduction rate is related to the seafloor-spreading rate, which in turn reflects the rate of heat loss from the Earth's interior, it is widely believed that the rate of continental-crust formation through time tracks the thermal history of the Earth (Lambert, 1980).

If continental crust were always preserved once it formed, the rate of crust formation could be found by determining a curve of crustal mass versus age, and differentiating with respect to time. If the mass of the continental crust of age greater than $\tau$ is represented by $M_c(\tau)$, and if the global heat loss per unit time at time $\tau$ is $\dot{Q}(\tau)$,

$$-\dot{Q}(\tau) \propto \frac{d}{d\tau} M_c(\tau) \qquad (5\text{-}3)$$

However, it is quite likely that some continental crust gets destroyed and returned to the mantle, also via the subduction process (Armstrong, 1981). If, for instance, the return of crust to the mantle is characterized by the time constant $\lambda_s(\tau)$,

$$-\dot{Q}(\tau) \propto \frac{dM_c(\tau)}{d\tau} \exp \int_0^\tau \lambda_s(\tau)\, d\tau \qquad (5\text{-}4)$$

The most extensive attempt to determine the crustal age distribution was by Hurley and Rand (1969). Based on the available data, they found a predominance of relatively young ages. They concluded that the formation of continental crust had accelerated through the past 4 Gy, and that peaks occurred in the production rate at about 2.8±0.2 Gy ago and at 0.5 to 1.5 Gy ago. The function $M_c(\tau)$ based on a smoothing of their published histogram, is shown in Fig. 5-7 as the dashed curve. If Eq. (5-3) were to hold, the data imply a complicated thermal history, with the Earth generally heating up with time. Some more recent data, especially some obtained using the Sm–Nd-isotope technique, suggest that the Hurley and Rand (1969) age estimates are generally somewhat too young. Insufficient data exist to compare all continents, but sufficient data are in existence for North America. Figure 5-6 shows two solid curves, one based on the Hurley and Rand data for North America alone, and a comparison curve based on inference from data summarized by Muehlburger (1980) and Sm–Nd isotopic data from McCulloch and Wasserburg (1978), DePaolo (1981a, 1981b), Farmer and DePaolo (1983), and Nelson and DePaolo (1982). The latter curve suggests that there is a significantly larger amount of older crust than was estimated by Hurley and Rand (1969). The time derivatives of these curves (approximate) shown in Fig. 5-6(B) are quite different. The inferred curve has more distinct episodicity and places the largest crust-forming episode at about $\tau = 2.8$ Gy. This episodicity is not easily accounted for by quasi-steady-state thermal evolution models.

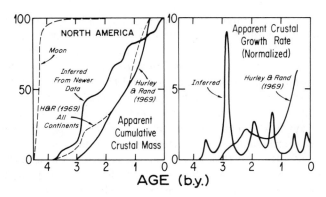

Fig. 5-7 Estimated mass of existing continental crust that is older than $\tau$, plotted against $\tau$, and normalized to present crustal mass. Solid lines are for the North American continent only. Dashed line is for all continents.

Although it is possible that the subduction term in Eq. (5-4) would substantially change the picture, it seems unlikely to be able to erase the episodicity since crust formation and crust destruction are coupled if both are related to subduction.

The curve shown in Fig. 5-7 is speculative, but through the use of the Sm-Nd method and the recently developed Lu-Hf method (Patchett and Tatsumoto, 1980; Patchett et al., 1981) in consort with age determinations by U-Pb and Rb-Sr, it should be possible in principle to determine such a curve with a fair degree of accuracy. To understand the meaning of the curve in relation to the Earth's thermal evolution requires, in addition, attention to the geology, petrology, and tectonics of the continents (see Moorbath and Windley, 1981, for a review). Assuming that, as on the Moon, crust of some sort formed as early as 4.5 Gy ago, it must be concluded that continental-type crust did not form to any significant degree at that time, but rather the initial crust must have been more like oceanic crust (Shaw, 1972; Condie, 1980; Smith, 1982) and has not been preserved *at all*. If this inference is correct, it would clearly (and not surprisingly) imply that the production of continental crust is dependent upon parameters other than the internal temperature of the Earth (e.g., the availability of condensed $H_2O$ at the Earth's surface to aid melting in subduction zones; see Ringwood, 1975).

It must be concluded that the details of the internal evolution of the Earth, as recorded by the chronology of continental-crust formation, are still uncertain. It is clear that the Earth's evolution has been progressing over its entire 4.5-Gy history with only barely discernible changes in long-term rates over the last 4 Gy. The internal energy still available in the Earth appears to be sufficient to keep the process going for several more billion years. The contrast with the Moon is striking indeed.

## 5-5 ATMOSPHERE AND HYDROSPHERE

The Earth's atmosphere and oceans probably formed as a result of the release of gases originally held in the interior (see Chapter 7 by Fanale; Rubey, 1951; Ringwood, 1979). The gases presumably escaped as a byproduct of magmatism. Magma, when it is under pressure deep in the Earth, can hold large amounts of volatile compounds in solution. These gases can be carried to the surface and then released to the atmosphere upon eruption of the lava (Burnham, 1979). Because magmatism is also intimately associated with the formation of continents, it has been proposed by some that the growth of continents and the outgassing of the Earth may be closely coupled (Rubey, 1951). However, we know that magmatism must have preceded the oldest continental crust known and that the rate of magmatic eruptions should be related to the internal temperature of the Earth. Thus, much of the outgassing could have taken place prior to the formation of the first permanent continental crust, when the Earth was probably hottest. Indeed it is likely that the process was underway during accretion before the Earth had completely formed (see Chapter 4 by Kaula).

Fragmentary evidence concerning the timing of the outgassing of the Earth is obtainable from the isotopes of the noble gases Ar and Xe, elements that should have been expelled from the Earth's interior along with the hydrogen, nitrogen, and water. Unfortunately, there are a large number of uncertain parameters, and for I-Xe isotopes (see Table 5-1), there are very few data. The abundance of the short-lived $^{129}$I (the parent of $^{129}$Xe) bears on this problem as shown schematically in Fig. 5-8. Because of its short half-life of 16 My, about 99 percent of the $^{129}$I originally incorporated into solid bodies in the solar system had decayed to $^{129}$Xe within about 100 My (about 99.99 percent had decayed after 200 My). If outgassing was delayed until after this time, all of the $^{129}$I would have already decayed. Thus, Xe gas leaving the mantle and entering the atmosphere and

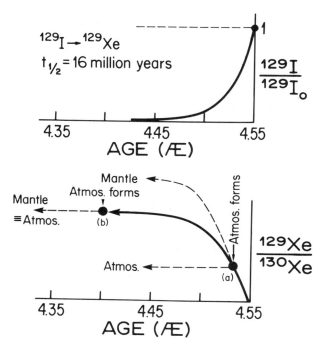

Fig. 5-8 Decay of $^{129}$I and growth of $^{129}$Xe abundance during the first 100 My of Earth history. Significant variations in $^{129}$Xe form only if I-Xe fractionation occurs during this time period.

any Xe left behind in the mantle would have the same relative amount of $^{129}$Xe. If this were the case, all Xe on the Earth would have the same fraction of $^{129}$Xe. However, if a substantial amount of the Xe had been released prior to approximately 4.45 Gy ago, atmospheric Xe would have a lower proportion of $^{129}$Xe (shown on the figure as a lower $^{129}$Xe/$^{130}$Xe ratio) than Xe that was held back in the Earth and released after all of the $^{129}$I had decayed (Thompson, 1980).

It has been observed that some of the Xe now coming from the Earth's interior (mantle Xe) is enriched in $^{129}$Xe (Hennecke and Manuel, 1975; Kyser and Rison, 1982; Staudacher and Allegre, 1982). This suggests that a substantial amount of outgassing took place in the first 100 My of Earth history, although because the I/Xe ratios are not known, and because there are so few data, no quantitative estimates can be made. Nevertheless, even if only 10% of the Xe was outgassed in the first 100 My, it seems likely that a large fraction could have been outgassed by the time the first continental crust formed 3.8 Gy ago.

The $^{40}$K-$^{40}$Ar system (see Table 5-1) does not yield information on very early outgassing, but it is useful for estimating the overall efficiency of the outgassing of the Earth. Again, quantitative estimates are difficult because the amount of K in the Earth cannot be estimated with sufficient accuracy (Hart et al., 1979). Following the initial demonstration by Gast (1960) that the Earth is depleted in K relative to chondritic meteorites, most recent estimates of the K content of the Earth are relatively low (about 120 to 170 ppm) (Anders, 1977; DePaolo, 1980). These estimates imply that 60 to 100 percent of the $^{40}$Ar of the Earth has been degassed, including the $^{40}$Ar in the continental crust (Hart et al., 1979). Furthermore, because most of the $^{40}$Ar was formed less than 4 Gy ago, this implies that there has been efficient degassing since that time also. However, variations of the $^{40}$Ar/$^{36}$Ar ratio trapped in glassy basaltic lava flows show that the degassing occurring between now and 4 Gy ago (at least) has been nonuniform. Some parts of the mantle are apparently much less degassed than others; presumably the less-degassed parts of the mantle are deeper (Hart et al., 1979).

Qualitatively, the existing data suggest that much of the Earth's degassing took place before the time the oldest continental crust formed, but that a substantial amount has also occurred since 3.8 Gy ago. The least that can be said is that the accumulation of degassed volatiles to form the atmosphere and hydrosphere took place on a shorter time scale than did the accumulation of sialic rock material to form the continental crust.

## 5-6 MANTLE STRUCTURE AND DYNAMICS

In principle, all of the isotopic systems listed in Table 5-1 may provide information on the structure of the mantle (in terms of chemically distinct reservoirs) and how and when that structure developed. In this regard, basaltic lavas have provided the most valuable means of probing the depths of the Earth's mantle. They bring to the surface samplings of the isotopic ratios in the mantle region from which they formed by partial melting. Variability in the isotope ratios $^{206}$Pb/$^{204}$Pb, $^{207}$Pb/$^{204}$Pb, $^{208}$Pb/$^{204}$Pb, $^{87}$Sr/$^{86}$Sr, $^{143}$Nd/$^{144}$Nd, $^{176}$Hf/$^{177}$Hf, $^{40}$Ar/$^{36}$Ar and others (Gast et al., 1964; Sun and Hanson,

1975; Tatsumoto, 1978; Hedge, 1970; Hoffman and Hart, 1978; O'Nions et al., 1977; DePaolo and Wasserburg, 1976; Patchett and Tatsumoto, 1980) all indicate the existence of domains within the mantle that are distinct in their chemical compositions (i.e., differing U/Pb, Th/Pb, Rb/Sr, Sm/Nd, Lu/Hf, and K/Ar ratios) and have been isolated against homogenization of these differences for time periods of more than 1 Gy. There are also a number of regularities in the geographic distribution of basaltic magmas coming from these different domains. Interpretations vary about how these data relate to the structure of the mantle. One that has gained considerable support holds that the mantle has two distinct layers—a shallow one that is "depleted," meaning it is thoroughly outgassed and has been greatly modified by having spawned the continental crust; and a deeper layer that is relatively more pristine, but was also probably degassed and modified slightly during the early history of the Earth. Variants of this model have been proposed by Schilling (1973), Sun and Hanson, (1975), Hart et al. (1979) and Wasserburg and DePaolo (1979). Figure 5-9, from the last reference, is a schematic representation of what

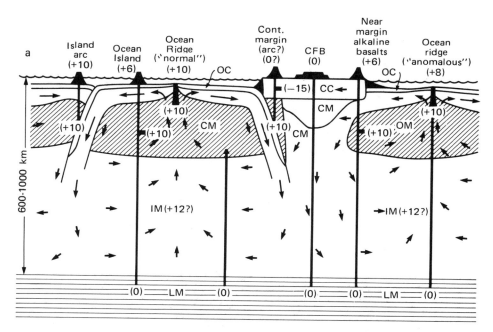

Fig. 5-9 Schematic model for the structure of the mantle based on neodymium isotopic measurements of basalt lava flows. The model depicts the mantle as having two layers. The upper layer is highly differentiated; the lower layer is much less differentiated. The lower layer retains some original volatiles and much of the radioactive heat-producing elements (K, U, and Th). The numbers in brackets are a representation (called $\epsilon_{Nd}$) of the isotope ratio $^{143}Nd/^{144}Nd$; positive numbers represent high ratios, negative numbers indicate lower ratios. The total Earth has a value of zero on this scale. The $\epsilon_{Nd}$ values shown at the top are those that are measured in the basalts. LM, lower mantle; CM, continental mantle; OM, oceanic mantle; OC, oceanic crust; CC, continental crust. (From Wasserburg and DePaolo, 1979.)

the mantle might look like in cross section, showing via the vertical stripes how different types of basaltic magmas come from different reservoirs characterized by different values of the parameter $\epsilon_{Nd}$ (the $^{143}$Nd/$^{144}$Nd ratio normalized to a standard). These reservoirs would also have distinct values of the other isotopic ratios listed above.

A somewhat peculiar aspect of this model is that the elements K, U, and Th (the radioactive elements that have provided the heat generated within the Earth since accretion stopped) would still be mostly buried deep in the mantle. The plausibility of this is the subject of current debate (Schubert and Spohn, 1981; Richter and McKenzie, 1981). Other models arrange the different reservoirs within the mantle in a less-organized fashion, as in Fig. 5-10 taken from Davies (1981). Such models do not require layering, and therefore allow larger-scale convection cells in the mantle (Hager and O'Connell, 1979). Although there is as yet no consensus on the details, the isotopic data provide a means of seeing into the mantle and studying its evolution. However, successful models must also take account of other relevant data, such as seismic wave velocities, the nature of phase transitions in the mantle, and plate tectonics.

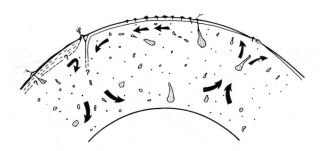

Fig. 5-10  An alternative model of mantle structure. In this case the parts of the mantle that are less differentiated are distributed as "blobs" within the more differentiated mantle, rather than forming a separate layer beneath it. (Reprinted by permission from Davies, 1981, *Nature, 290,* Copyright © 1981 Macmillan Journals Limited.)

Just as the crusts of the Earth and Moon are greatly different, the mantles of the two bodies also exhibit important differences. Data on Nd isotopes in lunar and terrestrial basalts (see Fig. 5-11) show that lunar basalts were highly variable in their $\epsilon_{Nd}$ values at a relatively early stage in the Moon's history (3 to 4 Gy ago). This indicates that the lunar mantle from which the basaltic magmas come, was heterogeneous very early. This heterogeneity probably formed during the early Moon-wide differentiation that also produced the crust, and it was frozen in at that time. The Earth's mantle, on the other hand, was apparently still rather homogeneous 3 Gy ago, as evidenced by the narrow range of $\epsilon_{Nd}$ values, all of which are near the zero line that represents a perfectly homogeneous, undifferentiated body. This relative homogeneity of terrestrial samples (there are measurable variations of $\epsilon_{Nd}$) probably is an indication of continuing vigorous mantle convection that tends to rehomogenize the chemical variability that forms.

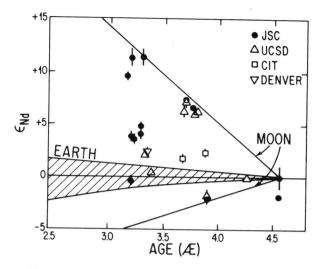

Fig. 5-11 Variability of $^{143}Nd/^{144}Nd$ ratio [represented as fractional deviations from the bulk-Earth value ($\epsilon_{Nd}$)] versus age, for basalts and related igneous rocks from the Earth and Moon. These variations represent variations in the mantles of the respective bodies. The data indicate that the lunar mantle was very heterogenous at an early point in time, consistent with early formation of the lunar crust (see Fig. 5-4). The Earth's mantle took much longer to develop the same-sized isotopic variations, consistent with the lesser age of the Earth's continental crust (see Fig. 5-7) in comparison with the lunar crust.

## 5-7 SUMMARY

Information about the internal evolution of the Earth and the Moon is provided by measurements of radioactive isotopes and their decay products in rocks and minerals. This information can be used to generalize about the evolution of the terrestrial planets. U–Pb isotopes indicate that the core of the Earth formed very early in the Earth's history, possibly contemporaneously with the Earth's accretion. This conclusion is consistent with data on the ages of lunar crustal rocks, which indicate that magma formation and differentiation of planetary bodies may begin when they are still much smaller than the present Earth. Age determinations on lunar rocks also indicate that small planets generally exhaust their energy for differentiation within about 1 Gy or so, in contrast to larger planets like the Earth, which are continuing their differentiation at the present time.

Age determinations and systematic variations of Sm–Nd and Lu–Hf isotopes indicate that the formation of continental crust on the Earth has been a continuing process over the past 3.8 Gy. There is evidence in the data for pronounced episodicity in this crust-formation process, which suggests that the internal evolution of a planet may not be simply a quasi-steady-state process. The Earth in particular appears to have experienced a particularly large crust-forming episode at approximately 2.8 Gy ago. This episode occurred long after the end of accretion and long after the time of formation of the core, and thus does not correlate with any other large-scale perturbation of the Earth's interior. The cause of this great event is currently unknown.

Isotopic evidence on the time of formation of the Earth's oceans and atmosphere is inconclusive. However, certain data, especially the few observations of excess $^{129}$Xe in some terrestrial rocks, suggest that degassing of the Earth's interior, and, by inference, the formation of a substantial fraction of the present mass of atmosphere and oceans, probably preceded the formation of much of the Earth's continental crust. Analogous information is inaccessible for the Moon, because of the apparent lack of primary volatile constituents in the lunar interior.

Isotopic variations in Pb, Sr, Nd, Hf, and Ar indicate that the differentiation of the Earth's interior has resulted in the formation of separate and isolated domains in the mantle with substantially different chemical and isotopic compositions. The deep interior appears to be less differentiated than the outer several hundred kilometers (also true for the Moon). Retention of heat-producing radioactive elements in the deep mantle may be a contributing factor in the prolonged geologic life of the Earth. The location and evolution of isotopically distinct reservoirs within the mantle provide information on the details of the Earth's internal evolution after core formation, but there is as yet no consensus on the proper interpretation of these data.

## 5-8 ACKNOWLEDGMENTS

Research by the author on some of the problems discussed in this paper has been supported by grants from the National Science Foundation, NASA, and the Petroleum Research Fund.

## 5-9 REFERENCES

Anders, E., 1977, Chemical compositions of the Moon, Earth, and eucrite parent body, *Phil. Trans. Roy. Soc. London A, 285*, 23-40.

Armstrong, R. L., 1981, Radiogenic isotopes: the case for crustal recycling on a near-steady-state no-continental-growth Earth, in S. Moorbath and B. F. Windley, eds., *The Origin and Evolution of the Earth's Continental Crust*, The Royal Society, London, p. 259-287.

Burnham, C. W., 1979, Magmas and hydrothermal fluids, in H. L. Barnes, ed., *Geochemistry of Hydrothermal Ore Deposits*, John Wiley, New York, p. 71-136.

Condie, K. C., 1980. Origin and early development of the Earth's crust, *Precamb. Res., 11*, 183-197.

Cox, A., 1973, *Plate Tectonics and Geomagnetic Reversals*, W. H. Freeman, San Francisco, 702 pp.

Davies, G. F., 1981, Earth's neodymium budget and structure and evolution of the mantle, *Nature, 290*, 208.

DePaolo, D. J., 1980, Crustal growth and mantle evolution—Inferences from models of element transport and Nd and Sr isotopes, *Geochim. Cosmochim. Acta, 44*(8), 1185-1196.

DePaolo, D. J., 1981a, Nd isotopic studies—Some new perspectives on Earth structure and evolution, *EOS, 62*(14), 137-140.

DePaolo, D. J., 1981b, Neodymium isotopes in the Colorado Front Range and crust-mantle evolution in the Proterozoic, *Nature, 291,* 193.

DePaolo, D. J., and G. J. Wasserburg, 1976, Inferences about magma sources and mantle structure from variations of $^{143}$Nd/$^{144}$Nd, *Geophys. Res. Lett., 3,* 743-746.

Farmer, G. L., and D. J. DePaolo, 1983, Origin of Mesozoic and Tertiary granite in the Western United States and implications for pre-Mesozoic crustal structure. 1. Nd and Sr isotopic studies in the geocline of the Northern Great Basin, *J. Geophys. Res., 88,* 3379-3401.

Faure, G., 1977, *Principles of Isotope Geology*, John Wiley, New York, 464 pp.

Gancarz, A. J., and G. J. Wasserburg, 1977, Initial Pb of the Amitsoq Gneiss, West Greenland, and implications for the age of the Earth, *Geochim. Cosmochim. Acta, 41*(9), 1283-1301.

Gast, P. W., 1960, Limitations on the composition of the upper mantle, *J. Geophys. Res., 65*(4), 1287-1297.

Gast, P. W., G. R. Tilton, and C. E. Hedge, 1964, Isotopic composition of lead and strontium from Ascension and Gough Islands, *Science, 145c,* 1181-1185.

Goldich, S. S., and C. E. Hedge, 1974, 3800-Myr granitic gneiss in south-western Minnesota, *Nature, 252,* 467-468.

Hager, B. H., and R. J. O'Connell, 1979, Kinematic models of large-scale mantle flow, *J. Geophys. Res., 84,* 1031-1048.

Hamilton, P. J., R. K. O'Nions, N. M. Evensen, D. Bridgwater, and J. H. Allaart, 1978, Sm–Nd isotopic investigations of Isua supracrustals and implications for mantle evolution, *Nature, 272,* 41-43.

Hart, S. R., J. Dymond, and L. Hogan, 1979, Preferential formation of the atmosphere-sialic crust system from the upper mantle, *Nature, 278,* 156.

Hedge, C., 1970, Variations in radiogenic strontium found in volcanic rocks, *J. Geophys. Res., 71*(24), 6119-6126.

Hennecke, E. W., and O. K. Manuel, 1975, Noble gases in an Hawaiian xenolith, *Nature, 257,* 778.

Hoffman, A. W., and S. R. Hart, 1978, An assessment of local and regional isotopic equilibrium in the mantle, *Earth Planet. Sci. Lett., 38*(1), 44-62.

Holmes, A., 1946, An estimate of the age of the earth, *Nature, 157,* 680-684.

Houtermans, F. G., 1946, Die Isotopenhäufigkeiten im natürlichen Blei und das Alter des Urans, *Naturwissenschaften, 33,* 185-219.

Hurley, P. M., and R. Rand, 1969, Pre-drift continental nucleii, *Science, 164,* 1229-1242.

Kaula, W. M., 1979, Thermal evolution of Earth and Moon growing by planetesimal impacts, *J. Geophys. Res., 84,* 999-1008.

Kyser, T. K., and W. Rison, 1982, Systematics of rare gas isotopes in basic lavas and ultramafic xenoliths, *J. Geophys. Res., 87,* 5611-5630.

Lambert, R. St. J., 1980, The thermal history of the earth in the Archean, *Precamb. Res., 11,* 199-213.

Lugmair, G. W., N. B. Scheinin, and K. Marti, 1975, Sm–Nd age of Apollo 17 basalt 75075—evidence for early differentiation of the lunar exterior, *Proc. Lunar Sci. Conf. 6th,* p. 1419-1429.

Mason, B., and C. B. Moore, 1982, *Principles of Geochemistry* (4th ed.): Wiley, New York.

McCulloch, M. T., and G. J. Wasserburg, 1978, Sm-Nd and Rb-Sr chronology of contental crust formation, *Science, 200*, p. 1003-1011.

Moorbath, S., R. K. O'Nions, R. J. Pankhurst, N. H. Gale, and V. R. McGregor, 1972, Further rubidium-strontium age determinations on the very early Precambrian rocks of the Godthaab district, West Greenland, *Nature Phys. Sci., 240*, 78-82.

Moorbath, S., and B. F. Windley, eds., 1981, *The Origin and Evolution of the Earth's Continental Crust*, The Royal Society, London.

Muehlburger, W. R., 1980, The shape of North America during the Precambrian, in *Continental Tectonics*, National Academy of Sciences, Washington, D.C., p. 175-183.

Nelson, Bruce K., and D. J. DePaolo, 1982, Crust formation age of the North American midcontinent, *Geol. Soc. Am. Abs. with Program., 14*, 575.

Nyquist, L. A., 1981, Radiometric ages and isotopic systematics of pristine plutonic lunar rocks, in *Papers Presented to the Workshop on Magmatic Processes of Early Planetary Crust*, Lunar and Planetary Institute, Houston, Tex.

O'Nions, R. K., P. J. Hamilton, and N. M. Evensen, 1977, Variations in $^{143}$Nd/$^{144}$Nd and $^{87}$Sr/$^{86}$Sr ratios in oceanic basalts, *Earth Planet. Sci. Lett., 34*(1), 13-22.

Papanastassiou, D. A., and G. J. Wasserburg, 1973, Rb-Sr ages and initial Sr in basalts from Apollo 15, *Earth Planet. Sci. Lett., 17*(2), 324-337.

Papanastassiou, D. A., and G. J. Wasserburg, 1976, Rb-Sr age of tractolite 76535, *Proc. Lunar Sci. Conf. 7th*, p. 2035-2054.

Papanastassiou, D. A., D. J. DePaolo, and G. J. Wasserburg, 1977, Rb-Sr and Sm-Nd chronology and genealogy of mare basalts from the Sea of Tranquility, *Proc. Lunar Sci. Conf. 8th*, p. 1639-1672.

Patchett, P. J., and M. Tatsumoto, 1980, Hafnium isotope variations in oceanic basalts, *Geophys. Res. Lett., 7*(12), 1077-1080.

Patchett, P. J., O. Kouvo, C. E. Hedge, and M. Tatsumoto, 1981, Evolution of continental crust and mantle heterogeneity—evidence from Hf isotopes, *Contr. Mineral. Petrol., 78*, 279-297.

Patterson, C. C., 1956, Age of meteorites and the Earth, *Geochim. Cosmochim. Acta, 10*(4), 230-237.

Peterman, Z. E., and Hedge, C. E., 1971, Related strontium isotopic and chemical variations in oceanic basalts, *Geol. Soc. Am. Bull., 82*, 493-500.

Richter, F. M., and D. P. McKenzie, 1981, On some consequences and possible causes of layered mantle convection, *J. Geophys. Res., 86* (By), 6133-6142.

Ringwood, A. E., 1975, *Composition and Petrology of the Earth's Mantle* McGraw-Hill, New York, 618 pp.

Ringwood, A. E., 1979, *Origin of the Earth and Moon*, Springer-Verlag, 295 pp.

Rubey, W. W., 1951, Geologic history of seawater—an attempt to state the problem, *Geol. Soc. Am. Bull., 46*, 1111-1147.

Schilling, J. -G. 1973, Icelandic mantle plume—geochemical evidence along the Reykjanes Ridge, *Nature, 242*, 565-571.

Schubert, G., and T. Spohn, 1981, Two-layer mantle convection and the depletion of radioactive elements in the lower mantle. *Geophys. Res. Lett., 8*(9), 951-954.

Shaw, D. M., 1972, Development of the early continental crust. Part 2: Prearchean, Protoarchean and later eras, in B. F. Windley, ed., *The Early History of the Earth*, John Wiley, New York, p. 33-53.

Smith, J. V., 1982, Heterogeneous growth of meteorites and planets, especially the Earth and Moon, *J. Geol., 90*(1), 1-48.

Solomon, S. C., 1979, Formation, history and energetics of cores in the terrestrial planets, *Phys. Earth Planet, Int., 19*, 168–182.

Stacey, J. S., and J. D. Kramers, 1975, Approximation of terrestrial lead isotope evolution by a two-stage model, *Earth Planet. Sci. Lett., 26*(2), 207–221.

Staudacher, T., and C. J. Allegre, 1982, Terrestrial xenology, *Earth Planet. Sci. Lett., 60*, 389–406.

Stettler, A., P. Eberhardt, J. Geiss, N. Grögler, and P. Maurer, 1973, $^{39}$Ar–$^{40}$Ar ages and $^{37}$Ar–$^{38}$Ar exposure ages of lunar rocks, *Proc. Lunar Sci. Conf. 4th*, p. 1865–1888.

Sun, S. S., and G. N. Hanson, 1975, Evolution of the mantle—Geochemical evidence from alkali basalt, *Geology, 3*(16), 297–302.

Tatsumoto, M., 1978, Isotopic composition of lead in oceanic basalt and its implication to mantle evolution, *Earth Planet. Sci. Lett., 38*, 63–87.

Taylor, S. R., 1977, Island arc models and the composition of the continental crust, in M. Talwani and W. C. Pitman, eds., *Island Arcs, Deep-Sea Trenches and Back-Arc Basins*, Am. Geophys. Union, Washington, D.C., p. 325–336.

Tera, F., and G. J. Wasserburg, 1974, U–Th–Pb systematics on lunar rocks and inferences about lunar evolution and the age of the Moon, *Proc. Lunar Sci. Conf. 5th*, p. 1571–1599.

Tera, F., D. A. Papanastassiou, and G. J. Wasserburg, 1974, Isotopic evidence for a terminal lunar cataclysm, *Earth Planet. Sci. Lett., 22*(1), 1–21.

Thompson, L., 1980, $^{129}$Xe on the outgassing of the atmosphere, *J. Geophys. Res., 85*(B8), 4374–4378.

Wasserburg, G. J., and D. J. DePaolo, 1979, Models of earth structure inferred from neodymium and strontium isotopic abundances, *Proc. Nat. Acad. Sci., 76*, 3594–3598.

Wasserburg, G. J., D. A. Papanastassiou, F. Tera, and J. G. Huneke, 1977, Outline of a lunar chronology, *Phil. Trans. Roy. Soc. London A., 285*, 7–22.

Wetherill, G. W., 1976, The role of large bodies in the formation of the earth and moon, *Proc. Lunar Sci. Conf. 7th*, p. 3245–3257.

York, D., and R. M. Farquhar, 1972, *The Earth's Age and Geochronology*, Pergamon Press, Oxford, Elmsford, N.Y., 178 pp.

Margaret G. Kivelson and Gerald Schubert
Department of Earth and Space Sciences
University of California
Los Angeles, California 90024

# 6

# ATMOSPHERES OF THE TERRESTRIAL PLANETS

# ABSTRACT

Venus, Earth, and Mars have vastly different surface temperatures. The differences are only partially explained by the decrease of solar-radiation flux with distance from the Sun. More significant effects arise from the variations in the degree to which the atmospheres act as absorbers of planetary thermal reradiation. Atmospheric circulation on a global scale also varies markedly among the three planets; many aspects of the large-scale winds can be understood by considering the forces arising from pressure variations and rotational motion. Some aspects of the winds observed at Venus have not been unambiguously interpreted.

## 6-1 INTRODUCTION

Atmospheric pressure at the surfaces of Venus and Mars is about 90 times greater and 140 times smaller, respectively, than the pressure of Earth's atmosphere at sea level. Surface temperature, on Earth about 288°K, is 730°K on Venus, hot enough to melt lead and zinc or to cause mercury and sulfur to boil, and only 218°K on Mars, cold enough to freeze water (carbon dioxide can also freeze in the polar regions of Mars and high in its atmosphere, where the temperatures are somewhat lower). Wind patterns representing large-scale atmospheric circulation differ markedly from one planet to another. How to account for these observed variations is the main purpose of this discussion. Some of the differences in the atmospheres of Venus, Earth, and Mars can be understood on the basis of a few fundamental principles. We explain these principles and apply them to the terrestrial planets with atmospheres (neither Mercury nor the Moon have atmospheres). We do not attempt to give explanations for all the observed physical characteristics of the terrestrial planets' atmospheres. Indeed, some aspects of the dynamics and circulation of Venus's atmosphere are not presently well understood and are the subject of ongoing research. In Chapter 7 by Fanale, questions of the origin and chemical composition of the atmospheres are addressed, so here we simply assume the presence of an atmosphere of known composition and seek to interpret its thermal structure and its circulation. Supplementary references that bear on the material of this chapter are Goody and Walker (1972), Leovy (1977), Pollack (1981), and Schubert and Covey (1981).

In the following sections, we identify the properties of the planets, such as size, spin rate, and distance from the Sun, that are important in understanding the characteristics of their atmospheres. Then we consider the rate at which the planets absorb energy from the Sun. Steady conditions result when energy is reradiated from the atmosphere at the same rate as it is absorbed, and we show how this condition establishes an "effective" temperature of the atmosphere. Surface temperatures are higher than the effective temperatures, because the atmosphere is opaque to thermal radiation from the surface and low levels of the atmosphere. We discuss how the atmospheric absorption leads to elevated surface temperatures through what is called the *greenhouse effect*.

To understand horizontal motions of atmospheres, we examine the forces that act on the atmosphere because of planetary rotation and pressure variations, and we explain why winds move along contours of constant pressure. The dominant pressure gradients result from nonuniform surface heating. The consequences for atmospheric circulation differ among the terrestrial planets (Earth, Mars, and Venus), and we briefly describe some major features of their large-scale circulations. For Venus, we point out a number of areas in which research is still active because the mechanisms responsible for the observed wind patterns are not fully understood.

## 6-2 SUMMARY OF PHYSICAL CONDITIONS AND GOVERNING LAWS

### 6-2-1 Orbital Parameters that Matter

The energy flux from the Sun falls off as the square of the distance ($R^{-2}$), so it is reasonable to expect atmospheric temperatures to decrease with distance from the Sun. Table 6-1 confirms this expectation, showing decreasing surface temperatures with increasing distance. (The atmospheres of terrestrial planets are warmed only by the Sun. The situation is more complicated for the outer planets, which are also warmed significantly by sources internal to the planets.)

TABLE 6-1 Surface Temperatures and Relevant Orbital Parameters

| Planet | Surface Temperature (°K) | Mean Distance from Sun (km) | Eccentricity | Obliquity |
|---|---|---|---|---|
| Venus | 730 | $1.08 \times 10^8$ | 0.007 | 177° |
| Earth | 288 | $1.50 \times 10^8$ | 0.017 | 23.5° |
| Mars | 218 | $2.28 \times 10^8$ | 0.093 | 24° |

The mean distance to the Sun is only one of the orbital parameters that affects the atmosphere of a planet. Also important is the eccentricity of the orbit (defined in Chapter 1), which determines the variation in the flux of solar energy that the atmosphere receives in each orbital period. For example, the distance between the Sun and Mars varies by approximately 19 percent during a Martian "year" (see eccentricity in Table 6-1) and the incident solar flux changes by about one-third. By contrast, the distance between the Sun and Earth changes by only approximately 3 percent and this effect is not very important.

The inclination of the plane of the equator to the orbital plane is called the *obliquity*. If the obliquity differs from 0° or 180°, the axis of rotation (approximately fixed in direction, as a gyroscope) tilts toward and away from the Sun as the planet moves around its orbit, thus producing seasons. From Table 6-1, it is clear that Mars has seasons like the Earth, but that there are no seasonal effects on Venus.

Our experience tells us that on Earth the atmosphere varies on a daily, as well as yearly, time scale. This means that the rate of rotation is important. Table 6-2 provides the rotation periods as well as the orbital periods. Also shown is the length of a solar day, which differs from a rotation period because of orbital motion. (If the planet moves around its orbit *without rotating*, an observer on the planet would still see the Sun rise

TABLE 6-2  Additional Parameters Important to Atmospheric Response

| Planet | Orbital Period (Days) | Rotation Period | Length of Solar Day | Overhead Motion of Sun |
|---|---|---|---|---|
| Venus | 224.7 | −243 days | 117 days | W to E |
| Earth | 365.3 | 23 h 56 min | 1 day | E to W |
| Mars | 687.0 | 24 h 37 min | 1 day | E to W |

and set once per year, so the number of solar days per year differs from the number of rotations.)

Mars and Earth spin at roughly the same rates. Venus spins much more slowly. Furthermore, the value of 177° for Venus's obliquity tells us that the spin axis is "upside down" relative to the others, so that the rotation is from east to west, or *retrograde*. At the surface of Venus, an observer would see the Sun rise in the west and, 58 days later, set in the east, heralding in night for the next 58 days. In a later section we discuss the dynamical consequences of this slow rotation.

## 6-2-2  Holding on to an Atmosphere

Planets acquire atmospheres either by retaining some gases as they condense initially or by outgassing light material from their interiors, for example, in volcanic emissions. Fanale in Chapter 7 discusses the question of how the atmospheres were formed.

Gases have a tendency to expand, so one may wonder why planets do not lose their atmospheres. Evidently, the confining effect of gravity opposes escape from the planet. The balance between the thermal pressure (upward) and the gravitational attraction (downward) determines how tightly the atmosphere is bound to the planet's surface and is described in terms of $H$, the scale height of the atmosphere. For an isothermal atmosphere $H = RT/Mg$, where $R$ is the universal gas constant per mole, $T$ is the temperature, $M$ is the average molecular weight (mol wt) of the atmospheric gases, and $g$ is the acceleration of gravity at the surface. The scale height is the vertical distance in which the pressure decreases by $e^{-1} \cong 0.37$. Thus small values of $H$ imply rapid decrease of atmospheric pressure with altitude. Table 6-3 gives the values of $g$ and $H$, as well as surface pressures, for the terrestrial planets. Note that $g$ is similar for Earth and Venus because they are close in size, but $g$ is much smaller for tiny Mars. All three scale heights are within a factor of two of one another. Variations in $T$, $M$, and $g$ among the terrestrial planets tend to compensate and yield scale heights that are roughly comparable.

TABLE 6-3  Atmospheric Surface Gravities, Scale Heights, and Surface Pressures

| Planet | Surface Gravity, $g$ (m · s$^{-2}$) | Scale Height, $H$ (km) | Surface Pressure (MPa) |
|---|---|---|---|
| Venus | 8.9 | 15 | 9.2 |
| Earth | 9.8 | 8.5 | 0.1 |
| Mars | 3.7 | 10.5 | 0.0007 |

### 6-2-3 Energy Input

The maximum rate at which a planetary atmosphere can absorb heat from the Sun depends on the flux of solar energy (i.e., energy per second per unit of area perpendicular to the Sun-planet direction) at the distance of the planet, a quantity called the *solar constant*. The values of the solar constant are listed in Table 6-4 for the planets of interest, and one can verify that they decrease with the square of the distance of the planet from the Sun, as previously noted. However, a cloudy atmosphere can reflect a substantial fraction of the incident energy flux, thus decreasing the rate of energy absorption. On Earth, clouds come and go, and typically 40% of the surface is cloud covered. On Mars, the cloud cover is variable. But on Venus, 100% of the surface is covered with clouds, and the planet reflects a large fraction of the radiation that falls on it. The reflectivity of a body is described quantitatively in terms of its *albedo*, and Table 6-4 indicates that the albedo of Venus is much larger than those of Earth and Mars. For this reason, the rate of energy absorption per unit area is smaller for Venus than Earth (see Table 6-4), even though the solar constant at Venus is almost double that of Earth. Surprisingly, Mars absorbs almost as much energy from the Sun as does Venus.

TABLE 6-4  Solar Energy and Atmospheric Absorption

| Planet | Solar Constant ($kW \cdot m^{-2}$) | Albedo | Absorbed Solar Flux ($kW \cdot m^{-2}$) | Altitude of Solar Energy Deposition |
|---|---|---|---|---|
| Venus | 2.62 | 80% | 0.132 | Clouds and above |
| Earth | 1.38 | 30% | 0.242 | Ground |
| Mars | 0.59 | 20% | 0.118 | Ground |

The rate at which an entire planetary atmosphere absorbs energy from the Sun can be found from the relation

$$\text{Solar energy absorbed per unit time} = (\text{solar constant})\, \pi R_p^2\, (1 - \text{albedo}) \quad (6\text{-}1)$$

In this expression, the factor $\pi R_p^2$ is the *area* of the outwardly flowing solar flux that is intercepted by a planet, that is, the area of a disk whose radius is $R_p$, the planetary radius.

### 6-2-4 Black-Body Radiation and Atmospheric Temperature

In the previous section, we considered the absorption of energy by planetary atmospheres. As energy is absorbed, the atmospheric temperature increases unless there is a way for the atmosphere to balance energy gain by energy loss. Since the atmospheres are in steady state, we know that such an energy loss takes place. In fact, the planet loses energy by reradiation into space.

For a body at temperature $T_e$, the energy radiated is a sensitive function of $T_e$. It is described by the Stefan-Boltzmann law, which says

$$\text{Energy flux} = \sigma T_e^4 \quad (6\text{-}2)$$

where $\sigma$ is the Stefan-Boltzmann constant, and its value is $5.67 \times 10^{-8}$ W · m$^{-2}$ · deg$^{-4}$. This flux is emitted from the entire surface of the planet, hence from an area $4\pi R_p^2$. Thus the total energy radiated per second is

$$(4\pi R_p^2) \sigma T_e^4 \tag{6-3}$$

Now we can describe the net rate at which the atmosphere gains energy as the difference between Eqs. (6-1) and (6-3), that is,

$$\text{Net energy gain per unit time} = (\text{solar constant})(\pi R_p^2)(1 - \text{albedo}) - (4\pi R_p^2)\sigma T_e^4 \tag{6-4}$$

Note that if $T_e$ is small, the net energy gain is positive. Then the atmospheric temperature increases, and, in turn, the rate at which energy is lost [Eq. (6-3)] increases. Warming continues until $T_e$ is large enough so that the two terms on the right side of Eq. (6-4) cancel, that is, energy loss and gain rates balance, giving a steady state.

The requirement of steady state allows us to calculate $T_e$, the effective temperature for the planetary atmospheres. For Venus, we find $T_e = 244°$K; for Earth, $T_e = 253°$K; and for Mars, $T_e = 216°$K. These temperatures differ from the surface temperatures of Table 6-1 by 486°K, 35°K, and 2°K, respectively. Why is Venus as hot as it is? The next section explains these differences.

## 6-2-5 The Greenhouse Effect

Surface temperatures at Venus and Earth exceed the effective radiation temperatures ($T_e$) because of what is called the *greenhouse effect*, which we examine in this section. First we must recognize that $T_e$ is the temperature of a layer sufficiently high in the atmosphere that it can radiate energy into space without further interaction with the atmosphere. Lower layers radiate as well, but the energy they radiate is reabsorbed within the atmosphere. If we require that in steady state the rates of energy loss and gain balance for each layer of the atmosphere, we can learn something about how temperature changes with altitude.

Our model atmosphere is illustrated in Fig. 6-1. We assume uniform composition. We also assume that the incident sunlight passes through the atmosphere without being absorbed and that it is absorbed by the ground. If the energy absorbed were reradiated as visible light, it would pass back through the atmosphere and the ground temperature would be $T_e$. However, the energy is reradiated as thermal radiation in the infrared portion of the spectrum, that is, at wavelengths longer than those of visible light. The atmosphere is not transparent to infrared wavelengths; rather, it absorbs and also emits such thermal (infrared) radiation. It is this atmospheric absorption that leads to ground temperatures greater than $T_e$. In our layered atmosphere model, we take each layer to be sufficiently thick to absorb all the thermal energy incident on it. The layer must, therefore, increase in thickness with altitude to compensate for the decrease in density of the absorbing gas.

As we noted, layer 1, the top layer, radiates into space, and the requirement that

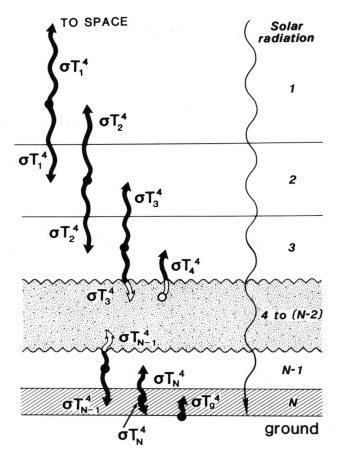

Fig. 6-1 Schematic of model used to calculate ground temperature. Layers are numbered to right. Heavy wavy lines represent energy flux, labeled to indicate magnitude. Arrow heads indicate regions of absorption. Circles indicate sources of radiation. Solar radiation is the thin wavy line. In this model, it is absorbed at ground level. Because there is no net change in energy, $T_1 = T_e$, and the net solar radiation flux $= \sigma T_1^4 = \sigma T_e^4$.

the atmosphere is neither heating nor cooling means that $T_1$ must equal $T_e$. But layer 1 radiates in two directions, so its energy-loss rate per unit area is $2\sigma T_1^4$. Layer 1 absorbs radiation from layer 2, and steady state requires that losses and gains balance, so $2\sigma T_e^4 = \sigma T_2^4$, or $T_2^4 = 2T_e^4$. Similarly, layer 2 loses energy at the rate $2\sigma T_2^4$ and absorbs energy from the adjacent layers at the rate $\sigma T_1^4 + \sigma T_3^4$. From the energy balance we can deduce $T_3^4 = 3T_e^4$, and evidently, the $n$th layer (temperature $T_n$) satisfies

$$T_n^4 = nT_e^4 \tag{6-5}$$

Finally, consider the layer $N$ (temperature $T_N$) just above the ground, whose temperature is $T_g$. Energy balance in layer $N$ gives

$$2\sigma T_N^4 = \sigma T_g^4 + \sigma T_{N-1}^4$$

or

$$T_g^4 = (1 + N)T_e^4 \tag{6-6}$$

The number of absorbing layers above the ground ($N$) is called the *optical thickness* of the atmosphere. For an optically thick atmosphere, $T_g$ can be quite large compared with $T_e$. Venus, for example, has a massive atmosphere. Its surface pressure (Table 6-3) is equivalent to the pressure at the depth of about 1 km in the ocean. The atmosphere contains $CO_2$, water vapor, $SO_2$, and cloud particles all of which absorb thermal radiation effectively. Equation (6-6) is satisfied for $T_e = 244°K$ at Venus if $N \cong 70$ and $T_g \geqslant 700°K$, and these are reasonable figures that fit with observations.

For Earth, whose atmosphere is predominantly $N_2$ and $O_2$, neither of which is a strong absorber of thermal radiation, the energy is largely absorbed by minor atmospheric constituents such as $CO_2$, water vapor, and water clouds. The optical depth is small but adequate to explain the difference between $T_e$ (253°K) and $T_g$ (288°K). Any significant addition of absorbing gases, such as $CO_2$, to the atmosphere increases the optical depth and, consequently, increases the temperature at the ground. This possibility has raised concern that major climate changes could result from relatively small changes of average ground temperature, a subject that is widely discussed. Unfortunately, large-scale climate models are extremely complex, and there is no consensus whether such climate changes are probable for realistic assumptions of the change in $CO_2$ content of the atmosphere.

Mars has $CO_2$ in its atmosphere sufficient, despite the low pressure, to account for the 2°K difference between $T_e$ and $T_g$.

## 6-2-6 Vertical Thermal Structure

In the model calculation of Sec. 6-2-5, we assumed that the solar radiation penetrates the atmosphere and is absorbed by the ground. This assumption, as well as the assumption of uniform composition, although a useful approximation, is not correct. Consequently, the vertical temperature profiles vary nonmonotonically with altitude. Figure 6-2 illustrates the variation of temperature with height for Venus (day side), Mars, and Earth. The greenhouse effect occurs in the *troposphere*, where the model of the previous section applies. At the highest levels, the *thermosphere* for all three planets has temperature increasing with altitude. At these levels the atmospheres are heated directly by absorbing a portion of the solar radiation (an effect not considered in the previous section). On Mars the temperature variation is also strongly affected by absorption of sunlight by dust suspended in the atmosphere. Earth's temperature profile is further complicated by the presence of

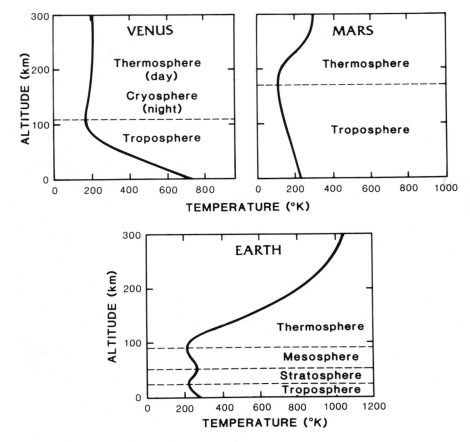

Fig. 6-2 Temperature variation with altitude for the planets Venus, Mars, and Earth.

a layer of ozone gas near an altitude of 50 km. The ozone absorbs ultraviolet radiation from the incoming solar radiation, and local heating results at the upper level of the stratosphere. Clouds on Earth (composed of liquid and solid water) are found in the troposphere. Venus clouds are composed of concentrated sulfuric acid; they are found between 50 and 60 km, with haze layers above and below. On Mars the clouds are composed of water and solid carbon dioxide and, when present, occur throughout the troposphere.

Comparing temperatures near 300 km with those near the ground, we see that the Earth is hot at the top and cold at the bottom, Venus is cold at the top and hot at the bottom, and Mars has a relatively uniform temperature throughout. These temperature profiles seem a bit paradoxical when we note that the bulk of the solar energy is deposited at the ground on Earth and Mars but in the clouds and higher at Venus; however, they are consistent with mechanisms for energy input and loss that we have described.

## 6-2-7 Motions of Atmospheres: Coriolis Force

Breezes, hurricanes, and jet streams all show that the atmosphere moves relative to the planet's surface. In this section and the next, we discuss the forces that produce horizontal atmospheric motions. The first to be considered arises only because of the planetary rotation; it is called the *Coriolis force*.

The reason that planetary rotation introduces a special force is illustrated in Fig. 6-3. Imagine a projectile (a parcel of air will do as well) moving radially away from the center of a rotating disk at a uniform velocity. A stationary observer would see motion in a straight line with no acceleration and would conclude that no forces are acting on the body. However, an observer rotating with the disk would observe the projectile curving

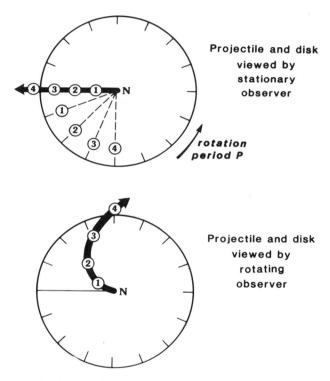

Fig. 6-3 Motion of a projectile and a rotating disk as viewed by observers in different frames. Above, the view of an observer at rest. The projectile moves to the left in a straight line and at a constant velocity. Positions are marked for equal increments of time (1 to 4). The disk rotates, and the observer sees the line on a disk along which the projectile was launched at $t = 0$ move to positions marked 1 to 4 at the corresponding times. Below, the view of a rotating observer at rest on the disk. This observer notes no motion of the line along which the projectile was launched, but sees the projectile moving on a curved path with positions at times 1 to 4 as marked.

away from the direction of its initial motion and would conclude that a perpendicular force is affecting the particle's motion, causing it to curve in its motion. This force is called the *Coriolis force*, and a rotating observer must include it to explain a particle's trajectory.

When we observe winds from the surface of a planet, we are rotating observers (rotating with the Earth, for example). Therefore, when we consider atmospheric dynamics, we must include the Coriolis force, whose magnitude per unit mass is

$$(4\pi/P)u \sin \phi$$

where $P$ is the planetary rotation period, $u$ is the velocity of the projectile (or parcel of atmosphere), and $\phi$ is the latitude angle. The force acts perpendicular to the direction of motion, to the right in the northern hemispheres of Earth and Mars (e.g., see illustration in Fig. 6-3) and to the left in the southern hemispheres of these planets (imagine looking at Fig. 6-3 from beneath). On Venus, these directions are exactly opposite, because the planet rotates in the opposite direction (retrograde) of Earth and Mars.

Both Earth and Mars rotate so rapidly that the Coriolis force is one of the two main forces controlling the winds. Venus rotates so slowly that the Coriolis force is unimportant in the dynamics of Venus's atmospheric circulation.

### 6-2-8 Motions of Atmospheres: Centrifugal Force

An object, such as a parcel of air, moving along a circular path experiences a force tending to push it radially outward from its center of motion. This is the well-known *centrifugal force*, whose

$$\text{magnitude per unit mass} = u^2/r$$

where $r$ is the radius of the object's circular path. When viewed from above the north pole, for example, see Fig. 6-4, the zonal winds in an atmosphere (winds parallel to latitude circles) are seen to be moving in circular paths about the polar axis. The radius of the circle is $a \cos \phi$ (see Fig. 6-4), where $a$ is the radius of the planet (we assume the winds are at an altitude that is small compared with the planetary radius). The centrifugal force per unit mass is $u^2/(a \cos \phi)$, and it points perpendicular to the polar axis, as shown in Fig. 6-4. The horizontal component of this force per unit mass points toward the equator and is

$$[u^2/(a \cos \phi)] \sin \phi = u^2 \tan \phi/a$$

The centrifugal forces acting on the global zonal circulations on Earth and Mars are not very significant, because the wind speeds are not high enough. On Venus, however, there is a rapid zonal circulation of the entire atmosphere, and the horizontal component of the centrifugal force is a dominant factor in the atmospheric dynamics. Centrifugal forces can also be important on Earth and Mars in small-scale flows such as tornadoes and dust storms.

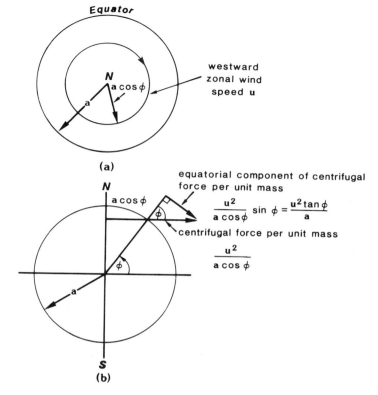

Fig. 6-4 View looking down on the north pole (a) of an atmosphere rotating to the west relative to a planet's surface. The westward horizontal wind speed (zonal wind speed) is $u$ at latitude $\phi$. (b) The rotation of the atmosphere produces a centrifugal force whose horizontal component is directed equatorward and has magnitude $(u^2 \tan\phi)/a$.

### 6-2-9 Motions of Atmospheres: Pressure Force and Geostrophic Balance

Another force that acts on a planetary atmosphere is the *pressure force*, directed from regions of high pressure toward regions of low pressure. Unless restrained by opposing forces, gases flow from high to low pressure, reaching equilibrium when the pressure is uniform.

For Earth and Mars, many aspects of atmospheric motion can be understood by considering only the pressure force and the Coriolis force. If a pressure gradient (from high to low pressure) is present on the surface of the planet, the atmosphere moves to reduce the pressure variation until the pressure force just balances the Coriolis force. A steady flow may then continue along contours of constant pressure, called *isobars*. The balance between the opposing forces with flow along isobars is called *geostrophic balance* and is illustrated in Fig. 6-5a for a limited region in which the Coriolis force is assumed constant in direction. Larger-scale views, of the types familiar from satellite images

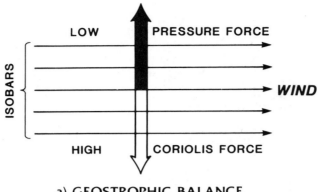

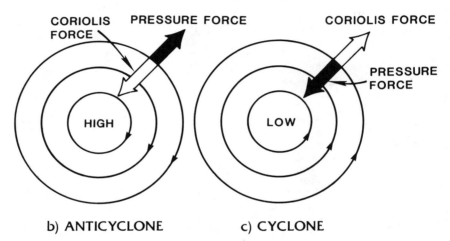

Fig. 6-5 Schematics of conditions for geostrophic balance. (a) The northward-directed force from high- to low-pressure balances the southward-directed Coriolis force. Wind flows along isobars. (b) Isobars and wind-flow directions around a northern hemisphere high-pressure region. (c) The same as (b) for a low-pressure region.

of cloud motions, are illustrated in Fig. 6-5b and c. In the vicinity of a high-pressure region in the northern hemisphere, the flow, clockwise as viewed from above, is anticyclonic (Fig. 6-5b). Correspondingly, a low-pressure region leads to cyclonic motion (Fig. 6-5c). The flow directions reverse if the high- or low-pressure regions are located in the southern hemisphere.

### 6-2-10 Motions of Atmospheres: Pressure Force and Cyclostrophic Balance

On Venus, the predominant horizontal-pressure gradient is north-south, with pressure decreasing toward the poles on longitude circles in both hemispheres. The westward zonal winds (the major circulation pattern of Venus's atmosphere) are kept moving along latitude circles by a balance of poleward pressure forces and the equatorward horizontal centrifugal force (see Fig. 6-6). This dynamical state is known as *cyclostrophic balance* and is, so far as we know, unique to Venus among all the planets and satellites in our solar system that possess atmospheres. As was the case with geostrophic balance, the zonal motion is along or parallel to isobars. Note that the dynamical state would be the same, that is, the forces would have the same directions, if Venus's atmosphere rotated from west to east instead of from east to west.

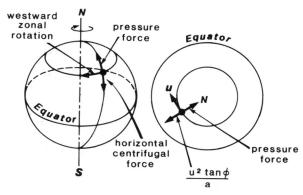

Fig. 6-6 Forces involved in the cyclostrophic balance of an atmosphere rotating to the west at high speed. The equatorward, horizontal centrifugal force is balanced by a poleward pressure force.

## 6-3 NONUNIFORM HEATING AND ATMOSPHERIC CIRCULATION

### 6-3-1 Latitude Dependence of Atmospheric Heating

In previous sections we have considered the energy balance for the planet as a whole, requiring that the net heat deposited from solar flux just balances the heat lost by radiation. In the discussion, we imagined that the planet's atmosphere radiates at a single "effective" temperature, $T_e$, and we ignored spatial variations of both solar heating and radiative cooling. Yet we know that Earth's atmosphere is warmer near the equator than at the poles, and warmer at midday than at midnight, so we must consider how the energy balance varies with position on the surface of the planet.

Let us first think about why the net energy deposited into the atmosphere varies with latitude. Figure 6-7 shows the absorbed flux of solar radiation and the flux of re-radiated energy as functions of latitude. At Earth, both are largest near the equator (0° latitude) and both drop off toward the poles (90° latitude). For the absorbed energy, the variation is primarily a geometrical effect arising from the relative orientation of the surface and the incoming sunlight. Near noon, for example, the sunlight falls directly on the surface near the equator, but merely grazes the surface near the poles. For the radiated energy, the decrease between equator and poles occurs because the poles are colder.

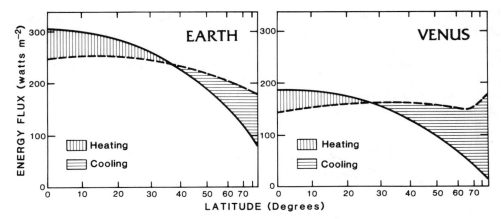

Fig. 6-7 Radiation balance for Earth and Venus. Solid curves represent energy flux absorbed from solar radiation as a function of latitude. Dashed curves represent energy flux reradiated into space. For Earth, these two processes alone produce net heating equatorward of about ~40° latitude and net cooling at higher latitudes. For Venus, heating occurs equatorward of about ~25° latitude. The latitude scale is drawn proportional to cosine (latitude), since only the component of the surface area perpendicular to the Sun's rays (the Sun is assumed to be directly over the equator) is effective in absorbing solar radiation. This trigonometric factor must be accounted for in the energy balance.

Equation (6-2) tells us that they radiate less energy. Near the equator, though, the rate of energy absorption exceeds the rate of energy loss, whereas poleward of approximately 40° latitude, the situation is reversed. This imbalance requires that the atmosphere itself carry heat poleward to compensate for the heat losses in the polar regions. On Venus, the same latitudinal imbalances exist (see Fig. 6-7) to drive circulation between equator and pole. On Mars, where the atmosphere is too tenuous to retain much heat, the heating and cooling rates both decrease from equator to pole and are believed to be almost in balance at all latitudes (though the relevant measurements have not been made).

### 6-3-2 Atmospheric Circulation: Earth

Earth's atmosphere, on average, rotates with the surface of the planet, but different regions of the atmosphere move in different directions relative to the surface. The resulting winds are called *zonal* if they blow east-west and *meridional* if they blow north-south.

Meridional winds exchange heat between equator and poles, as illustrated in Fig. 6-8a for the Earth's northern hemisphere. Near the equator, the air, heated by sunlight, rises and moves poleward. It is prevented from moving poleward beyond about 30° latitude by the strong effects of Earth's rotation. Instead, it sinks back to the surface and heads toward the equator, resulting in a large-scale circulation called a *Hadley cell*. Within the Hadley cell, the Coriolis force acts to divert the flow eastward as it moves north at high levels and westward as it moves south at low levels. Thus, near the surface steady northeasterly winds (winds are labeled by the direction from which they blow) appear. The dependable winds in these low latitudes were especially appreciated in the days of sailing ships when they acquired the name *trade winds*.

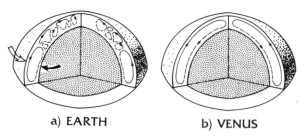

Fig. 6-8 Schematic of meridional circulations on Earth and Venus. Solid curves represent high-level flow. Dashed curves represent near-surface flow. On Venus, a single Hadley cell may exist between the equator and pole (see text). On Earth, the low-latitude Hadley cell is disrupted near 30° latitude by Coriolis forces. The directions of the Coriolis forces equatorward of 30° latitude are shown as an open arrow for near-surface winds and a filled arrow for high-level winds. At Venus, Coriolis forces are negligible.

Poleward transport at latitudes greater than 30°, where Coriolis forces are increasingly strong, is accomplished by the net effect of eddy motions forced by an instability known as the *baroclinic instability*.

Zonal winds are usually stronger than meridional winds. They vary in strength and direction relative to the surface with both latitude and altitude. In the era of air travel, the zonal jet streams at midaltitudes are particularly helpful.

### 6-3-3 Atmospheric Circulation: Mars

The circulation of the Martian atmosphere is similar to Earth's in many ways, but some things are quite different. In particular, the winter polar regions get so cold (about 150°K) that $CO_2$ freezes out on the surface. In spring and summer, the $CO_2$ sublimates away, thus creating a seasonal flow from one pole to the other, as contrasted with the flow between equator and poles that characterizes Earth and Venus. Day and night differ greatly in temperature (tens of degrees), and this difference produces strong *thermal tides*, with the atmosphere blowing from regions of heating to regions of cooling. The winds follow the motion of the Sun in the Martian sky.

A strong feedback mechanism links global dust storms and wind strength. The stronger the winds, the greater the entrainment of dust. More dust absorbs more heat from the Sun and enhances temperature contrasts around the planet. Greater temperature gradients produce stronger winds. Such a reinforcement keeps dust storms going for exceedingly long periods of time.

### 6-3-4 Atmospheric Circulation: Venus

Venus's atmospheric circulation stands in strong contrast to those of the Earth and Mars. First, the predominant circulation is a global east-to-west superrotation of the entire atmosphere. (*Superrotation* implies that the rotation of the atmosphere is faster than the rotation of the surface.)

We know about the superrotation from several observations. First, when Venus is observed in ultraviolet radiation, there are distinct cloud markings that rotate into and out of view periodically. Figure 6-9 shows a series of pictures obtained by the *Pioneer Venus Orbiter* on successive days in February 1979. The large-scale markings denote the famous "Y" (recumbent in the pictures), visible in image 1. In picture 2, the base of the "Y" appears, but the arms have rotated out of sight. The series of pictures continues until four days later (picture 5), when the planet appears again as in picture 1. This gives evidence that the atmosphere rotates in four days. The rotation is east to west (i.e., retrograde) at a rate 60 times faster than the surface. This speed corresponds to wind speeds of $100 \text{ m} \cdot \text{s}^{-1}$ at the cloud tops near the equator. The global retrograde superrotation of the atmosphere contrasts with Earth's zonal winds, which are strong only in jets and which rotate westward or eastward depending on latitude and season.

We have learned how Venus's atmospheric rotation varies with height from radio tracking of numerous Soviet and U.S. probes that have entered the atmosphere and have been carried along by the winds. Figure 6-10 illustrates the altitude profiles of the zonal

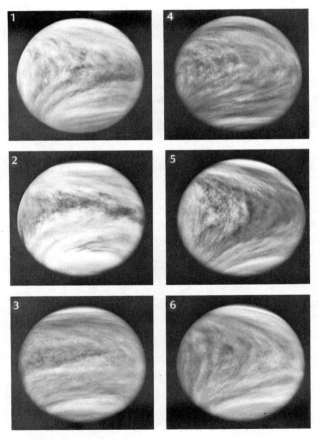

Fig. 6-9 Ultraviolet images of Venus, February 15, 1979, to February 20, 1979, from the *Pioneer Venus Orbiter*. The sequence is described in the text.

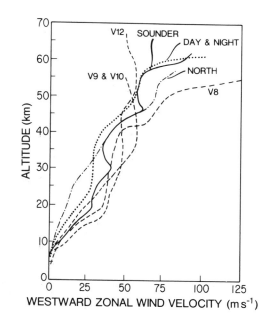

Fig. 6-10 Altitude profiles of east-to-west wind speed from spacecraft tracking (Schubert et al., 1980).

winds. The winds are westward at all heights, and wind speeds increase essentially monotonically with altitude.

Earlier we noted that the centrifugal force produced by the atmospheric rotation relative to the surface has a component that pushes equatorward and opposes the pressure force that is directed toward the poles, a state of cyclostrophic balance. Measurements of pressure gradients in the atmosphere by *Pioneer Venus* probes indicate that the balance is satisfied to a good approximation.

The dynamical mechanism responsible for the superrotation is not known. Several processes can contribute, any one of which can bring angular momentum up from the surface to the atmosphere. One thought is that the angular-momentum transport is carried by eddies induced by the overhead motion of the Sun, the so-called moving-flame mechanism. Other possible mechanisms, not to be described here, have been proposed; no consensus has yet emerged, and the problem remains an interesting one for future study.

In addition to the dominant rotation of the Venusian atmosphere, there are also meridional circulations that transport heat from equator to pole. Information on meridional winds comes from ultraviolet pictures, which allow cloud-top formations to be tracked, and from space probes. At cloud tops, the meridional winds are poleward. Near the base of the clouds, they are equatorward on all the probes. This is the level of the atmosphere where heat is deposited. Thus we believe that there is a cloud-level Hadley circulation of the type shown schematically in Fig. 6-8b. Because of the slow planetary rotation, the Hadley cells at Venus do not break up at high latitudes.

The actual meridional circulation at Venus may be more complex than indicated in Fig. 6-8b. In particular, the high-altitude Hadley cell stops at cloud level, but a low-altitude Hadley cell also seems necessary to transport poleward the small amount of solar energy absorbed by the near-equatorial surface. If such a near-surface cell exists, a third Hadley cell with reversed circulation would be required to separate the opposing flows from the high- and low-altitude Hadley cells. More information is needed to enable us to clarify the description of meridional flows on Venus.

## 6-4 SUMMARY

This brief review has pointed out some of the significant features of the large-scale structure and dynamical properties of the atmospheres of Venus, Earth, and Mars. To interpret the thermal structure, it was important to consider the planet's albedo and the effectiveness of the atmosphere as an absorber of thermal radiation. To explain global-circulation patterns, it was necessary to examine the variation of heating with latitude and to consider the rotation rate of the planet.

Both albedo and heating are strongly influenced by chemical composition, cloud cover, and atmospheric dust. If these features change with time, major climate changes including modified temperatures and circulation patterns can ensue. As we compare Earth with its neighbors in space, we gain insight into conditions that may have existed in the past and that show us ways in which our own atmosphere may evolve. Thus, the scientific challenges that make the study of planetary atmospheres attractive are augmented by the sense that we are also learning about our own planetary environment.

## 6-5 REFERENCES

Goody, R. M., and J. C. G. Walker, 1972, *Atmospheres*, Prentice-Hall, Englewood Cliffs, N.J., 150 pp.

Leovy, C. B., 1977, The atmosphere of Mars, *Sci. Am., 237*, 34–43.

Pollack, J. B., 1981, Atmospheres of the terrestrial planets, p. 57–70 in *The New Solar System* (J. K. Beatty, B. O'Leary, and A. Chaikin, eds.), Sky Publishing Corp., Cambridge, Mass.

Schubert, G., and C. Covey, 1981, The atmosphere of Venus, *Sci. Am, 245*, 66–74.

Schubert, G., C. Covey, A. Del Genio, L. S. Elson, G. Keating, A. Seiff, R. E. Young, J. Apt, C. C. Counselman, III, A. J. Kline, S. S. Kimaye, H. E. Revercomb, L. S. Sromovsky, V. E. Suomi, F. Taylor, R. Woo, and U. Von Zahn, 1980, Structure and circulation of the Venus atmosphere, *J. Geophys. Res., 85*, 8007–8025.

Fraser P. Fanale
Planetary Geosciences Division
Hawaii Institute of Geophysics
University of Hawaii
2525 Correa Road
Honolulu, Hawaii 96822

# 7

# PLANETARY VOLATILE HISTORY: PRINCIPLES AND PRACTICE

# ABSTRACT

The diversity of planetary surface-atmosphere volatile regimes can be understood and explained in terms of a corresponding planetary diversity in the following processes (and resulting states): (1) incorporation of volatiles during condensation of preplanetary material and planetary accretion (controls planetary bulk volatile inventory), (2) planetary energetics and internal thermal history (controls efficiency and chronology of degassing), (3) planetary volatile sinks, including space (controls current distribution, mass, and elemental mass balance of "available" surface volatile inventory), and (4) operation of external variables, for example, solar energy, on the transient, steady-state array of surface volatiles that are available at any point in time (controls molecular composition, atmospheric pressure, and climate change).

The net result of all these processes is a volatile history that is itself a controlling factor in overall planetary history: It affects interior history via melting curves and heat transfer; it affects most strongly the delicate environment at the surface of a planet and is the central factor in shaping conditions for the abiotic origination and sustenance of life.

Major current uncertainties in studies of each of these four processes or states are considered, and tentative conclusions drawn. Some examples of key issues for each process or state are (1) Does an apparent increase in bulk volatile content with increasing heliocentric distance—a trend to which most solid bodies appear to conform—reflect the influence of a central heat source during preplanetary condensation? If so, is the trend violated by a volatile-poor achondrite parent body, by a volatile-poor Mars, or by the Earth's Moon? This is probably not the case. (2) Concerning *planetary energetics*, were "magma oceans" common on both large and small objects 4.5 Gy ago? What were the energy sources? Did catastrophic degassing and initial supply of large surface-volatile inventories result for massive bodies with large gravitational fields, and did degassing of bodies with small gravitational fields cause the planets they later formed to be volatile-poor? Would degassing of a magma ocean be limited by solubility, and what would be the elemental mass balance of the evolved gas? (3) Concerning volatile sinks, in what forms are volatiles hidden in the Mars regolith? Did Venus ever possess an ocean? Did Io possess, degas, and lose a great deal of $H_2O$ and other volatiles early in its history? How has preferential hydrogen escape affected the chemistry of atmospheres? (4) How do astronomically induced variations in solar heating produce volatile redistribution and accompanying atmospheric pressure variations and climate changes on Mars and Earth? How does interaction with the Jovian magnetosphere affect the Galilean satellites' volatiles? Does the Earth have the only current liquid $H_2O$ ocean, or do subsurface liquid $H_2O$ hydrospheres exist on Europa, Mars, and other objects? Did early Mars experience more clement surface conditions, including rainfall?

Two of several pervasive "technique" issues affecting all these questions are the origin and significance of planetary primordial rare gas (i.e., planetary Ne, Ar, Kr, or Xe). The abundances of planetary primordial rare gases in planetary atmospheres have been used as a convenient index of the abundances and distribution of the major chemically reactive volatiles, since the former do not normally form chemical compounds. Studies of the processes by which the rare gases were incorporated into preplanetary and planetary material suggest that significant differences exist between those processes and the processes that caused incorporation of major volatiles. Although still an enormously useful tool, the full significance of the abundances of rare gases therefore requires further clarification. For example, the variation with heliocentric distance of planetary primordial rare gas contents cannot be predicted from (1) the postulate of equilibrium condensation (which requires an increase in bulk volatile content with heliocentric distance), and (2) the postulate that volatile content of rare gases should correlate well with general volatile content as is true for meteorites. The observed *decrease* of planetary primordial rare gas content with increasing heliocentric distance can be reconciled with the postulates above only by invoking an additional process in the evolution of terrestrial planets' atmospheres—namely, major early or initial atmospheric loss, with loss efficiency greater at Mars than Earth. Hence rare gas abundances should be regarded as only one of several lines of evidence (including isotopic studies of both rare gases and chemically active volatiles and nonvolatiles) that are best used in concert to decipher the history of planetary surface-volatile inventories.

## 7-1 PRINCIPLES OF VOLATILE HISTORY

### 7-1-1 Introduction

One of the many reasons why Rubey's (1951) paper on the geological history of seawater is recognized as a classic is that it properly identifies and defines a new and important subject—the origin and history of the Earth's *excess volatiles*. The identification of an excess volatile inventory immediately implies two key conclusions supported in that paper: (1) The Earth's surface volatile inventory far exceeds that which could be attributed to the weathering of rocks (hence the term *excess*) and (2) that the oceans, the $CO_2$ tied up in carbonates, the $N_2$ in the atmosphere, the sulphur in evaporites, and so forth, all constitute a coherent whole. Thus, it is unproductive to consider, for example, the degassing history of $CO_2$ in total isolation from that of $N_2$ or to consider the history of the oceans and the atmosphere as separate. With the advent of extensive planetary exploration from both Earth and spacecraft, a new dimension has been added to Rubey's analysis: We now compare excess surface volatile inventories of the various planets and satellites. Such a comparison, if done casually, results in the "zoo syndrome," that is, various planetary surfaces are dominated by different volatiles and each seems to have experienced a different history.

Despite the diversity, all planetary surface volatile inventories can be thought of as the products of the same four factors: (1) the condensation-accretion history of preplanetary material and its effect on planetary bulk-volatile concentrations; (2) the planet's energy history, leading to its degassing history, including the times, causes, and mechanisms of volatile transport to the surface-atmosphere environment; (3) the sinks available for volatiles. These sinks can cause temporary removal (such as condensation in a polar cap or regolith) or quasi-permanent removal (such as removal of $CO_2$ by deposit as carbonates which may be subducted and recycled) or truly permanent loss (e.g., removal of gaseous species from a planetary exosphere into space). The balance between the supply of volatiles (2) and their removal (3) continues throughout a planet's history. As it continues, this competition between supply and removal creates a transient steady-state population of volatiles at a planetary surface. It is (4) the operation of external *kneading processes* on this raw steady-state inventory of volatiles that determines the atmospheric pressure, the availability of liquid $H_2O$, and all the other climatic and chemical variables that control the environment at the planet's surface, including its ability to spawn and foster life. A good example of a kneading process is the *externally* (orbitally and axially) imposed variation in average surface insolation, or heating by sunlight, which on Mars moves quasi-available volatiles such as regolith-condensed volatiles and polar cap-condensed volatiles back and forth among the regolith, atmosphere, and the caps, causing periodic variations of a factor up to 100 in atmospheric pressure, corresponding variations in cap size, and weathering and transport mechanisms. This kneading of the $CO_2$ on Mars is also a good example of what might be called *comparative volatile studies*, since the Earth experiences analogous periodic climate variations driven with a similar period by the same cause.

The diversity of planetary volatile histories may be illustrated in the most simple terms by considering a single volatile, $H_2O$. We can (allowing ourselves to "megathink" for a moment) divide planets into categories based on their $H_2O$ abundance and distribution. Ignoring the current physical state of the water, we find that we can divide planetary objects into five categories: (1) planetary objects that are mostly $H_2O$ (all the large outer-planet satellites except Io and Europa); (2) those which are mainly composed of "rock" but have substantial $H_2O$ inventories and have sufficiently differentiated to transfer a large portion to the surface (Earth, Europa, and probably Mars); (3) those which were formed with substantial $H_2O$ content in bulk but failed to differentiate to create an $H_2O$ outer shell (probably most asteroids); (4) those which were formed with a substantial $H_2O$ inventory, transferred it to the surface but then lost it to space and interior reaction (probably Io and possibly Venus); and (5) those which never had a substantial $H_2O$ inventory (the Moon, Mercury, possibly Venus, and conceivably Io). A further interesting category comprises (6) those objects that are not only similar to the Earth in being rocky and degassing an $H_2O$ shell, but that presently contain liquid $H_2O$ zones in their crusts (possibly Europa and Mars). The uncertainties admitted to in the preceding list perhaps expose the shortcomings of such a "megathink" approach or, rather, the shortcomings of our knowledge. Nonetheless, it is an instructive exercise and yields some interesting results, especially the strange kinship, pursued elsewhere in this paper, between the Earth and the least studied of Jupiter's major moons—Europa (shown in Fig. 7-1). The main point, however, is that if such diversity exists for even a single volatile—$H_2O$—one can appreciate how a consideration of the wide variations in the abundance and distribution of all the volatiles throughout the solar system can engender what I have referred to as the "zoo syndrome." Nonetheless, it is the contention of this paper that the diversity can be entirely encompassed and explained within a framework consisting of the four processes listed earlier. We now consider these four processes in more detail and then indulge in the necessary body-by-body examination of our ability to explain the volatile inventories and volatile histories of individual objects using these four processes.

## 7-1-2 Planetary Formation and Bulk Volatile Content

To begin with, we consider the first factor: bulk volatile content, which results from planetary condensation-accretion processes. Specifically, an eminently reasonable starting hypothesis is that the differences in planetary bulk volatile content among objects is a

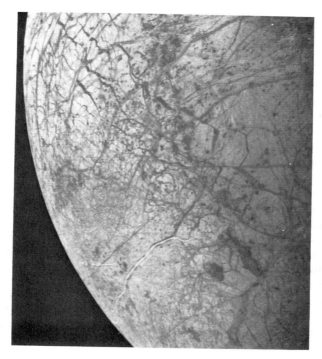

Fig. 7-1 Europa's icy crust. Owing to Jupiter's early "small star" history, Europa—like Io—formed as a rocky object (see text). Like the Earth, Europa has largely degassed its volatiles to form an $H_2O$ crust perhaps several kilometers to tens of kilometers thick. Sets of cracks record episodes in Europa's tectonic history. A key issue is whether or not tidal heating can maintain a liquid-$H_2O$ zone that might exist in the crust below a frozen outer layer. (NASA Photo P-21764C.)

function of the position within the preplanetary nebula (hence the temperature) at which the material comprising the object condensed (Lewis, 1972a, 1974; see Chapter 3 by Lin). In its simplest form, this *equilibrium condensation* hypothesis states that the center of the solar system represented a heat source that prevented condensation of volatile and semivolatile elements and compounds in its inner portions prior to gas dispersal. Thus a single compositional gradient radial to the Sun could describe the resulting array of solid planetary bodies, with the outer objects being more volatile-rich and chemically complete (Lewis, 1972b, 1974). In detail, this condensation could occur in such a way that each condensing component could continue to react with the remaining gas at lowering temperatures, or alternatively, could be buried away in the interior of a growing planet, which would grow with built-in refractories and a metal core, and with its volatiles already concentrated on the surface (Turekian and Clark, 1969). The first process is called *homogeneous accretion*, and the second, *heterogeneous accretion*. The testing of such hypotheses serves as the focal point for studies of planetary bulk volatile inventories. Let us now consider how well they explain the observations.

The surface of the Earth is covered with materials that have been processed through extensive magmatic-differentiation and chemical-fractionation processes, which were followed by extensive weathering and redistribution. Large portions of the crust, or even the crust as a whole, contain concentrations of some elements that are orders of magnitude larger than conceivable for the bulk Earth. Thus even in the case of our own planet, an understanding of its overall composition requires an indirect approach. One useful approach is to assume that planets, in bulk, have an overall composition similar to certain meteorites (called *chondrites*) that have never undergone extensive magmatic differentiation and weathering, but which are thought to be relics of the material that formed all the terrestrial planets. The fact that these meteorites have a composition similar to that of the Sun (except for the most volatile elements) lends credence to the hypothesis. However, the solid planets and satellites—even allowing for uncertainties in internal compression—clearly have greatly diverse densities and hence composition. The question is whether any equilibrium condensation scheme can explain most or all of this diverse array. In the simplest monotonic schemes, as described above, the best fit to densities is obtained if the chondrites condensed between the Earth and the asteroid belt (two to three times the Earth's distance from the Sun). Objects formed far inward of that heliocentric radius are somewhat volatile-depleted relative to these objects, whereas those further from the Sun are more volatile-rich.

The specifics of the homogeneous variant of the equilibrium hypothesis are shown in Fig. 7-2. The curve marked "adiabat" on the plot of temperature versus pressure is thought on dynamic grounds to represent radial variations in conditions in the preplanetary nebula when the planets formed. Although the choice of a particular adiabat is somewhat arbitrary, the ability of a *single* chosen adiabat (relating temperature and pressure) to provide an explanation for the densities and volatile contents of all these objects simultaneously is a fair if incomplete test of the hypothesis. The range of temperature and total nebula pressure (assuming solar-nebula composition), under which each of several listed major minerals is stable, is given, and the planets' positions—indicated by the symbols—determine what minerals they incorporate. This in turn determines both density and volatile abundances. How well does this hypothesis do? The chosen adiabat does a remarkable job of explaining the densities of the terrestrial planets. For a full discussion of the relationship between planetary densities and the stability fields in Fig. 7-2, see Lewis (1972a, 1974). The actual densities in grams per cubic centimeter—corrected for compression—are Mercury, 5.4; Venus, 3.9; Earth, 4.0; Moon, 3.3; and Mars, 3.7. Thus the planet nearest the Sun (Mercury) seemingly has the highest density because of the enrichment of metallic iron and its failure to incorporate its full share of corresponding lower-density silicates. Venus and the Earth are predicted to have their full share of silicates resulting in their lower density. Even the slightly lower density of Venus compared with that of the Earth can be explained as resulting from failure of Venus to incorporate its full share of sulfur (S), a volatile, but heavy, element. Mars' lower density results from a greater degree of oxidation of Fe and the conversion of higher-temperature silicates to lower-temperature forms. Furthermore, the model predicts condensation of hydrated minerals at the distance of the main asteroid belt, and meteorite studies and telescopic spectra of asteroids suggest that at least 80% of them consist of such assemblages (Johnson and Fanale, 1973; Gaffey and McCord, 1977). The model predicts dominantly icy objects at Jupiter's distance and beyond, and indeed all the main satellites of all the outer planets have densities approximately 1.0 to 1.7,

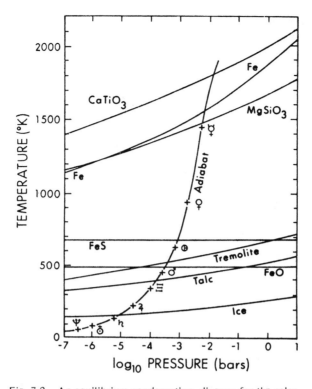

Fig. 7-2 An equilibrium condensation diagram for the solar system (Lewis, 1972b). This is a temperature versus pressure diagram on which the stability fields of some (but not all) probable key minerals in the condensation sequence of the preplanetary nebula are represented. The curved line is one of a family of adiabats, any one of which might be a reasonable representation of conditions as a function of heliocentric distance when the planets formed. If the planets condensed in solid-gas equilibrium under this set of conditions, their uncompressed densities could all be nearly explained in a self-consistent manner, and thus bulk-volatile contents predicted. (Objects from the right are Mercury, Venus, Earth, Mars, asteroids, Jupiter, Saturn, Uranus, and Neptune.) Alternative models, some of which involve nonsimultaneous formation of planets, disequilibrium processes, and preliminary degassing of planet-forming bodies are discussed in the text.

which suggests a dominance of $H_2O$ ice, except for Io and Europa. The latter two are explained as having been influenced by Jupiter as a "small star" heat source (see below). Thus there is some undeniable component of influence by an early central heat source in determining the bulk composition of solar-system objects. Clearly, the simple equilibrium-homogeneous-accretion hypothesis does a remarkable job of explaining in a semiquantitative way the evident main density trends and apparent compositional trends. True, there seem to be some possibly severe problems with the simplest form of this hypothesis, which are discussed below. However, if other hypotheses are offered to circumvent specific problems, care should be taken to examine these hypotheses with respect to their ability to *also* explain the *overall* trends in composition, which are explained so well by the equilibrium-condensation hypothesis.

By utilizing Fig. 7-2 and other similar, but more complete, condensation diagrams (e.g., Fegley and Lewis, 1980), we can forecast volatile contents. For example, $H_2O$ should be quite abundant on (or in) Mars, which forms at a temperature and pressure right in the middle of the tremolite field. However, it should be much less abundant in the Earth, which forms just outside the fields of hydrated minerals. Whether *Viking* observations support, refute, or bear at all on this key issue is discussed below. In general, planets whose preplanetary material condenses above 1200°K would be dominated by high-temperature silicates and metal and would be relatively free of volatiles. Those condensing between approximately 500°K and 1200°K would receive part or all of the available S, P, Cl, and F as sulfides, phosphides, halogen-containing silicates, and so forth (Fegley and Lewis, 1980). Farther out in the solar system, high-temperature silicates that have not already participated in accretion of a planet by the time temperatures of approximately 500°K are reached, are subjected to massive attack by $H_2O$ vapor, resulting in objects largely composed of amphiboles (a class of hydrated, igneous, rock-forming minerals) or serpentine and montmorillonite-like (sheeted or mica-like) structures that can contain huge amounts (more than 10%) of $H_2O$. Carbonaceous chondritic meteorites consist of this type of clay material plus a complement of organic compounds. The latter are expected as products of abiotic synthesis from a nebula of solar composition at temperatures slightly below about 350°K. Thus by the time $H_2O$ and other volatiles are added to the planetary mix as ices in regions of the nebula where and when the temperature falls to about 200°K prior to accretion, the nonicy portion is itself already exceedingly rich in volatiles because of "space weathering." If even 5 percent of the Earth consisted of such (nonicy but volatile-rich) material, it could more than supply the Earth's surface excess-volatile inventory, whether it was added homogeneously or as a veneer. The first step of icy addition (adding $H_2O$ ice to already hydrated rocky materials) might be expected to lower the density of the accreting object to approximately 1.6 (as in Jupiter's icy satellites), whereas the second step (adding $CH_4$ ice) would lower the density to approximately 1.0 to 1.2 (Lewis, 1972b). The final product would be a fair approximation to the most volatile-rich, solid objects in the outer solar system, such as Saturn's satellites, Uranus's satellites, and comets. (The final step in volatile addition would be to add the rare gases and hydrogen, resulting in a Jupiter- or Saturn-like object.

### 7-1-3 Energy History

A second key factor in the volatile history of any solid planetary object is its energy history. This is critical because it determines whether volatiles present in bulk can be transferred to the *surface volatile inventory* (e.g., on Earth this includes not only the atmosphere but the oceans, $CO_2$, and other volatiles chemically reincorporated into sediments, etc.). The planets' or satellites' energy history is, in turn, largely determined by the amounts of accretional (impact) energy added to the growing planetary nucleus, by energy from both short- and long-lived radionuclides, tidal energy, and so forth.

It is extremely critical to note that some of these energy sources have *inherent time scales* for heating associated with them (e.g., at least accretion heating and that from short-lived nuclides are skewed toward the time in early planetary history when short-lived nuclides were still available and the flux of accreting material was relatively high). In the most general terms, the size of a planetary object also determines its energy history; large objects with high gravitational fields can obviously accumulate a great deal more gravitational heat than small objects and also retain that heat. But small planetary objects with short *thermal-lag times* (the time needed to conduct heat from the interior to the surface) allow their small influx of heat energy simply to bleed out to the surface and to be radiated to space as fast as it is produced. This does not mean, however, that only large objects can ever differentiate or degas. A small object can have an active degassing and differentiation history (Wood, 1979) if it (1) incorporated short-lived nuclides from an immediately prior nucleosynthetic event when it formed (Lee et al., 1977), (2) if it is extremely rich in long-lived nuclides, (3) if it accretes from objects with high relative velocities, (4) if it has an unusually low melting point like icy objects, or (5) if it receives intensive, persistent tidal heating. A more workable general principle, however, is that, except for the tidal heating, continuing or increasing degassing after billions of years is more apt to characterize a large object than a small one. General discussions of thermal evolution of rocky objects are given by Hsui and Toksöz (1977) and Toksöz et al. (1978) and for icy objects by Parmentier and Head (1979).

## 7-1-4 Fate of Degassed Volatiles

Once degassed, planetary volatiles can react chemically (with or without the considerable aid of a liquid-$H_2O$ medium) with primary igneous rocks to form alteration products such as clays and carbonates. Or volatiles can be physically incorporated into planetary regoliths as hard, frozen permafrost or in the adsorbed phase (e.g., Fanale and Cannon, 1974; Fanale, 1976). Furthermore, planetary volatiles (especially those in the upper atmosphere, in tenuous atmospheres, or on the exposed surfaces of "airless" bodies) are constantly exposed to a whole host of escape processes, including thermal escape, dissociative recombination, photodissociation, dissociation by photoelectrons, charge exchange, sputtering, or ionization followed by field sweeping. Any of these processes can produce molecular velocities in excess of the escape velocity and cause permanent loss of volatiles if molecular trajectories are uninhibited by collisions. A comprehensive discussion of thermal processes is given by Hunten (1973) and a general discussion of escape processes by Pollack and Yung (1980). In Fig. 7-3 planetary escape is symbolized and shown in progress. Here the element S is escaping from Io, the innermost major moon of Jupiter. It is accompanied by escaping O, and both are probably derived from atmospheric or surface S and $SO_2$. In this case the loss is probably caused by atomic particle sputtering.

Working inward from the outer reaches of the solar system, we now consider the "zoo" of individual planetary volatile inventories as revealed by Earth-based and spacecraft observations and the ways in which the observations correspond to the general principles we have outlined.

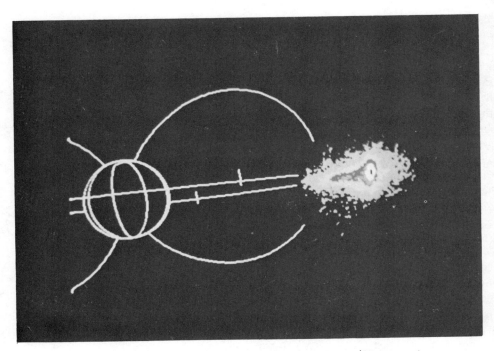

Fig. 7-3 Intensity contours of $S^+$ emission around Io (Pilcher et al., 1983). The closely parallel lines around Jupiter represent a symmetry plane for torus ions. The arcs issuing from Jupiter represent field lines, assuming a dipole Jupiter field. The escape of sulfur and oxygen (presumably derived from $SO_2$ and S) originates from Io as the result of interaction between the Jupiter magnetospheric particles (and field) and Io's atmosphere or surface S and $SO_2$ frost or adsorbate. There are many such nonthermal (and thermal) escape processes that effect escape of substantial quantities of volatiles from planets, causing major changes in atmospheric composition in the process. In the case of Io, Venus, and Mars, entire hydrospheres may have been lost as the result of $H_2O$ dissociation, H escape, and oxidation of the interior.

## 7-2 APPLICATION OF PRINCIPLES TO REAL WORLDS

### 7-2-1 Volatiles in and on Solid Objects in the Outer Solar System

First we may consider the comets, which probably originated at the most distant fringes of the solar system. Clearly, $H_2O$ is the dominant parent molecule of the ionic species streaming from most comets, and current models require comets to be mixtures of ice (dominantly $H_2O$) and dust. The presence of $CH_4$ (methane), $NH_3$ (ammonia), and $CO_2$ (carbon dioxide), as well as a variety of organic compounds, has also been inferred based on an assemblage of spectroscopic evidence and the physical behavior of comets in space. Spectra of cometary meteoroids and studies of what is believed to be cometary dust also suggest a silicate component as well as a carbonaceous component, both of which occur in carbonaceous chondritic meteorites (discussed below). Uncertainty exists concerning whether equilibrium was achieved between condensing particles and gas when comets formed, or whether disequilibrium paths (e.g., possibly indicated by the presence of C as $CO_2$ rather than as pure $CH_4$) were sometimes followed, possibly owing to rapid radial mixing, inhomogeneities in the nebula, and the low temperatures and pressures involved. Otherwise, comets do generally seem to represent the type of object envisioned as the lowest-temperature, most chemically complete, solid end product of the volatile-addition sequence described above. Comets may be important as direct contributors of volatiles to inner-planet atmospheres as well as representatives of the most finished product of the volatile-condensation sequence (see reviews by Whipple and Huebner, 1976; Donn et al., 1976; and Wyckoff, 1982).

Most outer solar-system satellites have been shown to have optical surfaces and deep interiors that are primarily composed of frozen volatiles. The Galilean satellites of Jupiter, the satellites of Saturn (Enceladus, Tethys, Dione, Rhea, and Hyperion), the satellites of Uranus (Titania and Oberon). Neptune's satellite Triton, and Pluto all fit into this category. Besides Io and Europa, notable exceptions are Jupiter's outer satellites (J6 to J14) and Saturn's satellite Phoebe. Phoebe and J6 to J14 are bodies of low albedo, have reflectance properties that are similar to carbonaceous chondrites, and are probably volatile-rich. They may also have icy surfaces darkened by a sprinkling of very fine, dark, opaque material. The satellites whose surfaces are dominated by frozen volatiles seem to fall into two groups:

1. bodies whose surfaces are characterized by varying mixtures of water, ice, and darker material presumed to be silicates
2. bodies with surfaces partially comprised of frozen methane that have extremely thin methane atmospheres. Triton (and Pluto?) appears to fall into this category, and Saturn's huge satellite, Titan, possesses a thick atmosphere (thicker than Earth's) rich in $CH_4$ and $N_2$. Surface and atmospheric observations cannot directly confirm the bulk "cosmic" $CH_4$ and $NH_3$ abundances suspected for these objects on theoretical grounds, but at least they are harmonious in broadest outline with the principles of condensation and accretion outlined above.

Several of the Saturnian satellites have densities close to unity, implying condensation of C- and N-containing ices following addition of two parts $H_2O$ ice to precondensed, low-temperature silicates. Additionally these very low densities may require simply leaving out much of the rocky portion.

The Galilean satellites of Jupiter have long been known to exhibit a wide range of densities, ranging from a rocky density of $3.5 \text{ g} \cdot \text{cm}^{-3}$ for the innermost, Io, to $1.6 \text{ g} \cdot \text{cm}^{-3}$ for the outer two, Ganymede and Callisto. The latter two objects have long been known to contain high percentages of $H_2O$ ice (but apparently not other ices) in their surface material based on identification of vibrational absorption features in their near-infrared telescopic spectra (Pilcher et al., 1972). Also the nonicy components of the surfaces are optically somewhat similar to the carbonaceous chondritic meteorite material described above that is dark and volatile-rich (Clark and McCord, 1980; Pollack et al., 1978). Thus what little information we have concerning the outer two major satellites' bulk composition is quite compatible (but not uniquely so) with the idea that they represent the stage of condensation in which two parts of $H_2O$ ice are added to one part of hydrated silicate material resembling carbonaceous chondritic meteorites but in which addition of the full complement of C- and N-containing ices has not occurred. As indicated, the inner two, Io and Europa, have *rocky* densities (Europa's is 3.1 g · cm$^{-3}$). The juxtaposition of large rocky satellites with icy objects apparently occurs because of the early thermal history of the primary, Jupiter, which approached stellar status but lacked the formational energy to initiate nuclear reactions in its interior. Nonetheless the early Jupiter was hot enough to prevent the cosmic complement of $H_2O$ ice from condensing from the nebula at Io's and Europa's distances, resulting in two rocky objects (Pollack and Reynolds, 1974; Cameron and Pollack, 1976; Pollack and Fanale, 1982).

*Voyager* has revealed Ganymede and Callisto to be rocky-icy objects; it has also revealed that at least Ganymede has experienced significant internally engendered ("endogenic") tectonic evolution. This is most dramatically represented in the so-called grooved terrain (Smith et al., 1979a). The link between this observation and the general history of volatiles in icy satellites seems clear: Since the lighter terrain is less cratered than the "dark cratered terrain" that complements it, the apparently extensional forces that produced the regional tectonics (see Chapter 8 by Bird) also caused removal of some initially present darker silicates (or refinement of ice). Furthermore, the extensional forces are thought to have resulted from net planetary expansion, which in turn resulted from a cycle of melting and recrystallization of the planet's interior $H_2O$ during its earliest history (Parmentier and Head, 1979; Cassen et al., 1980; Squyres, 1980). In general this implies overall planetary melting (except maybe the outer 100 km or so), internal ice-silicate separation, and development of a silicate core. This is in harmony with the general view that even "small" icy bodies (although Ganymede is the size of the planet Mercury) can manage to accumulate enough accretion energy and other sources of energy early in their history to enable planetary melting, mainly because to melt $H_2O$ ice–rich objects requires, even at Jupiter's distance, only about 150° temperature elevation from internally generated heat. The dominance of $H_2O$ ice on Ganymede and Callisto and even the greater postulated internal activity on Ganymede had been suspected from Earth-based telescopic observations and explained on a tectonic basis prior to *Voyager* (Fanale et al., 1977). Alternatively, it has been suggested that Callisto was contaminated by infalling dust (Pollack et al., 1978). Nonetheless it is not so easy even after *Voyager* to explain why Ganymede seems so much more active tectonically than Callisto, since their differences in terms of mass, silicate-to-ice ratio (only the silicate has meteoritic concentrations of heat-producing U, Th, and K nuclides), and so forth, are not overwhelming. The mass difference (approximately 50 percent) may have caused greater accretional heating and delayed mantle refreezing by about 0.5 Gy longer in the case of Ganymede, thus producing a reworked crust less "contaminated" by early high fluxes of primordial silicate.

Europa, a rocky Moon-sized object, has managed to cover its surface with $H_2O$ ice, presumably by planetary internal heating and degassing processes like those which degassed Earth's oceans. The paucity of craters on Europa's surface revealed by *Voyager* suggests that the exceptional purity of its surface $H_2O$ is due to a high level of internal activity as deduced in pre-*Voyager* studies of the Ganymede-Callisto-Europa triad (Fanale et al., 1977). Except for the Earth, Europa may be the only rocky object in the solar system whose surface is covered with $H_2O$.

Io is a special case in the Galilean system. Its surface is not thought to be substantially covered with $H_2O$ ice. Instead a major compound on its surface is frozen $SO_2$ (Fanale et al., 1979; Smythe et al., 1979). This does not mean that Io was $H_2O$-poor in bulk to begin with (most models of Io suggest it was $H_2O$-rich), but rather that it destroyed

The atmospheres of most outer solar-system objects are exceedingly tenuous (less than $10^{-6}$ bar). An exception is Titan, which has an atmosphere greater than 1 bar consisting of $CH_4$, $H_2$, and $N_2$. Titan is presumably unique among the Saturnian satellites in this regard, mainly because of its size, which is about that of Mercury. Ganymede and Callisto are also about the same size, but would be expected to have lacked large quantities of methane and ammonia ice when they formed for reasons given above. Ganymede, Callisto, and Europa may have up to approximately $10^{-6}$ bar of $O_2$ as residual atmospheres from the continual dissociation of the extremely small quantities of $H_2O$ vapor followed by the escape of the light hydrogen gas. But in the case of the Galilean satellites and perhaps the Saturnian satellites as well, any steady-state, tenuous atmosphere is constantly subject to various types of ionization, followed by field sweeping, which "erodes" the atmosphere and entrains the new ions in the ionized gas flowing by the satellite.

Sputtering of Io's surface and atmosphere is a major mechanism of material ejection for Io and is largely responsible for population of Io's neutral Na cloud (Matson et al., 1974) and may play a key role in maintaining Io's ionized torus as well (see Fig. 7-3 and Brown et al., 1983). But in Io's case, the volcanic-resurfacing rate greatly dominates the sputtering rate as do ordinary thermal processes of sublimation and condensation (over most of Io) because of the high vapor pressure of $SO_2$. In the case of the icy satellites, sputtering erosion is an important process (Lanzerotti et al., 1978). However the total erosion rate is probably less than 100 m per $10^9$ y (Haff et al., 1979). This could be enough to make it a significant modifier of landforms, but not capable of totally altering the global crustal structure. The process would be important for augmenting $H_2O$ transfer to the atmosphere and sustenance of any residual oxygen-rich atmospheres on the icy satellites (Haff et al., 1979; Kumar and Hunten, 1982).

Besides Titan, Io is also known to have an unusual atmosphere. Tidal heating (see Chapter 12 by Peale) supplied internal heat to Io at a rate over an order of magnitude greater than that from radioactive decay (Peale et al., 1979). As a result, Io exhibits extensive current volcanism and degassing (Smith et al., 1979a). Currently the dominant atmospheric gas, gaseous constituent of the plumes (Pearl et al., 1979), and surface condensate is $SO_2$, which was identified as a frozen surface constituent using Earth-based telescopic infrared spectra (Fanale et al., 1979; Smythe et al., 1979). This intensive degassing, continued throughout Io's history and coupled with magnetospheric sweeping and other planetary volatile-escape processes (e.g., see Kumar, 1979), has depleted Io of a large initial inventory of $H_2O$ and other volatiles (Pollack and Fanale, 1982). One consequence is that the surface volcanism involves flows of S allotropes and $SO_2$-rich plumes (Smith et al., 1979b). Another is that no $H_2O$ has even been detected, nor is there evidence for significant effusion of C- or N-containing compounds. The preceding information was obtained largely from ultraviolet (UV), visible, and infrared spectrometry including Earth-based, Earth-orbital, and *Voyager* studies. However it is augmented by *Voyager in situ* sampling of the magnetosphere, which is strongly contaminated near Io with material stripped from Io's atmosphere and surface. Sulfur and oxygen dominate the Io-supplied volatiles and semivolatiles in the magnetosphere (see Fig. 7-3), and significant carbon- and nitrogen-containing species have not been detected. Evidence pertaining to Io's surface-phase composition and atmosphere-surface interaction has been reviewed by Fanale et al. (1982a).

We prefer to believe that volatile-rock interaction and atmosphere stripping coupled with intense volcanism resulted in Io's loss of planetary volatiles, rather than that Io simply lacked the volatiles to begin with. One reason is that if Io accreted at such a high temperature as to prevent incorporation of substantial quantities of $H_2O$ and carbon-containing compounds, then the temperature even at Ganymede's position would still be too high for Ganymede to accrete most of its mass directly as $H_2O$ ice, which evidently it has (Fanale et al., 1977). Other arguments suggesting a dramatic whole-planet devolatization history for Io have been given by Pollack and Fanale (1982), and an analysis of the present atmospheric regimes of all the Galilean satellites has been given by Kumar and Hunten (1982). The nature of the initial surface environment of Io is a fascinating subject for speculation, since Io was presumably divesting itself of huge quantities of $H_2O$, while at the same time it was—according to current models—receiving as much energy from Jupiter (for about $10^6$ y) as the Earth receives from the Sun.

Many of the key questions concerning Ganymede, Callisto, and Europa have to do with the *state* of their $H_2O$. For example, how thick is Europa's $H_2O$ crust? Did Europa, Ganymede, and Callisto ever contain liquid-$H_2O$ zones below their icy crusts? It has been argued that liquid zones are unlikely because solid-state convection in the overlying ice would transport heat rapidly enough to freeze the liquids (Cassen et al., 1980). In other words, before the ice *ever* gets hot enough to melt, it starts circulating fast enough to carry to the surface and lose the internal heat that is needed for melting. However, the convective properties of ice are not well known, the role of salts has not been taken into account, and it is not at all clear whether tidal heating—although much less than that for Io—might not be able to maintain a liquid zone below a few kilometers of ice on Europa or deeper liquid zones in Ganymede or Callisto. Recent morphological studies indicate that some sections of Europa's crust have rotated, suggesting a possible liquid zone beneath. In the original analysis of Europa's energetics by Fanale et al. (1977), it was pointed out that the maximum crustal thickness was about 70 km. This is the value obtained if the density of the remaining material was as high as that of Io (it is probably lower). If Europa has a heat flow such as would characterize a planet made from primitive meteorites, and a crust more than 40 km thick, the heat flow could support a liquid zone in steady state (Fanale et al., 1979). This analysis did not include *either* the effects of cooling by solid-state convection *or* tidal-heat input. If indeed there is a liquid zone despite solid-state convection, it probably means that either (1) solid-state convection in such a thin crust is ineffective or (2) tidal heating is more effective than suggested by previous studies. These ideas have recently been pursued by Squyres et al. (1983), and their analysis also indicates a liquid $H_2O$ zone at depth is likely to result from overall planetary tidal heating.

### 7-2-2 Volatiles in and on Asteroids

Moving inward toward the Sun, we find indications that the main asteroid belt has a bulk composition probably harmonious with at least the broad outline of equilibrium-condensation theory. Recall that the stage of volatile incorporation immediately prior to direct condensation of $H_2O$ ice involved hydration of some higher-temperature phases by $H_2O$ vapor in the nebula, yielding hydrated phyllosilicates (mica-like sheet silicates or clays) and also the formation of some organic materials, yielding a mix not unlike that of the matrix material of carbonaceous chondritic meteorites (Lewis, 1972b). Material like these meteorites is abundant in the asteroids and is thought to represent primitive material. However "snowing out" of the full cosmic complement of $H_2O$ is not expected within 4.0 AU, which is consistent with the adiabat in Fig. 7-2. Indeed, most asteroids, including the largest, have optical surfaces and reflectance spectra resembling those of carbonaceous chondrites more than other known meteorite types (Johnson and Fanale, 1973; Zellner and Bowell, 1977). The largest asteroid, Ceres, not only has a spectrum generally resembling that of carbonaceous chondrites, but exhibits a deep 3.0-$\mu$m ($H_2O$-) absorption band in its infrared spectrum, suggesting its surface contains over 10% $H_2O$ (Lebofsky, 1978). But the spectral resolution available is not very good, so the form of the water is controversial (bound $H_2O$ versus ice). Asteroids appear to exhibit a great variety of surface compositions: Their spectra suggest that some of them have very metal rich surfaces or surfaces rich in high-temperature silicates (Gaffey, 1976; Gaffey and McCord, 1977). At least one asteroid, Vesta, is known to be covered with basalts; this is based on absorption bands in its infrared spectrum (McCord et al., 1970) and implies extensive internal differentiation. Also considering the complex collisional history probably experienced by the asteroids (Davis et al., 1979), the presence of many high-temperature surfaces in the belt should not be regarded as indicating that condensation ceased locally at high temperatures. Instead, revelation, caused by collision, of interior mantles and cores formed by magmatic differentiation seems more likely (see Chapter 8 by Bird).

Basaltic *achondritic* meteorites (which lack chondrules) are thought to be volatile-poor samples of basaltic material from some asteroid crust (possibly Vesta's). These meteorites have compositions and textures that suggest they are not primitive like chondrites, but reworked, like volcanic rocks; some even have vesicles like lava flows. The energy for differentiation of such a small body as an asteroid may have come from decay of short-lived nuclides (Lee et al., 1977; see Sec. 7-2-1).

A crucial test of models of solar-system volatile history arises here: Basaltic achondrites are volatile-poor, yet they appear to come from bodies in the main asteroid belt. This is not *in itself* a problem for equilibrium-condensation models, because the achondrites are differentiated igneous rocks and not the totality of the original condensed system. However, careful reconstruction of the composition of the *parent body* from which some achondrites must have been derived, based on numerous elemental correlations, suggests that the parent body was *also* very volatile-poor (Dreibus and Wanke, 1980) relative to the Earth and chondritic meteorites. This is particularly disturbing since, except for the depletion of moderately volatile and volatile elements, the composition is nearly that of chondritic meteorites, which is what it should be at 2 to 3 AU according to the adiabat shown in Fig. 7-1. This issue is unresolved, and it seems to pose a serious problem, at least for equilibrium condensation and homogeneous accretion models in their simplest form. The most probable explanation is that the Eucrite parent body was formed in the Earth-Venus region of the nebula and then dynamically displaced (Wasson and Wetherill, 1979).

## 7-2-3 Mars

Several types of fluid-formed channels exist on the surface of Mars (Sharp and Malin, 1975). Although water is generally thought to have played a role in the formation of most of them, the nature of that role is a matter of debate and probably varies among channel types. The largest outflow channels seem to have been formed more or less throughout Mars' history and are thought to possibly be the result of catastrophic outbreak of confined subsurface liquid $H_2O$ that broke through a confining layer, possibly of hard frozen permafrost, draining a huge aquifer (Carr, 1979). Since all the conditions for their formation could exist today, and since they seem scattered throughout Mars' history, there is no demonstrated need to imagine different past conditions at Mars' surface-atmosphere interface just to explain the larger channels. Intricate networks of smaller, very coarsely dendritic channels (sometimes called *gulleys*) exist on Mars as well. Figure 7-4 shows such dissected terrain. The process by which the gulleys formed is not exactly the same as that which formed the large channels, although headward erosion by sapping of ground ice or rainfall could have been involved. In any case, conditions different from those at present must have obtained when they formed, because (unlike the larger channels) they occur primarily on the oldest Martian terrain (about 4 Gy old) (Pieri, 1976). Whether these "different conditions" represented a dramatic difference in surface climate, allowing rainfall or merely some other change (e.g., a warmer subsurface thermal regime) is not known, but the channels are ubiquitous on the old terrain in low latitudes. To some scientists, this suggests a rainfall mechanism at least for global redistribution of the water, even if the channels' morphology suggests ground-water sapping more than surface runoff.

The relationship between atmospheric evolution and changes in surface-erosion style has been discussed by Pollack and Yung (1980). The channels and layered polar deposits lead us to examine possible causes of climate change. The best known is the obliquity variation (Ward et al., 1974). When Mars' axial inclination to the ecliptic is greatest, the poles are warmer (and the equator slightly cooler) than otherwise. Thus $CO_2$ cold-trapped in permanent caps at the poles could be transferred into the atmosphere periodically (every $10^5$ y or so), raising the atmospheric pressure. Even if some $CO_2$ remained trapped in permanent caps, the atmospheric pressure would be higher, because a higher average polar temperature would mean higher equilibrium-vapor pressures of $CO_2$. However, the north cap of Mars appears to be dirty $H_2O$ ice, not $CO_2$, and although there does seem to be a small $CO_2$ cap (or frost cover) at the south pole that has survived at least one or two summers, it may contain even less $CO_2$ than the atmosphere. Another, larger, reservoir is the $CO_2$ adsorbed on grain surfaces in the regolith (odoriferous compounds are adsorbed on charcoal in a refrigerator in the same way). The regolith is a deep rubble pile that thickens to hundreds of meters at high latitudes and to several kilometers near the poles. This probably contains 10 to 100 times the $CO_2$ that is present in the atmosphere (Fanale and Cannon, 1974). This $CO_2$ cannot exchange as easily with the atmosphere as that in any hypothetical massive, permanent $CO_2$ cap, but it is partly exchangeable nonetheless. It has been estimated that such degassing of adsorbed $CO_2$ would, together with cap growth at times of lowest obliquity, inevitably produce a periodic oscillation of atmospheric pressure around the present value of 7 mbar, from a low of less than 1 mbar at lowest obliquity to a high of approximately 20 mbar at highest obliquity (Fanale and Cannon, 1979; Fanale et al., 1982b). This is especially important

Fig. 7-4 A stream-dissected area on Mars in Thausmasia Fossae (43°S, 93°W). The large spectacular outflow channels in the vicinity of Chryse Planitia are thought to result from aquifer eruptions from below deep, hard, frozen permafrost layers. However, finely dissected terrain, as shown here, has a different geometry and occurrence. It occurs almost everywhere on (almost only) the oldest terrain. These streams are geometrically like terrestrial drainage systems caused by ice or groundwater sapping. A higher initial internal-heat flow may have played a role in making liquid available near the surface. However, their longitudinal omnipresence in the oldest terrain and concentration in lower latitude have led some investigators to suggest rainfall could have played a role in $H_2O$ global redistribution of water, even if the channels were not cut by ordinary runoff. (NASA Photo 532A16.)

from a comparative planetological point of view, since similar processes are responsible for periodic climate change on Earth. While the process is likely to have produced important variations in erosion and deposition rates and so forth, it seems an unlikely candidate for producing dramatically different climate—with much higher temperatures and pressures—in Mars' past. Indeed the presence of a huge adsorptive "capacitor" like a regolith would tend to thwart development of very high atmospheric pressures. Perhaps a significant greenhouse effect (see Chapter 6 by Kivelson and Schubert) might have obtained early in Mars' history caused by either a very high $CO_2$ pressure (greater than 1.0 bar) or by the presence of small amounts of especially effective atmosphere-warming components ($NH_3$?) in the atmosphere capable of stimulating a greenhouse effect, such as that which greatly warms Venus' atmosphere (Pollack and Yung, 1980). In any case, such different and nonrecurring conditions must have been engendered by nonrecurring events, for example, initial absence of an absorbing regolith and a rapid early outpouring of comparatively reduced gases associated with the planetary accretion process itself (Fanale, 1971b). Presumably such an early atmosphere would have been destroyed during Mars' very early history by hydrogen escape and storage of volatiles in a developing regolith (Fanale, 1971b). (See further discussion of early Mars and Earth atmospheric chemistry in Sec. 7-2-6.)

Where are Mars' volatiles stored? The regolith is the main storehouse for Mars' volatiles, but the forms of storage are several. As indicated above, a great deal more $CO_2$ is stored in the adsorbed phase on Mars' regolith than in its atmosphere and caps. Also, $CO_2$ may be stored as carbonates (Huguenin, 1976; Booth and Kieffer, 1978). It has been argued that vapor-solid interaction could produce carbonates from primary igneous rock even without the ion mobility that liquid $H_2O$ can offer (Huguenin, 1976). The *Viking Lander* gas chromatograph mass spectrometer (GCMS) did not detect life on Mars, but it did provide valuable data on soil volatiles (Biemann et al., 1977). A $CO_2$-release pattern was observed in heated Mars soil, which can apparently be better matched by the release of $CO_2$ warmed off the surfaces (desorbed) from clays than by release from carbonates easily broken down and degassed by heating (decrepitated). The possible added presence of those carbonates that would not decrepitate until higher temperatures than the GCMS could reach (more than 500°C) could not be eliminated, however. Water contents from 0.1 to 1.0 percent were measured in the soil at several sites by the GCMS. It is unlikely that this could be entirely attributed to $H_2O$ adsorbed on free surfaces. Chemically bound $H_2O$ or interlayer $H_2O$ in clays must have been involved along with bound $H_2O$ in hydrated salts such as found in the "duricrust" (a sulfate salt-cemented soil crust) (Anderson et al., 1967; Clark, 1978; Gibson et al., 1980). The existence of hydrated minerals including clays at the surface of Mars was anticipated for several years based on Earth-sited telescopic observations and *Mariner* orbiter observations in the infrared and mid-infrared region of the spectrum (Houck et al., 1973; McCord et al., 1970, 1978). Supporting evidence comes from other *Viking* experiments: The X-ray fluorescence experiment provided a partial chemical analysis compatible with the presence of a major clay component (Toulmin et al., 1977; Clark, 1978). Also, a continuous reflectance spectrum of the landing site materials, reconstructed from *Lander* camera pictures taken through several filters, has led investigators to suggest the presence of iron-rich clays. None of this evidence is unique, and alternative assemblages, including amorphous gels and palagonites (discussed later in this section) have been suggested (Soderblom and Wenner, 1978). Furthermore, it has recently been pointed out that smectite clays, where water and large ions are sandwiched between silicate sheets, are not an acceptable match for the visible and near-infrared spectrum of Mars and that amorphous gels are definitely to be preferred on this basis (Evans and Adams, 1980; Singer, 1981).

Still other volatiles are also probably stored chemically in the regolith. Sulfates are known to form a duricrust and to cement "instant rocks" in the *Viking* sites. The existence of nitrates in the soil has been speculated upon also, although the $N_2$ molecule is difficult to break down and it is not clear what type of processes would catalyze nitrate formation in the absence of life or liquid $H_2O$.

A probably significant sink for released S is sulfate salts, which appear to be prominent as the cement in the surface *duricrust* (Toulmin et al., 1977; Clark, 1978). The occurrence of UV-stimulated, vapor-solid weathering is not the only mechanism for volatile fixing available on Mars: When lavas flow over or intrusions are emplaced into hard, frozen soil on Earth, extensive interaction between preexisting soil, igneous rock, and remobilized volatiles is activated. This interaction sometimes produces volatile-rich melange soil called *palagonite*. Conditions on Mars seem favorable for such processes, and Earth palagonites have optical properties not greatly dissimilar from Mars' soils (Soderblom and Wenner, 1978). Possible mineral reservoirs of physically and chemically bound $H_2O$ and other volatiles on Mars have been reviewed by Clark (1978).

A very important sink for Mars' $H_2O$ is ground ice in the regolith or hard frozen permafrost (Anderson, et al., 1967; Fanale, 1976; Coradini and Flamini, 1979). This is important not only because of the huge potential capacity of the regolith for ground ice, but because such $H_2O$ is not locked away irreversibly, but rather is available for remobilization. Since pores make up a large fraction of the unconsolidated regolith, it follows that much ice can be stored this way. The bottom limit of the hard-frozen zone would be set by internal heat flow that warms the interior to $0°C$ or $-10$ to $-50°C$ (the melting points of some brines) at a depth of less than 500 meters near the equator and between 1 and 2 km near the poles (Fanale, 1976). The top extent of permanent ground ice is set by the depth to which seasonal summer heating causes net yearly sublimation of any shallow ground ice. Temperatures and atmospheric $H_2O$ contents on Mars are such that ground ice could exist in equilibrium with average basal atmospheric conditions anywhere poleward of about $40°$ latitude (Fanale, 1976), but ground ice "outliers" could well exist at lower latitudes (1) in another part of the obliquity cycle, (2) if compacted overlying soil prevented equilibrium with the atmosphere for geologic periods of time (Smoluchowski, 1968), or (3) if internal volcanic juvenile- or remobilized-$H_2O$ input from below created or maintained a regional ice deposit despite net sublimation. A controversial issue is whether the distribution of ground ice is affected by equilibration with the surface or whether relic ground ice can persist in disequilibrium with the atmosphere for billions of years. Much depends on whether one assumes that the largest pores are equal in size to the smallest grains, or whether larger channels for diffusion exist. Another issue is whether migrating $CO_2$ can aid in $H_2O$ transport.

The distribution of ground ice in three dimensions within the regolith is morphologically detectable, at least in principle. It can be revealed by unusually fluid or cohesive impact ejecta (compared with, say, the ejecta on the Moon) or by "thermokarst" topography (such as produced when ice melting causes ground collapse), patterned ground, and so forth (Carr and Schaber, 1977). A wide assemblage of such observations and tabulations (for example, crater size versus ejecta type versus latitude) are available (Rossbacher and Judson, 1981). Some investigations (especially those on ejecta blankets) suggest a greater availability of ground ice poleward of $\pm 40°$ latitude (Blasius and Cutts, 1980; Saunders and Johansen, 1980) as expected from equilibrium models (Fanale, 1976). However there are other landforms that show little latitudinal dependence (Rossbacher and Judson, 1981). There are also other variables (terrain type, age, and so forth) that need to be controlled (Cintala and Mouginis-Mark, 1980) before any consensus is approached on the important issue. Figure 7-5 shows such an ejecta lobe-morphology, which is definitely "nonlunar" or "non-Mercurian" and which may or may not reveal subsurface ice. An important point is that liquid $H_2O$ can also exist in equilibrium below the surface. Fanale (1976) pointed out that if Mars were chondritic, the $273°$ K isotherm would come within 1 km of the surface at the equator, and isotherms for brine melting, obviously, would be even shallower.

The assumption of a low thermal-conductivity regolith and $H_2O$ saturation with salts could bring the liquid interface to within less than 100 m of the surface at the equator. Perhaps some ejecta morphologies actually indicate the availability of subsurface liquid $H_2O$ (or perhaps impact energy played a role in heating and melting the ice). Certainly this would be consistent with regional aquifer activity.

The quantitative issues of Mars' bulk-volatile content and the mass of its present or past surface volatile inventories are not, however, easily addressed by examination of channels or hard frozen permafrost morphologies. Other possible approaches to such quantitative inventory taking of volatiles involve an analogy between bulk Mars material and meteorites. Such an analogy would allow estimation of the total degassed surface inventories of volatiles with complex histories (like C) to be "indexed" to more easily inventoried elements, notably the rare gases. The latter do not form chemical compounds and, in the most favorable case, might simply accumulate in a planetary atmosphere, once degassed.

Anders and Owen (1977) use a sophisticated version of this approach to estimate the bulk volatile endowment of Mars. Based on the $^{36}Ar$ and $^{40}Ar$ content of Mars' atmosphere, they deduce that Mars is depleted in bulk, in $^{36}Ar$ relative to $^{40}K$, the parent of

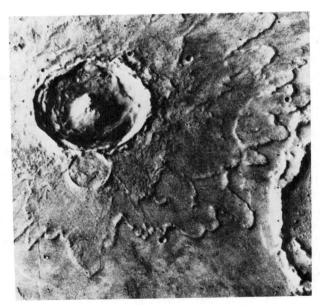

Fig. 7-5 The crater Yuty, as photographed by the *Viking Orbiter 1*. This crater, 18 km across, is one of several on Mars exhibiting types of ejecta patterns not observed on the Moon or Mercury and possibly attributable to a liquid component. This has led many—but not all—planetary geologists to conclude that abundant subsurface ice is present in vast regions on Mars. (NASA Photo P-16848.)

$^{40}$Ar. Thus they conclude that Mars is depleted in volatiles (indexed by $^{36}$Ar) relative to semivolatiles (indexed by K). They then use complex arguments based on an apparent correlation between size and semivolatile depletion for differentiated objects to infer that Mars is also depleted in K relative to other nonvolatile elements (such as U). The compound result of both inferred depletions is a huge inferred depletion in volatiles for bulk Mars. They conclude that Mars degassed an inventory of volatiles equivalent to only about 150 mbar of $CO_2$ (compared with 20 bars, or atmospheres-equivalent of $CO_2$ in Earth's carbonates) and that Mars degassed a global average of only 10 m of condensed $H_2O$.

A serious problem is the question of whether the $^{36}$Ar depletion, if as great as suggested, can reliably be used to index a comparably great bulk $-H_2O$, $-CO_2$, and $-N_2$ depletion for Mars. Bogard and Gibson (1978) point out that the Earth's surface volatile $^{36}$Ar:C ratio differs sharply from that of most meteorites. Also, although we are primarily discussing Mars, we must here consider rare gas results on Venus, and expect that any reasonable hypothesis might explain not only Mars, but the triad: Mars, Earth, and Venus. *Pioneer Venus* revealed that Venus' atmosphere contains 70 times as much $^{36}$Ar as does the Earth's, whereas it would be unreasonable to conclude that therefore Venus had lost the equivalent of 70 Earth oceans. We must anticipate Sec 2.4 on Venus somewhat in order to show the impact of Venus's high $^{36}$Ar content on our analysis of the bulk volatile endowment of Mars. Although Wetherill (1981) has suggested that Venus's extremely high $^{36}$Ar content resulted from a solar wind contribution essentially unique to Venus-forming material, Pollack and Black (1981) argue that the $^{20}$Ne:$^{36}$Ar ratio is far too low for this source and that, therefore, Venus's $^{36}$Ar had a similar origin to Earth's and that in meteorites. They go on to argue that the great discrepancy in the ratio of $^{36}$Ar to other volatiles between Venus and Earth indicates that $^{36}$Ar was incorporated into preplanetary material by rather different processes than C, N, and H. Therefore they argue that $^{36}$Ar is an unreliable index for chemically active volatile abundances. They use Mars' atmospheric $N_2$ instead to index $CO_2$ and $H_2O$. By this procedure, they calculate that Mars probably degassed the equivalent of 80 to 160 m of $H_2O$ on a global average—10 times that estimated by Anders and Owen (1977). Pollack and Black (1982) also conclude that the reason why $^{36}$Ar (and rare gas in general) is such a poor index of other volatiles is that rare gases were incorporated into grains by adsorption or chemisorption and thus the dependence of their degree of incorporation on temperature and pressure is different from those that are incorporated via chemical reactions. In fact, recent extensive laboratory analyses of phases of rare-gas carriers in meteorites and gas-release patterns for primordial rare gas (Yang and Anders, 1982) confirms that primordial rare gas was incorporated into preplanetary materials—and enriched in heavy rare gas relative to solar composition—by a process involving physisorption, chemisorption, and subsequent occlusion, as originally suggested by Fanale and Cannon (1972).

Yet another approach to estimating the degassed $CO_2$ and $H_2O$ inventory for Mars has been based on the *Viking* observation of $^{15}$N enrichment resulting from preferential $^{14}$N loss. Such preferential loss could occur because $^{14}$N—on the average—"stands above" $^{15}$N in the stratified uppermost layers of the atmosphere and hence is more exposed to certain essentially mass-independent loss processes (McElroy et al., 1976). The "reconstructed" $N_2$ concentration would imply a total $H_2O$ inventory equivalent to 20 m to 200 m, depending on what assumptions are made concerning sinks for nitrogen other than exospheric escape from the top of the atmosphere to space (e.g., nitrate formation). Also, the relative *non*enrichment of $^{18}O$ seems to require repeated flushing by the equivalent of 2 bar of some oxygen containing gas (McElroy et al., 1976). The regolith could store the amounts of $CO_2$ involved in most of these estimates—a few tens to hundreds of millibars—in the adsorbed phase alone and could exchange enough in response to obliquity variations to keep the $^{18}O$ enrichment suppressed (Fanale and Cannon, 1979). An interesting recent development results from the fact that the rate of $^{15}N$ enrichment is itself a function of $CO_2$ pressure owing to the differences in scale height between the molecules. Thus, Fox (1982) has shown that higher than present $CO_2$ pressures are required to keep the $^{15}N$ enrichment from being greater than observed, even if the lowest pressure parts of the obliquity cycle were ignored.

The most recent estimates for the Mars surface-volatile inventory are based upon chemical analogy to SNC (Shergottites-Nakhlites-Chassigny) meteorites which have many special characteristics, including the elemental and isotopic composition of rare gases, the nitrogen isotopic composition, and rare gas-to-nitrogen-to-carbon ratios, all of which very strongly point to an origin on Mars (Becker and Pepin, 1984). Unlike other (Eucritic) achondrites referred to above, these meteorites appear to have been derived from a parent body that is volatile rich—even more so than the Earth! Other characteristics, especially a strong depletion of chalcophile (sulfide-forming) elements, appear to suggest that the parent body (Mars?) accreted homogeneously and then differentiated thoroughly; this allowed chalcophile elements to be first incorporated into sulfides and later, still bound into sulfides, dragged out of the mantle into the core. If this is correct, it follows that Mars must have degassed an enormous $^{36}Ar$ inventory and then lost it to space. Currently, the favored atmospheric loss mechanism is *atmospheric cratering*, or removal of the atmosphere in the return jets of large impacts at the tail end of accretion (Watkins and Lewis, 1985). Further, the composition of the parent body mantle requires that almost all the water in the planet be used up to oxidize Mars' iron to create the current oxidized mantle.

Quantitative treatment of this problem by Wänke and Dreibus (1984) suggests a surviving surface $H_2O$ inventory equivalent to only 1-50m of water. The upper end of the range is only a factor of two below some of the estimates of Pollack and Black (1982) and McElroy et al. (1976) and probably sufficient to explain Mars' water-formed features.

In summary, the model of Anders and Owen (1977), which is based on analogy to meteorite chemistry, suggests that Mars was initially volatile poor in bulk. This in turn would indicate that special processes other than simple condensation from a particular portion of the nebula exert a major influence on bulk volatile endowment. The models of Pollack and Black (1979, 1982) and some of those of McElroy et al. (1976) are based on arguments decoupled from the atmospheric rare gas inventory and suggest a substantial degassed $H_2O$ inventory of perhaps 100 m. The model of Wänke and Dreibus (1984) simplifies matters greatly. Their model, based on analogy to certain meteorites for which a Martian origin is strongly suggested, forces the conclusion that almost all the $^{36}Ar$ has been lost from the Mars atmosphere. Once this conclusion has been accepted, it allows Mars to be regarded as having an original volatile-rich composition perfectly consistent with the Lewis condensation model. At the same time, the indexing of major volatiles to the rare gases in the planet-forming material remains applicable. The only mechanism necessary to preserve conceptually both the coupling between the rare gases and the other volatiles and the regularity of bulk composition in the nebula, and at the same time explain Mars' current paucity of $^{36}$Ar, is simple atmospheric loss. Thus the rare gases can be decoupled from major volatiles, during or following planetary accretion, by reaction of major volatiles followed by atmospheric loss of both rare gases and major volatiles. If correct, this important concept has implications for the other two members of the puzzling Mars-Earth-Venus triad as well.

## 7-2-4 Venus

In equilibrium-condensation schemes involving a nebula with a strong radial-temperature gradient, Venus might well be expected to form as a relatively anhydrous body, thereby explaining its present, exceedingly low, atmospheric $H_2O$ content (Lewis, 1974). However, it has been repeatedly pointed out that Venus might alternatively have lost the equivalent of Earth's oceans through a series of processes involving simultaneous emplacement of $H_2O$ in the atmosphere (as opposed to a "protected" oceanic reservoir), $H_2O$ dissociation, exospheric H escape, and oxidation of the Venusian interior. In fact, it has been pointed out that if the Earth were placed in Venus's position, the resulting increase in atmospheric $H_2O$ and $CO_2$ content due to its closer proximity to the Sun would cause still greater surface temperatures owing to enhanced trapping of solar radiation (the greenhouse effect), which would cause still greater atmospheric $H_2O$ and $CO_2$ contents, and so forth (the *runaway greenhouse* effect), until a new stable situation was achieved with essentially all the $H_2O$ and $CO_2$ in the atmosphere and a very high surface temperature like that of present Venus was reached (Rasool and DeBergh, 1970; Pollack, 1971). This would be followed by a loss of the $H_2O$. If the nebula models inspired by the high $^{36}$Ar content are correct, then Venus would have had to get rid of an ocean's worth of $H_2O$. Given the projected rates of hydrogen escape and $H_2O$ dissociation, this could have been accomplished in $10^8$ y (Walker, 1977). Exposing enough rocks to use up the oxygen thus produced might be more difficult, but even this is plausible. To simplify, there are three major possibilities: (1) Venus had only a fraction of an ocean's worth of $H_2O$ in bulk to start with, and that small $H_2O$ inventory has been eliminated by the processes described; (2) Venus had an $H_2O$ inventory like the Earth, degassed most of its $CO_2$ and $H_2O$ on accretion, quickly entered the runaway greenhouse mode, and has been similar to present Venus throughout most of its history; and (3) Venus has long possessed very different surface conditions, involving liquid $H_2O$ and an atmosphere like that of the Earth. In (3), although Venus was closer to the Sun, the solar output was some 30% lower in the geologic past, thus forestalling Venus's entry into the runaway greenhouse mode until recently (geologically speaking). Water on Venus could have been present in the liquid phase until the runaway greenhouse conditions were established. Since then, the atmospheric $H_2O$ was gradually lost by oxidation of interior rocks and escape of H.

To review and expand, the *Pioneer Venus Probe* found that Venus's atmosphere contained about two orders of magnitude more $^{36}$Ar than that of the Earth (see Fig. 7-6). Since Venus has about the same mass as the Earth and is supposed to be largely volatile-free in equilibrium-condensation models, this seemingly causes problems for those models as severe as the problems caused by the possible volatile poorness of both Mars and the achondrite parent body. Two explanations have been offered for this so far. In the first, the planets accreted under nearly isothermal conditions, but with pressure increasing greatly to the center (Pollack and Black, 1979, 1982). (Note that this does not imply that the *nebula* was actually isothermal; the planets need not have accreted simultaneously.) This is a radically different model of the condensation process, and it is not entirely clear how it would explain all the compositional data we have discussed above. However, it would explain the high $^{36}$Ar content of Venus, suggest a very interesting volatile history for that planet, and would also greatly affect our ideas about the volatile poorness of Mars as outlined above and the entire coupling between the rare gases and other volatiles. The other view of the $^{36}$Ar is that Venus merely partook of an extra component of solar wind-implanted rare gas that may not be dominant in the Earth and Mars, but that is the main component in certain meteorites (Wetherill, 1981). Normally one would dismiss this possibility because solar-wind rare gas has a Ne:Ar ratio 100 times that of planetary gas and much higher than that of Venus's atmosphere. However, the Ne/Ar ratio could have been changed by preferential neon loss during or after implantation but before planetary accretion (Wetherill, 1981). Pollack and Black (1982) argue that there is no evidence that such fractionation could have so greatly decreased the $^{20}$Ne:$^{36}$Ar ratio of any solar wind component on Venus. This (solar wind) hypothesis, if correct, would more or less decouple the rare gases from the $H_2O$, so Venus could have been always poor in $H_2O$. However, if the nearly isothermal condensation models of Pollack and Black are valid, then we must deal with either possibility number (2) or (3) above for Venus's history (Pollack and Yung, 1980; Phillips et al., 1981). The last is the most dramatic possibility. If true, how could it be detected?

Atmosphere-surface equilibrium (and atmospheric cold trapping) may control present Venus surface mineralogy. On a speculative note, if conditions had been very different in Venus's past, huge carbonate deposits might have been formed by the interaction of water, in equilibrium with intermediate-$CO_2$ partial pressures, with igneous rock. However, these deposits might have been long since decrepitated or subducted. If the massive $H_2O$ loss occurred recently and was accompanied by a sudden dehydration of the lithosphere, a sudden stiffening might result due to a higher viscosity for the higher melting point of the newly anhydrous lithosphere, coupled with a lower temperature resulting from the water loss (Phillips et al., 1981). Such a sharp change in the viscosity history might be reflected in the surface tectonics. Also, the relic effect on the current internal thermal profile might persist. The latter effects are exceedingly subtle, however, and perhaps the simple search for morphological structures relic from a more clement past might prove more fruitful. However with dynamic and chemically active processes occurring, it is not clear that survival of ancient morphologies is likely either. Finally, such extensive hydrogen escape could have resulted in an enormous D enrichment of any remaining Venusian $H_2O$. In fact there is a large D enrichment. However this does not *require* the loss of a huge $H_2O$ inventory, and only one-thousandth of the Earth's ocean mass need have been present to produce the observed enrichment. Thus the D enrichment (though an encouraging sign) does not provide quantitative evidence for loss of a Venusian hydrosphere or global ocean (Donahue and Pollack, 1983).

Perhaps the most appealing solution to the problem of the Mars-Earth-Venus triad is one indirectly based on studies of the SNC meteorites. In fact, if the Wänke and Dreibus (1984) Mars model is correct, it forces us to accept major atmospheric loss by impact or other loss processes to account for the paucity of $^{36}$Ar on Mars. Once the pertinence of

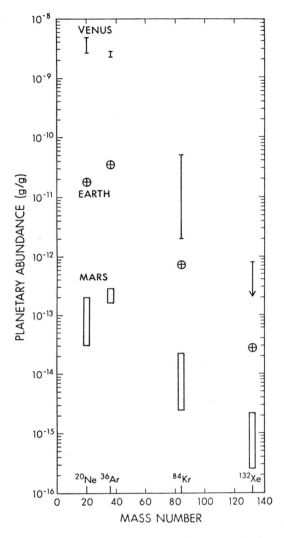

Fig. 7-6 Nonradiogenic rare gas abundances in the atmospheres of Venus, Earth, and Mars. Results are given in units of g/g, that is, the mass of each rare gas *in the atmosphere* divided by the mass of the planet. (Venus data from Donahue and Pollack, 1983; Mars data from Owen et al., 1977.)

this additional process is accepted, the possibility exists of a simple explanation for the peculiar three-way discrepancy in $^{36}$Ar among the Mars-Earth-Venus triad, which is shown in Fig. 7-6. As for Mars, the gradient in chemically active volatile and rare gas content implied in the Lewis scenario would be assumed applicable to all members of the triad, including Venus. Thus Venus would have been initially more volatile poor than the Earth, and its high $^{36}$Ar content at the present time would merely imply considerably less early atmospheric rare gas loss for Venus than for Earth, which in turn experienced less loss than did Mars. Differing atmospheric loss rates for Venus, Earth, and Mars can be understood if impact ejection of atmospheric volatiles (atmospheric cratering) is considered, since the effect of a thick, thermally supported atmosphere for Venus would be to decrease the efficiency of loss through atmospheric cratering, which could operate efficiently on a thin atmosphere, such as that of Mars. It has also been pointed out that the currently suggested origin for the Moon by the impact of a Mars-sized body on the accreted Earth might, by itself, explain not only the volatile depletion of the Moon, but the loss of initial $^{36}$Ar endowment from the Earth (relative to Venus) as well (Wänke and Dreibus, 1984). Again, such a scenario would leave the assumed monotonic bulk volatile gradient of the very early solar system intact and would require that rare gases and other volatiles decoupled only after planetary accretion.

## 7-2-5 Mercury and the Moon

Mercury, like the icy bodies of the outer solar system, appears to possess a bulk composition, based on its density, that can be understood in terms of the simplest equilibrium-condensation theory: Its density is 5.4 g · cm$^{-3}$, the same as the Earth's. However, after correcting for compression (see Table 4-1 listing reduced densities), Mercury is found to have a density 30% higher than that of the Earth. This requires that Mercury must possess a metallic core much larger than the Earth's, relative to its size. Mercury's larger metallic core suggests separation of the Mercury-forming material from nebular gas before temperatures were low enough even to allow condensation of all the high-temperature silicates (see Fig. 7-2). Thus Mercury should be expected, according to the simplest view, to be nearly devoid of planetary volatiles. Indeed, Mercury appears to lack a substantial atmosphere and to have a surface covered with essentially unaltered igneous rock. We do not fully understand, however, the efficiency with which atmospheric constituents may be lost from Mercury to space, perhaps by unfamiliar as well as familiar processes, and we must remember the example of Io, which may have lost a huge volatile inventory during its evolution. Nevertheless, the current lack of evidence of a substantial surface- or atmospheric-volatile inventory is significant, because even in steady state with all *known* escape processes, Mercury's degassing rate at present would have to be 10$^4$ times less than that of the Earth in order to prevent an accumulation of a substantial surface- or atmospheric-volatile inventory, or at least sustenance, in steady state, of an atmosphere of greater number density than observed (Kumar, 1976).

In the case of the Moon, the low, uncompressed density (3.3 g · cm$^{-3}$), together with chemical data on returned samples, suggests that moon material separated from any volatile reservoirs in the solar system before cooling even to the point at which metallic Fe could condense. Hence we would expect the Moon to be even more volatile-poor than Mercury in bulk. Moreover, isotopic studies of returned lunar samples suggest that for no longer than a few tens of millions of years could lunar materials have been in a well-mixed reservoir with even semivolatile-rich material. This makes it seem likely that the extreme paucity of volatiles evident in the *Apollo* samples applies to the Moon as a whole rather than just a volatile-depleted exterior. Neither the Moon nor Mercury lacks volatiles altogether, however; two sources of volatiles, the solar wind and radioactive decay, introduce some volatiles into even the most volatile-poor objects. Radioactive decay in the Moon produces $^{40}$Ar, which is released surprisingly efficiently from the Moon's interior and is measurable in its atmosphere, as well as He from decay of U and Th nuclides and daughters (Hodges, 1975). The solar wind implants H, He, and other gases in the Moon and Mercury, but less in the case of Mercury because of major shielding of Mercury from the solar wind by its magnetic field (see Chapter 14 by Russell). The hydrogen and He quickly reach equilibrium and escape as fast as implanted. The hydrogen can also combine to form $H_2$. Implanted C or N can also form $N_2$ and CO in the surface rocks, although in lunar samples some of these gases may be just terrestrial contamination (Gibson et al., 1980). Volatile escape from bodies such as the Moon and Mercury is primarily by ionization and field sweeping rather than thermal escape, so the volatile lifetimes in the Mercurian atmosphere are probably quite a bit longer than for the Moon. Thus, in neither the case of the Moon nor Mercury do we find striking and incontrovertible evidence against solar system-wide gradients in volatiles, since Mercury was expected to be volatile-poor and the Moon offers a special case hard to interpret owing to possible compound origins (e.g., fission or capture). In particular, the Moon may have formed by volatization and partial recondensation of Earth material as the result of impact of a Mars-sized body into the Earth (Kaula, 1979). In such a case the Moon's volatile-poor composition would not reflect nebular conditions. The interaction between sources such as a solar wind implantation and radioactive decay, with sinks such as ionization and sweeping to produce the very tenuous atmospheres of Mercury and the Moon, is discussed by Kumar (1976) and Shemansky and Broadfoot (1977).

## 7-2-6 The Earth

Throughout the previous sections, the Earth has been mentioned only as a point of comparison with various "denizens" of the planetary volatile "zoo." Rather than attempting to review the enormous literature on the history of the Earth's atmosphere and oceans, we consider the two most controversial issues, namely, (1) the time history, completeness, and mechanisms of Earth degassing, and (2) the degree of oxidation of atmospheric gases in earliest Earth history.

In Rubey's classic 1951 paper, the "excess volatiles" inventory was estimated as in Table 7-1. This table illustrates the relative abundances of volatiles, which clearly differs radically from the composition of our nitrogen- and oxygen-rich atmosphere. The total amounts can be appreciated by taking into account that the total $CO_2$ given (now primarily in carbonates) is about equivalent to 20 atm of pressure. Subduction may play a major role in hiding some volatile sinks quasi-permanently, so it has been suggested that perhaps this inventory should be increased by up to a factor of three. These volatiles were almost surely released due to the melting of the rocky materials (in which they were originally emplaced by processes operating on dispersed preplanetary material in the nebula). There is clear evidence that this is the case from the fact that the ratio of rare gases to other (chemically reactive) volatiles is much lower in the Earth's atmosphere (and all substantial solid-body atmospheres) than in the Sun. This, of course, signals that most of the volatiles were chemically combined or occluded in the pieces that made up the Earth, since if the Earth had simply gravitationally grabbed a primary atmosphere from the solar nebula, it would have taken a higher fraction of primordial Ne, $^{36}$Ar, and so forth. However, if one simply assembled the earth from ordinary, undifferentiated normal chondritic meteorites and degassed it (making some allowances for lower efficiency of degassing of chemically reactive or soluble gases versus the rare gases), this process would essentially produce all the rare-gas inventories and the entire Rubey volatile inventory with surprisingly few first-order deviations (Fanale, 1971a).

We now consider the controversial issue of the time history and mechanisms of supply of this secondary atmosphere-ocean-sediment volatile inventory. The issue is whether that inventory, "secondary" though it might be, was largely degassed in a catastrophic episode early in Earth's history (Fanale, 1971a). For example, an alternative is that it was supplied in a uniformitarian manner where the percentage of remaining interior $H_2O$ degassed per unit time remains about constant (Turekian, 1964). Another alternative is that the secondary degassing commenced well into Earth's history only when enough heat was accumulated from the decay of long-lived nuclides to trigger core formation. We first consider the arguments against catastrophic early degassing at the time of Earth's formation and then the arguments for it.

**Arguments against an early major catastrophic degassing event** One of the most important heat sources in the Earth is decay of long-lived nuclides, which took hundreds of millions of years to become a dominant heat source and which now serves as the main energy source for ongoing volcanism. This volcanism has a considerable amount of degassing associated with it, and its associated gas composition is not unreasonable for supplying the mass balance of the excess volatiles (Rubey, 1951).

It is hard to identify the small amounts of juvenile volatiles being added currently in a background of large amounts of recycled volatiles. Nonetheless, the fluxes could well be sufficient to explain the mass of the inventory, even assuming very low percentages of the juvenile component, providing that subtle sources of the flux such as thermal springs are taken into account (Rubey, 1951). Also, there is no evidence of the huge carbonate deposits one would expect to result if there was a massive initial $CO_2$ atmosphere (the latter would produce an exceedingly acid ocean and catastrophic weathering). It has also

TABLE 7-1 Excess Volatiles in the Earth's Surface and Atmospheric Inventory Not Accounted for by Rock Weathering[a]

| Volatile | Total Mass (g) |
|---|---|
| $H_2O$ | $1.66 \times 10^{24}$ |
| Total C (as $CO_2$) | $9.1 \times 10^{22}$ |
| Cl | $3.0 \times 10^{22}$ |
| N | $4.2 \times 10^{21}$ |
| S | $2.2 \times 10^{21}$ |
| H, B, Br, Ar, F, and so forth | $1.3 \times 10^{21}$ |

[a]If used to represent the total degassed volatile inventory of the Earth, these values would have to be raised to account for subduction of volatile-containing sediments (adapted from Rubey, 1951).

been argued that the $^{40}Ar$ content of the atmosphere could reasonably be supplied uniformly over the history of the Earth at a gradual pace consistent with the continuing emplacement and reworking of crustal rocks (Turekian, 1964; Wasserburg, 1964), which could be interpreted as implying a constant degassing rate for other volatiles. Finally, there is currently a substantial flux of primordial $^3He$ effusing from the Earth (Craig et al., 1978). This $^3He$ cannot possibly be recycled because it is chemically inert and also readily escapes to space. Also, it was not generated by decay throughout Earth's history, but was primordial. Its continued presence and effusion, therefore, could potentially place strict limits on the postulated efficiency of any early catastrophic degassing event. In fact, Tolstikhin (1975) has argued that the $^3He$ flux is so high that there are no valid models that can explain this flux, and also what is known of the composition of the Earth, and that additionally contain an episode of catastrophic degassing.

**Arguments favoring early catastrophic degassing** We now consider the arguments in favor of early catastrophic degassing of the Earth. First, there is evidence that even small bodies such as the Moon and the parent body of the Eucrites (Vesta or a still smaller body?) differentiated at the outset of their history (Lee et al., 1977). If so, the heat source for melting and differentiation must be identified. The energy from decay of the longer-lived nuclides like U, Th, and K would be "wasted" (i.e., lost to space) on small bodies, because the slow release of energy from their decay would be reradiated to space too rapidly to heat the small bodies to melting temperatures. However there were apparently short-lived nuclides, which heated small bodies rapidly enough to melt them and permit them to differentiate. In the particular case of the asteroids, this event would allow them to lose their volatiles to space before they accreted to create a large planet with a sufficient gravitational field for retention of light gases (Anders and Owen, 1977). Retention of short-lived nuclides could differentiate a large body such as the Earth at the outset, whereas at the same time the volatiles would be retained because of the large gravitational field of Earth. The greater likelihood of initial melting of the Earth over the Moon and the parent body of Eucrites becomes even more pronounced when, instead of short-lived nuclides, we consider the energy source provided by the gravitational energy of accretion, which for a body the size of the Earth exceeds that necessary for melting by over an order of magnitude (Fanale, 1971a). The amount of energy is so great that it is not necessary that it be efficiently converted to heat (10% to 25% seems likely) to melt the Earth. Nor is it necessary to assume that accretion occurs so rapidly that the heat produced is buried before it can be reradiated to space. Melting and degassing of each accreting radial increment would suffice. Accretion that takes place on too slow a time scale to bury the heat or that involves an array of preplanetary sizes can still (and probably did) melt even the Moon thoroughly at the outset of its history (Safranov, 1972; Wetherill, 1976), although the melting may have had a piecemeal character. If the Moon fissioned from the Earth, these calculations would not be applicable to the Moon but would still be applicable to the Earth, which could be melted by either accretion or fission. The likelihood of assembling the Earth without planetary melting seems slight and, moreover, the process would be highly self-aggravating once begun since exchange of core and mantle material in the gravitational field would produce yet another increment of energy (thoroughly buried), more than sufficient in itself to complete the process. (The accretion process and its relation to planetary heating are also discussed in Chapter 1 by Wasson and Kivelson.)

Second, Fanale (1971a) has argued that since every model for Pb-isotopic evolution yet proposed requires that global (though incomplete) U-Pb separation in the outer portions of the Earth (see Chapter 5 by DePaolo) occur as close to the outset of Earth history as permitted by the accuracy of the isotopic measurements; global volatile mobility is also required at the same time. In all cases this is within the first 200 My of Earth history, when accretion energy or short-lived nuclides were *decreasing* in availability faster than contributions from long-lived nuclides were *increasing*. This consideration pushes the likely time of planetary melting even closer to the outset of Earth history (Fanale, 1971a). Nor is it likely that processes other than ordinary igneous differentiation were responsible for core-mantle Pb separation; if nebula condensation processes (inhomogeneous accretion) were responsible, the Pb would be *more* skewed toward the outer portions than the U, which would be deep, whereas the reverse is true. Admittedly, migration of chemically reactive volatiles could be somewhat sluggish and inhibited in efficiency during the episode due to solubility effects (Fanale, 1971a). Also, if zone refining were involved, early hydrosphere formation would be somewhat limited by ongoing hydration simultaneous with the melting (Lange and Ahrens, 1982). Nonetheless, the Earth's entire surface-volatile inventory is still only a portion of its total volatile content in most compositional models (Fanale, 1971a), so supply of an initial atmosphere and hydrosphere seems assured. Catastrophic initial supply of a surface volatile inventory comparable in magnitude to the present one does *not* mean that the Earth is, in total, efficiently degassed (even $^3$He is not!) *or* that the volatile inventory had a composition similar to today's "Rubey's inventory" of excess volatiles. It should be mentioned that there exist mathematical models based upon rare gas isotopic composition which potentially could yield a unique solution for the degassing history of at least the rare gases themselves. I do not discuss these in detail because proper use of these models almost always requires input parameters that are poorly known at present. For example, those utilizing the current $^{40}$Ar:$^{36}$Ar ratio also require either knowledge of the K content of the mantle or knowledge of atmospheric $^{40}$Ar:$^{36}$Ar at one point in the distant past. So far, neither are well enough known to allow exercise of the models' potential.

I believe that these arguments are sufficient to indicate firmly that the Earth experienced a major catastrophic degassing event on formation. However, the arguments against such an event must still be answered in sequence. First, the $^{40}$Ar argument against catastrophic degassing (given above) is not relevant, since it merely indicates the occurrence of continuing activity and is irrelevant as evidence *against* a catastrophic degassing event because of $^{40}$Ar's growth function (parent's total half-life is 1.3 Gy). Thus it would not have been available for degassing initially. Analogous arguments apply to all evidence of continuing volcanism. Second, there is no reason to expect any simple, direct, easily recognizable evidence of massive carbonate deposits in view of the total absence of crustal rocks with crystallization ages more than 3.5 Gy. Third, although the $^3$He flux is juvenile and substantial, almost all the recent quantitative models of Holland (1983), which successfully relate the $^3$He flux to Earth degassing history, include and require major initial catastrophic degassing episodes. The retention of some of the $^3$He despite major U and Pb migration and global differentiation is not so mystifying as it might seem. Consider that the gravitational energy available for differentiation increases exponentially with the radius of the growing nucleus. This means that the central portion of the accreting Earth need not have melted immediately. Layer-by-layer zone refining of volatiles could have occurred in the outer layers as they accreted, and a magma ocean may have developed in the outer portion of a planet, a process that would be quite compatible with retention of volatiles in the central core. True, once differentiation started, it would propagate throughout the body, but by then the volatile migration paths required for total degassing of volatiles from the center of the Earth would be large, and degassing would be quite inefficient.

We now consider the degree of oxidation of the volatiles in the atmosphere of the Earth (and other planets) as a function of time. The current high degree of oxidation of the Earth's atmosphere (which essentially precludes further spontaneous abiotic organic synthesis *in the gas phase*) has resulted primarily from photosynthetic activity and secondarily from $H_2O$ dissociation coupled with exospheric hydrogen escape. However no igneous rock and no meteorite would be in equilibrium with the present atmospheric composition. At their melting points, most rocks and meteorites are in equilibrium with oxygen pressures of $10^{-8}$ to $10^{-15}$ bars, and this pressure is typically controlled by the $Fe^0:Fe^{2+}:Fe^{3+}$ ratios of the magmas (Kennedy, 1948; Lofgren et al., 1982). This limit on the oxygen activity, coupled with dissociation constants for $CO_2$ and $H_2O$ near the rock-melting temperatures, produces values of $CO/CO_2$, $H_2/H_2O$, and corresponding dry-gas compositions (if $H_2O$ is condensed) that have a considerable content (up to several percent) of free hydrogen and no free oxygen. Thus it is easy to see why, on theoretical grounds alone, planetary volatiles are not expected to be (at least prior to or in the absence of life) as oxidizing as is our current atmosphere. A natural transition from any initial reducing atmosphere on the Earth and Mars would occur fairly early in Martian or terrestrial history as the result of hydrogen escape, which, together with $H_2O$ dissociation, would inexorably produce a $CO_2$- and $N_2$-dominated atmosphere (Fanale, 1971b). Such an atmosphere would have persisted to the present on Mars and Venus (as is currently observed) but would have been altered by life on Earth to Earth's current atmosphere (i.e., photosynthetic production of $O_2$ from $CO_2$ by green plants). A more controversial issue is whether $CH_4$ or $NH_3$ atmospheres ever existed or whether atmospheric history tends to consist only of an early $CO_2$-rich phase. Such a $CO_2$- and $N_2$-rich atmosphere could have existed from the beginning of the history of Mars, Venus, and Earth and persisted until the present on Mars and Venus and until photosynthesis on Earth. The oxygen partial pressures indicated above, coupled with the high degree of dissociation at magmatic temperatures, force values of $H_2O:H_2$ to about 105 and $CO_2:CO$ to about 37 (Holland, 1964). In turn, this means that any volatile inventory removed from equilibrium with basaltic rocks near the melting points has a dry-gas composition with the C:O ratio greater than $\frac{1}{2}$ and the H:O ratio greater than 2, raising the possibility of *later* $CH_4$ or $NH_3$ synthesis (after release to the lower-temperature environment). Whether such synthesis occurs or not depends on conditions in the atmosphere and ocean, *not* in the magma chamber. The role of the magma chamber in producing any possible abiotic $CH_4$ or $NH_3$ is merely to supply the needed input mass balance to the atmosphere—a necessary but not sufficient condition for $CH_4$ or $NH_3$ synthesis—not the direct production of $CH_4$ or $NH_3$. In other words, development of the setting for abiotic organic synthesis (whether it occurs in a $CH_4$- and $NH_3$-rich environment or a $CO_2$-, $CO$-, and $N_2$-rich environment) has at least *two distinct stages*—(1) establishment of the system's elemental mass balance and (2) establishment of molecular equilibrium in the gas phase. These two steps occur under radically different sets of conditions (Fanale, 1971b).

Whether or not conditions ever favor $CH_4$ and $NH_3$ synthesis in the atmosphere (or the lithosphere), given the appropriate input balance, depends on several factors. Favoring $CH_4$ and $NH_3$ synthesis are lower gas-phase temperatures (versus magmatic temperatures) and condensation of $H_2O$ vapor. Both tend to drive the methane- and ammonia-synthesizing reactions (Fanale, 1971b). On the other hand, the favorable mass balance can be quickly destroyed by hydrogen escape. In order to sustain a reasonably high atmospheric-hydrogen abundance (a strict criterion in $CH_4$ and $NH_3$ synthesis), a correspondingly high flux of $H_2$ needs to be supplied from the interior to compensate for the $H_2$ outflux. Otherwise the $CH_4/CO_2$ ratio rapidly becomes negligible. Preliminary calculations indicate that such a high volcanic or impact outflux of gas with excess $H_2$ could be maintained only for a very short portion of the history of the Earth or Mars without requiring unreasonable degassing rates and accumulating an embarrassingly large total amount of surface volatiles (Fanale, 1971b). These calculations were done considering only thermal hydrogen escape. Most importantly, they neglect the diffusion limit on H escape (Hunten, 1973) and therefore perhaps should be redone. It has been argued that even if appropriate conditions for $CH_4$ and $NH_3$ synthesis were met, little of either compound would accumulate (especially $NH_3$) because of overwhelming rates of destruction by dissociation by solar photons (Pollack and Yung, 1980).

Is there really any compelling consideration that *requires* an early reducing atmosphere for Mars or Earth? On Earth, it has long been known that some significant progress toward abiotic organic synthesis can be made using mildly reducing $CO_2$-rich mixtures with some excess of $H_2$ and CO. Thus abiotic organic synthesis on Earth does not absolutely require a $CH_4$- and $NH_3$-rich atmosphere (Abelson, 1965). A second possible argument is that since solar output was 40 percent lower than at present in early solar-system history, a $NH_3$-rich atmosphere is required on the early Earth to produce enough greenhouse warming for there to be liquid $H_2O$, hence old limestones (Sagan and Mullen, 1972). However, Sagan and Mullen also show that even a few parts per million $NH_3$ would suffice, especially with a high $H_2$ pressure—an $NH_3$- and $CH_4$-rich atmosphere is not necessary. It has been argued by Pollack (1979) that it is difficult to sustain even trace concentrations of $NH_3$ against photolytic destruction. Also it is hard to imagine how atmospheric hydrogen could be stable against planetary escape. Owen et al. (1979) show that higher $CO_2$ pressures alone could account for the warming (but see below). Thus neither life on Earth nor the early existence of liquid $H_2O$ demand a $CH_4$- and $NH_3$-rich atmosphere. A third observation is the well-known fact that ancient Earth sediments are less oxidized than current analogous Earth sediments. This may suggest a much lower $PO_2$ than at present (Holland, 1978)—although even this has been debated (Burke, 1982)—but also does not require a $CH_4$- and $NH_3$-rich atmosphere.

Finally, a fourth possibility is that the very old "gulleys" on Mars discussed earlier in this chapter require an early $CH_4$- and $NH_3$-rich atmosphere. Certainly the ubiquitous distribution of these gulleys on the oldest Mars terrain and their essential absence on younger terrain suggest some profound global change (see Mars discussion earlier in Sec. 2-3). Also it is clear that high $CO_2$ pressures with low to moderate $NH_3$ pressures would account for substantial warming on Mars, but the stabilization of even trace concentrations of $NH_3$ is again a serious and perhaps fatal problem for such models (Pollack and Yung, 1980). Under favorable circumstances, a $CO_2$ atmosphere with a rather high pressure of approximately 1 to 2 bars could produce the warming needed to stabilize liquid $H_2O$ (Owen et al., 1979). However as pointed out by Rubey (1951), such high $CO_2$ pressures would imply fairly acidic waters (pH 5.0 to 6.0, depending on the degree of carbonate saturation of the water). This means that carbonate formation at the expense of any exposed igneous rocks would proceed very rapidly. Fanale et al. (1982b) argue on this basis that the rate of carbonate formation implied even for Mars would definitely be more than $10^{13}$ g · $y^{-1}$ based on analogy with current terrestrial rates, and that this means that *coexistence* of substantial bodies of water on Mars with such high carbon dioxide pressures would be limited to $10^7$ y. Perhaps this could be lengthened if the temperature were stabilized at approximately 0°C by a feedback loop in which rapid $CO_2$ removal always accompanied the occurrence of small bodies of water, but where the absence of such *small* bodies led to a rapid accumulation of $CO_2$, a rise in atmospheric temperature, and a reappearance of *small* bodies of $H_2O$. Also, no one has considered the possible effect of exchange between the atmosphere and a reservoir such as the ocean or regolith where available $NH_3$ could be protected against photolysis (Fanale et al., 1982b). Similar arguments can be made against the existence of large greenhouse effects on early Earth due to a huge $CO_2$ atmosphere that coexisted with large bodies of water.

Clearly, two (among many) critical questions in this area are (1) Could an atmosphere with at least a moderate $H_2$ content and trace contents of $CH_4$ and $NH_3$ be stabilized on the Earth or Mars against the effects of photolysis and hydrogen escape by atypically high early degassing rates *or any other means* and for how long? Otherwise, did organic synthesis occur on Earth in a $CO_2$- and $N_2$-rich atmosphere? and (2) What mechanisms could allow stabilization of atmospheres with $CO_2$ pressures high enough to stabilize a greenhouse effect coexisting with substantial bodies of water on Mars and the Earth despite the strong resulting tendency to form carbonates and thereby decrease the $CO_2$ pressures?

## 7-3 CONCLUSIONS

Planetary bulk volatile inventories are greatly affected by the distance from the preplanetary nebula center at which the preplanetary material accreted, with volatile contents increasing with increasing distance from the nebula center. Some objects appear to deviate from this trend, including probably the parent body of some (Eucritic) achondrites. This suggests some complexity may be overlaid on the simplest equilibrium-condensation hypothesis. This complexity may involve volatile loss from initially heated bodies large enough to accumulate rapidly supplied heat from short-lived nuclides, but still small enough to have low gravitational fields so they could lose volatiles to space. Inhomogeneous accretion may play a role, as may massive atmosphere loss to space but widespread radial transport of entire volatile inventories throughout the nebula seems contraindicated.

Our ideas concerning the degassing history of the planets have evolved, largely due to the space program and meteorite analysis, to include the following concepts: A hot origin, including "magma oceans," affected many solid objects, including some small ones. In addition to long-lived nuclides, short-lived nuclides, accretion energy, tidal heating, and other sources can heat bodies even if they are small and have fairly short thermal characteristic times. Only *persistent* heating continuing to the present requires (in the absence of strong tidal stresses) great planetary size. A major episode of catastrophic initial degassing characterized the early history of the Earth and probably many other solid planetary objects.

When volatiles are released to a surface environment, their elemental mass balance is strongly affected by the $Fe^0:Fe^{2+}:Fe^{3+}$ ratios in interior magmas. This "buffers" or controls the degree of oxidation of S, C, N, and so forth. Subsequently gas-gas and gas-rock reequilibration at low surface or atmospheric temperatures and pressures, together with this "frozen-in" mass balance supplied by the magmas, controls molecular evolution. The Earth, Venus, and Mars may have initially possessed atmospheres with significant $H_2$, $CH_4$, and $NH_3$ contents, but only for a very (geologically) short period, the length of which depends on the competition between early high degassing rates associated with accretion and very rapid preferential H loss from exospheres. ($NH_3$ is the least durable constituent owing to rapid photolysis.) Atmospheres rich in $CO_2$ and $N_2$ quickly resulted from H escape on all bodies. These persist to the present on Mars and Venus, but the Earth's atmosphere has become very $O_2$-rich owing to photosynthesis. A high $CO_2$ pressure on early Mars (with possibly some aid from minor amounts of reduced compounds) may have provided enough of a gas greenhouse effect to allow more clement conditions despite lower early solar output, thus accounting for the gulleys, which seem almost ubiquitous, on and restricted to, the oldest terrain in low latitudes. Alternatively, they might have been caused by an early accretion-engendered geothermal gradient that was much steeper than the current one, allowing for more efficient near-surface sapping processes. It is questionable whether high $CO_2$ pressure could coexist for long with *large* bodies of water on either the Earth or Mars, owing to the high erosion and $CO_2$ fixing rates that would result.

In the outer solar system, many objects are primarily $H_2O$ ice, and some contain large bulk contents of $NH_3$ and $CH_4$ as well. In these cases differentiation usually occurred despite weak heating, simply because of the low melting points of the volatiles themselves. Minor interior volatile contents also greatly affect melting points and heat transport in rocky objects. Some small bodies in the asteroid belt are clearly differentiated, and analogous differentiation may have resulted in bulk-volatile loss for many such small early objects and any planets to which they contributed. The rest of the asteroids have surfaces apparently dominated by hydrated silicates as predicted in equilibrium-condensation theory.

Io is the site of active S- and $SO_2$-driven volcanism, ultimately powered by tidal heating. Io probably was $H_2O$-rich but lost a potential early hydrosphere and other volatiles due to extensive degassing and a variety of escape processes, including thermal escape and magnetospheric ionization and field sweeping. (The latter processes are still major atmospheric loss or volatile redistribution processes on the Galilean satellites.) The early (water-rich) history of Io degassing is especially interesting since most Jupiter formational models suggest that Io initially received about as much energy from Jupiter as the Earth receives from the Sun. (This is why Io and Europa are rocky, not icy bodies in bulk.) Venus may also have possessed an early hydrosphere, destroyed by $H_2O$ dissociation, H escape, and interior oxidation. But the evidence for early hydrospheres is ambiguous for both Venus and Io. The Earth's current hydrosphere may not be unique; liquid $H_2O$ may possibly exist in the Mars regolith and Europa may have a substantial liquid-$H_2O$ zone maintained by tidal heating beneath an icy upper crust. Also Ganymede and Callisto may conceivably have more deeply located brine zones as well. The gradual freezing of $H_2O$ in objects in the outer solar system has also been a major cause of surface tectonics. Sputter erosion by magnetospheric ions has caused significant surface compositional and erosional effects on the icy satellites and both supplies and erodes their atmospheres.

Currently Mars' volatiles are lodged not only in the atmosphere and caps, but (mainly) in the regolith. Major forms include $H_2O$ as hard frozen permafrost and $CO_2$ as adsorbed $CO_2$. They also include C, S, and so forth, as salts and $H_2O$ as hydrated compounds such as amorphous gels and iron-rich clays. Some of these volatiles—including the adsorbed $CO_2$—are "available" for exchange among the atmosphere, polar caps, and regolith. This exchange is driven mainly by obliquity variations and results in major (hundredfold) periodic variations in atmospheric pressure and $CO_2$ cap size. However, no high $CO_2$ pressures (approximately 1 bar) such as may have characterized early Mars are achievable during most of Mars' history, and its water remains frozen, except perhaps within the regolith. The larger channels apparently result from catastrophic aquifer-like outbursts from beneath a permafrost cap. Given favorable conditions, liquid $H_2O$ can exist within the regolith near Mars' surface. The periodic climate change on Mars is physically quite analogous to obliquity-driven glacial epochs on Earth. The exact inventory of regolith volatiles on Mars is the subject of considerable debate. The debate centers around whether primordial rare gas can be reliably utilized to index the abundance of other atmosphere and surface volatile constituents. In fact, uncertainty concerning surface volatile inventories of the terrestrial planets generally results from uncertainties concerning the use of rare gas elemental concentrations in planetary atmospheres as tracers of the abundance and distribution of chemically active volatiles. While still a valid approach, use of rare gas tracers has been complicated by the growing realization that the processes that incorporated rare gases and those that incorporated chemically active volatiles into planet-forming material are significantly different. The so-called planetary component of the rare gases appears to have been incorporated by some combination of physical adsorption, chemisorption, and trapping processes, as opposed to simple condensation and chemical reaction for the chemically active volatiles. This means that isotopic compositions of the rare gases and surface elemental and isotopic compositions of chemically reactive volatiles and semivolatiles (not easily inferred from the current data set), together with additional studies of volatile distribution among phases in meteorites, may be needed before deductions concerning the bulk or surface volatile inventories of solid solar-system objects can be made with confidence.

Currently, the most serious anomaly in deciphering planetary history is that the primordial rare gas contents of the three objects—Mars, Earth, and Venus—increase dramatically with decreasing distance from the Sun, a trend which is opposite to that predicted by the equilibrium condensation scenario. However, the overall weight of the evidence argues strongly in favor of the equilibrium condensation hypothesis, implying initial bulk volatile endowments that increase with increasing heliocentric distance. At present, it appears that the reconciliation of these conflicting lines of evidence will leave both the equilibrium condensation hypothesis (and its heliocentric volatile-endowment gradient) and the putative correlations between primordial rare gas and chemically-reactive volatile contents more or less intact. The reconciliation would, however, involve differential early or initial atmospheric loss from the three objects, with Mars being the most affected, Venus least affected. The rare gases would, in this scenario, be decoupled from the major volatiles during or after planetary accretion by their greater vulnerability to atmospheric loss. Further decoupling would occur between the initial and present inventories of major volatiles as the result of destruction of $H_2O$ to oxidize iron, coupled with hydrogen escape.

## 7-4 ACKNOWLEDGMENTS

This work was performed under the auspices of grant NAS 7-100, National Aeronautics and Space Administration. I thank M. G. Kivelson, without whose unrelenting "encouragement" this paper would not have been written. This is Planetary Geosciences Publication No. 364.

## 7-5 REFERENCES

Abelson, P. H., 1965, Chemical events on the primitive Earth, *Proc. U.S. Natl. Acad. Sci., 55*, 1365–1370.

Anders, E. and T. Owen, 1977, Mars and Earth—origin and abundance of volatiles, *Science, 198*, 453–465.

Anderson, D. M., E. S. Gaffey and P. F. Low, 1967, Frost phenomena on Mars, *Science, 155*, 319–322.

Anderson, D. M., M. J. Schwarz and A. R. Tice, 1978, Water vapor adsorption by $N_2$ montmorillonite—modeling Martian water vapor adsorption, *Icarus, 34*, 638–644.

Becker, R. H. and R. O. Pepin, 1984, The case for Martian origin of the Shergottites: Nitrogen and noble gases in EETA 79001, *Earth and Planet. Sci. Lett., 69*, 225–242.

Biemann, K., J. Oro, P. Toulmin III, Le E. Orgel, A. O. Nier, D. M. Anderson, P.G. Simmonds, D. Flory, A. V. Diaz, D. R. Rushnech, J. E. Bitter and A. L. Lafleur, 1977, The search for organic substances and inorganic volatile compounds in the surface of Mars, *J. Geophys. Res., 82*, 4641–4658.

Blasius, K. R. and J. A. Cutts, 1980, Global patterns of primary crater ejecta morphology on Mars, *NASA TM 82385*, 147–150.

Bogard, D. D. and E. K. Gibson, Jr., 1978, The origin and relative abundance of C, N, and the noble gases on the terrestrial planets and in meteorites, *Nature, 271*, 150–153.

Booth, M. C. and H. H. Kieffer, 1978, Carbonate formation in Mars-like environments, *J. Geophys. Res., 82*, 1809–1815.

Brown, R. A., C. B. Pilcher and D. F. Strobel, 1983, Spectrophotometric studies of the Io torus, pp. 197–225, in *Physics of the Jovian Magnetosphere* (A. Dessler, ed.). Cambridge Univ. Press, New York.

Burke, K., 1982, Was the Earth's atmosphere anoxic between 3.8 and 2.0 Ga ago? *Proc. Planet. Volat. Conf.*, Lunar and Planet. Sci. Inst. Publication No. 488.

Cameron, A. G. W. and J. B. Pollack, 1976, On the origin of the solar system and of Jupiter and its satellites, pp. 61–84 in *Jupiter* (T. Gehrels, ed.), U. of Arizona Press, Tucson, Ariz.

Carr, M. H., 1979, Formation of Martian flood features by release of water from confined aquifers, *J. Geophys. Res., 84*, 2995-3007.

Carr, M. H. and G. G. Schaber, 1977, Martian permafrost features, *J. Geophys. Res., 82*, 4039-4054.

Cassen, P., S. J. Peale and R. T. Reynolds, 1980, On the comparative evolution of Ganymede and Callisto, *Icarus, 41*, 232-239.

Cintala, M. J. and P. J. Mouginis-Mark, 1980, Martian fresh craters' depths—More evidence for subsurface volatiles? *Geophys. Res. Lett., 7*, 329-332.

Clark, B. C., 1978, Implications of abundant hygroscopic minerals in the Martian regolith, *Icarus, 34*, 645-665.

Clark, R. N. and T. B. McCord, 1980, The Galilean satellites—new near-infrared spectral reflectance measurements (0.65-2.5$\mu$m) and a 0.325-5$\mu$m summary, *Icarus, 41*, 323-339.

Coradini, M. and E. Flamini, 1979, A thermodynamical study of the Martian permafrost, *J. Geophys. Res., 84*, 8115-8130.

Craig, H., J. E. Lupton and Y. Horibe, 1978, A mantle helium component in circum-Pacific volcanic gases: Hakone, the Marianas and Mt. Fassen, pp. 3-16 in *Terrestrial Rare Gases, Advances in Earth and Planetary Sciences, V3* (E. C. Alexander and M. Ozima, eds.), Japan Soc. Press, Tokyo, Japan.

Davis, D. R., C. R. Chapman, R. Greenberg, S. J. Weidenschilling and A. W. Harris, 1979, Collisional evolution of asteroids—populations, rotations and velocities, in *Asteroids* (T. Gehrels, ed.), U. of Arizona Press.

Donahue, T. M. and J. B. Pollack, 1983, Origin and evolution of the atmosphere of Venus, pp. 1003-1037 in *Venus* (D. M. Hunter, L. Colin, T. M. Donahue and V. I. Moroz, eds.) U. of Arizona Press, Tucson, Ariz.

Donn, B. et al., 1976, *The Study of Comets*, NASA SP-393.

Dreibus, G. and H. Wanke, 1980, The bulk composition of the Eucrite parent asteroid and its bearing on planetary evolution, *Zeit. f. Naturforschung, 35*, 204-216.

Evans, D. L. and J. B. Adams, 1980, Amorphous gels as possible analogs to Martian weathering products, *Proc. Lunar Planet. Sci. Conf., 11th*, 757-763.

Fanale, F. P., 1971a, A case for catastrophic early degassing of the Earth, *Chem. Geol., 8*, 79-105.

Fanale, F. P., 1971b, History of Martian volatiles—implications for organic synthesis, *Icarus, 15*, 279-303.

Fanale, F. P., 1976, Martian volatiles—their degassing history and geochemical fate, *Icarus, 28*, 179-202.

Fanale, F. P. and W. A. Cannon, 1972, Origin of planetary primordial rare gas—the possible role of adsorption, *Geochim. Cosmochemic. Acta, 36*, 319-328.

Fanale, F. P. and W. A. Cannon, 1974, Exchange of adsorbed $H_2O$ and $CO_2$ between the regolith and atmosphere of Mars caused by changes in surface insolation, *J. Geophys. Res., 79*, 3397-3402.

Fanale, F. P. and W. A. Cannon, 1979, Mars: $CO_2$ adsorption and capillary condensation on clays: significance for volatile storage and atmospheric history, *J. Geophys. Res., 84*, 8404-8414.

Fanale, F. P., T. V. Johnson and D. L. Matson, 1977, Io's surface and the histories of the Galilean satellites, pp. 756-782 in *Planetary Satellites* (J. Burns, ed.), Univ. of Arizona, Tucson, Ariz.

Fanale, F. P., R. H. Brown, D. P. Cruikshank and R. N. Clark, 1979, Significance of Io's IR reflectance spectrum, *Nature, 280*, 761-763.

Fanale, F. P., W. B. Banerdt, L. S. Elson, T. V. Johnson and R. Zurek, 1982a, Io's surface—its phase composition and influence on Io's atmosphere and Jupiter's magnetosphere, pp. 756-782 in *The Galilean Satellites* (D. Morrison, ed.), Univ. Arizona, Tucson, Ariz.

Fanale, F. P., J. R. Salvail, W. B. Banerdt and R. J. Saunders, 1982b, Mars—the regolith-atmosphere-cap system and climate change, *Icarus, 50*, 381-407.

Fegley, B., Jr. and J. S. Lewis, 1980, Volatile element chemistry in the solar nebular: Na, K, F, Cl, Br, and P, *Icarus, 41*, 439-455.

Fox, J., 1982, Nitrogen isotopic fractionation in the Mars atmosphere, *Proc Conf. Planet. Volat.*, Pub. Lunar and Planet. Inst., Houston, Tex.

Gaffey, M. J., 1976, Spectral reflectance characteristics of the meteorite classes, *J. Geophys. Res., 81*, 905-920.

Gaffey, M. J. and T. B. McCord, 1977, Asteroid surface materials—mineralogical characteristics and compositional implications, *Proc. Lunar Sci. Conf., 8th*, 113-143.

Gibson, E. K., M. A. Urbanic and F. F. Androawes, 1980, Volatile loss and thermal stability of Martian analog clay and sulfur minerals, *Reports Planet. Geol. Program, 1979-1980*, NASA TM 81776.

Haff, P. K., C. C. Watson and T. A. Tombrello, 1979, Ion erosion in the Galilean satellites of Jupiter, *Proc. Lunar Planet. Sci. Conf., 10th*, 1685-1699.

Hodges, R. R., Jr., 1975, Formation of the lunar atmosphere, *The Moon, 14*, 139-157.

Holland, H. D., 1964, On the chemical evolution of the terrestrial and cytherian atmospheres, in *The Origin and Evolution of Atmospheres and Oceans* (P. Brancazio, A. G. W. Cameron, eds.), John Wiley, New York.

Holland, H. D., 1978, The evolution of seawater, pp. 559-567 in *The Early History of the Earth* (B. F. Windley, ed.), John Wiley, New York.

Holland, H. D., 1983, *Origin and Evolution of Atmospheres and Oceans*, Princeton Univ. Press, Princeton, N.J.

Houck, J. R., J. B. Pollack, C. Sagan, D. Schaak and J. A. Decker, Jr., 1973, High altitude infrared spectroscopic evidence for bound water on Mars, *Icarus, 18*, 470-480.

Hsui, A. T. and M. N. Toksoz, 1977, Thermal evolution of planetary-size bodies, *Proc. Lunar Sci. Conf., 8th*, 447-461.

Huguenin, R. L., 1976, Mars: Chemical weathering as a massive volatile sink, *Icarus, 28*, 203-212.

Hunten, D. M., 1973, The escape of light gases from planetary atmospheres, *J. Atmosph. Sci., 30*, 1481-1494.

Johnson, T. V. and F. P. Fanale, 1973, Optical properties of carbonaceous chondrites and their relationship to asteroids, *J. Geophys. Res., 78*, 8507-8518.

Kaula, W. M., 1979, Thermal evolution of Earth and Moon growing by planetesimal impacts, *J. Geophys. Res., 84*, 999-1008.

Kennedy, G. C., 1948, Equilibrium between volatiles and iron oxides in igneous rocks, *Am. J. Soc., 246*, 529-549.

Kumar, S., 1976, Mercury's atmosphere: A perspective after *Mariner 10, Icarus, 28*, 579-591.

Kumar, S., 1979, The stability of an atmosphere on Io, *Nature, 280*, 758-761.

Kumar, S. and D. Hunten, 1982, The atmospheres of Io and other satellites, pp. 782-806 in *Satellites of Jupiter* (D. Morrison, ed.), Univ. Arizona, Tucson, Ariz.

Lange, M. A. and T. J. Ahrens, 1982, Accretion of water by the terrestrial planets, in *Proc. Conf. Planet. Volat.*, Pub. Lunar Planet. Inst., Houston, Tex.

Lanzerotti, L. J., W. L. Brown, J. M. Poate and W. M. Augustyniak, 1978, On the contribution of water products from Galilean satellites to the Jovian magnetosphere, *Geophys. Res. Lett., 5,* 155-158.

Lebofsky, L. A., 1978, Asteroid I Ceres—evidence for water of hydration, *Mon. Not. Roy. Ast. Soc., Short Comm., 182,* 17P-21P.

Lee, T., D. A. Papanastassiou and G. J. Wasserburg, 1977, Aluminum-26 in the early solar system—fossil or fuel? *Astrophys. J., 211,* L107-L110.

Lewis, J. S., 1972a, Low temperature condensation from the solar nebula, *Icarus, 16,* 241-252.

Lewis, J. S., 1972b, Metal/silicate fractionation in the solar system, *Earth Planet. Sci. Lett., 15,* 286-290.

Lewis, J. S., 1974, The temperature gradient in the solar nebula, *Science, 186,* 440-443.

Lofgren, G. E., et al., 1982, Petrology and chemistry of terrestrial, Lunar and meteoritic basalts, pp. 2-437 in *Basaltic Volcanism on the Terrestrial Planets* (W. M. Kaula, ed.) Pergamon Press, Elmsford, New York.

Matson, D. L., T. V. Johnson and F. P. Fanale, 1974, Sodium D-line emission from Io—sputtering and resonant scattering hypothesis, *Astrophys. J., 192,* L43-L46.

McCord, T. B., J. B. Adams and T. V. Johnson, 1970, Asteroid Vesta—Spectral reflectivity and compositional implications, *Science, 168,* 1445-1447.

McCord, T. B., J. H. Elias and J. A. Westphal, 1971, Mars—the spectral albedo ($0.3-2.5\mu$) of small bright and dark regions, *Icarus, 14,* 245-251.

McCord, T. B., R. Clark and R. L. Huguenin, 1978, Mars: Near-infrared spectral reflectance and compositional implication, *J. Geophys. Res., 83,* 5433-5441.

McElroy, M. B., Y. L. Yung and A. O. Nier, 1976, Isotopic composition of nitrogen: Implications for the past history of Mars' atmosphere, *Science, 194,* 70-72.

Owen, T., K. Bieman, D. R. Rushneck, J. E. Biller, D. W. Howarth and A. L. LaFleuer, 1977, The composition of the atmosphere at the surface of Mars, *J. Geophys. Res., 82,* 4635-4640.

Owen, T., R. D. Cess and V. Ramanathan, 1979, Early Earth—An enhanced carbon dioxide greenhouse to compensate for reduced solar luminosity, *Nature, 277,* 640-642.

Parmentier, E. M. and J. W. Head, 1979, Internal processes affecting surfaces of low density satellites—Ganymede and Callisto, *J. Geophys. Res., 84,* 6263-6277.

Peale, S. J., P. Cassen and R. T. Reynolds, 1979, Melting of Io by tidal dissipation, *Science, 203,* 892-894.

Pearl, J., R. Hanel, V. Kunde, W. Maguire, K. Fox, S. Gapta, C. Ponnaperuma and F. Rasilin, 1979, Identification of gaseous $SO_2$ and new upper limits for other gases on Io, *Nature, 280,* 755-758.

Phillips, R. J., W. M. Kaula, G. E. McGill and M. C. Malin, 1981, Tectonics and evolution of Venus, *Science, 212,* 879-893.

Pieri, D., 1976, Martian channels—distribution of small channels on the Martian surface, *Icarus, 27,* 25-50.

Pilcher, C. B., S. T. Ridgeway and T. B. McCord, 1972, Galilean satellites—identification of water frost, *Science, 178,* 1087-1089.

Pollack, J. B., 1971, A nongrey calculation of the runaway greenhouse—implications for Venus' past and present, *Icarus, 14,* 295-306.

Pollack, J. B. and D. C. Black, 1979, Implications of the gas compositional measurements of *Pioneer Venus* for the history of planetary atmospheres, *Science, 205,* 56-59.

Pollack, J. B. and D. C. Black, 1982, Noble gases in planetary atmospheres—Implications for the origin and evolution of atmospheres, *Icarus, 51*, 169-198.

Pollack, J. B. and F. P. Fanale, 1982, Origin and evolution of the Jupiter satellite system, pp. 872-910 in *The Satellites of Jupiter* (D. Morrison, ed.), Univ. Arizona, Tucson, Ariz.

Pollack, J. B. and R. T. Reynolds, 1974, Implications of Jupiter's early contraction history for the composition of the Galilean satellites, *Icarus, 21*, 248-253.

Pollack, J. B., F. C. Witteborn, E. T. Erickson, D. W. Strecker, B. J. Baldwin, and T. E. Bunch, 1978, Near-infrared spectra of the Galilean satellites—Observations and compositional implication, *Icarus, 36*, 271-303.

Pollack, J. B. and Y. L. Yung, 1980, Origin and evolution of planetary atmospheres, *Ann. Rev. Earth Planet. Sci., 8*, 425-487.

Rasool, S. I. and C. DeBergh, 1970, The runaway greenhouse and accumulation of $CO_2$ in the Venus atmosphere, *Nature, 226*, 1037-1039.

Rubey, W. W., 1951, Geologic history of sea water—An attempt to state the problem, *Bull. Geol. Soc. Am., 62*, 1111-1148.

Rossbacher, L. A. and S. Judson, 1981, Ground ice of Mars—inventory distribution and resulting landforms, *Icarus, 45*, 39-59.

Safranov, V. S., 1972, Evolution of the protoplanetary cloud and the formation of the Earth and planets, *NASA TTF-667*, Washington, D.C.

Sagan, C. and G. Mullen, 1972, Earth and Mars: Evolution of atmospheres and surface temperatures, *Science, 177*, 52-56.

Saunders, R. S. and L. Johansen, 1980, Latitudinal distribution of flow-ejecta morphology types on the ridged plains of Mars, *NASA TM, 82385*, 150-151.

Sharp, R. P. and M. C. Malin, 1975, Channels on Mars, *Bull. Geol. Soc. Am., 86*, 593-609.

Shemansky, P. E. and A. L. Broadfoot, 1977, Interaction of the surface of the Moon and Mercury with their exospheric atmospheres, *Rev. Geophys. Space Phys., 15*, 491-499.

Singer, R. B., 1981, Some observational constraints on iron mineralogy in Martian bright soils, *3rd Intl. Collo. on Mars*, Lunar Planet. Inst. Pub. 441, 236-238.

Smith, B. A. and the *Voyager* imaging team, 1979a, The Jupiter system through the eyes of *Voyager 1, Science, 204*, 951-972.

Smith, B. A., E. M. Shoemaker, S. W. Kieffer and A. F. Cook, III, 1979b, The role of $SO_2$ in volcanism on Io, *Nature, 280*, 738-743.

Smoluchowski, R., 1968, Mars: Retention of ice, *Science, 159*, 1348-1350.

Smythe, W. D., R. M. Nelson and D. B. Nash, 1979, Spectral evidence for $SO_2$ frost or adsorbate on Io, *Nature, 280*, 766.

Soderblom, L. A. and D. B. Wenner, 1978, Possible fossil $H_2O$ liquid-ice interfaces in the Martian crust, *Icarus, 34*, 622-638.

Squyres, S. W., 1980, Volume changes in Ganymede and Callisto and the origin of the grooved terrain, *Geophys. Res. Lett., 7*, 593-596.

Squyres, S. W., R. T. Reynolds, P. M. Cassen and S. J. Peale, 1983, Liquid water and active resurfacing on Europa, *Science*, (in press).

Toksöz, M. N., A. T. Hsui and D. H. Johnston, 1978, Thermal evolution of the terrestrial planets, *Moon and Planets, 18*, 281-320.

Tolstikhin, I. N., 1975, Helium isotopes in the Earth's interior and the atmosphere—a degassing model of the Earth, *Earth Planet. Sci. Lett, 26*, 88-96.

Toulmin, P., A. K. Baird, B. C. Clark, K. Keil, H. J. Rose, R. P. Christian, P. H. Evans and

W. C. Kellike, 1977, Geochemical and mineralogical interpretation of the *Viking* inorganic chemical results, *J. Geophys. Res., 82*, 4625-4634.

Turekian, K. K., 1964, Degassing of argon and helium from the Earth, pp. 74-82 in *The Origin and Evolution of Atmospheres and Oceans* (P. Brancazio and A. G. W. Cameron, eds.), John Wiley, New York.

Turekian, K. K. and S. P. Clark, Jr., 1969, Inhomogeneous accumulation of the Earth from the primitive solar nebula, *Earth Planet. Sci. Lett., 6*, 346-348.

Walker, J. C. G., 1977, *Evolution of the Atmosphere of Venus*, Macmillan, New York.

Wänke, H., and G. Dreibus, 1984, Volatiles on Mars, in *Proc. Workshop on Water on Mars*, Lunar and Planet. Inst., Houston, Texas.

Ward, W., B. C. Murray and M. C. Malin, 1974, Climatic variation on Mars 2: Evolution of carbon dioxide and polar caps, *J. Geophys. Res., 79*, 3387-3395.

Wasserburg, G. J., 1964, Comments on the outgassing of the Earth, pp. 83-85 in *The Origin and Evolution of Atmospheres and Oceans* (P. Brancazio and A. G. W. Cameron, eds.), John Wiley, New York.

Wasson, J. T. and G. W., Wetherill, 1979, Dynamical, chemical and isotopic evidence regarding the formation locations of asteroids and meteorites, pp. 926-974 in *Asteroids* (T. Gehrels, ed.), Univ. of Arizona, Tucson, Ariz.

Watkins, G. H. and J. S. Lewis, 1985, Evolution of the atmosphere of Mars as the result of asteroidal and cometary impacts, in *Proc. Workshop Evol. Martian Atmos.*, Pub. Lunar and Planet. Inst., Houston, Tex.

Wetherill, G. W., 1975, Late heavy bombardment of the Moon and the terrestrial planets, *Proc. Lunar. Sci. Conf., 6th*, 1539-1566.

Wetherill, G. W., 1976, The role of large bodies in the formation of the Earth and the Moon, *Proc. Lunar Sci. Conf., 7th*, 3245-3257.

Wetherill, G. W., 1981, Solar wind origin of $^{36}$Ar on Venus, *Icarus, 46*, 70-80.

Whipple, F. L. and W. F. Huebner, 1976, Physical processes in comets, *Ann. Rev. Astr. Astrophys., 14*, 143.

Wood, J. A., 1979, Cooling rates, pp. 849-889 in *Asteroids* (T. Gehrels, ed.), Univ. Arizona, Tucson, Ariz.

Wyckoff, S., 1982, Overview of comet observations, in *Comets* (L. Wilkening, ed.), Univ. Arizona, Tucson, Ariz.

Yang, J. and E. Anders, 1982, Sorption of noble gases by solids with reference to meteorites, III—sulfides, spinels, and other substances; on the origin of planetary gases, *Geochim. Cosmochim. Acta, 46*, 877-892.

Zellner, B. and E. Bowell, 1977, Asteroid compositional types and their distributions, pp. 185-197 in *Comets, Asteroids, Meteorites* (A. H. Delsemine, ed.), Univ. Toledo, Toledo, Ohio.

Peter Bird
Department of Earth and Space Sciences
University of California
Los Angeles, California 90024

# 8

# TECTONICS OF THE TERRESTRIAL PLANETS

# ABSTRACT

Because geologic events are not reproducible, the unraveling of past tectonic events is always difficult and controversial. This is especially so for other planets, where the available data provide even less complete knowledge than we had of the Earth in 1930. Because of these limitations, we cannot study planetary tectonics empirically, so we are limited to testing for the presence of particular styles suggested by theory and intuition. The major possibilities are nonuniform contraction, despinning, volcanism, isostatic adjustment, and several varieties of convection, including homogeneous, layered, plate-tectonic, delaminating, and plumose.

All of the five planets considered show volcanism and some isostatic adjustment in their early histories, supporting the concepts of hot accretion, early differentiation, and gradual cooling. On the smaller planets (the Moon and Mercury), a global lithosphere formed early and remained unbroken or was only slightly cracked by the small strains of nonuniform contraction and perhaps despinning. On Mars, the plains and highland provinces may record a phase of homogeneous convection, with drift of blocks of primary crust. Now there is a global lithosphere, disrupted by the Tharsis plume, which thinned it from below, and fed volcanism, which loaded the thick lithosphere until it was bent and cracked on a global scale. The Earth also has had a history of waning convection and growing plates, with two important changes in style. After the first eon, continents became too big and buoyant to be recycled through zones of downward flow. After the Archean Era, the convection probably changed from the symmetrical homogeneous to the asymmetrical plate-tectonic form. The retention of a hydrosphere here is responsible for most of the present volcanism, the formation of continental crust, the thickness of oceanic crust, their neat separation at two different elevation levels, and the asymmetry of subduction. Our expectation that higher surface temperatures on Venus should cause even greater tectonic activity is confounded by radar images showing circular features like ancient impact craters. This observation seems to require a Martian model with a global lithosphere, where the plateaus known as "*terrae*" are underlain by plumes. Such a model can be reconciled with rock mechanics if Venus has no significant granitic crust. However, if future data show large areas free from craters, and a granitic composition in the terrae, then Venus would more probably be a unique example of layered convection, with complementary overturning cells in both the crust and mantle. The lithosphere might then be thinner than the crust, and consist of a set of thin, brittle "rafts" which preserve some ancient surface without impeding the convection below.

## 8-1 INTRODUCTION

*Tectonics* is the science (or art) of using observations made across a two-dimensional surface during one "instant" of geologic time to infer a complete history of the three-dimensional strains and movements of the rocks. Clearly, the data are inadequate even when rocks are perfectly exposed and observed, so many constraining assumptions are needed. A popular guiding principle is that of greatest simplicity (Occam's razor), which holds that structures are assembled from a few homogeneous rock types and become more complex through time. However, believers in uniformitarianism are forced to accept complementary simplifying processes (e.g., erosion, melting). Once these are allowed into a reconstructed history, any appearance of a unique solution is destroyed. Clearly, a purely empirical approach to tectonics is not enough.

The empirical method is even weaker when the data are limited to what can be observed from orbit and a handful of landing sites (see Table 8-1). Collecting images with available light or active radar reveals the broad contours of the surface and such details as fault lines, lava flows, and craters. But little or nothing can be inferred about rock composition, and the fine textures by which a field geologist would infer the youth or age of a landform are not resolved. In fact, the technique of crater counting (see boxed material on p. 87) is the only way of dating rocks we have not sampled, and it does not work well for recent epochs when the meteorite influx was slow. Gravity measurements yield smoothed estimates of the minimum-possible stress and anomalous-mass distributions, but there is no assurance that the planet adheres to a minimum principle, or that the long-wavelength parts of its tectonics are the most important. A seismicity sample may be obtained, but over a week to a year the observed pattern is apt to be dominated by one or more aftershock swarms, rather than the true strain-rate distribution in the planet.

TABLE 8-1  Available Data on Planetary Tectonics

| Method | Reveals | Limitations |
|---|---|---|
| Imaging | Crater-count ages, faults, volcanic deposits | About 1 km resolution, only relative elevations |
| Radar altimetry | Major structures and loads | About 30-km resolution |
| Satellite gravity | Degree of isostasy | Wavelengths $\geq$ orbit height |
| Seismicity | Lower limit on fault slip rates | Short time sample, poor or no event locations |
| Heat flow | Lithosphere thickness | Questionable accuracy, limited to Lander sites |
| X-ray fluorescence | Major-element bulk chemistry of crust | Limited to Lander sites |

To make such limitations more graphic, consider what we would know of the Earth if we had investigated it from another planet with *Mariner* and *Viking* technology. We would probably extrapolate coastal topography beneath the oceans, never guessing their true depth or the existence of a second (basaltic) type of crust. We would detect subduction-zone volcanoes, a few (mostly inactive) faults, and only one impact crater. The most visible tectonic belts would be the Alpine-Himalayan continental-collision belt and the Rocky-Andes Mts. belt of the Americas, whose formation is somehow related to plate tectonics, but still not fully understood by our geologists on the ground. The landers sent down to smooth and featureless "safe" sites like the Amazon basin and Sahara might report such odd crustal compositions as hematite-bauxite or pure quartz, respectively. Neither lander would be likely to detect earthquakes with a short-period instrument on a windy site, and a Saharan lander might not even detect life! Even allowing for gravity and magnetic measurements from orbit, our knowledge of Earth would be far less than terrestrial geologists had available in 1930. Recall that it was another 35 y after that before the main (plate-tectonic) elements of Earth's orogenic cycle were discovered by empirical means.

Therefore, extraterrestrial tectonic models must still be motivated by theories more than by observations. By combining physical laws and laboratory observations on rock deformation, we can predict alternative classes of possible behavior of planetary surfaces. The suite of models can never be narrowed to a single prediction, because any theory will contain as parameters the temperature and chemical distributions in the interior, two things that are inherently unmeasurable even from the surface. However, a few scanty observations of actual faulting, gravity, and crater counting may then be put to powerful use as "filters" that reject whole classes of models and further focus our thoughts and future exploration.

The approach in this chapter is to summarize very briefly the relevant facts of rock mechanics and the classes of tectonic models to be found in the literature. For each, we emphasize specific predictions of observable features. Next we review the evidence on the terrestrial planets one by one and relate it to these models. In the end, we see what appears to be an excellent inverse correlation of tectonic activity with lithosphere thickness, which is in turn controlled by radius, age, and atmosphere.

However, the apparent orderliness of this approach does not preclude conclusions that are wrong. The theoretical approach may outrun the empirical, but it can also go astray if not all of the relevant physical processes have been considered. This has happened twice in the history of geology. First, William Thompson Lord Kelvin "proved" with a conductive heat-flow model that the Earth was only 20 to 40 My old; neither Lord Kelvin nor any other scientist of the nineteenth century knew of radioactive heat production, and few thought of convection as relevant. In this century, seismology "proved" that continental drift was impossible because the Earth's entire mantle was solid, and solid rocks could not flow; it was only a few decades ago that electron microscopes began to show us dislocations in crystals which prove that rocks do flow. In an audacious review such as this, there is almost certainly at least one fundamental error of that type!

## 8-2 ELEMENTS OF ROCK MECHANICS

To understand tectonics, it is essential to understand at least qualitatively the ways in which rocks can deform. This brief review is focused on physical mechanisms, rather than descriptive terms for the resulting structures. The most important mechanisms are (1) thermal expansion, (2) elasticity, (3) frictional faulting, and (4) dislocation creep. Each of these is a different type of *strain* (a word that is used here qualitatively to indicate the fractional change of shape of a body of rock). All four mechanisms may act at once under the same conditions, superimposing their strain effects. A related measure is the speed at which they operate to accumulate strain; we refer to this as the *strain rate* and measure it in units of (dimensionless) strain per unit time.

The causative factors leading to strain are primarily temperature and stress; some simple definitions regarding the latter are useful: *Stress* is the measure of how strongly equal and opposite opposed forces act on the opposite sides of internal planes in the rock. In the Systeme International (SI) metric system its unit is the Pascal (Pa), or associated unit megaPascal (MPa = $10^6$ Pa). Because these forces have a direction as well as an intensity, we distinguish further between *normal* stresses (those with force acting normal to the plane in question) and *shear* stresses (those with forces acting parallel to the plane in question). It is useful to remember that fluids will relax (by straining) until they have no shear stress within them and until the normal stress in all directions is equal (we then call it *pressure*). However, solids may be much stronger, so that the shear stresses (or the differences in normal stress across different planes) must reach high levels to cause visible strain.

### 8-2-1 Thermal Expansion

When subjected to an increase of temperature, rocks undergo a fractional expansion of all their dimensions of 5 ppm to 12 ppm per degree Celsius. If the rock is more or less rigidly constrained, this may change the stress state and cause other types of strain. When rocks are weakly confined, they expand, and the important effect is the decrease in their density, by 15 ppm to 36 ppm per degree Celsius. Rocks that are hotter than their surroundings tend to rise because a reduced density implies *relative* buoyancy; this effect is what causes convection. The high pressures of planetary interiors reduce these thermal expansions by 70% or more but cannot eliminate them. The fundamental thermal-expansion effect underlies all but one of the tectonic models proposed below.

### 8-2-2 Elasticity

*Elastic strain* is, by definition, proportional to stress; therefore it returns to zero when stress is removed. Elasticity of rocks is also universal, but not important in tectonics. This is because the shear stresses that rocks can support (less than $10^9$ Pa) give rise to strains of less than 1 percent, which are not visible or detectable by any type of remote sensing. Of course, the isotropic elastic compression of rocks by interior pressures affects planetary radii and densities, but these constant and spherically symmetric effects cause no surface deformations.

### 8-2-3 Faulting

The dominant deformation mechanism at the surface of any of the terrestrial planets is *faulting*, the sliding of one rock mass over another along the planar or curved surfaces known as *faults*. Faulting is independent of time and temperature, and its governing law states simply that the shear stresses ($\sigma_s$) acting on the fault must reach the critical value of $\sigma_s = \mu(\sigma_n - P)$ for fault slip to continue, where $\sigma_n$ is the stress normal to the fault plane, $P$ is the pressure of interstitial fluids (if any), and $\mu$ is the dimensionless coefficient of friction, approximately 0.8 ± 0.1. Remarkably, this law is valid regardless of rock composition (excepting hydrated clays), texture, strain rate, or temperature (up to 400°C), or pressure (up to $10^9$ Pa).

In this chapter, we describe the geometry of faulting loosely as belonging to one of three categories observed on Earth. The angle between the fault plane and the horizontal (called the *dip*) takes one of three values: 30°, 60°, or 90°. *Thrust* faulting is the overlapping of one block onto another by slip on a fault surface that dips gently (30°) into the interior; it is caused by horizontal compression in the direction of sliding. *Normal* faulting is the uncovering of one block by another, with slip on a fault surface dipping more steeply (60°); it is caused by horizontal tension in the direction of sliding. Finally, *strike-slip* faulting is purely horizontal movement across a vertical fault plane; it is caused by horizontal compression in one direction combined with horizontal tension at right angles to it. Because an old fault full of crushed rock has slightly lower strength than other adjacent planes, old fractures tend to keep sliding even if stress directions change slightly; this leads to large cumulative offsets with linear topographic expressions. On Earth, the presence of water causes faults to anneal and stick together with time, so that when they finally move they must "refracture" with a jerk, and seismic waves are radiated (Angevine et al., 1982). On other planets where water is absent, it is not clear that fault slip necessarily produces detectable seismic events!

## 8-2-4 Dislocation Creep

Ironically, the faulting phenomenon gives us visible evidence of tectonic activity, but no tectonics would be possible if the friction equation cited above were universally obeyed. This is because the increase of normal stress with depth would make the necessary shear stresses impossibly large. Therefore, we infer that in every planet there is an additional (weaker) strain mechanism: intracrystalline dislocation creep (Weertman and Weertman, 1975). This process strains rocks aseismically and without changing their volume. Strain continues to accumulate as long as shear stress acts, at a rate given by:

$$\dot{e}_s = A\sigma_s^3 \exp\left(\frac{-E - PV}{RT}\right)$$

where $\dot{e}_s$ is shear strain rate, $R$ is the gas constant, $T$ is absolute temperature, $P$ is pressure, and the three constants $A$, $E$, and $V$ depend upon the mineralogy. On Earth the important creeping minerals are quartz (in continental crust), which softens significantly above 400°C, and olivine and pyroxene (in the mantle), which creep significantly above 800°C. The activation energies ($E$) in each case are so high that strain rate (at constant stress) more than doubles for each 20°C rise in temperature. Conversely, the strain rate is at least halved for every 20°C drop, and becomes undetectable below the temperatures stated.

## 8-3 TECTONIC MODELS

Kaula in Chapter 4 has summarized the formation and structure of the terrestrial planets, and has concluded that each of the inner five has a thick mantle layer in which olivine or pyroxene is important. Therefore, we shall boldly assume that within each planet the same deformation mechanisms operate and that there is a transition at about 600 ± 200°C between the mechanism of faulting and that of creep. The layer above this transition is called the *lithosphere*, which consists of a single spherical shell, or a set of spherical caps (separated by faults); heat transfer within it is by conduction, and the vertical temperature gradient is large. Below the transition is the *asthenosphere*. With depth, this creeping layer rapidly loses strength until convection begins; the temperature gradient is thereafter small (nearly adiabatic) except in boundary layers that form around any barrier to vertical flow. The depth of the lithosphere-asthenosphere boundary is a function of the strain rate, sinking when the rocks are deformed quickly, and rising in quiescent periods. But in foreseeable cases, it would never rise to the surface or sink to the core. This basic division forms the conceptual basis for the following nine possible mechanisms that might deform or modify a planetary surface.

### 8-3-1 Nonuniform Contraction

Some overall temperature change is inevitable in the history of a planet, and except in the case of a special coincidence, thermal expansion (or contraction) of the interior does not match that of the surface. Within the interior, creep will relax any shear stresses quickly, leaving only a uniform anomaly in the pressure (with respect to lithostatic pressure). In turn, this interior pressure anomaly causes deviatoric stress in the thin, stiff lithosphere. These horizontal tensions (or compressions) in the lithosphere are greater in size than the pressure change, as the stress in a balloon exceeds the gas pressure within it. Stresses are amplified by approximately the ratio of planetary radius to lithosphere thickness. If that ratio is 100 (as in the Earth), any differential temperature change of 25°C between lithosphere and asthenosphere results in stresses exceeding the frictional-sliding limit of the lithosphere. The likely result would be a distributed set of faults through the lithosphere, in a variety of directions, with a spacing determined by the ability of the lithospheric blocks to withstand secondary internal stresses caused by the change in planetary radius (see Fig. 8-1a). If the contraction of the interior were greater than that of the lithosphere, these faults would be thrusts, dipping about 30° and pushing up ridges. The opposite situation would form normal faults, dipping 60° and dropping down linear blocks to form valleys. The height of these topographic features would depend upon the small difference in strength between faulted and unfaulted lithosphere and is thus very difficult to predict.

### 8-3-2 Despinning

The lithosphere of a planet may also cease to "fit" if the planet's rate of rotation changes, because this changes the degree of ellipticity required for a *lithostatic* (shear-stress-free) state. If inelastic tidal strains slow the rotation, the radius of curvature of the lithosphere will be forced to decrease near the poles and increase near the equator.

Melosh (1977) has analyzed this complex problem and found that when a planet has a thin lithosphere, the lithosphere strains elastically and does very little to retard the change in ellipticity. The resulting stresses are then dependent only on the ellipticity changes and not on lithosphere thickness or depth. This means that the upper parts fault, whether or not these faults cut through the entire lithosphere. The resulting faults are east–west-trending normal faults at high latitudes, intersecting strike-slip faults trending N60E and N60W in the midlatitudes, and (possibly) north–south-trending thrust faults near the equator (see Fig. 8-1b). Naturally, an admixture of nonuniform contraction can

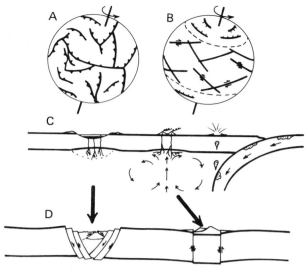

Fig. 8-1 Four types of planetary tectonics that are not primarily convective: (A) Nonuniform contraction produces a global fault network with no preferred orientation as the lithosphere is expanded or contracted to fit; (B) despinning produces conjugate strike-slip faults in the equatorial zone and normal faults near the poles; (C) volcanism can result from pressure-release melting after impact cratering (left) or during convection (center) or because crustal material or volatiles are convected back to the interior (right); (D) isostatic adjustment through faulting may involve slumping to fill a crater (left) or cylindrical faulting to adjust to volcanic loads (right).

change the pattern by expanding either the thrust-faulting or the normal-faulting regions. However, the predominance of east-west compression over north-south compression at all latitudes remains, and the resulting global orientation pattern of fault trends indicates that despinning has occurred.

It should be obvious that both of the mechanisms above operate very slowly, so their effects will not be noticed if the planet surface is modified by the faster processes below.

## 8-3-3 Volcanism

Melting of the interior is not something that a theoretician would predict, given a planet-sized lump of uniform, monomineralic rock. Instead, one would expect that slow heating from radioactivity and other causes (see Chapter 12 by Peale) would raise internal temperatures until the strong thermal activation of creep enabled convection to remove heat just as fast as it was produced. Similarly, it has been shown that steady fault slip cannot melt rocks, because of the thermal-softening effect (Yuen et al., 1978). Therefore, melting requires special circumstances, of which three classes are presently known.

The first is melting due to large meteorite impacts, which were probably far more common before 4 Gy. Contrary to intuition, the impact itself produces limited melting of the crater floor, because the total melt produced in escape-velocity impacts with terrestrial planets is only about twice the volume of the meteor (O'Keefe and Ahrens, 1977). Furthermore, most of this melt is splashed out along with the solid ejecta. The more important effect is that crater excavation immediately reduces the pressure on the mantle beneath the floor. Since the melting points of rocks are elevated by pressure, this sudden drop could melt asthenosphere that was already hot. According to this analysis, crater flooding may be taken as a rough indication that the lithosphere thickness at the time of impact was less than one crater diameter (see left-hand side of Fig. 8-1c).

The second melting mechanism relies upon volatiles (e.g., $H_2O$, $CO_2$) that are more soluble in melts than in solid rocks. If these are present (see Chapter 7 by Fanale), the melting point is depressed (without much effect on the creep strength), and the convective cycle may involve limited melting in the hot upwelling zones (see center of Fig. 8-1c). This is believed to be the cause of midocean-rise volcanism on Earth. Generally, the presence of such volatiles implies either incompleteness of planetary differentiation or convective overturning that returns volatiles to the interior from the atmosphere or ocean. The Earth presents an example of volatile recycling.

Third, a differentiated planet with a crust that was formed from a low-melting fraction of the mantle may be volcanic if convection draws small pieces of crust back into the interior. Although the geochemical evidence is mixed, it is possible that this effect contributes to subduction-zone volcanism on Earth (see right-hand side of Fig. 8-1c).

### 8-3-4 Isostatic Adjustment

The principle of *isostasy* is useful to use in approaching the study of vertical tectonic movements. It holds that any planet with a weak lithosphere (or none) tends to adjust itself until all its provinces "float" on the soft asthenosphere at their natural equilibrium level. These levels are lower where the crust is dense or thin, and higher where it is less dense or thicker. Technically, balance is achieved when all vertical columns everywhere have the same weight per unit area of base, as (hypothetically) measured at some common depth in the asthenosphere. It follows that every high mountain or plateau must be balanced by low-density (hot) mantle or an extra thickness of relatively low-density crust below it. These possibilities are illustrated in Fig. 8-4 in Sec. 8-4-3 later in this chapter. These low-density bodies at depth are referred to as the *compensation* of the surface topography, which is said to be *perfectly compensated* when isostasy holds exactly.

However, isostasy resembles a human law more than a natural law, in the sense that it is an ideal from which there are many departures. On Earth, for example, we often find that features either less than 30 km in width or less than $10^7$ y old are partially to totally uncompensated. The explanation is that a strong lithosphere can violate isostasy and hold topography above or below its isostatic level. An important part of tectonics is the study of the deformation of the lithosphere by vertical loads arising from lack of perfect compensation. These loads are often thought of as positive (as when volcanism adds mass to a planet's surface), but they may also be negative (as when cratering or erosion remove mass). Finally, a net load may be created if mantle convection changes the subsurface compensation without changing the topography.

In the absence of subsurface seismic-structure information, it is difficult and controversial to determine the true size of these loads. An integration of the topography fails to account for the anomalous subsurface densities that often reduce the net load, so it would seem that gravity would give a better indication. On the other hand, the load indicated by the integral of the gravity anomaly depends upon the part of the surface included in the integration and vanishes when this surface includes the whole planet. Hence any interpretation must be based on a geologic model and remains subject to revision. Specialists in this field attempt to distinguish between alternative models by studying how and why the ratio of gravity anomaly to topography varies with wave-length.

As soon as the load is imposed, dislocation creep at all levels begins to relax shear stress locally, so that it must be carried by shallower, colder levels of the lithosphere. This process decelerates but never ends, so theoretically all cases that are not disrupted go through an initial "elastic-bending" phase followed by faulting, collapse of the topography, and unbending. On Earth we have ancient shorelines establishing convenient datum surfaces with which to study the bending and infer lithospheric thicknesses. On other planets, only the second phase that produces visible faults is open to study.

These faults can be classed into two groups according to the direction of rock movement. The first consists of cylindrical faults with a vertical slip direction that surround the load and drop (or raise) the inner topography closer to its isostatic level. The second set occur only if the central region is hot and weak, a mechanical "hole" in the lithosphere. In this case, the anomalous pressure existing between the shallow load and its deeper compensation also acts horizontally to drive the lithosphere radially away from topographic highs (into lows). The associated map pattern of faults may include circumferential thrust (normal) faults, radial (normal) thrust faults, conjugate logarithmic spirals of strike-slip faults, or a combination, depending on local topography and temperature patterns (Fleitout and Froidevaux, 1982). Such fault patterns may be expected to spread radially outward at a decreasing rate until the central topographic or thermal anomaly disappears. However, if the lithosphere is thick, this process may take much longer than the present age of the solar system.

One final point about isostasy is that where there is volcanism, it may help a distant observer to estimate the thickness of the lithosphere or at least to put a lower limit on it. The argument is that the hot magma within a volcano is weak and will behave according to fluid laws. In particular, it will rise no higher than its isostatic level. If we take the top of the highest volcano to be the isostatic level of the magma, and the height of the surrounding plains to be the isostatic level of the cold lithosphere, then the difference between them ($\Delta h$) is just given by

$$\Delta h = T \left( \frac{\rho_L - \rho_M}{\rho_M} \right)$$

where $T$ is lithosphere thickness, $\rho_L$ is lithosphere density, and $\rho_M$ is magma density. If we make a further assumption that the magma and lithosphere are of similar chemical composition, we can apply terrestrial values of 4 to 10 percent for $(\rho_L - \rho_M)/\rho_M$ (Daly et al., 1965) to other planets. Then $T$ will be from 10 to 25 times the size of $\Delta h$.

## 8-3-5 Convection

Current theoretical models of the thermal histories of the terrestrial planets (Toksöz and Johnston, 1977) show that each should now have an asthenosphere in active convection, regardless of the planet's formation temperature. However, this review concerns only those convection systems that break and displace the surface, allowing remote sensing. Classical convection theory is of limited help, as it is just beginning to address the three major complications that may restrict planetary convection to the interior. One is the differentiation of a crustal layer whose compositional buoyancy overwhelms the thermal density anomalies tending to cause overturn. The other complications are the extreme temperature dependence of rock strength and the extreme nonlinearity of the creep law (i.e., lower effective viscosity at high stress); both of these make planetary convection more episodic and bistable than classical linear models. That is, substantial potential energy may accumulate before some large perturbation allows surface convection to begin; thereafter convection may be self-sustaining for some time by providing the heat and stress that keep surface viscosities low; and when it has dissipated the potential energy, it may "freeze up" and stop (or change form) with equal suddenness. Therefore, in searching for evidence of the following five forms of surface convection (see Table 8-2), we should not expect to see them in action throughout the whole surface or the whole history of a planet.

TABLE 8-2 Types of Planetary Convection

|  | Requires | Produces |
|---|---|---|
| Homogeneous | Hot formation or radio-activity | Young or mixed-age surface, lineated or polygonal topography |
| Layered | Global crust of low viscosity | Young or mixed-age surface, lineated or polygonal topography |
| Plate-tectonic | Strong boundary layer, fault-weakening | Young or mixed-aged surface, lineated topography with assymetrical trench-arc convergence zones, strike-slip faults |
| Delaminating | Crust with low-viscosity basal layer | Migrating regional uplifts (surface may be old) |
| Plumose | Heating from below (core, layered mantle) | Local uplift of spot or ramp (surface may be old) |

**Homogeneous convection** *Homogeneous convection* is the type observed in laboratory experiments on homogeneous fluids. It could only occur in a planet if the volume of crust were small and if the formation of a strong upper-boundary layer were suppressed (by a high surface temperature or some nonthermal weakening mechanism). Upwelling and downwelling are both locally symmetrical, so the flowlines may form closed loops in a pattern with long-term stability. Experiments show that a pattern of horizontal rolls converts to a pattern of hexagonal cells as convection gets more intense; however it can safely be predicted that even a weak surface lithosphere (which resists stretching more than bending) forces rolls to be dominant in the case of the planets. Thus, a lineated topography with alternating ridges and trenches in a young surface is produced (see Fig. 8-2A). The gravity field in this case is difficult to predict, because the primary convective mass anomalies cause deflections of the surface with an opposing gravity effect; these deflections have less excess mass per unit area but are closer to any observer. All that can be predicted is that the gravity anomaly is less than (or opposite from) that produced by the topography alone.

If a small amount of surface crust were added to this system, it would be swept into the downwelling depressions, but its relative buoyancy would presumably keep it on the surface. This would result in small patches of old surface surrounded by young surface.

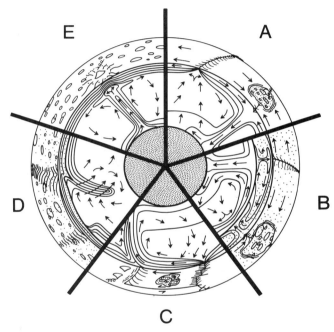

Fig. 8-2  Five types of planetary convection. Inner ring is the core, intermediate ring is a section through the mantle, and outer ring gives an oblique view of the surface. Solid curves are isotherms, and arrows indicate flow. Crustal material is dotted. Circle pattern on surface indicates impact craters and ancient surface. A = homogeneous, B = layered, C = plate-tectonic, D = delaminating, and E = plum convection. All cases except E are heated by radioactivity in the mantle; E is heated from the core below.

**Layered convection** *Layered convection* results when the amount of crust is increased to form a planetwide layer, but the lithosphere thickness is much less than the crustal thickness. Now the pattern of horizontal rolls appears in each layer, and the requirement of continuity of velocity across the boundary produces an out-of-phase pattern (see Fig. 8-2B). The upper, crustal layer is cooler, but does not necessarily have higher viscosities, since density, melting temperature, and creep-softening temperatures are correlated in rocks. Independent of viscosity, the thin upper layer would be less unstable than the thick mantle, and its convection would be "passive" (i.e., driven by shear stresses transmitted across the boundary); therefore the width of the cells would probably be determined by the mantle layer. At the surface, this case would look like homogeneous convection, except that the surface rocks would not be dense enough to account for the planet's known mass and inertia values without invoking an unusually large core. The prediction of gravity for this case is impossible without a number of constraints, because there are two layers of (opposed) thermal anomalies and two interfaces with (opposed) deflections to consider—almost *any* ratio of gravity to topography could result. As in the case of homogeneous convection, there might be "rafts" of lithosphere preserving patches of old surface above the downwelling zones.

**Plate-tectonic convection**  *Plate-tectonic convection* is a style no one would have predicted. As we know it on Earth, it involves the formation of stiff, cold upper thermal boundary layers (lithosphere) onto which the crustal layers are strongly bonded by high viscosity. Two special circumstances make it possible for downwelling to occur anyway. First, the low-density crust of the continents is distributed unevenly about the globe (and is cold and stiff enough to stay that way), so that many downwelling zones can operate for a long time without encountering this "indigestible" substance. When they do, they stop, and new sites are formed elsewhere. Second, the basalt crust of the ocean basins undergoes a phase change to become a garnet-bearing rock called *eclogite* at $10^9$ Pa pressure while it is being carried downward. This makes it even denser than the mantle below; hence it only hinders downwelling in the earliest stages and later promotes it (Ahrens and Schubert, 1975).

Plate tectonics shares the aspect of linear topographic features with homogeneous and layered convection. It differs in having long strike-slip faults along plate edges, which should be arcs of small circles. The general form of downwelling zones is less predictable, since such Earth-style features as volcanic arcs and fore-arc basins are not an essential part of the mechanism. There should, however, be a trench of some kind, with a consistent asymmetry caused by the fact that only one plate is being recycled into the interior (see Fig. 8-2C).

Significantly, no one has yet produced a laboratory model or computed simulation of convection in a homogeneous fluid that has the plate-tectonic style. In order to obtain great rigidity within lithospheric plates and also allow for relative movement of these plates, it is necessary to introduce heterogeneous bands of weak material (Schmeling and Jacoby, 1981). On Earth these bands are strike-slip faults and the thrust faults below the ocean trenches, which are many times weaker than plate interiors (Bird, 1978; Lachenbruch and Sass, 1980). The mechanical reason for this is not known, but most theories involve the physical or chemical effect of water in pores within the fault zones.

If such weak faults did not exist, the lithosphere on Earth would cease to move and would cool until it formed a single global shell capable of obscuring the convection below. However, there are still two ways in which such "buried" convection can be expressed at the surface—delamination and plume convection.

**Delamination**  *Delamination* can occur if there are two lithospheres (on top of the crust and mantle layers, respectively) separated by a lower crust which is hot and viscous. Then, there may be a localized and rapid overturn beneath the crust that drops away the cold thermal boundary layer of mantle lithosphere and replaces it with hot asthenosphere from the mantle interior (see Fig. 8-2D). In order for this process to begin, some force must rift the lithosphere so that the asthenosphere can intrude, and then the lower-crust viscosity must be low enough to let the intrusion spread faster than it cools (Bird, 1979). Once it is underway, the surface expression is an elevated plateau with straight or gently curved boundaries, ringed by a deep linear depression (Bird and Baumgardner, 1981). The lateral growth of this plateau is at a rate of centimeters per year. There would be no deformation of the (old) surface, and subsequent cooling would return that surface to its original elevation in $10^7$ to $10^8$ y, so the detection of this process on other planets is more a theoretical than a practical possibility.

**Plume convection** *Plume convection* is another consequence of the temperature dependence of viscosity, this time at the hot and fluid end of the scale. If a planet has heat sources in its core, or chemical stratification in its mantle, the upper-mantle layer is partially heated from below (as opposed to being heated only internally by radioactivity). In this case, theory and experiment both predict that the hot lower thermal boundary layer is unstable, giving off hot "plumes" that rise rapidly and independent of any general circulation. Studies show that plumes widen slightly as they rise, thus entraining any of the surrounding mantle that they have heated and maintaining long-term stability (Yuen and Schubert, 1976). The lowered viscosity inside the plume allows rapid upflow (as in a pipe) with negligible cooling; such flows are so stable that the surrounding circulation of the whole mantle can tilt them by more than 45° before they break down and form a new path to the surface.

Plumes are important in planetary tectonics because they are a concentrated heat source of great power. Where they meet the lithosphere from below, it is thinned. This produces a surface uplift in addition to any caused by the low density of the plume itself. Melting and volcanism are possible, as the very hot plume extends itself to lower-pressure environments. A narrow volcanic edifice upon a broad swell is the likely result (see Fig. 8-2E). However, mantle circulation may cause the "hot-spot" to move in jumps, as mentioned above. In that case there could be a chain of isolated volcanoes sitting on a broad linear "ramp" formed by the combination of plume movement and thermal subsidence of the old inactive "hot-spots."

## 8-4 ANALYZING THE PLANETS

Having (in theory, at least) reduced the possibilities for planetary evolution to combinations of nine basic mechanisms, let us examine each planet in turn to see which styles have dominated. This travelogue is arranged in the order of apparent tectonic complexity, which is almost the same as the order of planetary radii (see Fig. 8-3).

### 8-4-1 The Moon

Of the terrestrial planets other than Earth, we have the most information about the Moon. Because some of the *Apollo* missions included landings, it was possible to measure surface properties (e.g., moonquakes and rate of heat flow from the interior) and to return samples for chemical analysis and dating. These are extremely valuable supplements to remote observation.

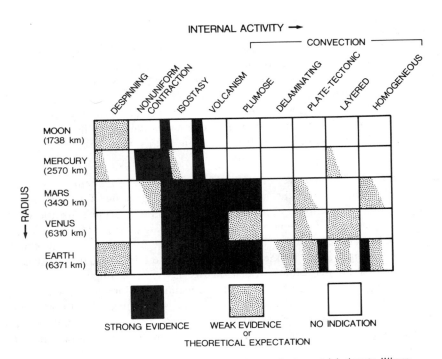

Fig. 8-3 Summary of inferred tectonic styles on the terrestrial planets. Where data permit, placement of shading and blackening in each box roughly indicates the portion of the planet's history concerned (left to right). Empty boxes do *not* indicate that a particular style is precluded by evidence, only that there is presently no indication that it has been operative.

The history of the Moon is actually less controversial than that of the Earth, because it involves relatively few provinces, they are perfectly exposed, and they are peppered with craters that give relative ages. The first known event is the segregation of a crust from an early "magma ocean" about 4.6 to 4.2 Gy (see Chapters 4 and 5 by Kaula and DePaolo, respectively). When this ocean began to freeze, the heavier minerals sank, but the lighter feldspar crystals (2.63 to 2.68 g · cm$^{-3}$) floated upward to concentrate in a global crustal layer about 60 to 100 km thick (Goins et al., 1977). Immediately the crust was subjected to heavy meteorite bombardment, not as a separate event but as a continuation of planetary accretion. It is likely that in the early stages this crust was hot enough at deep levels to creep and fill in the larger craters that cut deeply into it; this would explain the subdued form of the old craters on the side of the Moon away from Earth (e.g., the crater Mendeleev).

After the lithosphere had thickened to include the whole crust (about 3.9 Gy?) large impacts occurred, which made great holes in this crust to form the maria (see Chapter 5 by DePaolo). The first mystery about this event is why all the maria are on one side of the Moon. Did a single large planetesimal break up, in a pass close by the Earth, and then strike the Moon as a set of fragments? Did the Moon orbit close to the Earth with synchronous rotation, but with the present inner side formerly outward and exposed?

After these impacts, the craters were partly filled with low-viscosity basalt lavas, of density around 3.1 g · cm$^{-3}$. The principle of isostasy probably holds in the hot, disrupted region under any new crater, so the failure of this lava to completely fill the crater can be attributed to its density, which is significantly higher than that of the surrounding crust. The second problem is whether these basalts originated from pressure-relief melting under the craters, or whether a preexisting global layer of partial melt in the mantle (perhaps 200 km to 400 km down) was tapped to fill these holes. The first theory has difficulty explaining the 0.8-Gy duration of volcanism shown by crater patterns and radiometric ages (although it might take this long for an uplifted region of 200 km radius or more to cool down completely). The preexisting magma theory, on the other hand, has difficulty explaining just how the magma is moved and directed toward the crater at the proper time. Certainly, brittle tension cracks could not be opened at depths of more than a few tens of kilometers, due to the limited shear strength of all rocks.

Regardless of the magma source, the mare would initially have filled to an isostatic level. Later freezing and thermal contraction (volumetrically about 10 percent) would have lowered the mare surface, so that younger lavas (which continue to rise to the same absolute elevation) could continue flooding it with nonisostatic mass. Today we see *mascons* (concentrations of extra mass) over the maria, where the excess gravitational attraction of order 0.2 cm · sec$^{-2}$ implies an excess basalt thickness of order 1.5 km, or a total basaltic fill of roughly 15 km. If the lithosphere included the whole crust as we have assumed, it would already have been thick enough at the time of the last eruptions (about 3.1 Gy ago) to support this excess mass.

This discussion of the lack of perfect lunar isostasy does not imply that there was no isostatic compensation or deformation following cratering. For one thing, the large-diameter basins were almost certainly more than 15 km deep at the first instant; since their gravity anomalies are not larger, there must have been some isostatic adjustment to the excess mass of late basalt flows. Second, Orientale Basin is surrounded by three great circular ring faults, which are downthrown on the inside, between outward-tilted blocks; this is the pattern predicted by Melosh and McKinnon (1978) in their model of a lithosphere slumping radially into a hole. Subtle mare ridges in Mare Imbrium suggest a similar structure buried by lavas (Guest et al., 1979). If mantle upflow into the larger craters was a common process, then associated pressure-release partial melting could account for the basalts, and it would not be necessary to postulate a buried layer of partially melted mantle.

Solomon and Head (1980) have analyzed this process of incomplete isostatic adjustment of mare, in terms of the bending stresses created in an elastic lithosphere loaded by mascons and overlying a soft asthenosphere. They show that the radial stress becomes most tensile at a certain distance from the maria, which coincides with the concentric lunar rilles. *Rilles* are long, sinuous valleys that may be created by normal faulting; as the crust is stretched, a narrow central block bounded by two parallel faults drops lower than its surroundings. The distance of the datable rilles from their associated mare provides a measure of lithospheric thickness. This thickness seems to have varied geographically from about 25 km to 75 km during the time span of 3.8 Gy to 3.4 Gy before present. In contrast, it had been essentially zero at 4.2 Gy ago.

If this rate of thickening had continued to the present, the lunar lithosphere would be about 150 km to 450 km thick today. However, we would expect the rate of thickening to decrease, for two reasons. First, the thickening lithosphere itself slows the rate of heat flow outward to space, because it depresses convection to greater depths in the interior. Second, the decreasing total heat flow from the planet may eventually approach the rate of heat production by internal radioactivity, so that a steady state with a constant lithosphere thickness is asymptotically approached. In fact, from the two measured heat-flow values (average 18 mW · m$^{-2}$), we would estimate the Moon to have about a 100-km-thick global lithosphere today. The absence of distributed faulting of the nonuniform-contraction type indicates that lunar temperatures are now stable (Solomon and Chaiken, 1976). The lithosphere is strong enough to prevent internal convection from showing any surface effects. Since the Moon has lost any volatiles it had (see Chapter 7 by Fanale) and does not have a large core (or perhaps any core) to heat its mantle from below, there is no theoretical reason to expect melting and volcanism now or in the future. In terms of surface tectonics, it is a dead planet.

As a footnote, we should mention the exciting discovery of moonquakes at depths from 700- to 1000-km, whose location, timing, and occasional reversal of slip direction (!) all reveal them to be driven by monthly tidal stresses (Cheng and Toksöz, 1978). The nature of the mechanism that allows faults to slide at such high pressure is an important riddle in rock mechanics. However, these moonquakes do not indicate the extension of lithosphere (in the tectonic sense) to that depth, but only that the rocks in the asthenosphere have a Maxwell viscoelastic relaxation time of one month or more. That is, when stressed rapidly (by the monthly tides), these rocks do not have time to creep and behave as elastic or brittle. Yet under slow steady stresses they may well flow and convect. This is possible if their viscosities are above $10^{17}$ Pa-sec, a constraint easily satisfied if the asthenosphere of the Moon has the same temperature and viscosity as that of the Earth ($10^{21}$ Pa-sec).

In fact, the existence of heat-flow measurements allows an interesting comparison to be made between the interiors of the Earth and the Moon. Laboratory and computer models of convection have led to the development of a number of proportionalities or "scaling laws" of great generality, and these may be applied to planetary convection as well. The ratio of interior viscosities of two planets can be shown to be roughly the inverse cube of the heat-flow ratio, multiplied by the ratio of their gravities. This analysis gives a lunar mantle viscosity about four times that of the Earth; which in turn suggests a temperature only 70 to 190°C lower in the mantle of the Moon. Thus, we believe that the Moon's asthenosphere is convecting beneath its rigid outer lithosphere, although this cannot be directly observed or verified. This convection must be remarkably close to steady state, because even a very careful analysis finds no evidence of nonuniform expansion or contraction since 3.9 Gy (Golombek and McGill, 1983).

## 8-4-2 Mercury

We know very little of Mercury except its average density, weak magnetism, and what was seen of the surface in the two near approaches by *Mariner 10*. Through careful analysis of the overlapping landforms in these images, a tentative tectonic history has been assembled by Howard et al. (1974), Strom et al. (1975), Murray et al. (1975), and Dzurisin (1978). The major elements appear to be (1) differentiation of crust, mantle, and core; (2) cooling of a volcanic surface during heavy meteorite bombardment; (3) faulting of the lithosphere by despinning and contraction strains; (4) volcanic and isostatic (?) response to a major meteorite impact; and (5) quiescence, with minor meteorite erosion. Because much of this history is shared by the Moon, we here focus on the unique features of Mercury, which are its volcanic highlands, the Caloris Basin, and strong compression of the lithosphere.

On the Moon, the early crust of the highlands was cratered to saturation with little sign of continuing volcanism. Mercury's intercrater plains, which are the oldest surviving provinces, contain numerous "ghost craters" that were overlapped by flows of almost equal antiquity (Guest et al., 1979). The suggestion is that the early thin crust was denser than the lavas, allowing it to be isostatically depressed beneath them and recycled into the interior. For this to be true, any density difference resulting from the different compositions of the crust and the lavas would have to be smaller than that produced by the effect of temperature on density (up to 10%). In other words, the early crust and the lavas had similar compositions. This is the first major difference from the Moon, where a thick feldspar crust grew on the early magma ocean and later controlled the basaltic volcanism coming out of the mantle by blocking its ascent with a lower-density layer. On Mercury the situation was different: Either the lavas were derived by remelting of the crust, or else no crust ever formed.

The latter option is dubious at best. The albedo and color of Mercury's surface today have been interpreted as indicating a feldspar-rich composition (Hapke et al., 1975). Also, if comparative planetology has any merit, Mercury must have also had a magma ocean. (It is heavier, closer to the Sun, and has a large core.) So it would be hard to explain why feldspars never rose to choke the surface of Mercury. (Kaula, in Chapter 4, suggests that Mercury condensed with too little silica to form feldspars at all, which is a possible rebuttal to this argument.) However, it is far simpler to assume that the early lavas on Mercury came from remelting of even earlier crust, due to heating from strong convection in the mantle below. The overturn of the crust by volcanism would then be a form of layered convection early in the planet's history. When the lithosphere thickened to include the whole crust, this phase came to an end.

Toward the end of the heavy meteorite bombardment, Mercury suffered an especially large impact by a planetesimal, which created the huge impact scar of 1300-km diameter known as the Caloris Basin. Knobby and lineated terrains antipodal to Caloris are attributed to focusing of shock waves created by this impact (Strom, 1979). From the previous discussion (and drawing analogies to the Moon), we would expect such a large impact to make a hole through the entire thickness of the lithosphere. Indeed, the interior of Caloris today is largely a smooth plain interpreted as lava flows. These flows subsided as they cooled, and the decreased surface area of the outer skin (as it increased its radius of curvature within relatively rigid crater walls) led to wrinkling of the surface (Hapke, et al., 1975).

Because no late lava flows continued to erupt and smooth and level this surface (as on the Moon), it may be that the impact postdated the solidification of any global magma layer, and that the flooding occurred by the mechanism of pressure-relief melting discussed above. Unfortunately, the lack of late flows means that no uncompensated load should have formed within Caloris as it cooled; thus it cannot be used to test for the presence of isostatic-adjustment tectonics.

After the formation of a lithosphere and the Caloris impact, Mercury was deformed and faulted on a grand scale. Very spectacular "lobate scarps" up to 3 km tall and 500 km long are known to mark thrust faults because they cut through and shorten older craters (Strom et al., 1975). These scarps occur at all latitudes and, although sinuous, generally have trends within 45° of the North-South principal trend (Dzurisin, 1978). There may also be a preferred orientation of lineaments with trends N60°W and N60°E all around the planet (Dzurisin, 1978), which, because of their straightness and conjugate pattern, would probably be strike-slip faults. Both features could have been formed simultaneously in a stress field where east–west compression was the greatest and the vertical compression was the least. This would be evidence for despinning of Mercury (which now rotates only three times in every two orbits) combined with nonuniform contraction.

However, Strom (1979) argues that the lineations (if real) are older than the lobate scarps. If there were a clear division of the ages, this would invalidate the despinning model. But a combined model, with early despinning accompanied by nonuniform contraction (which continued after despinning) would be consistent with all our present knowledge. The necessary radius change has been estimated by Strom et al. (1975) to be 1 to 2 km. This could be achieved thermally by cooling the whole interior about 40° to 160°C, relative to any cooling of the lithosphere. Or the same contraction could have occurred in response to a smaller temperature drop, if the liquid-solid boundary in Mercury's large core migrated outward by some 10 to 20 km (Solomon and Chaiken 1976). Whatever combination of these two effects is responsible, the implied cooling is quite modest and does *not* preclude a substantial asthenosphere today.

### 8-4-3 Mars

In the exploration of the solar system, Mars gave us the first exciting look at another planet with active tectonics. Today the activity is modest, probably confined to volcanism, landsliding, and faulting in the Tharsis region. Yet there is reason to think that Mars underwent a more active early stage of horizontal lithospheric movements and "continental drift."

One of the remarkable facts about Mars is its division into two hemispheres of different elevation and age. The southern and higher terrain is the older, on the basis of crater counts—although not so old as the lunar highlands, if meteorite fluxes were equal. It has not yet been sampled by landers, as both *Viking* missions went to the safer northern plains. Still, we can determine two important things about the south by examining the large, circular impact basins known as planitae (Hellas, Argyre, and several more subdued planitae). First, the large basin relief of 2 km to 6 km (Christensen, 1975) suggests that the impacts blasted holes through an early crust that was much less dense than the mantle-derived lavas that quickly flooded the floors. If this crust did form in the same way as the lunar highlands, then it was probably once distributed uniformly over the whole planet, in that unrecorded (but hypothesized) time when the lithosphere was thinner than the crust. Second, we can infer from the ring-fault structures around basins as small as 120 km in diameter (Guest et al., 1979) that these craters penetrated through the lithosphere (Melosh and McKinnon, 1978), which therefore could not have been more than 60 km thick during these late impacts.

After most of the basins had formed, when the lithosphere was perhaps Earth-like in thickness, something happened to remove this crust from the northern half of the planet. Today the northern plains (Vastitas Borealis) are 5 km lower, less heavily cratered

(Condit, 1978), and appear to be floored by lava flows (Scott, 1978) of basaltic appearance (Binder et al., 1977) and density (Moore et al., 1977). It is inconceivable that these lavas are so much denser than the Martian mantle that they have depressed the early crust 5 to 10 km down and out of sight. It is equally unlikely that the early crust of the north was eroded away, because there is no possible place for storage of such a volume ($2 \times 10^9$ km$^3$) of light material. Therefore, it must have moved *laterally* to accumulate in the south.

The most plausible cause for this is that Mars had an early period of homogeneous convection involving the surface while the lithosphere was still thin. This would have produced drifting "continents" of low-density crust, whose trailing margins would preserve the normal faults and grabens formed when they rifted away from another continental mass. In fact, the "fretted terrain" at the highlands–north plains boundary looks very similar to seismic-reflection profiles of the Red Sea and Atlantic margins on Earth (after water and sediments are stripped away). Where continents coverged, such structures would have been crushed, and younger mountain chains uplifted, as the buoyant continents "choked" the downwelling zones and forced the convection to reorganize. Despite billions of years of cratering, erosion, and isostatic adjustment, the sympathetic eye can still find lineations in the topography of the southern highlands that may represent these old continental sutures (e.g., west and north of Hellas, north-east-trending; or west of Chryse, east-west-trending). Of course, detecting subtle lineations on Mars is an old, but not necessarily venerable, tradition!

In this view, the north plains are like seafloor on Earth, created by partial melting and volcanism at sites of upwelling. The reasons no topographic rises are visible today are that the system has been stopped for over 2 Gy and the lithosphere thickness has equalized. What stopped the overturn was apparently the collision of all the original crust into a southern "supercontinent," like Pangaea on Earth, which formed at 0.2 Gy. The present assembly is not necessarily the only or the first one; it would be the particular one that occurred after the lithosphere reached a critical thickness that prevented convection, once stopped, from reinitiating itself.

To gauge the lithosphere today, we can use the great active (or recently extinct) volcano Olympus Mons. Dividing its height of 25 km by the maximum plausible lava-rock density contrast of 10 percent, we conclude that the lava column must be confined in a strong lithosphere to at least 250 km in order to erupt. Such a thick lithosphere would have tremendous strength and could easily account for the confinement of planetary convection within the interior today. In fact, some have speculated that it was strong enough to support Olympus, the three volcanoes of the Tharsis Ridge, and the vast elevated terrain surrounding them for well over 1 Gy, while developing only minor surficial faults.

The great mound of Tharsis Montes corresponds to about 90° in longitude, and the mound rises above its surroundings to heights variously estimated as 9 km or 4 km (Downs et al., 1982). Sitting atop it are the three volcanoes Arsia, Pavonis, and Ascraeus Mons, which rise an additional 15 km above the mound. (Their collective mass is dwarfed by the excess material of Tharsis Montes.) Most early interpretations of this topography assumed that the surface of Tharsis is only thinly covered with lava flows and that the original flat surface of Mars has been warped upward by internal convection. Now, however, evidence is accumulating that volcanism in the Tharsis region has continued over most of the planet's history (Plescia and Saunders, 1982), so that the mound may be formed from a very thick accumulation of lava flows (Solomon and Head, 1982b; Willeman and Turcotte, 1982) and the original surface may actually be warped down beneath this load.

The support of the Tharsis bulge is the great tectonic problem of Mars, and (as the Introduction in this chapter warned) both the amount of load and the response are controversial. Figure 8-4 shows the major possibilities schematically in cross section. A valuable constraint is the gravity field, which shows a broad positive anomaly above Tharsis, but with only about 30 percent of the size it should have in the totally uncompensated case (Christensen and Balmino, 1979). This allows one to choose any model between the extremes of 70% shallow (crustal) compensation (with considerable stress in the lithosphere) and total compensation distributed deeply (from near the surface to a depth of 300 km to 400 km). Crustal compensation could be created simply by volcanic thickening of preexisting crust; for plausible density contrasts, some 20 km to 50 km of thickening would be required. The cause for deep compensation could be replacement of colder, denser lithosphere under Tharsis and Olympus by broad instrusions of hotter, lighter asthenosphere with the same composition (see Fig. 8-4). Such a "hot spot" could only be maintained by steady convective upwelling of a great plume or plumes beneath the region, which naturally would also explain why melting and volcanism are so severely restricted today. (Theoretical studies have not yet advanced far enough to predict whether such plumes under thick motionless lithosphere are stable for billions of years; however we can predict that if such a plume did move, the abandoned thermal dome would decay down to average elevation with a very long characteristic time of 1 to 2 Gy. There is no clear sign of this on Mars.)

One additional clue to the solution is the observation of radial rilles all around Tharsis, extending out at least 60° from its center. These rilles are thought to be downfaulted blocks caused by normal faulting and crustal stretching in the direction circum-

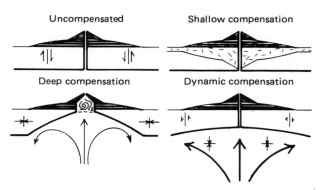

Fig. 8-4 Four hypotheses for the support of the Tharsis bulge and superjacent volcanoes on Mars. Heavy line indicates lithosphere-asthenosphere boundary; crust has random dashes. Paired arrows indicate stress, and single arrows show convective flow. Gravity and fault patterns seem to favor a mixed model, with part of the load uncompensated, and part compensated deeply by lithospheric thinning. The role of shallow compensation is uncertain.

ferential to Tharis (Carr, 1974). Their close relationship to the volcanic loading is shown by a mutual overlapping of rilles and flows extending far back in Martian history (Plescia and Saunders, 1982). Also, the great rift of Valles Marinerus, which is probably just the longest of these radial grabens, displays an intercutting pattern of scarps, landslides, and craters, which implies that it is still active (Blasius et al., 1977). A number of investigators (Banerdt et al., 1982; Solomon and Head, 1982b; Willeman and Turcotte, 1982), have attempted to compute models of a thick lithosphere subject to various vertical loads that will match these observations as well as Martian topography and gravity data. One clear consensus is that the uncompensated part of the Tharsis load is laterally supported by the strength of the surrounding lithosphere rather than by very fast (and viscous) convection in the asthenosphere. This is necessary to explain the great extent of the rilles and the activity of the Valles Marineris. Less obvious is whether the partial compensation is shallow (thickened crust), deep (heated and thinned lithosphere), or both. However, the simple fact of continuing volcanism implies a great and concentrated heat source, probably an upwelling mantle plume. Therefore deep compensation in the form of lithospheric thinning under Tharsis must play some role. Banerdt et al. (1982) have shown mathematically that some component of deep compensation is helpful in explaining the extension of the rilles into very small radii around Tharsis; in a pure bending model these would be suppressed at radii of less than 30°. Another factor that could be important is superposed nonuniform contraction, with the exterior lithosphere cooling more than the internal asthenosphere. This would tend to encourage normal faulting all over the surface of Mars (Solomon and Chaiken, 1976), reinforcing the effect of the weight of Tharsis.

## 8-4-4 Earth

The Earth today has a relatively thin lithosphere. Under the oceans it thickens as the square root of its age, from zero at spreading centers to 30 km in the oldest parts of the Pacific (Goetze and Evans, 1979). In continents, the situation is more complex. Earthquake locations and extrapolated creep laws both show a lithosphere-asthenosphere transition *within* the crust at 10- to 25-km depths (Sibson, 1982). However, the uppermost mantle just below the Moho is also a layer of great strength (in regions of low to average heat flow), because mantle olivine has a higher creep-softening temperature than crustal quartz and feldspars. This stiff upper mantle restricts the flow of the soft lower crust enough to maintain the integrity of continents. However, it must be remembered that when orogenies occur within continents, the upper crust and upper mantle generally decouple and deform with quite different styles (e.g., Bird and Rosenstock, 1984).

Plate tectonics as the dominant style of the Earth has been abundantly documented elsewhere (Le Pichon et al., 1976). Likewise, the existence of upwelling plumes beneath Hawaii, Iceland, Yellowstone, and a dozen other localities is no longer controversial. Together, these two mechanisms (and the associated isostatic adjustments) explain almost all of the topography, geology, and volcanic deposits forming today. A few remaining anomalies require other mechanisms from the theoretical list; the author has suggested delamination as the cause for migrating waves of uplift and volcanism within the western United States (Bird, 1979) and put forward a variant of layered convection with "invisible" subsurface mantle downwelling as an explanation for the seismic velocity anomaly beneath the Transverse Ranges of California (Bird and Rosenstock, 1984). Despinning of the Earth is well documented (Munk and McDonald, 1960), but associated faulting has never been confidently identified amid the welter of larger deformations produced more rapidly by plate tectonics. Similarly, the fragmentation of Earth's lithosphere into many changing plates makes it unlikely that nonuniform contraction effects will ever be identified, although it has almost certainly occurred (Schubert et al., 1980). This mechanism was once the last resort of theoreticians (e.g., Bucher, 1933, p. 123) for all unexplained orogenies, before the understanding of plate tectonics. This should serve as a caution that our present synthesis may also be subject to drastic revision.

The present synthesis of Earth tectonics must also be qualified by saying that it may only apply to the last billion years; this is the greatest age of rocks obviously formed by sea-floor spreading or subduction. The student of early Earth history is in some ways much poorer in data than the student of the Moon: although all the desired rock samples are available, there is absolutely no information about the topography, gravity, heat flow, seismicity, crustal thickness, planetary radius, or rotation rate prevailing at the time they formed. The rocks of the Archean Era (i.e., those from the first half of Earth history) available for study show an association of "greenstone" belts (metamorphosed basalt flows and sandstones) and grey tonalitic gneisses with younger granite intrusions. These bodies of rock have small dimensions (5 to 50 km) and ambiguous structural relations (in West Australia the greenstones surround the gneisses; in the Canadian shield, vice versa). Stratigraphic relations show that the gneisses rose above sea level and the greenstone belts formed below (Kummel, 1961). This observation shows that sea level was high enough to cover some, but not all of the area of continental crust (as it is today); therefore we can infer that the ratio of Archean crust to ocean was similar to the present one.

With regard to the global tectonic regime, there are two other useful constraints: The surface temperature has remained in the general area of 0°C to 100°C during the last 3 Gy (to permit the survival of algal life), and the heat flow has declined by a factor of at least three in the same time due to radioactive decay and planetary cooling (Schubert et al., 1980). If we apply the scaling laws of convection theory, the joint implication of these facts is that at 3 Gy ago, the mantle was about 200°C hotter and the speed of convection about five times as great.

The effect that this would have on surface tectonics depends on how much continental crust and ocean had already formed, but this is not known. One extreme possibility is that all continental crust differentiated immediately and has merely been reworked and remelted ever since. In that case, the higher Archean heat flow would have led to excellent crust-mantle decoupling in the soft lower crust and the development of layered convection. The granite bodies would mark sites of upwelling in this model, while the greenstone belts would represent crustal downwelling sites that continuously engulfed and recycled sediments. (No obvious source for the basalt flows is implied by this model.) Conversely, if the amounts of Archean crust and ocean were small, then erosion and deposition would have redistributed the crust into thin patches. The temperature at the base of the crust would have been low despite the high heat flow, and the crust would have been firmly attached to the convecting mantle. We would then interpret the granitic blocks as microcontinents assembled and sutured together in the downwelling zones of a homogeneous or plate-tectonic convection system arrested by buoyancy, and interpret the greenstone belts as remnants of lost oceanic crust that once separated them.

We have no record at all of events on Earth with ages exceeding 3.8 Gy, because no rocks have survived unaltered. Presumably the Earth is at least as old as the Moon, and a significant period of time passed when crust was either not differentiated or not preserved here. The argument that it never formed is weakened by evidence of early crustal layers on the three smaller (and cooler) planets we have already examined; therefore destruction or reheating is indicated. Under the layered-convection model proposed above, it could be argued that it took 1 Gy for the top of the crust to form a lithosphere strong enough to resist recycling through the crustal convecting layer, with consequent resetting of radiometric ages. Or, in the microcontinent model, it would be argued that the size and number of crustal blocks had to grow to a critical size before their buoyancy was great enough to prevent their being dragged down into the mantle as a part of the circulating noncontinental lithosphere. A decision between these models will probably have to be made on the basis of geochemical arguments constraining the history of the total mass of differentiated crust. Preliminary indications from neodymium-isotope ratios (DePaolo, 1981) are that crustal fractionation has been gradual (see Chapter 5 by DePaolo); this favors the microcontinent model of homogeneous or plate-tectonic style.

The homogeneous convection style, in which downwelling is symmetrical, is a better choice for the Earth's first eon, because it avoids two objections that have been raised to early plate tectonics. These are the absences of (1) Archean "blueschists," to indicate subduction, and (2) basalts intruding older basalts, to indicate sea-floor spreading (Kröner, 1979). Blueschists of the recent plate-tectonic age form when sediments are caught in a zone of recirculating flow between one subducting and one stationary plate (Cloos, 1982) where the ratio of temperature to pressure is especially low. But sediments caught between two plates that were both sinking would not have been returned to the surface (except in an adjacent upwelling zone that would be too hot to form blueschist). Ocean floor, on the other hand, may have been formed in the Archean Era, but not preserved. Even today the uplift of ocean-floor basalts onto continents is a rare event and is probably caused by abortive plate-tectonic subduction of a buoyant continent beneath an oceanic arc. Without asymmetrical subduction, these rocks would never have been lifted high enough to be visible on land today.

As the Earth gradually cooled and its lithosphere thickened, the forces required to bend downwelling lithosphere through a tight curve down into a trench would have increased roughly at the cube of its thickness (Goetze and Evans, 1979). (Specifically, the bending moment per unit length of trench is proportional to the cube of thickness.) On the other hand, the convective density anomalies that gravity acts upon to create these bending moments would have increased more slowly as the square of lithosphere thickness. This means that the convecting system would have been thrown out of balance by the thickening of the lithosphere, and bending at trenches would come to consume more and more of the energy available. There would have been a decrease in the stability of the old homogeneous convection with respect to plate tectonics, in which one plate is not deformed and the other bends very gently through a large arc. The resulting change of style apparently occurred around 1 Gy ago. The new style of plate tectonics would not have been possible without an ocean, because it is the pore water dragged into the fault between the converging plates that reduces friction and makes the sliding possible (Bird, 1978). If Earth had been as dry as Mars, its horizontal tectonics might also have come to an end eons ago!

In comparing Earth to other planets, it is essential to realize how much our hydrosphere controls the phenomena we regard as normal. Water not only lubricates subduction, but is directly responsible for the generation of magmas in downwelling zones, despite their anomalously low temperatures. Many believe that the present continental crust is not primary, but was formed by differentiation of basaltic magmas in these arcs (Wyllie, 1982) and by further sorting during weathering and sedimentation. Likewise, "oceanic" crust (i.e., that found between the granite "continents") could be different on a dry Earth, because the present mantle temperature is probably too low for convection alone to produce any melting in the absence of volatiles. The major physiographic characteristic of the Earth, the neat separation of oceanic and continental crust into two levels, is also a physical result of the ocean. Without it, the upper level would vanish as erosion and isostatic rebound distributed any crust more uniformly around the planet. In short, our planet might have a "frozen" global lithosphere, no horizontal tectonics, no continents or ocean basins, but only a monotonous thin covering of sand on an irregular ancient surface, occasionally relieved by plume volcanism and vertical tectonics. This exercise in imagination is a very good introduction to our neighboring planet and closest analogue.

### 8-4-5 Venus

Venus is the most poorly known of the five planets because of its thick cloud cover, and thus the interpretation of its tectonics is still very primitive. Four successful *Venera* soft landings were made east of the elevated Beta Regio, but the chemical analyses returned are inconclusive; either a granitic or a basaltic crustal composition can be inferred. (Without knowing whether the rocks are igneous, metamorphic, or sedimentary, we can infer at least local differentiation and volcanism from the fact that either rock composition would yield the wrong total mass if extrapolated throughout the planet.) Next in terms of the resolution of view obtained are the Earth-based (Arecibo) radar observations reported by Campbell and Burns (1980). Despite coverage of almost half the surface at 5-km to 20-km resolutions, no evidence for a grid of linear faults has yet been found. Thus there is no reason to think that despinning and nonuniform contraction have been important in shaping the present surface of Venus. Isostatic readjustment cannot be properly studied without detailed topographic data over unambiguous volcanoes or impact craters, but we do have data on global topography from satellite altimetry (Masursky et al., 1980), and global gravity from satellite orbits (Sjogren et al., 1980), which can be compared. Although the two are correlated, the gravity anomalies over broad features are only 15 to 35 percent of the attraction of the topography (Sjogren et al., 1980), so at least partial isostatic compensation is required. Phillips et al. (1981) have made a strong case that the remaining gravity anomalies probably result from deep compensation at 85- to 150-km depths, rather than from uncompensated topography supported by a strong lithosphere.

The crucial remaining question is whether the surface of Venus is a global lithosphere, or whether it convects. Kaula and Phillips (1981) have searched the topographic data for any elevated linear features corresponding to plate-tectonic spreading centers; the result was that the "best candidates" have inconsistent elevations, a disorganized and disconnected map pattern, and a shape different from that predicted by convection theory. They took this to mean that homogeneous and plate-tectonic convection are absent, or at best they are slow and unimportant in transmitting the total heat flow. Additional arguments for a global-lithosphere model are based on some 50 quasi-circular features in the radar images, with diameters of 20 to 1200 km (Campbell and Burns, 1980). If these are interpreted as impact craters, and if meteorite-influx histories of Venus and the Moon are comparable, this implies that at least part of the crust on Venus is over 1 Gy old and cannot be involved in convection (Masursky et al., 1980). On this basis, Phillips et al. (1981) have presented a Venus model in which plume convection dominates. In their view, hot upwelling plumes lie beneath each of the major highlands (Ishtar Terra, Aphrodite Terra, Beta Regio) and provide both a magma source for construction of continental-sized volcano complexes, and the thermal buoyancy needed to isostatically support them. A difficulty with this model is that it cannot explain the deep curvilinear troughs (Artemis, Dali, Devana, and Diana Chasmae) adjacent to these highlands; they must be attributed to an earlier, arrested stage of surface convection. It is also difficult to explain why these have not been filled with sediment, considering that dense Venusian winds should be able to transport sand and gravel up to 1 cm in diameter (Warner, 1982).

It must be admitted that the global-lithosphere or "Martian" model *is* consistent with known rock mechanics, even though Venus is much hotter than Mars. Consider the steep slopes at the edge of the terrae, such as the "Vesta Rupes" southwest of Ishtar, which slopes down 3 km in 350 km. Using approximate values of gravity and density, we find the shear stress beneath this slope must increase with depth at a rate of $0.24$ MPa $\cdot$ km$^{-1}$ down to about 200-km depths. Meanwhile, temperature increases from the surface value of 460°C at a gradient of roughly 15°C $\cdot$ km$^{-1}$ (Phillips et al., 1981). Combining these conditions, we find that the strain rate from dislocation creep becomes significant ($10^{-8}$ y$^{-1}$) at only 28-km depth with an olivine or gabbroic flow-law, or at only 9 km with a granitic flow-law (Tullis and Yund, 1981). In other words, a thin lithosphere must act alone to support the horizontal normal stress that keeps Ishtar from spreading and collapsing. This stress (approximately = 3 km height $\times$ 8.9 m $\cdot$ s$^{-2}$ gravity $\times$ 3000 kg $\cdot$ m$^{-3}$ density $\times$ 100 km depth-to-compensation $\div$ lithosphere thickness) becomes 280 to 900 MPa, depending on the composition at these depths. The lower figure is consistent with the limits imposed by frictional sliding, and the upper figure is not. That is, Ishtar Terra *could* be a static feature *if* it contains less than 9 km of granitic crust. A purely gabbroic crust is quite possible, considering the speculations of the previous section on the control of planetary differentiation by the presence or absence of water. Perhaps the 3-km height of Ishtar is determined by the fact that gabbroic crustal roots (with a 10 percent density contrast) cannot extend more than 30 km down and still remain part of the lithosphere.

Although this "Martian model" is self-consistent, the linear troughs are a perplexing puzzle and lead us to entertain alternatives. Perhaps some of the circular features are not impact craters, for they fail to show the expected central depression (Masursky et al., 1980) or circular scarps (Campbell and Burns, 1980). (Solomon et al. (1982) have pointed out that significant viscous relaxation of crater forms is to be expected on Venus; this type of argument can be used either to excuse the peculiar shape of these "craters" or, conversely, to argue that they must be very young.) On a planet that is so hot, we can equally well speculate that these are volcanic features, such as collapsed calderas or eroded intrusion domes on Earth. Or, with more data we may find that these *are* ancient impact craters, but that they are confined to areas of ancient surface surrounded by mobile belts. In that case crustal convection is possible (especially with a granitic composition). Solomon and Head (1982a) have already shown that all the other arguments against convection are inconclusive.

It was already mentioned above that layered convection can produce any finite ratio of gravity to topography, including the one observed on Venus. Likewise, a wide variety of topographic forms is possible. The spreading centers in the crustal layer would necessarily be the highest points, but could have different absolute heights depending on the thickness of the mantle boundary layers they overlay. Furthermore, the colder, more brittle upper crust could fault and deform in a chaotic fashion in mantle-convergence-crustal-divergence zones, where mantle heat flow is low (see Fig. 8-5). Conversely, the thickening of crustal upper-boundary layers could be arrested or reversed by high heat flow near mantle-divergence zones, producing more ductile flow and gentler topography. Additional complications could result if the crust was insufficient to cover the whole planet: Where mantle upwelling swept aside the crust, the topography could be W-shaped in profile (Tellus Regio?), and where mantle convergence was unobscured by the crust, there would be symmetrical outer rises about a central trough (Dali Chasma?). (Alternatively, the chasms could indicate grabens formed by crustal extension above mantle downwelling.) Such steep asymmetrical scarps as the Vesta Rupes could be formed by intracrustal thrust faults, like the Himalayan front on Earth, that maintain steep slopes by constantly sweeping up a soft viscous crust.

These widely divergent possibilities for Venus could be constrained by just a little more data. A clear view of the surface (which will soon be obtained by synthetic-aperture radar imaging from orbit) might clear up the crucial questions of origin and distribution

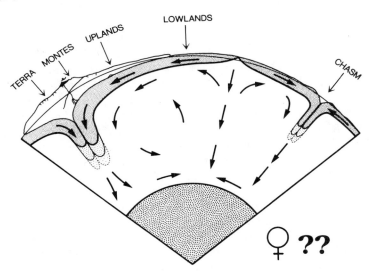

Fig. 8-5 Cartoon of a layered-convection model for Venus, with considerable vertical exaggeration. Crust and asthenosphere are white, and mantle lithosphere is shaded. This model is only applicable if Venus has a large amount of silicic crust, which is mobile at near-surface temperatures. It predicts large amounts of young surface, hence could be invalidated if circular features on Venus prove to be ancient impact craters and are uniformly distributed.

of the small circular features. Wider sampling of crustal composition by landers would be very helpful, as would technological innovations allowing a heat flow experiment to survive on the surface. The exploration of Venus (see Chapter 17 by Kivelson) remains one of the most exciting and accessible problems in space.

## 8-5 CONCLUSIONS

A detailed survey of the inner five terrestrial planets for tectonic styles suggested by theory reveals possible examples of every type (see Fig. 8-3). Each planet has had volcanism at least in its early history; on the Moon this may have been a passive response to crater excavation, but on the other planets it seems to indicate a retention of volatiles that depressed the melting point enough to match the temperatures in upwelling convection.

On Mercury, the lavas may have been remelted crust, recirculating in the upper level of an early layered-convection system. Each planet also shows evidence of at least partial isostatic compensation of larger and older features, confirming that the theoretical concept of a thickening lithosphere over a convecting interior is universally valid for stony planets. In each case, the gradual loss of primordial heat of accretion and differentiation and the decay of radioactive elements have decreased the planetary heat flow and thickened the lithosphere, leading to a decline of tectonics. Mercury and the Moon have lithospheres that were never broken by internal convection, but only by the small strains of isostatic adjustment, nonuniform contraction, and perhaps despinning. Mars probably had homogeneous convection in its early history, which collected the ancient (granitic?) crust in the South and formed waterless basaltic "oceans" in the North; but today horizontal motions have stopped, and even the plumes that rise to the surface at Tharsis do not wander. The Earth has probably differentiated its oceans and continental crust gradually over time. The simultaneous thickening of the crust and the lithosphere forced two changes of convective style, from steady-state homogeneous convection to "jerky" homogeneous convection that was locally arrested by buoyant continents, and then from homogeneous convection to the asymmetric plate-tectonic style. It is likely that surface convection on Earth today is only possible because of water lubrication of subduction zones. Venus presents a dilemma, because its greenhouse effect has boiled the water off a surface that would otherwise be much like the Earth's. If this happened early, then Venus probably has only basaltic crust of great strength and high melting point, so that its tectonics are like those of Mars despite a 500°C difference in surface temperature. However, if granitic crust formed before the water was lost, it would be highly mobile today. The continent-sized *terrae* would then have to be dynamic features, probably overturning piles of buoyant crust swept together by mantle convergence in the lower level of a layered-convection system.

## 8-6 ACKNOWLEDGMENTS

As a newcomer to the field of extraterrestrial geology, I must stress that none of the data and few of the concepts in the review are original. The authors of the references cited (which are only representative and not exhaustive) were the original developers of the models presented here, excepting only delamination and layered convection. Very valuable discussions that helped to form my ideas were held with W. M. Kaula about Venus, with D. G. Sandwell and P. R. Mullen about Mars, and with W. G. Ernst and S. Peacock concerning the early history of the Earth. W. M. Kaula and S. C. Solomon each reviewed the manuscript and suggested a number of significant improvements.

## 8-7 REFERENCES

Ahrens, T. J., and G. Schubert, 1975, Gabbro-eclogite reaction rate and its geophysical significance, *Rev. Geophys. Space Phys.*, *13*, 383-400.

Angevine, C. L., D. L. Turcotte, and M. D. Furnish, 1982, Pressure solution lithification as a mechanism for the stick-slip behaviors of faults, *Tectonics*, *1*, 151-160.

Banerdt, W. B., R. J. Phillips, N. H. Sleep, and R. S. Saunders, 1982, Thick shell tectonics on one-plate planets—applications to Mars, *J. Geophys. Res.*, *87*, 9723-9734.

Binder, A. B., R. E. Arvidson, E. A. Guinness, K. L. Jones, E. C. Morris, T. A. Mutch, D. C. Pieri, and C. Sagan, 1977, The geology of the *Viking Lander 1* site, *J. Geophys. Res.*, *82*, 4439-4451.

Bird, P., 1978, Stress and temperature in subduction shear zones—Tonga and Mariana, *Geophys. J. Roy. Astr. Soc.*, *55*, 411-434.

Bird, P., 1979, Continental declamation and the Colorado Plateau, *J. Geophys. Res.*, *84*, 7561-7571.

Bird, P., and J. Baumgardner, 1981, Steady propagation of delamination events, *J. Geophys. Res.*, *86*, 4891-4903.

Bird, P., and R. W. Rosenstock, 1984, Kinematics of present crust and mantle flow in southern California, *Bull. Geol. Soc. Am.*, *95*, 946-957.

Blasius, K. R., J. A. Cutts, J. E. Guest, and H. Masursky, 1977, Geology of the Valles Marineris: First analysis of imaging from the *Viking 1 Orbiter* primary mission, *J. Geophys. Res.*, *82*, 4067-4092.

Bucher, W. H., 1933, *The Deformation of the Earth's Crust*, Princeton Univ., Princeton, N.J., 518 pp.

Campbell, D. B., and B. A. Burns, 1980, Earth-based radar imagery of Venus, *J. Geophys. Res.*, *85*, 8271-8281.

Carr, M. H., 1974, Tectonism and volcanism of the Tharsis region of Mars, *J. Geophys. Res.*, *79*, 3943-3949.

Cheng, C. H., and M. N. Toksöz, 1978, Tidal stresses in the Moon, *J. Geophys. Res.*, *83*, 845-853.

Christensen, E. J., 1975, Martian topography derived from occultation, radar, spectral, and optical measurements, *J. Geophys. Res.*, *80*, 2909-2913.

Christensen, E. J., and G. Balmino, 1979, Development and analysis of a twelfth degree and order gravity model for Mars, *J. Geophys. Res.*, *84*, 7943-7953.

Cloos, M., 1982, Flow melanges—numerical modeling and geologic constraints on their origin in the Franciscan subduction complex, California, *Bull. Geol. Soc. Am.*, *93*, 330-345.

Condit, C. D., 1978, Distribution and relations of 4- to 10-km-diameter craters to global geologic units on Mars, *Icarus*, *34*, 465-478.

Daly, R. A., G. E. Manger, and S. P. Clark, Jr., 1965, Density of rocks, in Handbook of Physical Constants, *Memoir Geol. Soc. Am.*, (S. P. Clark, Jr., ed.,) *97*, 19-26.

DePaolo, D. J., 1981, Nd isotope studies—some new perspectives on Earth structure and evolution, *EOS Trans. AGU, 62,* 137-140.

Downs, G. S., P. J. Mouginis-Mark, S. H. Zisk, and T. W. Thompson, 1982, New radar-derived topography for the Northern Hemisphere of Mars, *J. Geophys. Res., 87,* 9747-9754.

Dzurisin, D., 1978, The tectonic and volcanic history of Mercury as inferred from studies of scarps, ridges, troughs, and other lineaments, *J. Geophys. Res., 83,* 4883-4906.

Fleitout, L., and C. Froidevaux, 1982, Tectonics and topography for a lithosphere containing density heterogeneities, *Tectonics, 1,* 21-56.

Goetze, C., and B. Evans, 1979, Stress and temperature in the bending lithosphere as constrained by experimental rock mechanics, *Geophys. J. Roy. Astr. Soc., 59,* 463-478.

Goins, N. R., A. M. Dainty, and M. N. Toksöz, 1977, The deep seismic structure of the Moon, *Proc. Lunar Planet. Sci. Conf., 8th,* 471-486.

Golombek, M. P., and G. E. McGill, 1983, Grabens, basin tectonics, and the maximum total expansion of the Moon, *J. Geophys. Res., 88,* 3563-3578.

Guest, J., P. Butterworth, J. Murray, and W. O'Donnell, 1979, *Planetary Geology,* John Wiley, New York, 208 pp.

Hapke, B., E. Danielson, Jr., K. Klaasen, and L. Wilson, 1975, Photometric observations of Mercury from *Mariner 10, J. Geophys. Res., 80,* 2431-2443.

Howard, H. T., et al., 1974, Mercury's surface—preliminary description and interpretation from *Mariner 10* pictures, *Science, 185,* 169-179.

Kaula, W. M., and R. J. Phillips, 1981, Quantitative tests for plate tectonics on Venus, *Geophys. Res. Lett., 8,* 1187-1190.

Kröner, A., 1979, PreCambrian crustal evolution in the light of plate tectonics and the undation theory, *Geol. Mijn. 58,* 231-240.

Kummel, B., 1961, *History of the Earth,* Freeman, San Francisco, 610 pp.

Lachenbruch, A. H., and J. H. Sass, 1980, Heat flow and energetics of the San Andreas fault zone, *J. Geophys. Res., 85,* 6185-6222.

Le Pichon, X., J. Franchetau, and J. Bonnin, 1976, *Plate Tectonics,* Elsevier North-Holland, New York, 311 pp.

Masursky, H., E. Eliason, P. G. Ford, G. E. McGill, G. H. Pettengill, G. G. Schaber, and G. Schubert, 1980, *Pioneer Venus* radar results—geology from images and altimetry, *J. Geophys. Res., 85,* 8232-8260.

Melosh, H. J., 1977, Global tectonics of a despun planet, *Icarus, 31,* 221-243.

Melosh, H. J., and W. B. McKinnon, 1978, The mechanics of ringed basin formation, *Geophys. Res. Lett., 5,* 985-988.

Moore, H. J., R. E. Hutton, R. F. Scott, C. R. Spitzer, and R. W. Shorthill, 1977, Surface materials of the *Viking* landing sites, *J. Geophys. Res., 82,* 4497-4523.

Munk, W. H., and G. J. F. McDonald, 1960, *The Rotation of the Earth: a Geophysical Discussion,* Cambridge Univ., Cambridge, 323 pp.

Murray, B. C., R. G. Strom, N. J. Trask, and D. E. Gault, 1975, Surface history of Mercury—implications for terrestrial planets, *J. Geophys. Res., 80,* 2508-2514.

O'Keefe, J. D., and T. J. Ahrens, 1977, Impact-induced energy partitioning, melting, and vaporization on terrestrial planets, *Proc. Lunar Sci. Conf., 8th,* 3357-3374.

Phillips, R. J., W. M. Kaula, G. E. McGill, and M. C. Malin, 1981, Tectonics and evolution of Venus, *Science, 212,* 879-887.

Plescia, J. B., and R. S. Saunders, 1982, Tectonic history of the Tharsis region, Mars, *J. Geophys. Res., 87,* 9775-9792.

Schmeling, H., and W. R. Jacoby, 1981, On modelling the lithosphere in mantle convection with non-linear rheology, *J. Geophys., 50*, 89-100.

Schubert, G., D. Stevenson, and P. Cassen, 1980, Whole planet cooling and the radiogenic heat source contents of the Earth and Moon, *J. Geophys. Res., 85*, 2531-2538.

Scott, D. H., 1978, Mars, highlands-lowlands—*Viking* contributions to *Mariner* relative age studies, *Icarus, 34*, 479-485.

Sibson, R. H., 1982, Fault zone models, heat flow, and the depth distribution of earthquakes in the continental crust of the United States, *Bull. Seism. Soc. Am., 72*, 151-163.

Sjogren, W. L., R. J. Phillips, P. W. Birkeland, and R. N. Wimberly, 1980, Gravity anomalies on Venus, *J. Geophys. Res., 85*, 8295-8302.

Solomon, S. C., and J. Chaiken, 1976, Thermal expansion and thermal stress in the Moon and terrestrial planets: clues to early thermal history, *Proc. Lunar Sci. Conf., 7*, 3229-3243.

Solomon, S. C., and J. W. Head, 1980, Lunar mascon basins—lava filling, tectonics, and evolution of the lithosphere, *Rev. Geophys. Space Phys., 18*, 107-141.

Solomon, S. C., and J. W. Head, 1982a, Mechanism for lithospheric heat transport on Venus—implications for tectonic style and volcanism, *J. Geophys. Res., 87*, 9236-9246.

Solomon, S. C., and J. W. Head, 1982b, Evolution of the Tharsis province of Mars: the importance of heterogeneous lithospheric thickness and volcanic construction, *J. Geophys. Res., 87*, 9755-9774.

Solomon, S. C., S. K. Stephens, and J. W. Head, 1982, On Venus impact basins—viscous relaxation of topographic relief, *J. Geophys. Res., 87*, 7763-7771.

Strom, R. G., 1979, Mercury—a post-*Mariner 10* assessment, *Space Sci. Rev., 24*, 3-70.

Strom, R. G., N. J. Trask, and J. E. Guest, 1975, Tectonism and volcanism on Mercury, *J. Geophys. Res., 80*, 2478-2507.

Toksöz, M. N., and D. H. Johnston, 1977, The origin and development of the Earth and other terrestrial planets, *Proceedings of the Soviet-American Conference on the Cosmochemistry of the Moon and Planets*, (J. H. Pomeroy and N. J. Hubbard, eds.), NASA, Washington, D. C., 295-327.

Tullis, J., and R. A. Yund, 1981, An experimental study of the rheology of crustal rocks, in *Summaries of Technical Reports of the National Earthquake Hazards Reduction Program, U.S.Geol. Survey Open File Report 81-833*, 511-514.

Warner, J. L., 1982, Lunar and planetary science conference—Earth and Venus compared, *Geotimes, 27*, 16-19.

Weertman, J., and J. R. Weertman, 1975, High temperature creep of rock and mantle viscosity, *Ann. Rev. Earth Planet. Sci., 3*, 293-315.

Willeman, R. J., and D. L. Turcotte, 1982, The role of lithospheric stress in the support of the Tharsis Rise, *J. Geophys. Res., 87*, 9793-9801.

Wyllie, P. J., 1982, Subduction products according to experimental prediction, *Bull Geol. Soc. Am., 93*, 468-476.

Yuen, D. A., and G. Schubert, 1976, Mantle plumes—a boundary layer approach for Newtonian and non-Newtonian temperature-dependent rheologies, *J. Geophys. Res., 81*, 2499.

Yuen, D. A., L. Fleitout, and G. Schubert, 1978, Shear deformation zones along major transform faults and subducting slabs, *Geophys. J. Roy. Astr. Soc., 54*, 93-119.

James L. Gooding
Code SN2/Planetary Materials Branch
Solar System Exploration Division
NASA/Johnson Space Center
Houston, Texas 77058

# 9

# PLANETARY SURFACE WEATHERING

## ABSTRACT

Weathering is the collection of processes that, through atmosphere-surface interactions, leads to the decomposition or alteration of rocks, minerals, or mineraloids. Physical weathering can occur by frost riving, involving the destructive expansion of ice crystals grown in rock voids, by mineral riving, involving similar destructive growth of secondary mineral crystals, and possibly by other processes. Chemical weathering involves chemical reactions between minerals or mineraloids and planetary volatiles (usually $H_2O$, $O_2$, and $CO_2$) through oxidation, hydration, carbonation, or solution processes. Venus, Earth, and Mars all possess permanent atmospheres such that weathering should be expected to significantly affect their respective surfaces. In contrast, Mercury and the Moon lack permanent atmospheres but conceivably could experience surface weathering in response to transient atmospheres generated by volcanic or impact cratering events. Weathering processes can be postulated for other rocky objects including Io, Titan, asteroids, and comets. Analyses of weathering products formed on various planetary objects can be used to deduce environments of weathering and, hence, the histories of atmosphere-surface interactions and the incorporation of volatiles into planetary regoliths.

## 9-1 INTRODUCTION

*Weathering* refers to the collection of processes that, through interactions between a planetary surface and atmosphere, leads to the mechanical decomposition or chemical alteration of rocks, minerals, or mineraloids. As its name implies, weathering represents a natural attack by the elements of weather on objects that are exposed to them. On Earth, rain, ice, and oxygen in the air are the principal agents of weathering, which, through processes commonly aided by sunlight, are to blame for familiar outdoor annoyances such as the deterioration of paint and the rusting of iron-based metal objects. Although the processes of weathering may be different on other planets, we should expect that, in general, any rocky planetary surface exposed to a chemically active atmosphere for any length of time should experience weathering.

Among Earth-oriented geoscientists, weathering is acknowledged in various ways, ranging from its indictment as a pervasive nuisance that degrades important rock outcrops, to its veneration as the provider of soils from which most of humanity's food supply is ultimately derived. As exploration of the planets continues, igneous petrologists will undoubtedly adopt the former view, whereas future planners of *terraforming* and colonization may be drawn toward the latter view. The present article aims simply to review weathering as a general planetary phenomenon which, along with other geologic processes, must be understood before planetary histories can be fully deciphered.

By definition, the terrestrial-type planets of greatest interest in studies of weathering are Venus, Earth, and Mars. However, as will be discussed, even objects such as Mercury and the Moon, which do not possess permanent atmospheres, may still experience incipient weathering from gas clouds of volcanic or impact origin. In addition, smaller objects such as asteroids and comets may similarly experience short-lived weathering events. Finally, less familiar objects such as Jupiter's satellite Io and Saturn's satellite Titan may each, through interactions between its peculiar surface and atmosphere, be said to *weather*.

Figure 9-1 attempts to place weathering in proper planetary context. Weathering is an integral part of the surface geochemical cycle through which rocks degrade into soils and sediments, which, in turn, may become other rocks available for reprocessing by the same cycle. The role of weathering in planetary surface evolution may also be better appreciated by reference to the operational definitions given in Table 9-1. These

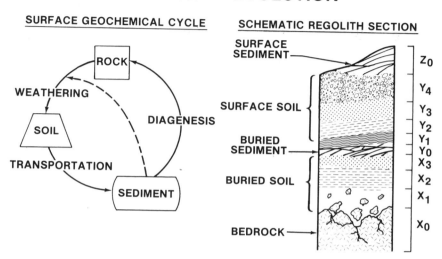

Fig. 9-1 The role of weathering in planetary regolith evolution. As part of the surface geochemical cycle, weathering leads to the physical and chemical decomposition of rocks and contributes to the formation of soils. In general, soils may form from either rocks or sediments, and a given section of regolith may contain more than one soil profile. For example, soil profile $X$, developed directly on bedrock, underlies more recent sediment which, after deposition, weathered to form soil profile $Y$. Very recent sediment $Z$ occurs on the surface and may later evolve into yet another soil.

TABLE 9-1 Important Definitions for Planetary Surface-Weathering Studies

| Term | Definition |
| --- | --- |
| Regolith | The unconsolidated layer of soils, sediments, and rock fragments that overlies bedrock |
| Soil | Unconsolidated material that forms in place by the comminution, decomposition, or alteration of rocks or material derived from rocks |
| Sediment | Unconsolidated material which forms by comminution or decomposition of rocks but which is transported from its place of origin and deposited elsewhere |
| Weathering | The collection of processes which, through interactions between surface and atmosphere, leads to the decomposition or alteration of rocks, minerals, or mineraloids and the possible formation of new phases |

definitions represent the author's attempt to formulate a consistent set of terms that reflect geological reasoning but which have general planetary applicability.

It should be noted that the term *soil* is not synonymous with *regolith*, although the distinction is not always adequately made. Similarly, not all loose, fine-grained material on a planetary surface may be soil. If it can be shown that the material did not form in place, then the material must have been transported from elsewhere and, thus, is more properly called a *sediment*. A commonly observed characteristic of soils on Earth is zonal structure that preserves some identifiable relationship to the underlying parent material (see right-hand side of Fig. 9-1). This structure should be expected in some form within true soils on other planets. Such zonation may be only weakly developed but, if present, may serve to distinguish soils from sediments. In those cases where soil cannot be unambiguously distinguished from sediment, more noncommittal descriptions such as *surface fines* or *surface material* might be preferable.

From the definitions in Table 9-1, it should be clear that weathering is not the only phenomenon that can generate soils. As on the Moon, soils and sediments may be produced almost exclusively by impact cratering. Also, weathering is defined in a manner that allows but does not require the involvement of chemical or biological processes. By so defining *regolith, soil, sediment,* and *weathering*, serious prejudices that were acquired during the birth of planetary geoscience on Earth can be substantially removed. Properly defined, these terms should be equally applicable to all terrestrial-type planets, including Earth, such that the goal of studying weathering as a planetary phenomenon becomes feasible.

## 9-2 METHODS OF STUDY

Weathering can be studied by three principal approaches:

1. Field and laboratory study of weathering environments and products
2. Laboratory simulations
3. Theoretical modeling

Method 1, ideally, should be the starting point for all studies but, in any case, it is the vehicle for testing ideas and predictions formulated by other methods. Unfortunately, we are not yet able to properly visit and sample all the planetary environments and materials of interest. Although the unmanned *Viking* and *Venera* missions (to be discussed later in this article) have provided important information about the weathering environments on Mars and Venus, respectively, we do not have documented Martian or Venusian samples available for study. Thus, method 1 is not yet practical for planets other than Earth but must eventually be included in the complete study plan for each object.

In the absence of samples for direct study, environments of interest can be experimentally simulated and their effects on various materials determined. Thus, for effects measurable within the lifetime of a patient observer, method 2 can provide very important information. However, natural processes that may be "rapid" relative to the lifetime of a planet may still require hundreds or thousands of years to produce measurable results and are, therefore, effectively beyond the reach of laboratory simulations.

Using available information about a given weathering environment, method 3 works from basic principles to define the set of possible processes and products which is likely to include those of real importance on the planet of interest. The possibility that a process operates can thereby be considered as an issue separate from that of the rate at which the process might occur. Obviously, rates are better studied by methods 1 or 2. As observational knowledge of the environment of interest improves, the theoretical models can become more specific and useful as guidelines for the design of new experiments to be performed by methods 1 or 2. Thus, theoretical modeling should be viewed as an iterative aid to observations rather than as a substitute for them.

## 9-3 TYPES OF WEATHERING

### 9-3-1 Physical

Physical weathering leads to the mechanical breakdown of rocks but without change in material composition. Given known differences in thermal expansivities among minerals, repeated cycles of rapid heating and cooling might be expected to create significant grain-boundary stresses that could lead to fracturing of the host rock. Although Grabau (1960, p. 31-34) summarized the conventional arguments for thermally induced rock weathering on Earth, its importance as a general planetary phenomenon remains poorly understood. Laboratory experiments by Griggs (1936) failed to produce noticeable degradation of granite even after simulated diurnal heat-cool cycles equivalent to many years of desert exposure on Earth. It is possible, though, that the rate of degradation was simply too slow to be noticed in Griggs' experiments and there remains some evidence that solar heating can mechanically decompose rocks on Earth (Ollier, 1963).

Upon first consideration, it might seem that the surface of the Moon might favor thermally induced fracturing of rocks. However, *Apollo* expeditions to the Moon found no convincing evidence for such breakdown of lunar rocks. Although the surface of the Moon may reach 384°K during lunar days and 102°K during lunar nights, the diurnal cycle is 27 days long (compared with 24 hours for Earth), so that heating and cooling of rock surfaces are relatively slow and, apparently, not destructively stressful.

Venus represents another extreme environment in which diurnal temperature variations are insignificant compared with the prevailing atmospheric temperature. Demonstrable evidence for physical weathering of rocks on Venus might indicate that sustained heat, as well as heating cycles, can be an effective weathering agent.

More important than simple hot-cold cycling in the physical weathering of rocks is the growth of foreign crystals along hairline fractures or grain boundaries in rocks. The freezing of water that has percolated into voids in a rock exerts a powerful force that can cause widening of the voids, and, ultimately, fragmentation of the rock. This process of *frost riving* derives its power from the peculiar physical properties of water, which include a specific volume increase (i.e., volumetric expansion) during the water-ice transition. Nitrogen, oxygen, carbon dioxide, and other common planetary volatile components all exhibit specific volume decreases (i.e., volumetric contraction) during freezing. Thus, water is the only common planetary volatile compound that should be expected to perform frost riving.

The work done by ice crystals in rock splitting is sometimes matched by that done by salt crystals or by other mineral crystals in suitable environments. The percolation of aqueous salt solutions into voids in rocks, followed by chemical reactions and evaporation of the water, can lead to similarly forceful volumetric increases as secondary (weathering product) mineral crystals (commonly halides, sulfates, or carbonates, or silicates), nucleate, and grow (Johnston, 1973; Whalley et al., 1982). The weathering process is then known as *mineral riving*. Recognizing the unusually broad solvent properties of water, it seems that no other liquid would facilitate mineral riving as does water. Thus, mineral riving is probably important only on planets that possess liquid water.

Both frost riving and mineral riving can undoubtedly contribute to the distinctive, nearly concentric patterns of fracturing and peeling, known as *exfoliation*, which characterize some mechanically weathered rocks (Ollier, 1971). Therefore, when exfoliated rocks are found on a planetary surface, physical weathering should be strongly suspected as a possible explanation and the possible role of water should not be overlooked.

### 9-3-2 Chemical

Chemical weathering occurs when minerals or mineraloids chemically decompose or react to form new phases. Because weathering processes are, by definition, surface-atmosphere interactions, chemical compositions of planetary atmospheres exert strong influence on the nature of chemical weathering. As shown in Table 9-2 the three most chemically reactive gases common to the atmospheres of Venus, Earth, and Mars are $O_2$, $H_2O$, and $CO_2$. Accordingly, three important varieties of chemical weathering reactions are

TABLE 9-2  Near-Surface Atmospheres of Terrestrial-Type Planets

|  | *Venus*[a] | *Earth*[b] | *Mars*[c] |
|---|---|---|---|
| TEMP. (°K) | 750 (477°C) | 288 (15°C) | 250[d] (−23°C) |
| PRESSURE (Pa) | $9.7 \times 10^6$ (96 atm) | $1.0 \times 10^5$ (1 atm) | $8.0 \times 10^2$ (0.01 atm) |
| % $CO_2$ | 96.4 | 0.03 | 95.3 |
| % $N_2$ | 3.4 | 78.1 | 2.7 |
| % $O_2$ | <0.004 | 21.0 | 0.13 |
| % CO | 0.003 | $<10^{-6}$ | 0.07 |
| % $SO_2$ | 0.02 | $<10^{-4}$ | − |
| % $H_2O$ | 0.14 | 0.8[e] | 0.03[e] |
| ppm $O_3$ | − | <0.1 | 0.03 |
| ppm Ne | 4 | 1800 | 2.5 |
| ppm Ar | 19 | 9300 | 16,000 |
| ppm Kr | <0.2 | 100 | 0.3 |
| ppm Xe | − | 8 | 0.08 |

[a] Hoffman et al. (1980); Marov (1978); Oyama et al. (1979).
[b] ICAO standard sea-level atmosphere, unpolluted, at 50% relative humidity.
[c] Hess et al. (1977); Owen et al. (1977).
[d] Northern Hemisphere summer maximum.
[e] Typical value; known to vary.

*oxidation, hydration,* and *carbonation* (see Table 9-3). When abundant liquid water is available, *solution* may represent a fourth important type of chemical weathering reaction (Table 9-3). Solution is distinguished from hydration (the incorporation of water into the structure of a solid) as the process by which a solid actually dissolves.

TABLE 9-3  Examples of Chemical Weathering Reactions

1. OXIDATION
   (1-1)  $4\ CaFeSi_2O_6 + O_2 \longrightarrow 2\ Fe_2O_3 + 4\ CaSiO_3 + 4\ SiO_2$
   (1-2)  $3\ CaFeSi_2O_6 + H_2O \longrightarrow Fe_3O_4 + 3\ CaSiO_3 + 3\ SiO_2 + H_2$
   (1-3)  $3\ CaFeSi_2O_6 + CO_2 \longrightarrow Fe_3O_4 + 3\ CaSiO_3 + 3\ SiO_2 + CO$
2. HYDRATION
   $3\ CaMgSi_2O_6 + 4\ H_2O \longrightarrow Mg_3Si_4O_{10}(OH)_2 + 3\ Ca(OH)_2 + 2\ SiO_2$
   $CaAl_2Si_2O_8 + 3\ H_2O \longrightarrow Al_2Si_2O_5(OH)_4 + Ca(OH)_2$
3. CARBONATION
   $CaMgSi_2O_6 + 2CO_2 \longrightarrow CaMg(CO_3)_2 + 2\ SiO_2$
   $CaAl_2Si_2O_8 + CO_2 \longrightarrow CaCO_3 + Al_2SiO_5 + SiO_2$
4. SOLUTION
   (4-1)  $CaMgSi_2O_6 + 4\ H_2O + 2H^+_{(aq)} \longrightarrow Ca^{2+}_{(aq)} + Mg^{2+}_{(aq)} + 2\ H_4SiO_{4\,(aq)} + 2\ OH^-_{(aq)}$
   (4-2)  $CaCO_3 + H_2O \longrightarrow Ca^{2+}_{(aq)} + HCO^-_{3\,(aq)} + OH^-_{(aq)}$

Among the four major categories of chemical weathering, oxidation is particularly noteworthy in that it can be driven by either $O_2$, $H_2O$, or $CO_2$, given suitable pressures and temperatures. Furthermore, oxidation, in the geologic context, is a highly selective process that is powerless over some phases but that readily destroys others. Phases may be subject to oxidation if they contain certain geologically important elements with variable oxidation states that can be controlled by weathering processes. The elements most notable in that regard are C, P, S, V, Mn, Fe, Co, Ni, and Cu. However, the phases most commonly and distinctively vulnerable to attack are those containing abundant iron either as $Fe^0$ or $Fe^{2+}$; $Fe^{3+}$ is the ultimate oxidation product.

In contrast with oxidation, the processes of hydration and carbonation are straightforward. Namely, only water can perform hydration and only carbon dioxide can perform carbonation. In the planetary geological context, solution refers specifically to water as the solvent so that solution reactions are also relatively unambiguous except that the rates and products of dissolution are commonly determined by the crystal chemistries of individual mineral surfaces (e.g., Berner et al., 1980).

The physical states of reactants and products are not specified for oxidation, hydration, and carbonation reactions in Table 9-3 because, in general, two different conditions might occur. Gas-solid reactions would involve the direct interaction of gases with minerals or mineraloids, whereas liquid-solid reaction would proceed with the gases dissolved in a suitable solvent, namely, water. Gas-solid and liquid-solid reactions may differ greatly in their thermodynamic and kinetic aspects, with liquid-solid reactions expected to occur faster and to greater degrees of completion. However, gas-solid reactions, though perhaps ponderously slow, may dominate in those environments that lack liquid water. Obviously, the fourth category of reaction, solution weathering, does not occur at all without liquid water.

The reactions listed in Table 9-3 use pyroxenes and a feldspar as common reactants to be expected among primary igneous rocks on the surfaces of terrestrial-type planets. The reactions so exemplified are chosen merely to illustrate the types of chemical interactions that are to be expected. More complex reactions and reactions involving other reactants (e.g., olivine or basalt glass) could easily be written and can be expected to occur in some weathering environments.

## 9-4 STYLES OF WEATHERING ON INDIVIDUAL PLANETS

### 9-4-1 Earth

As with all aspects of planetary science, our collective understanding of weathering phenomena began on our home planet, Earth, the reference point against which all other planetary weathering environments are compared. The complex and varied environments on Earth lead to equal complexity and variety among weathering processes and products. It is probably fair to say that, with few exceptions, the weathering phenomena to be found on other planets are analogous to or are subsets of those phenomena that occur on Earth. Accordingly, the styles of weathering on various planetary objects, to be discussed below, can be compared with that on Earth by a "scorecard" of the type given in Table 9-4.

TABLE 9-4 Styles of Weathering on Planetary Bodies Through Geologic Time

| Body | Physical Weathering By | | Chemical Weathering By | | | | |
|---|---|---|---|---|---|---|---|
| | Frost Riving | Mineral Riving | Permanent Atmosphere | Transient Impact Atmosphere | Transient Volcanic Atmosphere | Liquid Water | Biosphere |
| Mercury | | | | ± | ± | | |
| Venus | | | + | ± | ± | ? | |
| Earth | * | * | * | ± | ± | * | * |
| Moon | | | | ± | ± | | |
| Mars | ± | ± | + | ± | ± | + | ? |
| Asteroids | | | | ± | ± | ? | ± |
| Comets | ? | | + | ± | ? | ± | |
| Io | | | ? | ± | ± | | |
| Titan | | | ? | ± | ? | ? | |

Key: *known importance, + probable importance, ± possible importance, ? unknown importance

The principal determinant of weathering on Earth is the abundance of liquid water, a fact that cannot be overemphasized. Although Earth possesses many extremely arid deserts, no place on Earth's surface can have completely escaped visitation by liquid water through geologic time or on even shorter time scales. Even Earth's most arid spots must occasionally experience water as light rain, dew, or melted frost. Thus, when care is taken to separate mechanism from rate, it is clear that weathering on Earth is driven and styled by liquid water. Observational evidence demonstrates that frost riving and salt riving are active in deserts and that, in all climatic zones, liquid-solid reactions dominate chemical weathering. Thus, it can reasonably be expected that, where it occurs, liquid water will also dominate weathering on other planets.

All three analytical approaches described in Sec. 9-2 of this chapter have been successfully applied in studies of weathering on Earth. Method 1 is exemplified by the classical study of Goldich (1938), whereas method 2 is represented by the instructive works of Griggs (1936) and Correns (1963). The principles underlying method 3 can be found within the broader treatment of aqueous geochemistry by Garrels and Christ (1965). General aspects of weathering on Earth have been reviewed by Loughnan (1969), Ollier (1969) and Carroll (1970). Weathering of volcanic rocks to form soils, a topic of special importance in planetary science, has been reviewed by Ugolini and Zasoski (1979). Only key points digested from the above and other literature sources are summarized here.

Basically, the relative contributions of physical and chemical weathering are largely determined by topography and climate, with composition of the material being weathered having somewhat less importance. A terrane having steep surface slopes, occurring at high elevation, and exposed to a dry climate is most affected by physical weathering and yields relatively coarse-grained regolith rich in fragments of the source materials. In contrast, the same terrane constrained to low surface slopes, low elevation, and wet climate suffers much greater attack by chemical weathering, producing regolith rich in fine-grained material and secondary minerals or mineraloids. Naturally, the wide variation of topography and climate across Earth's surface produces wide variation in the ratio of physical to chemical weathering.

Earth's complexity assures that the surface-atmosphere interactions producing weathering can take many forms. Thus, because atmospheric precipitation is the ultimate source of ground water and ground ice, water-driven alteration processes within the regolith can properly be regarded as "weathering." In fact, it is well known that retention of pore water in subsurface soils on Earth facilitates weathering that is more rapid and complete than is generally found at the actual surface-atmosphere interface.

The surface geochemical cycle (see Fig. 9-1) is born by weathering of igneous rocks and, on terrestrial-type planets, the most abundantly exposed igneous rocks are probably mafic volcanic rocks akin to basalt. Chemical weathering of their constituent primary phases (olivine, pyroxene, feldspar, ilmenite, magnetite, glass, and sulfides) on Earth produces various secondary phases including oxides, phyllosilicates, and salts. Iron oxides (e.g., hematite), hydrous iron oxides (e.g., goethite), and clay minerals (especially kaolinites and smectites) are common weathering products of crystalline rocks. Glassy basalts commonly alter to poorly crystalline mineraloids such as palagonite and allophane and a variety of accessory minerals, including zeolites.

Throughout these alteration processes, chemical and isotopic changes accompany the mineralogical changes. Water-soluble species are leached from the parent material and transported away, ultimately to be deposited elsewhere as salts or other secondary minerals. Furthermore, the stabilities of various primary phases relative to weathering leads to nonuniform alteration with mafic phases commonly being altered faster than felsic phases. Furthermore, petrogenetically important isotopic ratios such as $^{18}O/^{16}O$ and $^{13}C/^{12}C$ can be changed by weathering reactions that tend to enrich some weathering products in the heavier isotopes as functions both of mineral type and of temperature. Also, strong sorbents such as clay minerals and zeolites tend to strongly bind certain ions while rejecting others so as to produce trace-element concentrations in weathering products that significantly differ from the trace-element concentrations in the original parent minerals. Consequently, the composition of weathering products derived from a polymineralic rock, in general, cannot be used as a reliable indicator of the bulk composition of the source rock because many fractionations may have occurred. However, by carefully analyzing the weathering behavior of many different basalts, for example, the chemical and isotopic compositional variety to be expected among basalt weathering products can be established, so that, given a new sample of basalt weathering products, important information about the particular weathering environment that produced it can be reconstructed.

## 9-4-2 Mars

Most of our collective knowledge about Mars was obtained from the tremendously successful spacecraft missions of *Viking 1* and *Viking 2*, which included the soft landing of two unmanned scientific stations on the Martian surface in 1976. We now know that the Martian permanent atmosphere is thin but oxidizing in nature and contains a variable but significant concentration of water vapor (see Table 9-2). Hence, weathering on Mars might be expected to resemble that on Earth. However, at all but the lowest elevations and except during times of maximum solar heating, the atmospheric pressure at the Martian surface is below that required to preserve liquid water as a thermodynamically stable phase. Consequently, the intensity of water-driven weathering is probably much lower in the current Martian surface environment than it is on Earth. Rates and mechanisms of aqueous alteration on Mars might be strongly influenced by the thicknesses (usually a few molecular layers) of films of "unfrozen" water that are known to occur at subfreezing temperatures in porous rocks and soils on Earth (Anderson et al., 1967).

The possible importance of salt riving on Mars was suggested prior to *Viking* (Malin, 1974). It was further reiterated by *Viking* results, suggesting high salt contents in Martian surface fines, as inferred from the measured abundances of Cl and S in those materials (Clark and Van Hart, 1981). Furthermore, the surface frosts formed during winter at the *Viking 2* landing site have been interpreted as combined $H_2O$ and $CO_2$ condensates (Wall, 1981). Thus, both frost riving and salt riving are probably viable processes on Mars, and it is not surprising that many of the rocks at the two *Viking* landing sites show evidence of physical weathering (Garvin et al., 1981). Examples of Martian rocks that may have experienced physical weathering are shown in Fig. 9-2.

Despite the fact that attempts to scrape material from Martian rock surfaces using *Viking* surface samplers showed the rocks to be strong and scratch-resistant (Moore et al., 1979), Strickland (1979) used enhanced color images obtained by the *Viking* landers to argue for the existence of weathered rinds on Martian rocks. If present, the rinds must be very thin and tightly bound to the rocks. In any case, the reddish-brown color of Martian surface fines and the visible to near-infrared reflectance spectra of reddish-brown regions on Mars (Singer et al., 1979) are consistent with the occurrence of abundant $Fe^{3+}$-bearing phases, probably produced by oxidative weathering of mafic igneous rocks.

The *Viking* lander experiments provided our closest approach to the use of study method 1—material sampling and analysis—in the investigation of weathering on Mars. However, they gave us no samples for direct study and, indeed, were never designed to study Martian weathering. Those limitations, and the crude similarities between the sur-

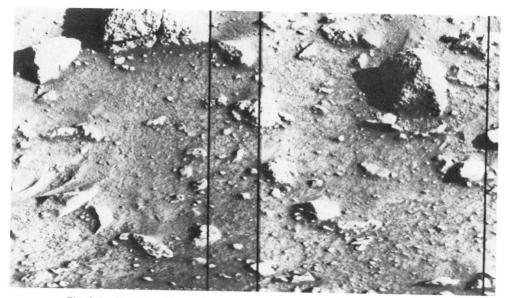

Fig. 9-2 Possible evidence for physical weathering of rocks at Chryse Planitia on Mars. The rock at the upper left may have been fractured by frost riving or salt riving. The rock at upper-right may have shed an arcuate array of fragments by exfoliation in response to riving by similar processes. The vertical black strips in the image are lines of missing data that were probably lost during transmission from Mars to Earth. (*Viking* Lander 1 image 12A081/012), NASA photograph.)

face environments of Mars and Earth, have led some researchers to pursue a modified version of method 1, namely, the field and laboratory study of Mars-like environments on Earth. Two of the most promising areas for study are the Dry Valleys of Antarctica (Morris et al., 1972; Gibson et al., 1983) and the summit area of the Mauna Kea, Hawaii, shield volcano (Ugolini, 1976; Japp and Gooding, 1980). Although such analog studies provide no direct information about Martian weathering, they provide a necessary base of experience for the interpretation of Martian weathering processes and products in proper geological context. The value of such studies will probably be realized only when we are eventually faced with the task of analyzing and interpreting returned Martian samples.

Method 2, experimental simulation, has been utilized in Martian weathering studies by a number of investigators since Huguenin's (1974) advocacy of solar ultraviolet radiation-stimulated photochemical oxidation as the dominant chemical weathering process on Mars. This suggestion might seem reasonable in view of the expected high flux of solar ultraviolet radiation transmitted through the thin Martian atmosphere. Unfortunately, careful laboratory testing of the putative processes, performed by Morris and Lauer (1980, 1981), failed to verify Huguenin's results and, instead, provided strong evidence that his originally reported high weathering rates were artifacts of thermally driven reactions rather than products of the assumed photocatalyzed reactions. Thus, the importance of ultraviolet-stimulated weathering on Mars remains to be demonstrated.

Method 3, theoretical modeling, has also been applied to the problem of chemical weathering on Mars (O'Connor, 1968; Gooding, 1978; Gooding and Keil, 1978). Among the most important and easily demonstrated results of these studies is the fact that $Fe^{3+}$ should be the thermodynamically stable oxidation state of iron on Mars, as it is on Earth, regardless of the abundance of water on the planet. Beyond that, the theoretical conclusions become very dependent on assumptions and fragmentary evidence about the abundance and phase behavior of water in the Martian surface and near-surface environments. Given the observed composition of the Martian atmosphere and granted a supply of liquid water, possibly from melted ground ice or as films of "unfrozen" interstitial water, we can expect that chemical weathering reactions on Mars should generally be similar to liquid-solid reactions on Earth. However, the much higher partial pressure of $CO_2$ on Mars, relative to Earth, should cause the water to be more acidic than on Earth, possibly accelerating rates of weathering. By observation, though, liquid water must be extremely rare to nonexistent at the actual surface-atmosphere interface, so that gas-solid reactions should dominate the Martian surface under current climatic conditions. As a consequence, very few hydrous secondary minerals should be currently forming on the surface of Mars. If formed by liquid-solid reactions in subsurface environments or in more water-rich surface paleoenvironments, a number of hydrous secondary phases could theoretically survive at the surface if kept at very low temperatures (less than $240°K$). Most clay minerals, though, would probably remain susceptible to slow chemical attack by $CO_2$, which would tend to decompose them into simpler phases. Thus, the survival of demonstrably unstable secondary phases on Mars to the present day could allow eventual study of "fossil" weathering products.

### 9-4-3 Venus

Most of our knowledge about surface conditions on Venus has come from the soft-landed, unmanned *Venera* spacecraft, particularly *Venera 9* and *10*, which landed in 1975, and *Venera 13* and *14*, which landed in 1982. In addition, the multiprobe spacecraft portion of the *Pioneer Venus* mission in 1978 provided valuable information on the composition of the Venusian atmosphere.

None of the *Venera* experiments were designed specifically to study weathering. However, the gamma-ray and X-ray fluorescence spectrometers provided the first elemental analyses of Venusian surface materials and indicated generally "basaltic" compositions. Materials at the sites of *Venera 13* and *14*, in particular, are apparently akin to alkaline and tholeiitic basalts, respectively (Surkov et al., 1983).

Acquisition of imagery of the Venusian surface has been very limited due to the understandably short working lifetimes of the *Venera* spacecraft in the hostile Venusian environment (about 2 to 3 hours each for *Venera 13* and *14*). In the few images returned to Earth by *Venera 9, 10, 13,* and *14*, the evidence for physical weathering of rocks is ambiguous, although the loose, fine-grained materials that occur at least at the landing sites of *Venera 9, 10,* and *13* can reasonably be interpreted as weathering products. Furthermore, the planar fissility of the surface at the *Venera 14* site (see Fig. 9-3) can be interpreted either as unusual, thermally driven exfoliation or as thin bedding in fine-grained weathering products that have been indurated by chemical processes (Florensky et al., 1983).

Intuitively, it would seem that the high temperature and pressure of the Venusian atmosphere (see Table 9-2) would favor chemical alteration of rocks at rates that should

Fig. 9-3 Possible evidence for rock weathering on Venus. Planar structures in the surface might represent exfoliative weathering or bedding in chemically indurated, fine-grained weathering products (Florensky et al., 1983) in this *Venera 14* panorama. (Photograph provided by Dr. V. L. Barsukov of the USSR Academy of Sciences). Copyright 1983 by American Association for the Advancement of Science and reprinted here with permission.)

be readily reproduced in laboratory experiments. It is surprising, though, that few such experiments have been reported. In a series of exploratory experiments, however, Gooding (1981b) found that powdered rocks exposed to $CO_2$ under conditions approaching those on Venus darkened significantly over time periods as short as a few days. The darkening can be tentatively attributed to the oxidation of iron-bearing minerals in the rocks (Gooding, 1982) by reactions analogous to Eq. (1-3) in Table 9-3. Although better experiments are needed to confirm the observation and its speculative explanation, it is interesting to note that, at least at the *Venera 10* site, the loose, fine-grained materials are photometrically darker than the adjacent rocks. Furthermore, the black mineral magnetite, $Fe_3O_4$, should be the stable form of iron on the Venusian surface if oxidation is controlled by $CO_2$-CO reactions (see Fig. 9-4), a prediction generally consistent with the Lewis and Kreimendahl (1980) chemical model for the Venusian surface. Thus, oxidative darkening may be a weathering effect that can be sought on Venus as a direct test of theoretical predictions.

Chemical interactions between the atmosphere and surface of Venus have also been modeled theoretically by Khodakovsky et al. (1979) and Barsukov et al. (1980). Secondary (weathering product) minerals predicted to form by alteration of basalts include glaucophane, annite, epidote, and magnetite. In addition, sulfur is predicted to occur in

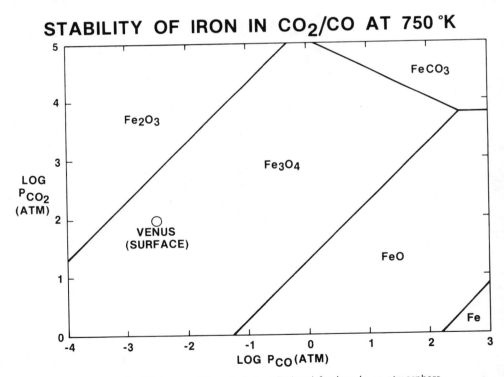

Fig. 9-4 Equilibrium-stability diagram calculated for iron in an atmosphere of $CO_2$ with minor CO at 750 K. The atmospheric composition of Venus (pressures of $CO_2$ and of CO in atmospheres), as determined by *Pioneer Venus* and *Venera*, is shown by the circle. Theoretically, $Fe_3O_4$ should be the stable form of iron on Venus if oxidation is controlled by $CO_2$-CO reactions.

altered surface materials as pyrite and anhydrite. These predicted mineral assemblages imply that significant amounts of water could be stored in Venusian weathering products (as the hydrous minerals glaucophane, annite, and epidote) and that sulfur compounds could be available to act as agents of cementation in weathered materials. These predictions can be tested directly if the sulfur and water contents of Venusian surface materials can be measured. However, the required chemical analyses should be accompanied by diagnostic mineralogical analyses.

Nozette and Lewis (1982) pointed out that chemical weathering on Venus may vary with elevation in view of the fact that atmospheric composition, pressure, and temperature change significantly with altitude. Those authors suggest that Ca- and Mg-rich weathering products formed at high elevations may be blown by wind to low elevations where they further react to form new phases and, thereby, possibly exert some control over the abundances of H-, C-, S-, and O-based gases in the Venusian atmosphere. These predictions are also subject to testing if comparative mineralogical and chemical analyses of highland and lowland materials on Venus can be made in the future.

## 9-4-4 Moon and Mercury

Although Earth's Moon and the planet Mercury differ substantially in their bulk physical properties, both objects have experienced surface evolution through similar processes, strongly dominated by impact cratering and volcanism (Murray et al., 1975). Neither body possesses a permanent atmosphere, though both are relentlessly bombarded by the solar wind (the stream of ions emitted by the Sun), and there is reason to believe that transient atmospheres generated by impact or volcanic events could lead to short-lived weathering events.

Both volcanic and explosive impact events on the Moon should produce transient clouds of gas or vapor, including neutral monatomic Na and K as species (Naughton et al., 1972). It has been shown experimentally that K vapor, in particular, can physically decompose both terrestrial and lunar crystalline basalts under simulated lunar conditions (Naughton et al., 1971). Glassy basalts, though, are apparently attacked much more slowly or not at all. Although the mechanism of this K-vapor weathering remains unclear, some evidence from lunar sample studies suggests that it may have operated on the Moon (Naughton et al., 1971), possibly contributing to soil and sediment formation.

Although the vast majority of lunar rocks are remarkably fresh in appearance (by Earth standards), the observation of "rust" on Ni-Fe metal particles in *Apollo 16* lunar rocks inspired a debate about whether it formed naturally on the Moon or by contamination after the astronauts had collected the samples. Detailed analytical and laboratory simulation studies led most investigators to conclude that the rusting was not lunar in origin but resulted, instead, from the action of water vapor that leaked into the sample containers during spacecraft transportation to Earth or prior to processing in the lunar sample curatorial laboratory (Taylor et al., 1974). However, Haggerty (1978) presented some thought-provoking arguments for the lunar origin of at least some of the rusty particles. It might also be significant that akaganéite, $\beta$-FeO(OH, Cl), the dominant phase in lunar rust (Taylor et al., 1974), is commonly not the dominant phase among Fe-oxide weathering products of Ni-Fe metal in stony meteorites (Gooding, 1981a). Indeed, a recent opinion allows that at least some lunar rock rust may have formed by attack of volcanic- or impact-generated Cl-$H_2$O-rich vapor on indigenous lunar metal particles under conditions distinctively different from those characteristic of surface weathering on Earth (Taylor et al., 1981). Only further lunar exploration and sample collections can finally settle this lingering uncertainty and, indeed, it can be argued that the search for rusty or otherwise weathered rocks on the Moon should be an important aspect of "prospecting" for naturally occurring volatile-rich lunar materials that would be important resources for future colonization and utilization of the Moon.

Gibson (1977) pointed out that bombardment by the solar wind, which is known to induce the formation of volatile H-C-N-O compounds on the Moon, may form similar compounds in the regolith of Mercury and perhaps at a rate even greater than that on the Moon, due to Mercury's close proximity to the Sun. Thermal evaporation of such compounds from Mercury's hot (illuminated) hemisphere to its cold (dark) hemisphere might result in chemical weathering of regolith materials, especially if $H_2O$ was one of the compounds involved. In addition, the volcanic- and impact-induced weathering processes outlined above for the Moon can also be reasonably expected to apply to Mercury. Thus, the study of weathering on both the Moon and Mercury may still be a valid and fruitful pursuit.

## 9-4-5 Io

Io is unquestionably the most dynamic and, hence, most interesting satellite of Jupiter. The discovery in 1979 of several active volcanoes on Io was one of the most exciting scientific yields of the *Voyager 1* and *2* spacecraft missions. As a consequence of its volcanism, Io possesses a tenuous but quasi-permanent atmosphere of gaseous $SO_2$ (Pearl et al., 1979; see also Chapter 7 by Fanale). In addition, a cloud of monatomic Na vapor around Io was discovered from Earth-based telescopic measurements made even before the *Voyager* missions (e.g., Matson et al., 1978). Thus, Na, and $SO_2$, and possibly other chemically reactive vapor species are available as weathering agents on Io.

If crystalline rocks exist on Io, they may be subject to attack by Na vapor by the process described above in Sec. 9-4-4. However, the lesser effectiveness of Na, relative to K, in performing such weathering (Naughton et al., 1971), combined with the lower mean surface temperature on Io relative to the Moon, might reduce the potency of the process on Io. Still, the possibly longer-term duration of Na clouds on Io, relative to the Moon, might actually increase the likelihood of alkali weathering on Io as a significant surface process.

The very low limit set for the abundance of water vapor on Io (Pearl et al., 1979) argues against frost riving by water ice as a significant process on Io. It is not known whether $SO_2$ frost could perform riving, although various sulfur salts that may exist on Io (Fanale et al., 1979; Nash and Nelson, 1979) might perform salt riving if salt crystallization occurred in rock voids as a consequence of $SO_2$ attack on the rocks. In this regard, it is interesting to note that Booth et al. (1982), in an experimental study intended to test weathering reactions on Mars rather than on Io, found that olivine and olivine tholeiitic basalt react with $SO_2$ in low-temperature, low-pressure atmospheres containing water vapor. It is at least conceivable that similar rocky materials could be attacked by dry $SO_2$ on Io. The reaction products would probably be sulfites or sulfates of Fe, Mg, Ca, or Na, depending on the chemical and mineralogical composition of the rocks being weathered. Thus, chemical weathering might account for the formation of some salts on Io, in addition to those that may be condensates from the volcanic plumes.

### 9-4-6 Titan

Titan is the largest and most dense (hence, most rocky) of Saturn's satellites and is the only one to possess an atmosphere. In fact, *Voyager 1* determined that Titan's atmosphere exists at a pressure about 1.6 times that of Earth's atmosphere and is probably composed mostly of $N_2$ (Tyler et al., 1981). In addition, Titan's atmosphere is densely clouded by aerosols or small solid particles inferred to be condensates of $CH_4$ and, possibly, more complex C-H-N compounds. The location of the rocky surface and the abundance of water on Titan remain unknown. However, the very low surface temperature of Titan would require that all water be effectively trapped as ice and has led to the speculation that any freeze-thaw phenomena on Titan would involve $CH_4$ rather than $H_2O$ (Tyler et al., 1981). It is not known whether $CH_4$ could perform frost riving, although we might speculate that the crystallization of C-H-N-based salts from liquid $CH_4$ might perform salt riving. However, a recent analysis of *Voyager* data suggests that condensed (liquid and solid) $CH_4$ may not be globally abundant on Titan (Eshleman et al., 1983; Flasar, 1983), so that the significance of frost riving and salt riving on Titan remains highly conjectural.

### 9-4-7 Comets and Asteroids

Comets are operationally distinguished from asteroids by their characteristic comas, as observed telescopically, and by their commonly more eccentric or highly inclined orbits. Beyond those behavioral differences, though, comets and asteroids may share deeper evolutionary threads, and both types of objects may be sources of meteorites, the solid objects that fall onto planetary surfaces from space.

Comets are known to have volatile-rich nuclei that probably include ices of $H_2O$, $CO_2$, and other species (Delsemme, 1975). Under the pressure and temperature conditions predicted for cometary nuclei, $H_2O$ ice is expected to exhibit phase transitions, including crystallization from amorphous forms (Klinger, 1980), such that frost riving might occur. In addition, various chemical reactions might occur between the solid dust particles of the nuclei and either the gaseous atmospheres generated by vaporization of the ices or films of "unfrozen" pore water. Thus, both physical and chemical weathering involving $H_2O$, $CO_2$, and other species can be expected to occur on or in cometary nuclei. In the absence of more detailed information about the starting materials and reaction conditions, it is not possible to predict the weathering products, although it might be reasonable to expect hydrous silicates and salts to be prominent among them. In fact, interplanetary dust particles that are thought to be debris shed by comets, exhibit infrared absorption spectra that suggest the presence of hydrated silicates and carbonates (Sandford et al., 1982).

Because they do not display comas, asteroids are inferred to contain much less volatile ice than do comets, although Feierberg et al. (1981) presented spectroscopic evidence for mineralogically bound water on at least three asteroids. The existence of mineral-structural water on these objects would be consistent with the inference that such asteroids are the parent bodies of certain meteorites, namely, C1 and C2 carbonaceous chondrites, which contain abundant hydrous phyllosilicate and salt minerals thought to have formed by aqueous alteration of olivine, pyroxene, and other phases (Kerridge and Bunch, 1979). Providing that such effects ultimately arose from interactions of solids with a variety of volatile-element "atmospheres," such features can be rightfully interpreted as weathering products. Temperatures that prevailed during aqueous alteration of carbonaceous chondrites remain unknown so that neither "hydrothermal" nor "hydrocryogenic" alteration can yet be excluded from consideration and, indeed, both may have been important (Gooding, 1984). Given the apparently high abundance of water in the attacking fluids and the salt minerals deposited from them, both frost riving and salt riving may have occurred in concert with chemical weathering.

## 9-5 SIGNIFICANCE OF WEATHERING IN PLANETARY EVOLUTION

As a collection of surface-atmosphere interactions, weathering is significant in two major aspects of the evolution of terrestrial-type planets. These aspects are (1) weathering products as indicators of weathering environments, and (2) atmosphere-regolith interactions as factors in planetary volatile cycling and storage. Aspect 1 has been pursued on Earth in the study of "fossil" soil profiles as keys to deducing paleoclimates (McLaughlin, 1955; Nahon et al., 1980). Although well-preserved products of physical weathering (e.g., frost-rived rocks) might be useful in reconstructing some types of paleoenvironments, the most useful records should be expected among chemical weathering products. Clearly, hydrous phyllosilicates and hydrous salts require water for formation and, hence, if found on a planetary surface devoid of water under current climatic conditions, such phases can be reasonably interpreted as relics of past environments that were more rich in water. Furthermore, analyses of stable isotopic ratios in the weathering products might reveal the temperature at which weathering occurred. Given a thorough theoretical, experimental, and observational understanding of their stabilities in a wide variety of environments, such mineral assemblages, upon further study, might yield indications of whether they formed in acidic or alkaline solutions and whether the solutions were hot or cold. Such an investigational approach is almost certainly necessary for a complete understanding of weathering products to be found on Mars, for example. Eventual study of weathering products among returned Martian samples can be expected to yield important information about the climate history of Mars.

Aspect 2 is important because, through geologic time, atmosphere-regolith interactions can significantly alter the compositions of both the atmosphere and regolith of a planet. Earth and Venus, as examples, are very similar in their bulk physical properties, yet Venus possesses a dense $CO_2$ atmosphere, whereas Earth has a more modest atmosphere rich in $N_2$ and $O_2$. Although abundant plant life is presumably responsible for the production of our $O_2$-rich atmosphere on Earth, could the primitive atmospheres on Venus and Earth have been initially similar in the ancient past? That question remains to be answered, although it should be clearly kept in mind, as pointed out by Rubey (1951), that Earth, too, would have a dense $CO_2$ atmosphere if the $CO_2$ currently stored as $CaCO_3$

in Earth's voluminous deposits of limestone were to be released. Most of Earth's $CO_2$ inventory has been immobilized as carbonate minerals that formed as a consequence of chemical weathering.

In the absence of oceans as storage bins for $H_2O$ or marine limestones as storage bins for $CO_2$, the regolith of a terrestrial-type planet may become the storage site for that portion of the planet's volatile inventory not contained in its atmosphere. On a cold planet such as Mars, $H_2O$ and $CO_2$ may be stored physically as ices, as gases adsorbed on regolith materials, or as hydrous phases or carbonates formed by chemical weathering (Fanale, 1976). On a hot planet such as Venus, though, chemical weathering may be the only effective mechanism for incorporating volatiles into the regolith. In either case, though, it is important to understand weathering as an aid to understanding the abundances and locations of volatiles on these planets. For example, the occurrence of abundant phyllosilicates on Mars or Venus would imply that significant portions of the $H_2O$ budgets for these two planets have been, in effect, irreversibly locked into their respective regoliths. Conversely, the paucity of phyllosilicates on these same two planets would imply that their respective $H_2O$ inventories are confined to the atmosphere (in the case of Venus) or to an atmosphere-ground ice system (in the case of Mars). Without the analysis of weathering products, though, these vital parts in the respective evolutionary puzzles for Mars and Venus would remain forever missing.

It should now be apparent that planetary science is not complete unless studies of planetary surface weathering are included. Some type of physical or chemical weathering can be expected to occur on all terrestrial-type planets, rocky planetary satellites, comets and asteroids, regardless of whether the attendant atmospheres are permanent or transient. Only the quantitative, not qualitative, significance of weathering varies with availability of atmospheres. Therefore, the study of weathering as a general planetary phenomenon is an essential ingredient in the study plans for all rocky planetary objects.

## 9-6 ACKNOWLEDGMENTS

The *Venera 14* image was obtained through Dr. M. B. Duke from a collection of photographs generously provided by Dr. V. L. Barsukov of the USSR Academy of Sciences. Preparation of this article was supported by the NASA Planetary Geology Program and the Johnson Space Center.

## 9-7 REFERENCES

Anderson, D. M., E. S. Gaffney, and P. F. Low, 1967, Frost phenomena on Mars, *Science, 155*, 319-322.

Barsukov, V. L., V. P. Volkov, and I. L. Khodakovsky, 1980, The mineral composition of Venus surface rocks—a preliminary prediction, *Proc. Lunar Planet. Sci. Conf., 11th*, 765-773.

Berner, R. A., E. L. Sjöberg, M. A. Velbel, and M. D. Krom, 1980, Dissolution of pyroxenes and amphiboles during weathering, *Science, 207*, 1205-1206.

Booth, M. C., E. K. Gibson, Jr., and R. Kotra, 1982, Chemical weathering on Mars—interactions of sulfur dioxide with olivine and olivine tholeiite in simulated Martian environments, in *Lunar Planet. Sci. XIII*, Lunar and Planet. Inst., Houston, Tex., 55-56.

Carroll, D., 1970, *Rock Weathering*, Plenum Press, New York, 203 pp.

Clark, B. C., and D. C. Van Hart, 1981, The salts of Mars, *Icarus, 45*, 370-378.

Correns, C. W., 1963, Experiments on the decomposition of silicates and discussion of chemical weathering, in *Clays and Clay Minerals, Proc. Tenth Nat. Conf.* (A. Swineford, ed.), Macmillan, New York, 443-459.

Delsemme, A. H., 1975, The volatile fraction of the cometary nucleus, *Icarus, 24*, 95-110.

Eshleman, V. R., G. F. Lindal, and G. L. Tyler, 1983, Is Titan wet or dry?, *Science, 221*, 53-55.

Fanale, F. P., 1976, Martian volatiles—their degassing history and geochemical fate, *Icarus, 28*, 179-202.

Fanale, F. P., R. H. Brown, D. P. Cruikshank, and R. N. Clarke, 1979, Significance of absorption features in *Io's* IR reflectance spectrum, *Nature, 280*, 761-763.

Feierberg, M. A., L. A. Lebofsky, and H. P. Larson, 1981, Spectroscopic evidence for aqueous alteration products on the surface of low-albedo asteroids, *Geochim. Cosmochim. Acta, 45*, 971-981.

Flasar, F. M., 1983, Oceans on Titan?, *Science, 221*, 55-57.

Florensky, C. P., A. T. Basilevsky, V. P. Kryuchkov, R. O. Kusmin, O. V. Nikolaeva, A. A. Pronin, I. M. Chernaya, Yu. S. Tyuflin, A. S. Selivanov, M. K. Naraeva, and L. B. Ronca, 1983, *Venera 13, Venera 14*—sedimentary rocks on Venus?, *Science, 221*, 57-59.

Garrels, R. M., and C. L. Christ, 1965, *Solutions, Minerals, and Equilibria*, Freeman, Cooper, San Francisco, 450 pp.

Garvin, J. B., P. J. Mouginis-Mark, and J. W. Head, 1981, Characterization of rock populations on planetary surfaces—techniques and a preliminary analysis of Mars and Venus, *Moon and Planets, 24*, 355-387.

Gibson, E. K., Jr., 1977, Production of simple molecules on the surface of Mercury, *Phys. Earth Planet. Interiors, 15*, 303-312.

Gibson, E. K. Jr., S. J. Wentworth, and D. S. McKay, 1983, Chemical weathering and diagenesis of a cold desert soil from Wright Valley, Antarctica—an analog of Martian weathering processes, *J. Geophys. Res. (Proc. Thirteenth Lunar Planet Sci. Conf., 13th, Part 2), 88*, Supplement, A912-A928.

Goldich, S. S., 1938, A study in rock-weathering, *J. Geol., 46*, 17-58.

Gooding, J. L., 1978, Chemical weathering on Mars—thermodynamic stabilities of primary minerals (and their alteration products) from mafic igneous rocks, *Icarus, 33*, 483-513.

Gooding, J. L., 1981a, Mineralogical aspects of terrestrial weathering effects in chondrites from Allan Hills, Antarctica, *Proc. Lunar Planet. Sci. Conf., 12B*, 1105-1122.

Gooding, J. L., 1981b, Alteration of rocks in hot $CO_2$ atmospheres—preliminary experimental results and application to Venus, *Rep. Planet. Geol. Program—1981*, NASA Tech. Memo 84211, 460-462.

Gooding, J. L., 1982, Alteration of rocks in hot $CO_2$ atmospheres: further experimental results and application to Mars, *Rep. Planet. Geol. Program 1982*, NASA Tech. Memo 85127, 327-329.

Gooding, J. L., 1984, Low-temperature aqueous alteration in the early solar system: Possible clues from meteorites weathered in Antarctica, *Lunar Planet. Sci. XV*, Lunar Planet. Inst., Houston, Tex., 308-309.

Gooding, J. L., and K. Keil, 1978, Alteration of glass as a possible source of clay minerals on Mars, *Geophys. Res. Lett., 5*, 727-730.

Grabau, A. W., 1960, *Principles of Stratigraphy*, Vol. 1, Dover, New York, 581 pp.

Griggs, D., 1936, The factor of fatigue in rock exfoliation, *J. Geol., 44*, 783-796.

Haggerty, S. E., 1978, Apollo 16 deep drill—a review of the morphological characteristics of oxyhydrates on *Rusty* particle 60002,108 determined by SEM, *Proc, Lunar Planet. Sci. Conf. 9th*, 1861-1874.

Hess, S. L., R. M. Henry, C. B. Leovy, J. A. Ryan, and J. E. Tillman, 1977, Meteorological results from the surface of Mars—*Viking 1* and *2, J. Geophys. Res., 82*, 4559-4574.

Hoffman, J. H., V. I. Oyama, and U. von Zahn, 1980, Measurements of the Venus lower atmosphere composition—a comparison of results, *J. Geophys. Res., 85*, 7871-7881.

Huguenin, R. L., 1974, The formation of goethite and hydrated clay minerals on Mars, *J. Geophys. Res., 79*, 3895-3905.

Japp, J. M., and J. L. Gooding, 1980, Size-fraction analyses of Mauna Kea, Hawaii, summit soils and their possible analogy with Martian soils: *Rep. Planet. Geol. Program—1980*, NASA Tech. Memo. 82385, 212-214.

Johnston, J. H., 1973, Salt weathering processes in the McMurdo Dry Valley regions of South Victoria Land, Antarctica, *New Zealand J. Geol. Geophys., 16*, 221-224.

Kerridge, J. F., and T. E. Bunch, 1979, Aqueous activity on asteroids—evidence from carbonaceous meteorites, in *Asteroids* (T. Gehrels, ed.), Univ. Ariz., Tucson, Ariz., 745-764.

Khodakovsky, I. L., V. P. Volkov, Yu. I. Siderov, M. V. Borisov, and M. V. Lomonosov, 1979, Venus—preliminary prediction of the mineral composition of surface rocks, *Icarus, 39*, 352-363.

Klinger, J., 1980, Influence of a phase transition of ice on the heat and mass balance of comets, *Science, 209*, 271-272.

Lewis, J. S., and F. A. Kreimendahl, 1980, Oxidation state of the atmosphere and crust of Venus from *Pioneer Venus* results, *Icarus, 42*, 330-337.

Loughnan, F. C., 1969, *Chemical Weathering of the Silicate Minerals*, Elsevier North-Holland, New York, 154 pp.

Malin, M. C., 1974, Salt weathering on Mars, *J. Geophys. Res., 79*, 3888-3894.

Marov, M. Ya., 1978, Results of Venus missions, *Ann. Rev. Astr. Astrophys., 16*, 141-169.

Matson, D. L., B. A. Goldberg, T. V. Johnson, and R. W. Carlson, 1978, Images of Io's sodium cloud, *Science, 199*, 531-533.

McLaughlin, R. J. W., 1955, Geochemical changes due to weathering under varying climatic conditions, *Geochim. Cosmochim. Acta, 8*, 109-130.

Moore, H. J., C. R. Spitzer, K. Z. Bradford, P. M. Cates, R. E. Hutton, and R. W. Shorthill, 1979, Sample fields of the *Viking* landers, physical properties and aeolian processes, *J. Geophys. Res., 84*, 8365-8377.

Morris, E. C., H. E. Holt, T. A. Mutch, and J. F. Lindsay, 1972, Mars analog studies in Wright and Victoria valleys, Antarctica, *Antarctic J., 7*, 113-114.

Morris, R. V., and H. V. Lauer, Jr., 1980, The case against UV photostimulated oxidation of magnetite, *Geophys. Res. Lett., 7*, 605-608.

Morris, R. V., and H. V. Lauer, Jr., 1981, Stability of goethite ($\alpha$FeOOH) and lepidocrocite ($\delta$-FeOOH) to dehydration by UV radiation—implications for their occurrence on the Martian surface, *J. Geophys. Res., 86*, 10893-10899.

Murray, B. C., R. G. Strom, N. J. Trash, and D. E. Gault, 1975, Surface history of Mercury: Implications for terrestrial planets, *J. Geophys. Res., 80*, 2508-2514.

Nahon, D., A. V. Carozzi, and C. Parron, 1980, Lateritic weathering as a mechanism for the generation of ferruginous ooids, *J. Sed. Petrol., 50,* 1287-1298.

Nash, D. B., and R. M. Nelson, 1979, Spectral evidence for sublimates and adsorbates on Io, *Nature, 280,* 763-766.

Naughton, J. J., J. V. Derby, and V. A. Lewis, 1971, Vaporization from heated lunar samples and the investigation of lunar erosion by volatilized alkalis, *Proc. Lunar Sci. Conf., 2nd, 1,* 449-457.

Naughton, J. J., D. A. Hammond, S. V. Margolis, and D. W. Muenow, 1972, The nature and effect of the volatile cloud produced by volcanic and impact events on the Moon as derived from a terrestrial volcanic model, *Proc. Lunar Sci. Conf., 3rd, 2,* 2015-2024.

Nozette, S., and J. S. Lewis, 1982, Venus—chemical weathering of igneous rocks and buffering of atmospheric composition, *Science, 216,* 181-183.

O'Connor, J. T., 1968, Mineral stability at the Martian surface, *J. Geophys. Res., 73,* 5301-5311.

Ollier, C. D., 1963, Insolation weathering: Examples from Central Australia, *Am. J. Sci., 261,* 376-387.

Ollier, C. D., 1969, *Weathering,* Edinburgh, Oliver and Boyd, 304 pp.

Ollier, C. D., 1971, Causes of spheroidal weathering, *Earth-Sci. Rev., 7,* 127-141.

Owen, T., K. Biemann, D. R. Rushneck, J. E. Biller, D. W. Howarth, and A. L. Lafleur, 1977, The composition of the atmosphere at the surface of Mars, *J. Geophys. Res., 82,* 4635-4639.

Oyama, V. I., G. C. Carle, F. Woeller, and J. B. Pollack, 1979, Venus lower atmospheric composition—analysis by gas chromatography, *Science, 203,* 802-805.

Pearl, J., R. Hanel, V. Kunde, W. Maguire, K. Fox, S. Gupta, C. Ponnamperuma, and F. Raulin, 1979, Identification of gaseous $SO_2$ and new upper limits for other gases on Io, *Nature, 280,* 755-758.

Rubey, W. W., 1951, Geologic history of sea water: An attempt to state the problem, *Bull. Geol. Soc. Amer., 62,* 1111-1147.

Sandford, S. A., P. Fraundorf, R. Patel, and R. M. Walker, 1982, Laboratory infrared spectra of interplanetary dust, *Meteoritics, 17,* 276-277.

Singer, R. B., T. B. McCord, R. N. Clark, J. B. Adams, and R. L. Huguenin, 1979, Mars surface composition from reflectance spectroscopy—a summary, *J. Geophys. Res., 84,* 8415-8426.

Strickland, E. L. III, 1979, Martian soil stratigraphy and rock coatings observed in color-enhanced *Viking Lander* images, *Proc. Lunar Planet. Sci. Conf. 10th,* 3055-3077.

Surkov, Yu. A., L. P. Moskalyeva, O. P. Shceglov, V. P. Kharyukova, O. S. Manvelyan, V. S. Kirichenko, and A. D. Dudin, 1983, Determination of the elemental composition of rocks on Venus by *Venera 13* and *Venera 14* (preliminary results), *J. Geophys. Res. (Proc. Lunar Planet. Sci. Conf., 13th, Part 2), 88,* Supplement, A481-A493.

Taylor, L. A., H. K. Mao, and P. M. Bell, 1974, β-FeOOH, akagenéite, in lunar rocks, *Proc. Lunar 5th Sci. Conf., 1,* 743-748.

Taylor, L. A., R. H. Hunter, and H. Y. McSween, 1981, The significance of rust in lunar rocks and meteorites, *Meteoritics, 16,* 391-392.

Tyler, G. L., V. R. Eshleman, J. D. Anderson, G. S. Levy, G. F. Lindal, G. E. Wood, and T. A. Croft, 1981, Radio science investigation of the Saturn system with *Voyager 1*—preliminary results, *Science, 212,* 201-206.

Ugolini, F. C., 1976, Soils of the high elevations of Mauna Kea, Hawaii, an analogy to Martian soils?, *Intl. Colloq. Planet. Geol. Proc., Geol. Romana, 15,* 521-524.

Ugolini, F. C., and R. J. Zasoski, 1979, Soils derived from tephra, in *Volcanic Activity and Human Ecology*, Academic, 83-124.

Wall, S. D., 1981, Analysis of condensates formed at the *Viking 2 Lander* site—the first winter, *Icarus, 47,* 173-183.

Whalley, W. B., G. R. Douglas, and J. P. McGreevy, Crack propagation and associated weathering in igneous rocks, *Z. Geomorph. N.F., 26,* 33-54.

Andrew P. Ingersoll
Division of Geological and Planetary Sciences
California Institute of Technology
Pasadena, California 91125

# 10

# JUPITER AND SATURN

Reprinted with revisions from *The New Solar System* Second Edition (1982).

# ABSTRACT

Observations and theories of bulk planetary properties; interior structure and evolution; and atmospheric composition, temperature, and winds are reviewed. Jupiter and Saturn are both fluid masses of hydrogen, helium, and other elements in approximately solar proportions. Both planets radiate stored internal energy. Their atmospheres exhibit lightning and auroras, colored cloud bands, strong zonal winds, and long-lived oval storms that last for years or centuries. Saturn differs from Jupiter in that it has now cooled to the point where precipitation of helium in the metallic core has partially depleted the atmosphere of helium. Saturn's lower gravity and consequently greater depth of atmosphere and deeper nonconducting fluid layers may account for the greater widths and speeds of Saturn's zonal jets. Jupiter has more vigorous eddies and more spots in all size ranges. A major unknown in models of atmospheric structure, composition, and dynamics is the interaction between the surface layers and the deep fluid interior. It is hoped that *Voyager* data and other data on surface phenomena will help resolve these major issues.

## 10-1 INTRODUCTION

Jupiter and Saturn are the Sun's principal companions. Together they account for 92 percent of the extrasolar mass of the solar system. The most basic statistics about them emphasize their enormity: Jupiter possesses 318 times the mass of the Earth, or about 0.1 percent that of the Sun; Saturn is 95 times more massive than our planet. Both are some 10 times Earth's diameter and about one-tenth that of the Sun. Clearly, these bulk physical properties place them in a class midway between the Earth and the Sun.

At wavelengths of the visible spectrum, Jupiter and Saturn shine by reflected sunlight (see Figs. 10-1 and 10-2) and are among the brightest objects in the night sky despite

Fig. 10-1   Jupiter, as viewed by *Voyager 1* from a distance of $54 \times 10^5$ km.

Fig. 10-2  *Voyager 1* recorded Saturn and its rings from a distance of $18 \times 10^5$ km. Note that cloud features of Saturn are fewer and lower in contrast than on Jupiter.

their great distance from us. They also glow brightly at infrared wavelengths, partly because like all planets they reradiate to space a fraction of the energy received from the Sun. In addition, however, Jupiter and Saturn emit their own stored energy, generated long ago during the gravitational contraction of the nebulae from which they condensed. Had these planets been 10 times more massive, their internal temperatures might have

Fig. 10-3  The night hemisphere of Jupiter displayed surprising evidence of lightning (bright patches below center) and auroras (curved arcs at top) in this *Voyager 1* image. The planet's north pole lies roughly midway along the auroral arc.

risen high enough to trigger nuclear fusion. Then our solar system would have contained a multiple star, and conditions on all planets would have been considerably different from those observed today. Even though the gas giants never reached that stage, enough energy was stored to last until the present.

Jupiter and Saturn are of great interest to planetary scientists for several reasons. By virtue of their size and mass, they occupy a distinct place among objects in the solar system. They and their satellite families constitute two miniature "planetary" systems, whose character and evolution can be compared with our perceived history of the solar system proper. These giants also exhibit a wide range of atmospheric phenomena that includes multicolored clouds arranged in parallel bands, huge circulating storms which can last for decades or centuries, lightning and giant auroras (see Fig. 10-3), and swift-moving cloud currents that approach supersonic speeds relative to the planets' interiors (see Fig. 10-4).

Thus, at one extreme, we can compare Jupiter and Saturn with the Sun and stars, and their satellites with our own solar system. At the other extreme, we can compare the atmospheric phenomena observed on these planets with those on Earth. Our understand-

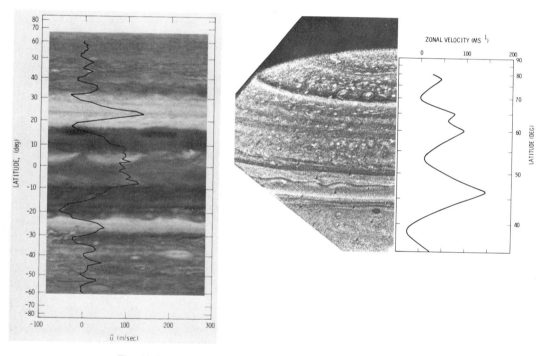

Fig. 10-4 A comparison of wind speed with latitude for Jupiter (left) and Saturn. Wind velocities are eastward, measured with respect to each planet's internal rotation period (9 h 55.5 min for Jupiter, and 10 h 39.4 min for Saturn), which are based on observations of the planet's periodic radio emissions. Negative velocities are, therefore, winds moving westward with respect to these reference frames. The *Voyager* imaging team has combined these results with images of each planet projected so that the latitude scales match the data.

ing of terrestrial atmospheric phenomena was inadequate to prepare us for what we have found recently about the gas giants, but by testing our theories against observations of these planets, we can understand the Earth in a broader context. This chapter, therefore, delves into their atmospheres, interiors, and possible internal histories.

## 10-2 OBSERVATIONS

Our knowledge of Jupiter and Saturn is derived from Earth-based telescopic observations begun 300 years ago, modern observations from high-altitude aircraft and orbiting satellites, and the wealth of recent findings from interplanetary spacecraft. Between 1973 and 1981, four unmanned probes flew by Jupiter (*Pioneer 10* and *11, Voyager 1* and *2*), and three of those by Saturn.

Basic characteristics like mass, radius, density, and rotational flattening were determined during the first era of telescopic observation. Galileo's early views revealed the Galilean satellites which bear his name. Newton estimated the mass and density of Jupiter from observations of those satellites' orbits. Other observers, using improved optics, began to perceive atmospheric features on the planet. The most prominent of these, the Great Red Spot (Fig. 10-5), can be traced back 300 y with near certainty, and it may be older still. Beginning in the late nineteenth century, astronomers made systematic measurements of Jovian winds by tracking features visually with small telescopes. Photographic observations with larger telescopes later augmented these early efforts. Most recently, tens of thousands of features on Jupiter and Saturn have been tracked with great precision using the *Voyager* imaging system. The constancy of these currents over the long time intervals spanned by classical and modern observations is one of the truly remarkable aspects of Jovian and Saturnian meteorology.

Determination of chemical composition in the atmospheres began in the 1930s with the identification of methane ($CH_4$) and ammonia ($NH_3$) in the spectra of sunlight reflected from their clouds. About 1960, molecular hydrogen ($H_2$) was detected. Because hydrogen is a simple, symmetrical molecule, its vibrational and rotational absorptions are weak. Fortunately, hydrogen occurs above the clouds in such abundance that its absorption lines are nevertheless detectable in the spectra of both planets. It was quickly verified that the proportions of hydrogen, carbon, and nitrogen in Jupiter's atmosphere were consistent with a mixture of solar composition. Similar inferences have been made for Saturn, although actual observations of these compounds (especially ammonia) are extremely difficult. During the 1970s, as infrared detectors improved, observers recorded absorptions by other gases present in extremely minute amounts (one part per million or even per billion by volume). Ethane ($C_2H_6$), acetylene ($C_2H_2$), water, phosphine ($PH_3$), hydrogen cyanide (HCN), carbon monoxide (CO), germane ($GeH_4$), and compounds with deuterium ($^2H$) and isotopic carbon ($^{13}C$) all were detected in Jupiter's atmosphere in this way (see Table 10-1). The list for Saturn is shorter than for Jupiter, partly because Saturn is a fainter object and partly because it is colder; many of these compounds freeze at the level of its cloud tops and become harder to detect.

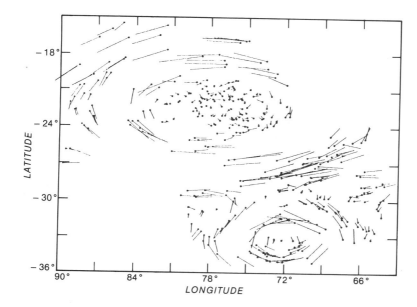

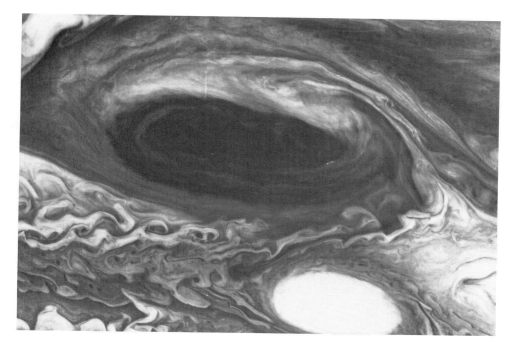

Fig. 10-5 High-resolution mosaic of Jupiter's Great Red Spot and a white oval (known as BC since its formation more than 40 years ago). The winds, determined from changes observed over a 10-h period, are counterclockwise around both oval features and reach speeds of 100 m · s$^{-1}$.

TABLE 10-1  Fractional abundances (by number of molecules) for a solar-composition atmosphere and for Jupiter's and Saturn's atmospheres near the cloud tops. The list contains seven of the ten most abundant elements in the Sun (and the universe). The other three—Si, Mg, and Fe—are believed to reside in the cores of the giant planets. Blanks indicate unobserved compounds. All numbers are uncertain in the least significant figure.

| Molecule | Sun | Jupiter | Saturn |
| --- | --- | --- | --- |
| $H_2$ | 0.89 | 0.90 | 0.94 |
| He | 0.11 | 0.10 | 0.06 |
| $H_2O$ | $1.0 \times 10^{-3}$ | $1 \times 10^{-6}$ | |
| $CH_4$ | $6.0 \times 10^{-4}$ | $7 \times 10^{-4}$ | $5 \times 10^{-4}$ |
| $NH_3$ | $1.5 \times 10^{-4}$ | $2 \times 10^{-4}$ | $2 \times 10^{-4}$ |
| Ne | $1.4 \times 10^{-4}$ | | |
| $H_2S$ | $2.5 \times 10^{-5}$ | | |

Helium, the second most abundant element in the Sun, is presumably an important constituent of Jupiter and Saturn as well. Unfortunately, it has no detectable absorptions to make its presence known. However, instruments aboard *Pioneer* and *Voyager* noticed the effect of collisions between helium and hydrogen molecules on the latter's infrared absorptions, giving us a kind of "back-door" identification of helium. And we can also combine infrared measurements with radio occultation data (obtained as the spacecraft passed behind the planets) to determine the molecular weight of the mixtures in their atmospheres. *Voyager* experimenters refined both of these methods to yield helium abundances with an uncertainty of only 2% to 3% by volume. The values they derived (10 percent for Jupiter and 6 percent for Saturn) are roughly consistent with solar composition, but the evidence suggests that some helium has been depleted from the upper atmosphere of Saturn, a fact which provides intriguing clues to its internal structure and history.

The temperatures and energy budgets of the gas giants have been studied from Earth for decades, but the best determinations have once again come from spacecraft. Both infrared and radio-occultation experiments have probed the cloud tops (in the upper troposphere) up to where the temperature increases with height (the lower stratosphere). Ultraviolet instruments probe the much greater altitudes of the thermosphere and ionosphere. We find that pressures near the cloud tops range from hundreds of millibars (mb) to 1 bar, as on Earth, but the temperatures there are much colder—about 125° K and 95° K based on infrared emissions from Jupiter and Saturn, respectively (see Fig. 10-6). From these observations, we infer that Jupiter and Saturn radiate between 1.7 and 1.8 times the amount of heat they absorb from the Sun. These excesses provide us with additional insight into internal structures and histories.

The latitudinal distributions of temperatures and heat flux for Jupiter and Saturn (see Fig. 10-7) indicate how effective the winds are in redistributing heat from equator to pole. We would expect to see thermal gradients if this took place near the cloud tops. However, the planets exhibit almost complete lack of thermal contrast, when averages

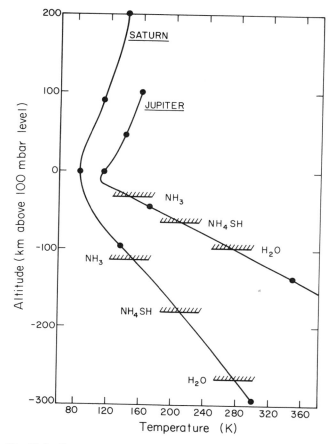

Fig. 10-6 Pressure-temperature profiles for the upper atmospheres of Jupiter and Saturn, as determined by both ground-based and spacecraft measurements at radio and infrared wavelengths. Each dot marks the point where atmospheric pressure is 10 times greater than the next dot above it; the total range indicated runs from 1 mb at top to 10 bars near the bottom. Shaded bands show the altitudes at which various clouds should form, based on a gaseous mixture of solar composition. Temperatures are lower on Saturn because it is farther from the Sun, and the range of cloud altitudes is broader because Saturn's weaker gravity allows a more distended atmosphere than on Jupiter.

are taken with respect to longitude, time of day, and season. This fact, combined with their latitudinal banding, suggests that the heat transport occurs at deeper levels.

We have inferred the upper atmospheric structure of these worlds by observing the effect of free electrons on spacecraft radio signals occulted by the planets, by comparing ultraviolet emissions on the day and night side, and from watching the Sun in the ultraviolet as it sets behind the planet as seen from the *Voyager* spacecraft. Temperatures can be deduced from changes in electron density with altitude. The distribution of various

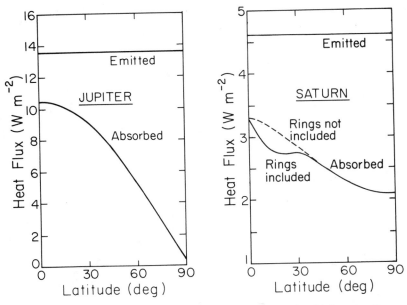

Fig. 10-7 A comparison of absorbed solar energy and emitted infrared radiation for both planets, averaged with respect to longitude, season, and time of day. Saturn is farther from the Sun and colder, so its heat fluxes are lower (the rings also shadow a large band around Saturn's equator). However, Saturn's seasonal cycle introduces a large uncertainty in the curves shown. Both planets emit more infrared radiation than the amount of sunlight they absorb, implying an internal heat source. The emitted radiation is also more uniform, which suggests heat transport across latitude circles at some depth within the planet. These results are derived largely from *Pioneer* and *Voyager* infrared observations, as well as from those in visible light. Irregularities in the energy flux associated with belts and zones are not shown.

emissions over the disk and limb are clues to composition and excitation mechanisms. Excitation stems mainly from charged particles (causing auroras) and energetic solar photons (airglow) striking the atmosphere from above. These processes can alter the chemical composition of deeper layers by causing stable compounds to form that become mixed downward.

## 10-3 ATMOSPHERIC STRUCTURE AND COMPOSITION

Temperatures, pressures, gas abundances, cloud compositions—all as functions of altitude and horizontal position—are the principal components of atmospheric structure. Atmospheric dynamics, which is discussed in the next section, concerns the wind and its causes and effects. However, the distinction between structure and dynamics is not always clear. Atmospheric circulation alters the structure by carrying heat, mass, and chemical species from place to place. The structure, in turn, affects wind patterns by controlling the absorption of sunlight, emission of infrared radiation, and release of latent heat, all of which lead to heating, expansion of the gas, and pressure gradients which drive the wind.

The vertical temperature structure is perhaps most fundamental to understanding a planetary atmosphere. This is normally given as temperature versus pressure (see Fig. 10-6), the latter derived from the altitude using a relationship known as hydrostatic equilibrium. The thickness, in kilometers, of a given pressure interval is less for Jupiter than for Saturn, owing to the more intense Jovian gravity. Temperatures at the deepest levels measured, which correspond to nearly 1 bar of pressure, tend to follow an adiabatic gradient (this is how temperature would change with pressure for a parcel moving vertically at a rate such that there is no heat exchange with the surroundings). An adiabatic gradient is usually a sign that the atmosphere is well mixed by convective currents. We would expect an adiabatic gradient to extend down almost indefinitely because heat from the interior cannot be carried off by infrared radiation, which is blocked by the opacity of the gases there. Thus convection must carry the internal heat up to levels where radiation to space occurs. For Jupiter and Saturn this occurs near pressure levels of 100 to 300 mb. At altitudes higher than this, the temperature increases with height due to the absorption of sunlight.

The rise in temperature with altitude above the 100-mb level probably requires more heating that can be accounted for by sunlight striking a gaseous atmosphere. Some scientists attribute this to a layer of fine black dust, produced photochemically, but other evidence of its existence has been hard to find. Farther up, the atmosphere is so thin that temperature is extremely sensitive to small energy inputs. *Voyager* discovered auroral emissions from Jupiter, especially intense at latitudes where magnetic field lines from the orbit of the satellite Io intersect the planet. The energy producing these auroral emissions is equivalent to about 0.1 percent of the total incident sunlight (a large input at these altitudes). Thus the charged-particle flux from Io seems capable of causing the high temperatures (1000 to 1300°K) observed near the top of Jupiter's atmosphere, since solar photons produce temperatures of only 200°K. The dissipation of upward-propagating waves could probably also account for the high temperatures, but little evidence for these waves exists.

The vertical cloud structure is inferred by several means. First, one can compute the altitudes above which each atmospheric constituent can condense (see Fig. 10-6). The procedure is to take a uniform solar composition mixture (Table 10-1), or whatever mixture is appropriate, and compute the partial pressures of each gas at each level. We compare these with the saturation vapor pressures, which are determined from the temperatures at each level. Condensation occurs where the computed partial pressure exceeds the saturation vapor pressure, and since the latter falls rapidly with temperature, clouds will form at the coldest layers. Above the cloud-forming levels, particle fallout may reduce the abundance of condensates and prevent clouds from extending above the top of the convective zone. When these calculations are worked for Jupiter and Saturn, we find three distinct cloud layers (see Fig. 10-6). The lowest is composed of water-ice or possibly liquid water droplets. Next are crystals of ammonium hydrosulfide ($NH_4SH$), which is basically a compound of ammonia ($NH_3$) and hydrogen sulfide ($H_2S$). At the top we expect an ammonia ice cloud.

These cloud structures assume solar composition throughout. But the real situation is more complex. Vertical mixing could carry particles from the lower clouds upward, thereby changing the composition of the upper clouds. Moreover, hydrogen sulfide has not been detected on either Jupiter or Saturn, possibly because it precipitates out below the cloud tops, and possibly because it changes to sulfur when exposed to sunlight. The existence of water clouds is also uncertain. Water vapor apparently occurs only at 1 ppm by volume, even though solar composition would imply abundances 1,000 times greater. The paradox is that we can see through holes in the clouds (see Fig. 10-8) to warm levels with temperatures of at least $300°K$, where both $H_2O$ and $H_2S$ should be abundant. Perhaps these "hot spots" are the deserts of Jupiter, where dry air, stripped of its condensable gases, is descending into the well-mixed interior. The observations nevertheless cast some doubt on the solar composition model.

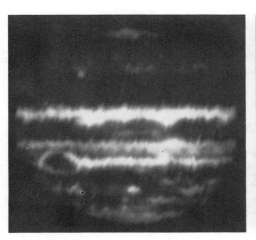 

Fig. 10-8  Near-simultaneous infrared (left) and visible-light views of Jupiter. Holes in the upper cloud deck allow radiation to escape from warmer layers underneath; these spots appear bright in the infrared. The infrared image was taken at Palomar Observatory by Richard Terrile.

The abundance ratios of various gases provide clues to chemical processes that are occurring on the two planets. If the atmospheres are in chemical equilibrium, with hydrogen the dominant species, virtually all the carbon should be tied up in methane, all the nitrogen as ammonia, all the oxygen as water, and so on. But in Jupiter's case, we have identified gases such as $C_2H_6$, $C_2H_2$, and CO, which imply disequilibrium. Both $C_2H_6$ and $C_2H_2$ are relatively easy to account for, since they are formed in the upper atmosphere as byproducts of methane photodissociation. In these reactions, solar ultraviolet photons at wavelengths less than 1600 Å knock hydrogen atoms off the methane ($CH_4$) molecule, leaving free radicals that can react with other $CH_4$ molecules to form $C_2H_6$ and $C_2H_2$. The existence of CO is more interesting, since oxygen is not readily available in the upper atmosphere—water, the principal oxygen-bearing molecule, tends to condense in the lower clouds. Several explanations have been proposed. One is that oxygen ions, injected into the magnetosphere by the $SO_2$ volcanoes on Io, enter the upper atmosphere from above. Another explanation is that CO is created at depth in Jupiter's atmosphere, and is mixed upward before it is destroyed. But phosphorus should also react with oxygen at depth in this way, so the existence of phosphine ($PH_3$) is something of a mystery. A critical unknown in these theories is the vertical mixing rate as a function of altitude.

The spectacular colors of Jupiter and more muted colors of Saturn provide further evidence of active chemistry in the atmospheres. Every observable feature in the *Voyager* photographs corresponds to clouds of various colors and brightness. Infrared images show that cloud color correlates with altitude (see Fig. 10-8). Blues have the highest brightness temperatures (calculated from emissions at specific wavelengths), so they must lie at the deepest levels and are only visible through holes in the upper clouds. Browns are next highest, followed by white clouds, and finally red clouds, such as those in the Great Red Spot, which is a very cold feature judging from its infrared brightness. The trouble is that all cloud species predicted for equilibrium conditions are white. Color must come when chemical equilibrium is disturbed, either by charged particles, energetic photons, lightning, or rapid vertical motion through different termperature regimes. The most likely coloring agent is probably elemental sulfur, which forms a variety of colors depending on its molecular structure. Sulfur is definitely present on Io, which has many of the same colors as Jupiter (see. Fig. 10-9). Some scientists believe phosphorus explains the Great Red Spot's color, and organic (carbon) compounds have been proposed by others to explain almost all of the colors.

Thus coloration is a subtle process, involving disequilibrium conditions and trace constituents. The correlation with altitude presumably reflects the processes that cause chemical reactions to occur. For example, higher altitudes receive more sunlight and a higher charged-particle flux. Certain regions may contain more lightning (see Fig. 10-3). And other regions may be sites of intense vertical motion. A different question is why these processes should be organized into large-scale patterns that last for years and sometimes for centuries. This question involves the dynamics of the atmospheres, to which we now turn.

Fig. 10-9 Io transits the southern hemisphere of Jupiter in this *Voyager 2* image with a resolution of about 200 km. The color of the planet's clouds is similar to that of Io's sulfur-covered surface.

## 10-4 ATMOSPHERIC DYNAMICS

The dominant observable dynamical features in Jupiter and Saturn's atmospheres are the counterflowing eastward and westward winds (see Fig. 10-4). Instead of one westward current at low latitudes (the trade winds) and one eastward current at high mid-latitudes (the jet stream) as on Earth, Jupiter has five or six of each kind of current in each hemisphere. Saturn seems to have fewer of such currents than Jupiter, but more than our planet; its currents are also stronger than Jupiter's. In fact, the eastward wind speed at Saturn's equator is about 500 m $\cdot$ s$^{-1}$, about two-thirds the speed of sound there. As on Earth, these winds are measured with respect to the rapidly rotating interiors. For Jupiter and Saturn, which have no solid surfaces, the rotation of the interior is deduced from that of the magnetic field, which is generated in the metallic core. Even supersonic wind speeds would be small compared with the equatorial velocity due to the planet's rotation.

These currents, the zonal jets, apparently have not changed their latitudinal positions during 80 years of modern telescopic observations. During the four months between the two *Voyager* encounters of Jupiter, the zonal velocities changed by less than the measurement error, in this case about 1.5 percent. This is remarkable for several reasons. First, although the zonal jets on Jupiter correlate with the latitudinal positions of the colored bands, the bands often change their appearance dramatically in a few years, while the jets do not change. Second, as revealed in *Voyager* photographs at 30 times the resolution of Earth-based images, there is an enormous amount of eddy activity in addition to the zonal jets. The time required for structures to be sheared apart by the eastward and westward jets is only about 1 to 2 days. This is also the lifetime of small eddies that suddenly appear in the zonal shear zones. Larger eddies, including the long-lived white ovals and the Great Red Spot (see Fig. 10-5), manage to survive by rolling with the currents. *Voyager* repeatedly observed smaller spots encounter the Red Spot from the east, circulate around it in about 6 days and then partially merge with it (see Fig. 10-10). How the zonal jets and the large eddies can exist amidst such activity is something of a mystery.

An important fact about the eddy motions was learned from statistical analysis of Jupiter's winds. During one 30-h period of *Voyager 1* atmospheric observations and another 30-h period for *Voyager 2*, many thousands of 100-km features were tracked, usually at 10-h intervals. Averaged over the planet, the typical (root mean squared) zonal wind was found to be 50 m $\cdot$ s$^{-1}$. Departures from the zonal mean wind (eddy winds) at each latitude were moving from 10 to 15 m $\cdot$ s$^{-1}$. And a positive correlation was seen between the northward and eastward velocity components at those latitudes where the zonal wind increases with latitude. A negative relation was found in the opposite situation. This correlation is consistent with the idea that the zonal winds are continuously pulling the eddies apart (see Fig. 10-11). The kinetic energy of the eddies is not lost, however, but goes into the zonal jets. Thus the eddies help maintain the zonal jets, not vice versa. The eddies presumably get their energy from buoyancy, and are driven either by heat from Jupiter's interior or by solar heat absorbed at low latitudes.

A similar process helps maintain the Earth's jet streams. However, on Earth this energy transfer averaged over the globe is only one-thousandth of the total energy flowing through the atmosphere as sunlight and infrared radiation. On Jupiter the transfer of energy from eddies to mean zonal flow is more than one-tenth of the total energy flow. In other words, Jupiter is able to harness the thermal energy flow 100 times more effi-

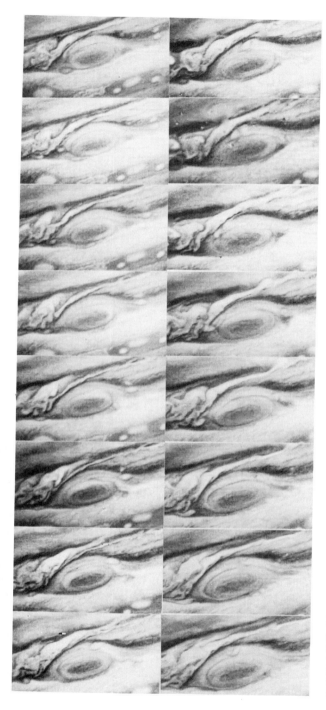

Fig. 10-10 A blue-light image of the Great Red Spot was taken every other rotation of Jupiter over a period of about 2 weeks in this sequence, which begins at upper left, continues down each column, and ends at lower right. Note the small, bright clouds that encounter the Red Spot from the east, circle counterclockwise around it, and partially merge with it along the southeast boundary.

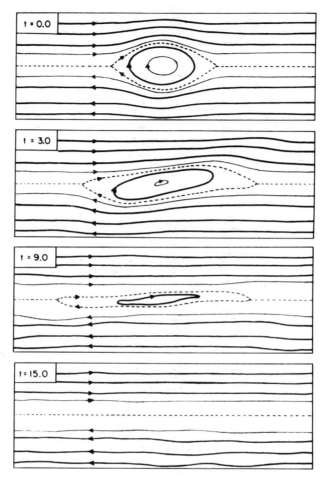

Fig. 10-11  A computer model of an unstable vortex in a zonal shear flow. Here the initial vortex is spinning too slowly to remain intact and becomes pulled apart. Its energy is transferred into the zonal jets, since total energy is conserved in this model. On Jupiter and Saturn, as on Earth, vortices like this one get their energy from buoyancy (Ingersoll and Cuong, 1981).

ciently than our own atmosphere. One possible explanation is that the Earth's eddies—the midlatitude cyclones and anticyclones—get their energy from the sideways transfer of heat due to temperature differences from equator to pole. Jovian eddies get half their energy by vertical convection of heat from the interior. The details have not been worked out, but the latter process is likely to be more efficient than the former.

The constancy of Jupiter's zonal jets still must be explained. If the eddies and zonal jets occupy the same thickness of atmosphere, the eddies could double the kinetic energy of the jets in about 75 days at the observed rate of energy transfer. If the zonal jets extend much deeper than the eddies, the doubling time will be much longer. One explanation for the more than 80-y lifetime of the zonal jets is that the mass involved in the jet motions is at least several hundred times greater than that of the eddies. If the eddies were confined to cloudy layers with pressures no greater than about 5 bars, the jets would have to extend down to pressures of 1000 bars or more. Such behavior is not as unlikely as it first seems. In a rapidly rotating sphere with an adiabatic fluid interior, the small-amplitude motions (relative to the basic rotation) are of two types. The first are waves with periods near the planet's rotation period. The second are steady zonal motions on coaxial cylinders (see Fig. 10-12), each cylinder moving about the planet's rotation axis at a unique rate. The observed zonal jets, as far as we know, could be the surface manifestations of these cylindrical patterns.

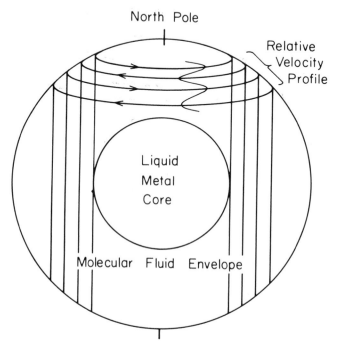

Fig. 10-12 The possible large-scale flow within the molecular fluid envelopes (see Fig. 10-15) of Jupiter and Saturn. Each cylinder has a unique rotation rate; zonal winds (see Fig. 10-4) may be the surface manifestation of these rotations. The tendency of fluids in a rotating body to align with the rotational axis was observed by Geoffrey Taylor during laboratory experiments in the 1920s and was applied to Jupiter and Saturn by F. H. Busse in the 1970s. Such behavior seems reasonable for Saturn and Jupiter if their interiors follow an adiabatic temperature gradient.

The problem with such speculation is that we lack information about winds below the visible cloud tops. A totally different approach is to treat Jupiter simply as a larger version of the Earth. Computer models designed for the Earth's atmosphere have given realistic zonal wind patterns when applied to Jupiter (see Fig. 10-13). This is somewhat surprising, since the models assume an atmosphere less than 100 km thick with a rigid lower boundary (the Earth's surface) and no internal heat flow. On the other hand, the process of eddies driving zonal flows is a very general one and seems to occur in a wide

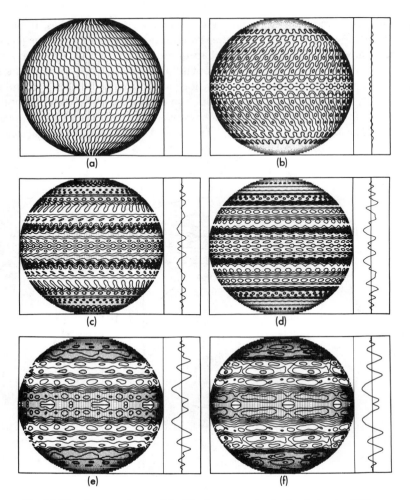

Fig. 10-13 A computer model of Jovian atmospheric circulation by Williams (1978). Williams assumes that all of the clouds' energy exchanges occur within a narrow layer of the atmosphere (as is the case on Earth); the eddies get their energy from sunlight. His model first produces small-scale eddies (A), which become unstable (B) and give up their energy to zonal jets (C, D) that eventually dominate the flow (E, F). Eddies driven by internal heat demonstrate the same behavior, so remaining questions center on the depths of both the eddies and the zonal flow.

range of situations on rotating planets. In fact, several years before *Voyager* measured it, atmospheric scientists proposed eddy-to-mean-flow energy conversion as an explanation of Jupiter's zonal flow, largely on the basis of these meteorological models.

Clearly, the lack of data on vertical structure has allowed a wide range of models to develop explaining the zonal jets of both Jupiter and Saturn. Some progress may be made by examining the secondary predictions of each model, and comparing these with actual observations.

## 10-5 THE GREAT RED SPOT

There is a similarly wide range of theories concerning the long-lived oval structures, well known on Jupiter and discovered by *Voyager* on Saturn. As already mentioned, Jupiter's Great Red Spot (GRS) is probably more than 300 y old. It covers 10° of latitude, and so is about as wide as the Earth. The three white ovals slightly to the south of the GRS first appeared in 1938; other spots have lasted for months or years. These long-lived ovals, which drift in longitude but remain fixed in latitude, tend to roll between opposing zonal jets. The circulation around their edges is almost always counterclockwise in the southern hemisphere and clockwise in the northern hemisphere, indicating that they are high-pressure centers (see Fig. 10-5). They often have cusped tips at their east and west ends. And at the time of the *Voyager* encounters, both the GRS and the white ovals had intensely turbulent regions extending off to the west.

Any theory of the GRS and white ovals must explain their longevity and their isolated nature. Longevity involves two problems. The first concerns the hydrodynamic stability of rotating oval flows. If these spots were unstable they would last only a few days, which is roughly the circulation time around their edges (or the lifetime of smaller eddies that are pulled apart by the zonal jets). The second problem concerns energy sources. A stable eddy without an energy source will eventually run down, although on Jupiter this dissipative time scale could be several years.

How do theorists account for the long lifetimes of Jupiter's ovals? The "hurricane" model postulates that these structures are giant convective cells extracting energy from condensing gases below (latent heat release). The "shear-instability" model postulates that they draw energy from the zonal currents in which they sit. Still another model postulates that they gain energy from the smaller, buoyancy-driven eddies, much as the zonal jets gain energy. It is also possible that the large ovals gain energy by absorbing smaller eddies.

One model that explains most features of the observed flow fields is the *solitary-wave theory*. Assuming certain conditions, notably a very stable density stratification, there is a class of waves in the visible clouds in which the east-west flow lines would be displaced slightly to the north and south, creating a wave. But unlike ordinary waves in which the displacement is periodic, this wave has just one crest, and is therefore called a solitary wave, or *soliton*. When two solitons collide, they pass through each other unscathed, suffering only a net displacement during the event. Thus the lifetime of a soliton at a given latitude band is limited only by the lifetime of the zonal currents that define the band. One criticism of the solitary-wave theory is that it only describes structures that have a long east-west dimension; short wave disturbances are neglected, yet in reality these can grow to large amplitude and perhaps affect the behavior of the long-wavelength solitons.

An alternate view of Jovian vortices assumes that the effective density stratification is small. This assumption rules out the kinds of waves discussed above. But if there is a basic zonal flow that extends from the top of the clouds well into the interior, the flow lines can be permanently displaced when there is a discontinuity of a certain quantity called *potential vorticity* (related to angular momentum). Regions of anomalous potential vorticity can exist as closed, circulating spots. These vortices or spots exist in the slightly less dense cloud zone and do not affect the zonal flow of the deeper atmosphere. When two such vortices collide they merge, unlike the solitary waves. After merging there is a short transient phase during which the new vortex ejects material, usually westward and equatorward. The new larger vortex finally settles down to a steady state until it encounters another vortex. A computer model can expose this behavior (see Fig. 10-14), but it is also typical of Jovian spots. The merging provides a natural explanation for how large Jovian spots maintain themselves against dissipation—they devour smaller transient spots produced by buoyancy.

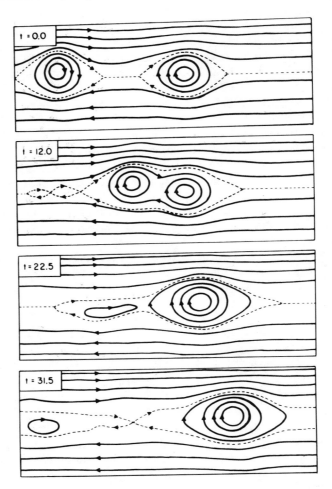

Fig. 10-14  This computer model demonstrates the collision and merging of two stable vortices (first two frames), followed after a short time by the ejection of material (last two frames). Each vortex by itself, including the resulting larger vortex, could last indefinitely. This merging behavior agrees with observations (see Fig. 10-10), and contrasts with the solitary-wave behavior, discussed in the text, in which disturbances pass through each other without interacting. Merging provides a means by which large oval spots could maintain themselves, given a supply of smaller buoyancy-driven eddies (Ingersoll and Cuong, 1981).

The above might be called a *modon theory* of Jovian spots, after a concept proposed by oceanographers to explain circulating rings of water that break off from the Gulf Stream and other major currents. A criticism of the modon theory is that it requires a zonal jet flow in the deep atmosphere beneath the clouds, and we have no knowledge of whether such a flow exists or not. The cylindrical pattern in Fig. 10-12 is theoretically possible and would support the modon theory. We are now studying how this flow might be maintained against dissipation and determining whether it is hydrodynamically stable. The observed zonal velocity profiles on Jupiter and Saturn (see Fig. 10-4) exhibit a fair degree of north-south symmetry, but there are also significant departures from symmetry, especially in the peak amplitudes of zonal jet pairs in the north and south. Differences in internal structure (see Fig. 10-15) could account for the difference between Jupiter's and Saturn's wind profiles. Further theoretical work and further analysis of *Voyager* data are needed; our data may ultimately limit the unknown parameters, so that the correct model will emerge.

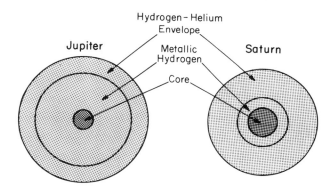

Fig. 10-15 The interiors of Jupiter and Saturn are shown schematically. The cloud-zone thickness is only 0.1 to 0.3 percent of the radius. Beneath that, clear atmosphere extends downward and gradually transforms into the liquid of the molecular fluid envelope. An abrupt transition occurs where this liquid becomes metallic. A second abrupt transition marks the interface of the hydrogen-helium zone with an ice-silicate core. Saturn's metallic zone is smaller than Jupiter's—a consequence of Saturn's lower mass, gravitational field strength, and internal pressure.

## 10-6 HEAT BUDGET AND INTERNAL STRUCTURE

As mentioned earlier, both Jupiter and Saturn radiate more heat than they absorb. This fact leads to several questions: What is the source of the internal heat? Can it be accounted for in any simple way, such as by the slow cooling of an initially hot object created 4.6 Gy ago? What is the effect of this excess heat on the atmospheric structures and circulation? What is its effect on the internal structure?

We consider the internal structure first. A crucial step in modeling the interior is knowing that temperature increases with depth along a given adiabatic profile. Here we assume that only by convection could the quantities of heat observed be carried from the interior to the cloud tops, and convection leads to an adiabatic temperature gradient. Knowing that the interior is warm, we can deduce that there is essentially no solid surface, only a gradual transition from gas to liquid followed by an abrupt transition from molecular liquid to metallic liquid (in which the molecules have been stripped of their outer electrons), and finally to an ice-rock core at the center (see Fig. 10-15 and Table 10-2). Also, knowing the effect of temperature on density, we can derive the overall hydrogen-helium ratio from the bulk density and moment of inertia. These calculations are generally consistent with the solar composition models for both planets. Finally, knowing the temperature distribution we can estimate the amount of stored heat both now and in the past, and see if it could have lasted 4.6 Gy.

TABLE 10-2  Properties of Jupiter's and Saturn's interiors are given at four distinct levels: (1) the 1-bar level in the atmosphere, (2) the transition from molecular to metallic hydrogen, (3) the outer boundary of the (hypothetical) rock-ice core, and (4) the center of the planet.

|  | (1) | (2) | (3) | (4) |
|---|---|---|---|---|
| **JUPITER** | | | | |
| Fractional radius | 1.0 | 0.76 | 0.15 | 0.0 |
| Radius ($10^3$ km) | 71 | 54 | 10 | 0 |
| Pressure ($10^6$ bar) | $10^{-6}$ | 3.0 | 42 | 80 |
| Density (g · cm³) | $2 \times 10^{-4}$ | 1.1/1.2 | 4.4/15 | 20 |
| Temperature (Kelvin) | 165° | 10,000° | 20,000° | 25,000° |
| Fractional mass | 1.0 | 0.77 | 0.04 | 0.0 |
| Mass ($10^{30}$ g) | 1.90 | 1.46 | 0.08 | 0.0 |
| **SATURN** | | | | |
| Fractional radius | 1.0 | 0.46 | 0.27 | 0.0 |
| Radius ($10^3$ km) | 60 | 28 | 16 | 0 |
| Pressure ($10^6$ bar) | $10^{-6}$ | 3.0 | 8 | 50 |
| Density (g · cm³) | $2 \times 10^{-4}$ | 1.1/1.2 | 1.9/7 | 15 |
| Temperature (Kelvin) | 140° | 9,000° | 12,000° | 20,000° |
| Fractional mass | 1.0 | 0.43 | 0.26 | 0.0 |
| Mass ($10^{30}$ g) | 0.57 | 0.24 | 0.15 | 0.0 |

Attempts to model this gradual cooling have yielded an interesting difference between Jupiter and Saturn. One starts with an initially warm planet. How warm is not important, since very little time is spent in the warm stage. As the planet cools, it contracts. The rate of energy loss is followed to the present, where the model result is compared with observation. For Jupiter, the calculated and observed heat fluxes are consistent for Saturn, they are not quite consistent: The observed internal heat flux is greater than the calculated value. Either Saturn is only half as old as Jupiter, an unlikely possibility or else another energy source besides cooling and contraction is contributing to Saturn's heat output.

A source which would operate on Saturn but not on Jupiter, as required by the observations, has been proposed. Because Saturn is smaller than Jupiter and farther from the Sun, its surface and interior are colder. Calculations suggest that Saturn's interior temperatures are too low for helium to be uniformly mixed with hydrogen throughout the metallic zone. Instead, helium should be condensing at the top of this zone, and raindrops of helium should be falling toward the center of Saturn, converting their gravitational energy into heat. This process began about 2 Gy ago when temperatures first dropped to the helium condensation point. On Jupiter this point can only have been reached recently, and the process is not yet generating significant amounts of heat.

This hypothesis requires that the helium in Saturn should be concentrated toward the center. The outer edge of the metallic zone lies 45% of the distance from the center to the cloud tops (see Fig. 10-15). Helium that condenses at the top of this zone is derived from the entire gaseous envelope which surrounds the metallic zone. This is due to the fact that the envelope is a convecting fluid and its compositional differences become homogenized quickly compared with the age of the solar system. In order to explain Saturn's excess heat flux, the envelope should have lost about one-half its original helium. Two types of observation support this hypothesis.

First, as already mentioned, the helium-to-hydrogen ratio in the atmospheres of both planets has been estimated from spacecraft observations. *Voyager* results are more accurate, and they imply just about the right amount of helium depletion if Saturn started with a uniformly mixed atmosphere of solar composition. Second, the mass of the envelope relative to Saturn's mass has been determined from measurements of the gravity field. The envelope responds more to centrifugal forces than the core does, so the amount of gravitationally induced oblateness, or flattening, is a measure of the relative mass of the envelope. Both satellite orbits and spacecraft trajectories have been used to fix this oblateness, and we find that the envelope has less mass than expected if the planet were well mixed. These observations are consistent with the idea that one-half of the heavier helium has settled out toward the core. In similar studies performed for Jupiter, no significant helium depletion is found. Jupiter's helium-to-hydrogen ratio could be equal to the solar value throughout the planet. Thus it appears that we can account for both Jupiter's and Saturn's internal heat by postulating that both planets started hot 4.6 Gy ago with essentially uniform solar abundance mixtures of hydrogen and helium.

Resolving these questions is important because it helps us acquire confidence in our models of the interiors and early histories of these planets. These models say that large bodies like the Sun, Jupiter, and Saturn give off large amounts of excess energy shortly after forming. This early high-luminosity phase is the basis of our current understanding of the differences between the inner and outer solar system, and between the inner and outer Galilean satellites. According to our scenario, volatile material orbiting the Sun in the inner solar system was either lost by evaporation or otherwise failed to be incorporated into the planets as they were forming. The result is a general increase of volatile content and decrease of density as we move away from the Sun. The same reasoning seems to account for the density differences and water-to-rock ratios of the Galilean satellites. The innermost satellite, Io, apparently has no water, whereas the outer ones, Ganymede and Callisto, seem to have approximately solar ratios of water to rock.

However, the *Voyager* results on the satellites of Saturn do not show the same trend. In fact, the Saturnian satellites are all low-density objects, showing little evidence of volatile loss. The most likely explanation seems to be that Saturn's early high-luminosity phase was weaker than Jupiter's, owing to the smaller mass of Saturn. Thus no water, even close to the planet, was driven out of the Saturnian system. (This explanation does not account for the small, irregular variations of density in the Saturnian system, or the irregular distribution of surface ages as revealed by crater statistics.)

In closing this discussion of Jupiter and Saturn, let us speculate on their internal heat flux and its possible role in the dynamics of the atmosphere and interior. There are no direct measurements of the zonal velocity profile with depth. However, there are limits to the rate at which velocity can change with depth. According to the thermal wind equation of meteorology, this change is proportional to the temperature gradient with latitude. However, no significant temperature gradients were found on either Jupiter or Saturn by the *Pioneer* and *Voyager* instruments at the deepest measurable levels. Therefore, the velocity change with depth must be extremely small, and consequently the winds measured at the cloud tops must persist below. A rough estimate is that the eastward wind speeds cannot fall to zero (relative to the internal rotation rate) at pressures less than about 10,000 bars—about 1,000 times the pressure at the cloud base (see Fig. 10-6). This is far too deep to be affected by solar energy, which is mostly absorbed in the clouds. This leaves internal heat as the obvious source for these deep motions.

The internal heat flux may also provide the mechanism for maintaining the poles and equator at the same temperature. Convection is normally very effective in producing an adiabatic temperature distribution, especially when the heat sources are located below the heat sinks. This is the arrangement for Jupiter's and Saturn's interiors. Once an adiabatic state is reached, it is maintained by the buffering effect of the internal circulation. For Jupiter and Saturn the heat is derived from the cooling of the planet, and heat sources are distributed in a spherically symmetric manner throughout the interior. The heat sinks are in the atmosphere; their total influence is defined by the infrared emission to space minus the sunlight absorbed. The net heat loss is therefore greatest at the poles and least at the equator, although it is positive at all latitudes (see Fig. 10-7). We expect that small departures from adiabaticity have arisen in the interior to drive the internal heat flow poleward. These departures are too small to be observed or to drive significant lateral heat fluxes in the atmosphere. The atmosphere is effectively short-circuited by the interior; there is no large-scale temperature gradient to disrupt the banding in the atmosphere.

The situation is very different on Earth. Here the poleward heat transport must take place in the atmosphere and oceans. Absorption of sunlight by the oceans and subsequent latent heat release from water in the atmosphere lead to a net heat gain in the tropical atmosphere. Radiation to space leads to a net heat loss in the polar atmosphere. These combine to create appreciable temperature gradients with latitude, making the atmosphere decidedly nonadiabatic. Our atmosphere and oceans show obvious evidence of mixing across latitude circles.

Again we find ourselves speculating on reasons for a pronounced difference between the gas giant planets and the Earth. And again the cause of the difference—in this case the bandedness of Jupiter and Saturn—may lie in the fluid interiors beyond our reach. Some of our speculations may turn out to be wrong. The entire circulation may be taking place in the atmosphere. Further analysis of *Voyager* data and theoretical work should resolve many of our questions.

## 10-7 REFERENCES AND ADDITIONAL READINGS

Gehrels, T., (ed.), 1976, *Jupiter*, Univ. Arizona, Tucson, Ariz., 1254 pp.

Gehrels, T., and M. S. Matthews, eds., 1984, *Saturn*, Univ. Arizona, Tucson, Ariz., 968 pp.

Ingersoll, A. P., 1982, *Jupiter* and *Saturn*, pp. 117-128 in *The New Solar System*, Second Edition, (J. K. Beatty, B. O'Leary, and A. Chaikin, eds.), Sky Publishing Corp., Cambridge, Mass.

Ingersoll, A. P., and P. G. Cuong, 1981, Numerical model of long-lived Jovian vortices, *J. Atmos. Sci.*, 38, 2067-2076.

Stone, E. C., and A. L. Lane, 1979, *Voyager 1* encounter with the Jovian System, *Science*, 204, 945-1008.

Stone, E. C., and A. L. Lane, 1979, *Voyager 2* encounter with the Jovian System, *Science*, 206, 925-996.

Stone, E. C., and E. D. Miner, 1981, *Voyager 1* encounter with the Saturnian System, *Science*, 212, 159-243.

Stone, E. C., and E. D. Miner, 1982, *Voyager 2* encounter with the Saturnian System, *Science*, 215, 499-594.

Stone, E. C., 1981, *Voyager* missions to Jupiter, *J. Geophys. Res.*, 86, 8123-8841.

Stone, E. C., 1983, *Voyager* mission to Saturn, *J. Geophys. Res.*, 88, 8625-9018.

Williams, G. P., 1978, Planetary circulations—1. Barotropic representation of Jovian and terrestrial turbulence, *J. Atmos. Sci.*, 35, 1399-1429.

David J. Stevenson
Division of Geological and Planetary Sciences
California Institute of Technology
Pasadena, California 91125

# 11

# ORIGIN, EVOLUTION, AND STRUCTURE OF THE GIANT PLANETS

# ABSTRACT

Jupiter and Saturn are adiabatic spheres of hydrogen and helium, with central concentrations of heavier elements of the order of 10 Earth masses. Uranus and Neptune are ice-rich by comparison and have masses comparable to the cores of Jupiter and Saturn. It is argued that these planets started out substantially hotter than at present and have since been gradually cooling. It is also argued that the high-density cores of Jupiter and Saturn are primordial and might be the nuclei that promoted gaseous accretion and led to the current large masses of these planets. In this review, emphasis is placed on the basic ideas needed to construct static and evolving models, and an assessment is made of the relationship between these models and current ideas for the formation and evolution of the solar system.

## 11-1 INTRODUCTION

How can we possibly know what a giant planet is made of? The corresponding problem for Earth is still only partially resolved despite the enormous amount of information available from seismology. Techniques for "sounding" the giant planets are much less developed or focused, and we have to rely on a small number of global measures, such as a few low harmonics of the gravity field. Fortunately, we are able to deduce from their average densities alone that Jupiter and Saturn are predominantly hydrogen. This can be asserted with great confidence, because there is no other material with a density as low as that of hydrogen for a given temperature. Unfortunately, no similar simplicity is offered by nature for the more distant large planets, Uranus and Neptune.

In this chapter we focus on the better understood giant planets, Jupiter and Saturn, although speculations concerning the more icy bodies (Uranus and Neptune) are also offered. The important role of hydrogen is emphasized by Sec. 11-2, which is devoted to how a self-gravitating ball of hydrogen behaves. Increasing levels of complications are then introduced, as demanded by the observations, eventually leading to some important assertions about the origin and evolution of giant planets. It is demonstrated that the observed internal heat flow of these planets implies that they started out much hotter than they are at present and have since been gradually cooling. It is also argued that the high-density (rock or ice) cores of these planets are primordial and are probably the nuclei that promoted gaseous accretion and led to giant-planet formation. This has important implications for the early evolution of the solar system. This chapter does not attempt to be a comprehensive review of the subject; for that, the reader should consult Zharkov and Trubitsyn (1978) or Stevenson (1982) or Hubbard and Stevenson (1984). Rather, the intent is to provide a very basic understanding of the issues posed by giant-planet structure and the procedures used to understand them.

## 11-2 SPHERES OF HYDROGEN

Consider a self-gravitating sphere of pure hydrogen. Suppose, first, that the temperature $T = 0°K$, so that the hydrogen is entirely solid. If the mass is sufficiently small, the hydrogen solid is not subject to high pressure and remains in its familiar molecular form.

However, if the pressure is high enough so that the electron clouds of each molecule begin to overlap, we can expect a nonnegligible population of conduction electrons. The electron distribution responsible for the preservation of $H_2$ molecules is then disrupted, and a transition to a monatomic, metallic state occurs. This state, called *metallic hydrogen*, is closely analogous to the conventional alkali metals, and is amenable to precise theoretical computation.

We can estimate the pressure at which this occurs by noticing that the work done in moving energy levels from an insulating to a conducting configuration is of the order 10 eV per molecule, and this work is approximately equal to $Pv$, where $v \simeq 20\, a_0^3$ ($a_0 \equiv$ first Bohr radius $= 0.529 \times 10^{-8}$ cm) is roughly the volume per molecule in this dense configuration. We then find $P \simeq 5$ Mbar (megabars) (1 Mbar $\equiv 10^6$ bars $\equiv 10^{12}$ dynes $\cdot$ cm$^{-2}$). More accurate estimates are usually in the range of 2 to 5 Mbar (Ross et al., 1983). This transition is likely to be first order at $T = 0°K$, with a density change of approximately 20 percent. The density at which the transition occurs is approximately 1.0 g $\cdot$ cm$^{-3}$ in molecular hydrogen.

The internal pressure $P_{int}$ within a homogeneous, self-gravitating, hydrostatic body is approximately $\bar{\rho}gR$, where $\bar{\rho} \simeq M/R^3$ is the average density, $M$ is the mass, $R$ is the radius, and $g = GM/R^2$ is the surface gravitational acceleration with $G$ as the universal gravitational constant. Inserting the values for mass and radius for Jupiter and Saturn, we find $P_{int} \simeq 20$ Mbar and 10 Mbar respectively. The detailed models, discussed further below, confirm that both of these planets are massive enough to have large amounts of metallic hydrogen. Uranus and Neptune probably do not have any metallic hydrogen (although they might have in the deep interior a multicomponent mixture that includes protons and free electrons).

If $T \neq 0°K$, then most of the conclusions are qualitatively unaltered. The reason is that the giant planets are *cold:* The thermal energy per particle ($\simeq kT$, where $k$ is Boltzmann's constant) is much less than the internal energy per particle (which is about equal to the magnitude of the gravitational energy per particle, according to a basic argument called the *Virial theorem*). In other words, throughout most of the interior, the fractional change of volume with temperature is very small. Under these circumstances, an insulator-metal transition must still exist within Jupiter and Saturn, although it will occur in a *liquid* rather than a solid.

We can deduce that the planets are "cold" by making use of the fact that for a homogeneous, fluid planet the temperature increase with depth cannot be too large. If it were, then convective motions would rapidly redistribute the heat, transporting it from the deep interior to the outside. Convection stops when the temperature varies with pressure no more rapidly than in what is called an *adiabatic state*. The argument is consistent if the adiabatic state is fluid throughout, since in those circumstances, viscosity does not inhibit the fluid motions. (In contrast, a solid planet such as the Earth has high viscosity and large deviations from adiabaticity.) We can proceed by trial and error as follows: In the observable atmosphere of Jupiter, the temperature is about 170°K at $P = 1$ bar. As we proceed deeper, the temperature can increase no more rapidly than $T \propto P^n$, where $n$ is the appropriate adiabatic index. In the ideal gas regime, $n \simeq 0.3$, and in metallic hydrogen $n \simeq 0.3$ again (a coincidence). The intermediate, dense molecular-hydrogen regime is more complicated, but the value of $n$ is nevertheless remarkably constant. If we assume $n = 0.3$ throughout the planet, we find $T \simeq 30{,}000°K$ at $P \simeq 40$ Mbar. A more detailed calculation predicts $T \simeq 20{,}000°K$. This is believed to be roughly the central temperature of Jupiter, because the interior is convective and adiabatic (see Sec. 11-4). This temperature corresponds to a thermal energy per proton of order $kT \simeq 2$ eV; by comparison, the gravitational energy is $GMm_p/R \simeq 20$ eV, where $M$ and $R$ are the mass and radius of Jupiter, respectively, and $m_p$ is the proton mass. All that remains is to check that the calculated temperatures exceed the melting point of hydrogen. This is indeed the case, as the phase diagram in Fig. 11-1 demonstrates. We have thereby confirmed that thermal energies are relatively unimportant in modifying bulk properties of the planet. Of course, the temperature is important in other respects, especially in processes of planetary evolution. We return to this later.

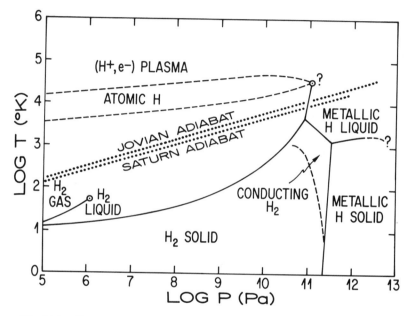

Fig. 11-1  Phase diagram of hydrogen. The pressure units are pascals ($10^5$ Pa $\equiv$ 1 bar). Solid lines indicate first-order phase transitions (except that the line ending in a question mark and separating fluid $H_2$ from fluid metallic H is uncertain). The conducting $H_2$ region is hypothetical. The dotted adiabats for Jupiter and Saturn are lines of constant entropy fixed by the atmospheric temperature at 1 bar, given in Table 11-1. (Reproduced, with permission, from the *Annual Review of Earth and Planetary Sciences*, 10, 257–295. © 1982 by Annual Reviews Inc.)

In the limit of very high pressure, the unbound, degenerate electrons (see text in box page 258) provide most of the pressure and $P \propto n^{5/3}$, where $n_e$ is the electron density. If all the electrons are unbound and degenerate, the mass density $\rho \propto An_e/z$, where $A$ is the atomic weight and $z$ is the nuclear charge. It is significant that $A/z = 1$ for hydrogen but is approximately twice as large for all other elements. This causes hydrogen to have a significantly lower mass density than other elements for a given electron density. However, the giant planets are not at high enough pressures so that $P \propto \rho^{5/3}$, the relation that would

> **Degeneracy of Electrons**
>
> The laws of quantum mechanics require that at most one electron can occupy any quantum mechanical state in which its basic properties (energy, angular momentum, etc.) are fully specified. An electron gas is said to be *degenerate* when only the lowest states are occupied, subject to the exclusion rule stated above. In the giant planets, the requirement for degeneracy is $T$ less than or of order $10^5\,°K$ and is satisfied.

apply for an ideal, degenerate electron gas. Coulomb effects (electron-ion interactions, electron exchange) are important and tend to increase the exponent, so that $P \propto \rho^2$ is a better approximation. In fact, the simple relation $P = K\rho^2$ turns out to work remarkably well even for molecular hydrogen. It is instructive, therefore, to analyze the implications of this approximate relation.

If we insert $P = K\rho^2$ in the equation of hydrostatic equilibrium for a nonrotating body (which requires that the radial gradient of pressure at $r$ be balanced by the gravitational attraction of the material inside $r$), we find

$$\frac{dP}{dr} = -\rho g = -\rho \left(\frac{G}{r^2}\right) \int_0^r 4\pi x^2 \rho(x) dx \tag{11-1}$$

The solution that is finite at the origin has the form

$$\rho = \rho_c \left(\frac{\sin kr}{kr}\right) \tag{11-2}$$

$$k \equiv \left(\frac{2\pi G}{K}\right)^{1/2}$$

Since the density must be positive, $R = \pi/k$ defines the outer radius of the planet. Note that $R$ is independent of the mass and depends only on $K$. If we choose $K = 2.7 \times 10^{12}$ cm$^5 \cdot$ g$^{-1} \cdot$ s$^{-2}$ which is the best fit to the "exact" equation of state, then $R = 7.97 \times 10^9$ cm. For comparison, the average radius of Jupiter is $6.98 \times 10^9$ cm. Significantly, the average radius of Saturn is only $5.83 \times 10^9$ cm. This is *not* because Saturn is less massive, but because it is further removed from a pure hydrogen composition than in Jupiter. If it had the same composition as Jupiter, then it would have about the same radius as Jupiter.

The simple density profile of Eq. (11-2) can also be used to estimate the moment of inertia, $I$, about an axis. One finds $I/MR^2 = (\frac{2}{3})(1 - 6/\pi^2) \simeq 0.26$. The "observed" value of Jupiter is about 0.25, and the value for Saturn is about 0.23. We return later to how these are deduced and what they imply. For the moment, it suffices to notice that these values are lower than for a homogeneous planet, implying a central concentration of heavier material.

Because the estimates above are roughly consistent with observations, it is interesting to extend this model slightly by including a central, dense core. The general formula for the density in the region external to the core is then

$$\rho = A\left(\frac{\sin kr}{kr}\right) + B\left(\frac{\cos kr}{kr}\right) \qquad (11\text{-}3)$$

and, to lowest order in $B$, the planetary radius

$$R = \frac{\pi}{k}\left(1 - \frac{B}{\pi A}\right)$$

and the planetary mass

$$M = \frac{4AR^3}{\pi}\left(1 + \frac{B}{\pi A}\right).$$

In order to satisfy hydrostatic equilibrium just outside the dense core, $B \simeq GM_c k/2K$, where $M_c$ is the mass of the core. (The *radius* of the core does not matter, provided it is small.) If we seek to explain why the reduced radius of Jupiter is reduced relative to that of a pure hydrogen sphere, then $M_c \simeq 0.16\, M_{\text{Jupiter}}$. For a reasonable choice of average core density (approximately 20 g·cm$^{-3}$), this yields $I/MR^2 \simeq 0.235$, too low compared with observations. Thus we deduce that the radius of Jupiter is less than that of a pure hydrogen sphere, not just because of a dense core, but because the envelope is not pure hydrogen. A similar conclusion applies for Saturn.

## 11-3 SPHERES OF OTHER ELEMENTS AND MIXTURES

The simple model of a sphere of pure hydrogen is a remarkably good first estimate for Jupiter and Saturn, yet it is nevertheless clearly inadequate since it predicts a planet substantially larger than either of these bodies. The presence of other constituents is also evident from observations of the atmosphere. One way to proceed is by analysis of self-gravitating spheres of heavier elements: helium, oxygen, carbon, and so forth (chosen because of their cosmic abundance). This can be done using a combination of experimental data at low pressure and theoretical analysis (the Thomas-Fermi-Dirac model) at very high pressure (see Salpeter and Zapolsky, 1967). A somewhat more useful approach is pursued here for the giant planets and involves consideration of appropriate cosmochemical mixtures.

Since fractionation of helium from hydrogen cannot be achieved by any plausible process prior to giant-planet formation, it is reasonable to suppose that the average helium fraction is approximately the same as that in the interstellar medium or the primordial Sun, about 20 to 25 percent by mass. A theoretical analysis of the influence of helium on the equation of state shows that $p = K\rho^2$ is still a quite good approximation, with $K \simeq (2.7 \times 10^{12})(1 - Y/2)^2$ cm$^5$·g$^{-1}$·s$^{-2}$ where $Y$ = helium mass fraction. It follows that $R = (7.97 \times 10^9)(1 - Y/2)$ cm, which provides the observed Jovian radius if $Y \simeq 0.25$

and the observed Saturnian radius if $Y \simeq 0.5$. This result is satisfactory for Jupiter, but not reasonable for Saturn. The lower moment of inertia factor, $I/MR^2$, for Saturn had already warned us that Saturn was not likely to be so close to cosmic composition as Jupiter.

In many respects, the addition of helium does not change the general character of the internal structure. The hydrogen still metallizes at several megabars, and the helium electronic states probably remain bound until somewhat higher pressures before these electrons are gradually released to the conduction sea. However, there is one very interesting feature of the hydrogen-helium mixture that may be very important to giant-planet structure and evolution: It is very likely that helium has only limited solubility in metallic hydrogen. The analogous laboratory behavior of noble gases dissolved in metals is well known. For example, helium gas is highly insoluble in liquid or solid aluminum metal (chosen here as an example because of its high electron density). The problem with extrapolating this behavior to the giant planets is that at the lowest pressures for which metallic hydrogen exists, the structure of the helium atoms begins to change. Nevertheless, detailed calculations (Stevenson, 1979) indicate that helium has limited solubility in metallic hydrogen for $T \lesssim 10,000°K$ and $P$ about equal to a few Mbars. Limited solubility is also expected in the very high pressure limit, where all the electrons are no longer bound to the nuclei (Stevenson, 1975), but this is a subtle effect and its magnitude has been questioned (MacFarlane and Hubbard, 1983). This limited solubility could allow formation of a layered planet with the outer region being helium-poor and the deeper regions being enriched in the more dense helium. Current observations support this possibility for Saturn but not for Jupiter. This is consistent with the expectation that since Saturn has a colder atmosphere than Jupiter and both planets are likely to be adiabatic, all levels within Saturn are colder than the corresponding pressure levels in Jupiter. If helium redistribution occurs, it has important implications for thermal evolution and planetary magnetism. We return to this issue later. It is important to stress, however, that redistribution of helium does not solve the problem of Saturn's small radius or moment of inertia if the total helium content is cosmic. Heavier material is required. (It also turns out that heavier material is required in Jupiter, but this can be deduced only by detailed modeling and is not evident at the crude level of understanding sought in this discussion.)

After hydrogen and helium, the next three most abundant elements are oxygen, carbon, and nitrogen. (Neon is about as abundant as nitrogen but less interesting since it does not condense and has no chemistry.) In the hydrogen-rich environments encountered in Jupiter and Saturn, at least, the preferred dominant forms of these three elements are water, methane, and ammonia, respectively. This does not necessarily mean that they were originally incorporated into the planet in these forms; in particular, CO and $N_2$ are possible primordial reservoirs of carbon and nitrogen. It is also likely that the molecules $H_2O$, $CH_4$, and $NH_3$ cease to be well-defined entities at megabar pressures. It has long been suspected that water dissociates at high pressure ($P \gtrsim 200$ kbar) and temperature into an ionic melt $H_3O^+OH^-$, isoelectronic (same number of electrons) with $NH_4^+F^-$ (Hamann and Linton, 1966), and it is likely to metallize above several megabars. Methane, like all hydrocarbons, is expected to decompose at extreme pressure (Ree, 1979) and perhaps even unmix to form elemental carbon (diamond or metal) and hydrogen, but this depends on complex aspects of chemical reactions (see discussion in Stevenson, 1982). Ammonia is likely to form $NH_4^+$ ions in the presence of dissociated water and excess hydrogen. As with helium, the behavior of the ices ($H_2O$, $NH_3$, and $CH_4$) in the giant planets must depend on the chemical environment, and an understanding of individual pure members may be a grossly inadequate basis for planetary model construction. Nevertheless, it is the current basis for models, together with the assumption of *volume additivity*, which states the rules for adding volumes of the constituents and weighting the volumes appropriately.

The pressure-density relationships for the three ices are quite well known up to pressures of approximately 1 Mbar because of recent shock-wave experiments and their interpretation (Ross et al., 1981). At extreme pressures ($\gtrsim 10^2$ Mbars), a theoretical approach is expected to work (Salpeter and Zapolsky, 1967). The important "intermediate" pressure range is poorly understood at present. The ices are much less compressible than hydrogen yet substantially more compressible than rock, and all have densities much greater than 1 g · cm$^{-3}$ at megabar pressures. It is important to remember this when trying to understand the meaning of the average densities of the giant planets, all of which are too low to be explainable by a pure ice composition.

A major unsolved problem with ices concerns their mixing properties with hydrogen. It is known, for example, that hydrogen and water mix in all proportions above a temperature of approximately 650 K, at pressure less than or about equal to several kilobars (Seward and Franck, 1981). Complete mixing at higher pressures is also suspected for the temperatures expected in Jupiter and Saturn (Stevenson and Fishbein, 1981). The extent of mixing in Uranus and Neptune remains unclear at present. This is an important issue for all the giant planets because they have dense cores that may consist, in part, of ice. The failure of this material to homogenize with the overlying hydrogen has some interesting implications for the formation and evolution of these planets, another point to which we will return later.

The remaining important group of constituents expected to be important in giant planets is loosely describable as "rock." It is the material out of which the terrestrial planets (e.g., Earth) formed and is refractory (i.e., condenses from the vapor phase at a much higher temperature than ices condense). By mass, rock consists of comparable amounts of iron, magnesium, silicon, and oxygen. Numerous other elements are present, of course, but less important. The ways in which these elements are combined in Earth comprise the sciences of petrology and high pressure geochemistry, but for our present concerns, these details are neither important nor well understood. At the pressures encountered in terrestrial planets, an iron core surrounded by a silicate mantle is expected. At the much higher temperatures and pressures encountered in giant planets, the phase assemblage is not known. For example, silicates may be metallized and soluble in metallic iron. It is also not known whether this material is liquid or solid at the centers of Jupiter and Saturn.

The pressure-density relationship for rock can be constructed in much the same way as that for ice, by combining low pressure ($p \lesssim 3$ Mbar), shock-wave data with an interpolation through intermediate pressures (3 Mbar $\lesssim p \lesssim$ a few hundred megabars), which merges with the theoretical (Thomas-Fermi-Dirac) limit. The high-pressure limit is insensitive to the phase assemblage chosen, but the density in the low-pressure regime is uncertain to ±10% because of ignorance concerning the appropriate phase assemblage.

Figure 11-2 summarizes the conclusions of the discussion presented in this section and the previous section. It shows the dependence of radius on mass for planets com-

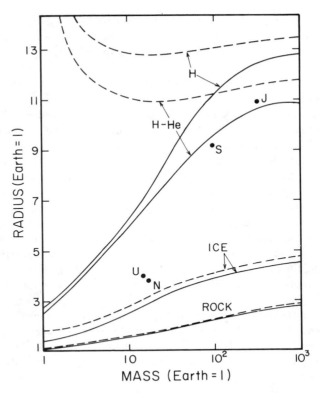

Fig. 11-2 The mass-radius relationship for self-gravitating bodies of various labeled compositions. The solid lines are for cold matter ($T = 0$ K); the dashed lines correspond to adiabatic models (see text). The positions of the giant planets are labeled by J, S, U, and N. (Reproduced, with permission, from the *Annual Review of Earth and Planetary Sciences*, 10, 257–295. © 1982 by Annual Reviews Inc.)

posed of four different possible materials: pure hydrogen, a cosmic hydrogen-helium mixture, ice ($H_2O$, $CH_4$, and $NH_3$ in relative cosmic abundance), and rock. The curves labeled H-He essentially represent cosmic composition, since all other elements are so much less abundant than either hydrogen or helium that their presence in cosmic proportion changes the density very little. Both isothermal ($T = 0°$K) and adiabatic cases are shown. (The chosen adiabat for H and H-He is that appropriate to Jupiter and passes through $T = 170°$K at $P = 1$ bar. The chosen adiabats for ice and rock are appropriate for the deep interior of Uranus and pass through $2000°$K at 200 kbar.) The H and H-He curves show that except in the cases of low-mass "gasballs," the isothermal and adiabatic results are very similar, supporting our earlier claim that these planets are cold, to a first approximation. Notice, also, that the radius of an adiabatic gasball is insensitive to mass, consistent with our simple analysis of $p \propto \rho^2$ in Sec. 2.

Figure 11-2 enables a preliminary assessment of the compositions of the giant planets, each of which is designated by the first letter of its name. We see that Jupiter is closest to an adiabatic cosmic mixture, with a radius that is perhaps only 5 percent too low. In accordance with our discussion at the end of Sec. 2, this discrepancy can be explained by the presence of an approximately 20-Earth-mass core of rock. This choice is also compatible with the moment of inertia. The radius of Saturn is about 15 percent too low; this can be explained by a rock core that is also about 20 Earth masses (but a much larger fraction of the total mass than in the case of Jupiter). Uranus and Neptune are clearly very different, and the figure does not provide a clear guide to their composition, except to suggest strongly that ice is important.

## 11-4 HOW GIANT PLANET MODELS ARE CONSTRUCTED

We now dispense with the generalities (generic models) and proceed to the specifics of constructing models for individual planets. Clearly, we would like to formulate an inverse problem in which the observational and theoretical constraints lead to a unique model. In practice, only a limited inverse problem can be formulated in which simplifying assumptions allow the number of unknowns to be matched to the number of independent, relevant observables. The biggest uncertainty in this procedure is the validity of the simplifying assumptions.

The relevant observables are given in Table 11-1. They are all external in the sense that there are no observations, comparable to terrestrial seismology, which provide almost direct information on the interior. However, the gravity field provides integral constraints on internal densities, and atmospheric measurements provide important thermal and compositional boundary conditions.

Consider, first, the gravity field of a planet. If the planet were not rotating and were in exact hydrostatic equilibrium, the external gravity field would be spherically symmetric and convey only one bit of information about the planet: its total mass. Fortunately,

TABLE 11-1  Planetary Properties[a]

|  | *Jupiter* | *Saturn* | *Uranus* | *Neptune* |
|---|---|---|---|---|
| Mass (Earth = 1) | 318.05 | 95.147 | 14.58 (±0.1) | 17.23 (±0.08) |
| Average radius of $P = 1$ bar surface ($10^9$ cm) | 6.980 (±0.001) | 5.83 (±0.003) | 2.55 (±0.02) | 2.45 (±0.03) |
| Average density (g · cm$^{-3}$) | 1.334 (±0.006) | 0.69 (±0.01) | 1.26 (±0.07) | 1.67 (±0.1) |
| $J_2$ | $(14{,}733 \pm 4) \times 10^{-6}$ | $(16{,}479 \pm 18) \times 10^{-6}$ | $(3354 \pm 5) \times 10^{-6}$ | $(41 \pm 4) \times 10^{-4}$ |
| $J_4$ | $(-587 \pm 7) \times 10^{-6}$ | $(-937 \pm 38) \times 10^{-6}$ | $-3 \times 10^{-5}$ | — |
| Rotational period | $9^\text{h}55^\text{m}29.7^\text{s}$ | $10^\text{h}39.4^\text{m}$ | $(15.6 \pm 1.2)^\text{h}$ | $\sim 18^\text{h}$? |
| $I/MR^2$ | 0.253 (±0.003) | 0.227 (±0.006) | 0.16 (±0.02) | 0.15? |
| Temperature at $P = 1$ bar (°K) | 170 ± 20 | 135 ± 15 | 75–80? | 70–75? |
| Excess luminosity (watts) | $(4 \pm 1) \times 10^{17}$ | $(1.3 \pm 0.3) \times 10^{17}$ | $\leqslant 1.5 \times 10^{15}$ | $(2 \pm 1) \times 10^{15}$ |

[a] See Stevenson (1982) for detailed references on data.

planets rotate and are oblate. The external gravitational potential $\Phi$ can then be expressed as

$$\Phi = -\frac{GM}{r}\left[1 - \sum_{n=1}\left(\frac{R_e}{r}\right)^{2n} J_{2n} P_{2n}(\cos\theta)\right] \qquad (11\text{-}4)$$

where $R_e$ is the equatorial radius, $\theta$ is the colatitude (the angle between the rotation axis and the radial vector **r**), $P_{2n}$ are polynomial functions of their arguments called the *Legendre polynomials*, and the dimensionless numbers $J_{2n}$ are known as the *gravitational moments*. The $J_{2n}$ can be measured (for low $n$) by tracking flyby spacecraft or by observing the orbital precession of several satellites or rings. The second-order moment $J_2$ is most easily measured and is crudely related to the planetary moment of inertia (see Chapter 4 by Kaula, especially boxed text on p. 80). An approximate algebraic formula can be derived (see Stevenson, 1982) in the form

$$\frac{I}{MR^2} = \frac{\frac{2}{3} J_2}{J_2 + \frac{\Omega^2 R^3}{3GM}} \qquad (11\text{-}5)$$

where $\Omega$ is the planetary rotational angular velocity, but in reality the relationship between $J_{2n}$ and the internal density distribution is determined by a complicated integro-differential equation. Nevertheless, this simple formula illustrates in a semiquantitative fashion the relationship between an internal property ($I/MR^2$) and external observables ($\Omega$, $J_2$). For example, it is clear that $J_2$ must be large if $\Omega$ is large and $I/MR^2$ is not small. Since the giant planets are rapid rotators, this aids accurate measurement. Other factors being equal, it is also clear that $J_2$ is small if the planet is centrally condensed (low $I$). This is physically reasonable since the gravity field of a centrally condensed object should more closely approach that of a point object, *despite* the oblateness of the outer, low-density envelope. Table 11-1 shows that $J_2$ is known to high accuracy for all the giant planets except Neptune. The Jupiter and Saturn values are derived from spacecraft flybys (primarily *Pioneer 11*), the Uranus value comes from observed precession of an eccentric ring, and the Neptune value is derived from observing its moon Triton. Since the interpretation of $J_2$ requires accurate knowledge of $\Omega$, only Jupiter and Saturn have adequate constraints on interior structure. The rotation period for Uranus is *derived* from the observed $J_2$ and the observed oblateness, assuming hydrostatic equilibrium, and is consequently rather uncertain because of the relatively large error in the oblateness.

The next gravitational moment, $J_4$, is not related in a clear way to any particular integral measure of the planetary density distribution but depends primarily on the properties of the region exterior to the sphere of radius equal to the polar radius (i.e., the equatorial bulge). Unlike the deep interior, this region is sensitive to temperature. Consequently, $J_4$ is useful as a constraint on internal temperature *provided* the values of $J_2$ and average density have already provided good constraints on composition. In practice, the interpretation of $J_4$ tends to be somewhat ambiguous even for Jupiter where it is known to high accuracy. The higher-order moments ($n \geqslant 3$) are not yet known or constrained to useful accuracy.

We now turn to the atmospheres of these planets. There is no direct relationship between the temperature or composition of a terrestrial planetary atmosphere and the temperature or composition of that planet's interior, so it is not immediately apparent why such a relationship should be expected for the giant planets. The relationship exists because, for hydrogen, which is observed to be the dominant constituent in each giant-planet atmosphere, at the temperatures and pressures of interest there is no gas-liquid or gas-solid phase transition. In other words, these planets have bottomless atmospheres. The other important factor is the convective nature of these atmospheres. It is known from the combined analysis of infrared and visible observations that these planets (except, possibly, Uranus) emit more energy than the solar energy they absorb. This internal heat flux is too great to be transported by microscopic transport processes (conduction, radiation) if the temperature gradient is stable (i.e., subadiabatic). The heat, therefore, must be transported by convection. In the absence of phase transitions, this should provide homogeneity. In other words, the atmospheric composition should be a good measure of composition deep within the envelope, except for those constituents that are condensible. Unfortunately, the exceptions include water, a likely major constituent.

Perhaps the most interesting compositional variable is the ratio He/H. This can be measured indirectly by analyzing the strength of the pressure-induced infrared absorption of molecular hydrogen, which depends on collisions between hydrogen and other atoms or molecules, including helium. In Jupiter's atmosphere, the helium abundance is thus found to be 19 ± 4% by mass, consistent with cosmic abundance. In Saturn's atmosphere, the helium abundance is 11 ± 4%, almost certainly lower than either the Jovian value or cosmic abundance (Gautier and Owen, 1983). As previously noted, this suggests unmixing in the interior of Saturn because of the limited solubility of helium in metallic hydrogen. The evolutionary implications of this are discussed in Sec. 11-6.

The next most abundant observable constituent (after hydrogen and helium) is methane. Surprisingly, methane is enhanced relative to cosmic abundance in all the giant planets. The enhancement is twofold in Jupiter (Gautier et al., 1982), perhaps fivefold in Saturn (Courtin et al., 1983), and probably even larger in Uranus (Wallace, 1980). This is surprising for Jupiter and Saturn because it is unlikely that these planets formed in a region where the temperature was so low that methane could condense as an ice and be preferentially incorporated into the planet as solid (grains or icy planetesimals). The required temperature for condensation is approximately 20 K ($CH_4$ ice) and approximately 40 K ($CH_4 \cdot 7H_2O$; methane clathrate). One unlikely explanation is that because of some internal-phase transition, the central portions of the planets are depleted in $CH_4$ and the outer portions are enriched. A more likely explanation is the addition of solid material (comets) that formed far out in the solar system (Uranus or beyond) and then was scattered gravitationally into the region of Jupiter and Saturn, where some of it accreted onto these planets long after they formed (approximately $10^8$ y). This is at least qualitatively consistent with the postulated massive inner Oort cloud of comets that has received considerable recent attention (see, for example, Shoemaker and Wolfe, 1984 and Chapter 15 by Neugebauer). In any event, the presence of a methane enhancement is circumstantial evidence for enhancement of other ices and provides justification for models in which the outer envelope is substantially enriched in heavier material relative to a cosmic abundance mixture. There is no clear evidence either for or against substantial enhancements of $H_2O$ or $NH_3$.

Infrared observations or occultation observations provide limited information on the atmospheric temperature profile, a boundary condition for interior models. This is usually expressed as the temperature at the 1-bar level (see Table 11-1), which happens to be roughly at an optical depth of unity in Jupiter and Saturn (see Chapter 6 by Kivelson and Schubert). More importantly, it is a level at which both observation and calculation indicate an adiabatic temperature profile. As one proceeds deeper, the temperature continues to be adiabatically related to the pressure, characteristic of highly subsonic convection in a low-viscosity medium. This conclusion seems firm for at least Jupiter and Saturn and probably Neptune, based on the observed intrinsic (i.e., excess) heat flows in Table 11-1, together with theoretical calculations demonstrating the inadequacy of microscopic heat-transport processes. (For a detailed discussion of this, see Stevenson and Salpeter, 1977). There is only an upper bound on the Uranian heat flow, but the actual heat flow would have to be substantially less than this before radiation or conduction became adequate heat-transporting processes deep within the planet.

One other observational constraint, the magnetic field, is not listed in Table 11-1. It has a less useful role in the construction of planetary models because of the absence of a quantitative theory of planetary magnetism (but see Stevenson, 1983). The presence of a large field argues for the existence of a hydromagnetic dynamo, which requires a large, electrically conducting region in substantial nonuniform motion (see Chapter 13 by Levy). In Jupiter and Saturn, convection in the metallic-hydrogen core is more than sufficient to guarantee a dynamo. In Uranus and Neptune, even ionic water may be sufficiently conducting, and metallic material is likely to be present at still deeper levels. It may eventually become possible to use magnetic field observations in a more quantitative fashion, but for the present they do not serve as strong constraints on the models.

We are now in a position to understand how a static planetary model can be constructed. ("Static" means that the model is designed to describe the present state of the planet but does not attempt to explain how it evolved to that state.) The ingredients are

1. The equation of hydrostatic equilibrium
2. The equations of state that relate pressure, density, and temperature for a specified composition
3. The thermal structure $T(P)$, which is adiabatic in all the models discussed here and consistent with the outer boundary condition (see Table 11-1)
4. The observed gravitational moments ($J_2$, $J_4$) and the theory of planetary figures that relates these moments to the internal-density distribution
5. The outer compositional boundary condition (a rather weak constraint, since the enrichment of the envelope in water, for example, is not known directly)

Let us see how we might construct a planetary model using a body like Jupiter as an example. Starting from the outside (known mass and radius), we integrate inward using the equation of hydrostatic equilibrium, the known adiabat, and an assumed composition. Assuming homogeneity, we continue inward until the remaining interior mass and radius are consistent with a rock or rock-ice core. (This almost invariably happens unless the envelope is highly enriched in heavier elements.) At that point, we postulate the presence of this core and obtain a model that is consistent with the observed external mass and radius. Most models generated in this way would not reproduce the observed $J_2$ and $J_4$, however. The usual procedure, therefore, is to adjust the envelope composition *and* temperature to satisfy these moments. The motivation for adjusting temperature is that the "observed" 1-bar boundary condition is actually subject to significant (about

10 percent) uncertainty, and this is a large enough effect to influence interior models. An alternative and possibly preferable approach acknowledges the uncertainty in the equations of state (especially in the pressure range 0.1 Mbar to 3.0 Mbar) and adjusts the equation of state (within bounds set by experiment and theory), together with envelope composition, to satisfy both $J_2$ and $J_4$.

Clearly this is not a unique procedure, since it is possible to have a variety of central cores or central enhancements of heavy material, or to postulate inhomogeneities (layering) of the envelope. The conventional procedure in published models is governed more by Occam's razor than by compelling reasoning. Nevertheless, some features of the models are remarkably robust. In particular, the presence of a central core or concentration of heavy material appears to be mandatory in all the giant planets and to be 10 Earth masses to within a factor of 3 in all cases.

## 11-5 TYPICAL MODELS: A GUIDED TOUR OF GIANT-PLANET INTERIORS

Figures 11-3, 11-4, and 11-5 are pie diagrams representing likely interior structures of Jupiter, Saturn, and Uranus, respectively. (Neptune is not accorded a separate figure because it is not sufficiently distinguishable from Uranus in our present state of ignorance.)

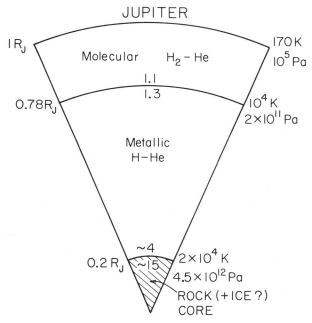

Fig. 11-3 Schematic representation of a typical (no particular) Jupiter model. Numbers adjacent to boundaries are densities in g · cm⁻³. Pressures are given in pascals ($10^5$ Pa ≡ 1 bar). (Reproduced, with permission, from the *Annual Review of Earth and Planetary Sciences*, 10, 257-295. © 1982 by Annual Reviews Inc.)

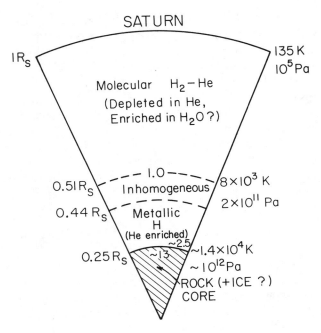

Fig. 11-4 Schematic representation of the present-day Saturn, showing a partially differentiated structure because of downward helium migration. Numbers adjacent to boundaries are densities in g · cm$^{-3}$. Pressures are given in pascals ($10^5$ Pa ≡ 1 bar). (Reproduced, with permission, from the *Annual Review of Earth and Planetary Sciences*, 10, 257-295. © 1982 by Annual Reviews Inc.)

Suppose we were to travel down into the interior of Jupiter in a "submersible." What would we observe? As we proceed downward, the temperature rises rapidly in accordance with an adiabat, initially that for an ideal gas. Water clouds are encountered at approximately 5 bars pressure level and clouds of other condensibles (e.g., silicates) several hundred kilometers deeper. At only 700 km below the cloud deck, which we see from Earth ($NH_3$ cirrus), the temperature has reached 2000°K. No phase transitions are encountered (except those associated with cloud formation of minor constituents). The gas becomes progressively more nonideal as we descend, until the effects of intermolecular interactions gradually begin to exceed the ideal gas pressure (at $\rho \simeq 0.05$ g · cm$^{-3}$, $T \simeq 3000°$K). This dense, molecular fluid is undergoing convection, perhaps with convective velocities of order 1 cm · s$^{-1}$. Eventually at $P \simeq 2$ Mbar ($\rho \simeq 1$ g · cm$^{-3}$ and $T \simeq 10^{4°}$K), the molecular hydrogen dissociates and ionizes to form a fluid, metallic state (Coulomb plasma). This might be an abrupt transition (as it would probably be at 0°K), but it is more probably gradual at the ambient temperature. This region, at 75 to 80 percent of Jupiter's radius, may represent the outer extent of dynamo generation of the Jovian magnetic field, although significant dynamo action could conceivably occur in the molecular region. Incidentally, the attenuation distance for light deep in the molecular hydrogen envelope (but not so deep that the photons can excite electrons across the bandgap of the semiconducting $H_2$) is probably of order 1 cm. The metallic fluid is, of course, com-

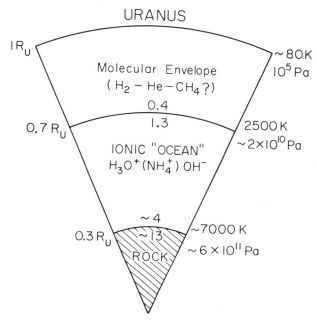

Fig. 11-5 Schematic representation of Uranus (but Neptune may be very similar), based primarily on Hubbard and MacFarlane (1980). See text for other possibilities. Numbers adjacent to boundaries are densities in g · cm$^{-3}$. Pressures are given in pascals ($10^5$ Pa ≡ 1 bar). (Reproduced, with permission, from the *Annual Review of Earth and Planetary Sciences, 10,* 257–295. © 1982 by Annual Reviews Inc.)

pletely opaque. This region is the largest component by mass in Jupiter, extending downward another order of magnitude in pressure (to approximately 40 Mbar) before a rock (and ice?) core is encountered near the planet center. The precise nature and mass of this core are not known. It is likely to be soluble in the overlying metallic hydrogen but has persisted because upward convective mixing is inefficient (Stevenson, 1982). This would imply that the core is primordial, a possibility that we return to in the next section. At the center of Jupiter, the pressure is roughly 100 Mbar, the density is approximately 20 g · cm$^{-3}$, and the temperature is in excess of 20,000°K (but probably not greater than 30,000°K). It is important to realize that the large uncertainties in these values arise because of the uncertain nature and size of the core. In contrast, the conditions immediately above the core in the metallic-hydrogen zone are known to within 10 percent if the heavy-element enhancement is not very large.

Our guided tour of Saturn has some similarities but also some striking differences. If the envelope is helium-depleted, then the presence of other heavier material (most probably water) seems to be required as partial compensation since the Saturnian $J_4$ does not indicate a low-density envelope. However, this heavier material is entirely consistent with the observed fivefold methane enhancement in the atmosphere. (The C/O ratio in the source material for this enhancement is not known, but it is likely that the water enhancement is even greater.) Dense, extended water clouds are likely at 10 bars to 30 bars, and temperature rises rapidly (just as in Jupiter) as we proceed downward a few hundred kilometers. The fluid becomes highly nonideal at $\rho \gtrsim 0.05$ g · cm$^{-3}$ and eventually transforms to the metallic phase at a similar pressure and density to Jupiter's transition region (but at a *lower* temperature because Saturn's adiabat is colder than Jupiter). The transition occurs at approximately 50 percent of Saturn's radius, much deeper than the transition within Jupiter because of Saturn's lower mass. Unlike Jupiter, the temperature at this level may be low enough for helium raindrops to form. In these circumstances, the region immediately below the molecular-metallic-hydrogen transition is expected to follow the phase boundary defining the pressure- and temperature-dependent solubility of helium. This implies a *stable* compositional gradient linking the helium-depleted envelope to the helium-enriched core. This inhomogeneous rain-forming region is perhaps 5000 km thick (see Fig. 11-4) and inhibits dynamo generation (Stevenson, 1980). The underlying, helium-rich, metallic-hydrogen core is much smaller than the metallic core of Jupiter and is bounded below by a central rock (and ice?) core of roughly 20 Earth masses.

Our guided tour of Uranus or Neptune is necessarily speculative, and the particular configuration shown in Fig. 11-5 is only one possibility. As in Jupiter and Saturn, temperature is likely to rise rapidly as we proceed downward below the observable atmosphere, reaching about 1000°K at less than 2000 km deep. At this level, the mixture may be highly enriched in the ices, relative to cosmic abundance. At a level corresponding to about 70 percent of the radius ($P \simeq 200$ kbar, $\rho \simeq 0.4$ g · cm$^{-3}$, $T \sim 2000$ to 2500°K), it is possible that an "ionic ocean" is present ($H_3O^+OH^-$, with dissolved $NH_3$; density $\simeq 1.3$ g · cm$^{-3}$). This corresponds to the model of Hubbard and MacFarlane (1980). However, it is also possible that water enrichment is gradual or that the water is even uniformly mixed below a level of cloud formation at the outer extremities of the planet. Uncertainties in the moment of inertia and equations of state currently exclude firmer statements. In any event, a central rock core of a few Earth masses also seems to be required.

## 11-6 FORMATION AND EVOLUTION OF THE GIANT PLANETS

The internal heat source of the giant planets is almost certainly gravitational in origin. It could either be the gradual escape of primordial heat that was generated by the gravitational energy release accompanying the formation of the planet, or it could be derived from previous or ongoing differentiation (e.g., downward migration of helium, which is more dense than hydrogen). Other potential sources of internal heat are less probable or clearly insufficient: fission (the dominant energy source within terrestrial planets) would supply only about 0.1 percent of Jupiter's heat flow if the core consists of chondritic material; fusion of deuterium could never have occurred since it requires a temperature of approximately $5 \times 10^5$ °K; chemical energy sources are conceivable in principle but no plausible source has been proposed; and other exotic explanations (e.g., monopoles) are ad hoc.

The simplest model we can consider is an adiabatic giant planet that begins at a high temperature and then undergoes gradual cooling at a rate dictated by the atmospheric opacity. Suppose that $T_e$ is the effective (radiating) temperature of the planet, $T_0$ is the equilibrium-effective temperature in the absence of an internal heat source (i.e., caused by insolation), $\bar{C}_v$ is the average specific heat per gram, and $T_i$ is an appropriately defined average internal temperature. The luminosity (total thermal energy output from the interior) is then

$$L \equiv 4\pi R^2 \sigma(T_e^4 - T_0^4) \simeq -\frac{d}{dt}(M\bar{C}_v T_i) \tag{11-6}$$

where $\sigma$ is Stefan-Boltzmann's constant. A detailed discussion of the meaning and limitations of this equation can be found in Hubbard (1980). Throughout most of the evolution, $R$ is essentially constant because the planet is cold (thermal expansion is negligible through most of the interior, as discussed in Sec. 11-2). Since the planet is essentially adiabatic, $T_i \propto T_e$. If the opacity of the atmosphere is constant during the evolution, it follows that the elapsed time $\tau$ for a planet to cool from an initial high temperature to the observed present value is

$$\tau \simeq \frac{(0.25)(\text{Present heat content})}{(\text{Present excess luminosity})} \tag{11-7}$$

(The result is insensitive to the initial temperature because the strong $T^4$ dependence of the luminosity guarantees that any early high-temperature phase lasts only a short time and contributes little to $\tau$.) Planets that evolve homogeneously should have $\tau \simeq 4.6 \times 10^9$ y, since there is no reason to suspect that the age of a giant planet differs significantly from the conventionally defined age of the solar system (e.g., the age of the oldest meteorites).

The "homogeneous age" of Jupiter, thus evaluated from Eq. (11-7), is $(5 \pm 1.3) \times 10^9$ y, consistent with the age of the solar system. This implies that the observed heat output of Jupiter is consistent with gradual cooling from an initial state that was at least twice as hot as the present state. This interpretation is also consistent with all the other observable characteristics of Jupiter. The situation with Saturn is less clear, since the evaluated age is $(3.5 \pm 1) \times 10^9$ y, based on the most recent heat-flow estimate (Hanel et al., 1983; the age of $2.8 \times 10^9$ y or less quoted in earlier papers was based on an earlier, higher heat-flow estimate). Saturn is marginally consistent with homogeneous cooling. The observations suggest a helium depletion in the Saturnian atmosphere, however; so it is important to understand the evolutionary consequences of helium rainfall. It is found (Stevenson, 1980; Hubbard and Stevenson, 1984) that the energy release from helium differentiation is comparable or larger, per unit drop in internal temperature, than the energy release from primordial, stored heat. However, the effects are not strictly additive, since the planet self-regulates (i.e., cools less rapidly if differentiation is proceeding). As a consequence, the model illustrated in Fig. 11-4 is consistent with the observed heat output, provided differentiation began about $2 \times 10^9$ years ago. If the onset of differentiation were very recent, we would be in the unlikely position of living at a special time. The possible antiquity of the differentiation onset is compatible with our limited knowledge of the helium-hydrogen phase diagram and the absence of pronounced differentiation in Jupiter. Nevertheless, the most compelling argument for differentiation is the atmospheric composition, since the heat-flow measurement is equivocal.

The interpretations of the Neptune heat flow and the Uranian upper bound are even less well understood. The analysis of Hubbard and MacFarlane (1980) suggests that the heat flow may be *lower* than expected (i.e., the time scale $\tau$ substantially exceeds the age of the solar system). Two possible interpretations are (1) that these planets were not much hotter primordially than they are now or (2) that they are not adiabatic.

In any event, a hot start seems to be required for Jupiter and Saturn. Their present cores of approximately 20 Earth masses are also likely to have already been in place as these planets formed, since the core material is likely to be soluble in metallic hydrogen and could not "rainout" from above. What does this imply about the formation of the giant planets? Two formation scenarios have been suggested. In the scenario of the giant, gaseous protoplanet (e.g., Cameron et al., 1982), the primordial solar nebula undergoes gravitational instability, forming gaseous bodies of order the mass of Jupiter. Although partial rainout of condensibles can occur to form a dense core at this stage, the sequence of steps leading from this state to the observed terrestrial giant planets remains unclear (but see Cameron, 1984). In particular, the heavy-element enhancement manifest in Jupiter and Saturn has no simple explanation. The alternative scenario involves the conventional view in which there are no instabilities in the gas phase, but condensation, sedimentation, and coagulation of solid material, leading to kilometer-sized planetesimals that grow by collision to Moon-sized and (eventually) Earth-sized solid bodies. In the models formulated by Mizuno (1980), a solid body of several Earth masses nucleates collapse (radial inflow) of gas and eventual formation of a giant planet. This is an attractive theory because it may explain the similar masses of the giant-planet cores. The situation remains unclear, however, because the collapse may occur for a much lower central mass (Stevenson, 1984).

## 11-7 CONCLUDING COMMENTS

We have much to learn about the giant planets, but the conclusions outlined above demonstrate that our current, limited knowledge is sufficient to reach important conclusions: the predominance of hydrogen and helium in Jupiter and Saturn; the existence of a central, primordial, dense core, probably composed of rock; approximately adiabatic thermal structures with temperatures exceeding $1 \times 10^{4\circ}$K (Saturn) and $2 \times 10^{4\circ}$K (Jupiter); internal heat flows that require that Jupiter and Saturn, at least, were once much hotter than now; and evidence for ongoing differentiation of helium from hydrogen in Saturn.

Future progress lies primarily in three areas: observation, experiment, and theory. In the long term, *in situ* observations (atmospheric probes) are essential for all these planets. Jupiter will be probed by *Galileo*, but probes into the other giant planets may not eventuate this century. Useful information may require probes that can reach approximately $10^2$ bars and transmit back information from that level. Experimental work on relevant cosmically abundant materials, including mixtures, will continue to play an essential role. Many experiments have not yet been attempted (e.g., $H_2$–$H_2O$ mixtures at high pressure). Both shock-wave and static data are useful. Theory will also continue to play an essential role, since it is usually not possible to simulate exactly the thermodynamic conditions encountered within a giant planet. The biggest uncertainties here are long-standing issues such as the nature of the molecular-metallic-hydrogen transition.

What do we gain from these endeavors? The giant planets comprise 99.5 percent of the planetary mass in our solar system, and it would be foolish indeed to attempt any understanding of solar system's origin and evolution that does not take full account of the properties and inferences described in this chapter. Perhaps almost as important, at least to a physicist, is the fascination about what the environment is like and how materials behave deep within a giant planet. It is both tantalizing and frustrating to reflect on the fact that the most common metal in the solar system (and perhaps the universe) has yet to be encountered or studied on Earth.

## 11-8 REFERENCES

Cameron, A. G. W., 1985, in *Protostars and Planets* (D. Black, ed.), Univ. Arizona, Tucson, Ariz., preprint.

Cameron, A. G. W., W. M. DeCampli, and P. Bodenheimer, 1982, Evolution of giant, gaseous protoplanets embedded in the primitive solar nebula, *Icarus, 49*, 298-312.

Courtin, R., D. Gautier, A. Marten, and B. Bezard, 1983, The composition of Saturn's atmosphere at temperate northern latitudes from *Voyager* IRIS spectra, *Bull. Am. Astr. Soc., 15*, 831.

Gautier, D., and T. Owen, 1983, Cosmological implications of helium and deuterium abundances on Jupiter and Saturn, *Nature, 302*, 215-220.

Gautier, D., B. Bezard, A. Marten, J. P. Baluteua, N. Scott, A. Chedin, V. Kunde, and R. Hanel, 1982, The C/H ratio in Jupiter from the *Voyager* infrared investigation, *Astrophys. J., 257*, 901-912.

Hamann, S. D., and M. Linton, 1966, Electrical conductivity of water in shock compression, *Trans. Faraday Soc., 62*, 2234-2241.

Hanel, R. A., B. J. Conrath, V. J. Kunde, J. C. Pearl, and J. A. Pirraglia, 1983, Albedo, internal heat flux and energy balance of Saturn, *Icarus, 53*, 262-285.

Hubbard, W. B., 1980, Intrinsic luminosities of the Jovian planets, *Rev. Geophys. Space Phys., 18*, 1-9.

Hubbard, W. B., and J. J. MacFarlane, 1980, Structure and evolution of Uranus and Neptune, *J. Geophys. Res., 85*, 225-234.

Hubbard, W. B., and D. J. Stevenson, 1984, Interior structure of Saturn, in *Saturn* (T. Gehrels, ed.), Univ. Arizona, Tucson, Ariz., p. 47-87.

MacFarlane, J. J., and W. B. Hubbard, 1983, Statistical mechanics of light elements at high pressure. V. Three dimensional Thomas-Fermi-Dirac theory, *Astrophys. J., 272*, 301-310.

Mizuno, H., 1980, Formation of the giant planets, *Progr. Theor. Phys., 64*, 544-557.

Ree, F. H., 1979, Systematics of high pressure and high temperature behavior of hydrocarbons, *J. Chem. Phys., 70*, 974-983.

Ross, M., H. C. Graboske, Jr., and W. J. Nellis, 1981, Equation of state experiments and theory relevant to planetary modeling, *Phil. Trans. Proc. Roy. Soc. (London), A303*, 303-313.

Ross, M., F. H. Ree, and D. A. Young, 1983, The equation of state of molecular hydrogen at very high density, *J. Chem. Phys., 79*, 1487-1494.

Salpeter, E. E., and H. E. Zapolsky, 1967, Theoretical high pressure equations of state, including correlation energy, *Phys. Rev., 158*, 876-886.

Seward, T. M., and E. U. Franck, 1981, The system hydrogen-water up to 440°C and 2500 bar pressure, *Ber. Bunsenges. Phys. Chem., 85*, 2-7.

Shoemaker, E. M., and R. F. Wolfe, 1984, Evolution of the Uranus-Neptune planetesimal swarm, *Lunar Planet. Sci. Abstracts, 15*, 780-781.

Stevenson, D. J., 1975, Thermodynamics and phase separation of dense, fully ionized hydrogen-helium fluid mixtures, *Phys. Rev., 12B*, 3999-4007.

Stevenson, D. J., 1979, Solubility of helium in metallic hydrogen, *J. Phys. F: Metal Physics, 9*, 791-800.

Stevenson, D. J., 1980, Saturn's luminosity and magnetism, *Science, 208*, 746-748.

Stevenson, D. J., 1982, Interiors of the giant planets, *Ann. Rev. Earth Planet. Sci., 10*, 257-295.

Stevenson, D. J., 1983, Planetary magnetic fields, *Rep. Prog. Phys., 46*, 555-620.

Stevenson, D. J., 1984, On forming the giant planets quickly (superganymedean puffballs), *Lunar Planet. Sci. Abstracts, 15*, 822-823.

Stevenson, D. J., and E. Fishbein, 1981, The behavior of water in the giant planets, *Lunar Planet. Sci. Abstracts, 12*, 1040-1042.

Stevenson, D. J., and E. E. Salpeter, 1977, The phase diagram and transport properties for hydrogen-helium fluid planets, *Astrophys. J. Suppl., 35*, 221-237.

Wallace, L., 1980, The structure of the Uranus atmosphere, *Icarus, 43*, 231-259.

Zharkov, V. N., and V. P. Trubitsyn, 1978, *Physics of Planetary Interiors*, Pachart Publ., Tucson, Ariz., 338 pp.

S. J. Peale
Department of Physics
University of California
Santa Barbara, California 93106

# 12

# CONSEQUENCES
# OF TIDAL EVOLUTION

## ABSTRACT

A simple explanation of the tidal distortion of a solar-system body from the differential gravitational attraction by an orbiting companion is given. This is followed by an explanation of how energy dissipation caused by time variations in the tidal bulge leads to changes in the orbits and spins that can be large over the age of the solar system. Consequences of this orbital and spin evolution are pointed out for all solar-system bodies for which the effects of tidal dissipation are significant.

## 12-1 INTRODUCTION

Everyone who has lived near the ocean is very much aware of the variation in the level of the ocean water, which we call the *tide*. There are two high tides and two low tides each day, and the astute observer will notice that these extreme tides occur about fifty minutes later on the average than they did the day before. The fact that the Moon reaches its maximum elevation above the horizon with this same average 50-min delay on each successive day makes it logical to suspect the Moon as the cause of our observed ocean tide.

In contrast to their awareness of the ocean tide, few people know that the solid body of the Earth also experiences a tide with an amplitude of about 30 cm. This motion is not at all obvious, as all objects on the solid surface of the Earth move together, just as the ocean tide would not be obvious to an observer on a ship in midocean. But the solid-body tide is detectable with sensitive instrumentation.

The Moon interacts with the Earth through gravitational force, and one can generalize the idea that the moon is responsible for the Earth tide to every solar-system object: Each planet or satellite in the solar system experiences tidal distortions due to all the other objects attracting the planet or satellite gravitationally. As the first test of this hypothesis, we expect the Sun to also raise a tide on the Earth, and we observe that it does. The Earth tide due to the Sun is not as large as that due to the Moon, but it is quite noticeable when one considers that the highest tides occur when the Sun and Moon are nearly lined up—a time when they both pull on the Earth in the same or opposite directions. The newspapers in any coastal city publish the times and heights of the local tidal extremes, and it is easy to see that these extremes are maximal at new and full Moon.

In the remainder of this chapter, we show how tidal deformation of a body is generated, develop a simple physical model with which we can describe the essential features of the tidal distortion, and indicate the conversion of tidal energy into heat as the tide oscillates. It is this change of mechanical energy into heat that allows slow changes in the orbits and rotations of many objects in the solar system, and we show what the consequences of such tidal evolution have been for each planet-satellite system.

## 12-2 THEORY OF TIDES

A gravitational field of force (see text in box) of a solar-system object leads to a distortion of another massive object that is in the field because the gravitational force decreases as the distance from the source of the field increases. This means that different parts of the

## The Force of Gravity

We are all familiar with gravity as the force that attracts us and all other objects to the Earth. An object dropped near the Earth's surface accelerates toward the center of the Earth at the rate of 980 cm/sec². However, the Moon is accelerated toward the Earth only at the rate of 0.27 cm/sec², as this is the acceleration necessary to keep it moving in its almost circular orbit. From these two numbers, we find that the acceleration of gravity is proportional to $1/r^2$, where $r$ is the separation of the centers of the two bodies that are being attracted to each other—either the Earth and Moon or the Earth and the object we drop. We also know that if we double the mass of an object, that is, double the amount of material in it, the force of gravity on the object is doubled. It is twice as heavy, but since the acceleration of an object is inversely proportional to its mass, the acceleration in the Earth's field remains the same. But we see that the force of gravity on a body must also be proportional to its mass. By observing the orbital motions of the satellites about the planets (just as we observed the Moon about the Earth), we can infer that the force of gravity must be proportional to the mass of the attracting object as well, leading to $F = Gm_1 m_2/r^2$, where $G$ is the constant of proportionality called the *gravitational constant*.

object in the field experience different gravitational forces. This is illustrated in Fig. 12-1, where the forces on small-mass elements due to the satellite $m_s$ are represented by arrows at three points on a planet $m_p$. The separation of $m_p$ and $m_s$ is represented by $r$, and the

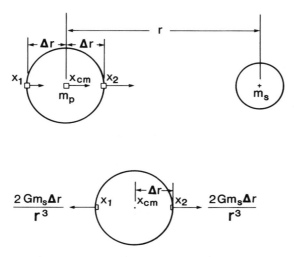

Fig. 12-1   Differential gravitational forces across a body of finite size.

radius of $m_p$ is $\Delta r$. If $\Delta m$ is the mass of each element indicated on $m_p$, the gravitational force due to $m_s$ on that element is given by

$$\Delta F = \frac{Gm_s \Delta m}{x^2}$$

where $G$ is the gravitational constant, and $x$ is the actual distance of $\Delta m$ from the center of $m_s$. The fact that $\Delta F$ decreases as $x$ increases from $r - \Delta r$ to $r + \Delta r$ is indicated in Fig. 12-1 by the decrease in the length of the arrows from points $x_2$ to $x_{cm}$ to $x_1$. Here the subscript $cm$ indicates the center of the mass $m_p$. As the acceleration of each $\Delta m$ is $\Delta F/\Delta m$, we see that the side of $m_p$ closest to $m_s$ tends to accelerate more toward $m_s$ than the center, and the center tends to accelerate more than the far side. If the masses ($\Delta m$) of the elements at $x_1$, $x_{cm}$, and $x_2$ were completely free from one another, $x_2$ would move away from $x_{cm}$ toward $m_s$, and $x_1$ would be left behind by $x_{cm}$. Relative to $x_{cm}$, the two points $x_1$ and $x_2$ would accelerate in opposite directions at rates indicated in the bottom part of Fig. 12-1. These differential accelerations tend to stretch $m_p$ into a prolate ellipsoid, similar to the shape of an egg.

But the $\Delta m$'s are not free, since they are attracted gravitationally to one another and to the remaining mass in $m_p$. The accelerations toward the center of $m_p$ due to self-gravity acting on the $\Delta m$'s at $x_1$ and $x_2$ are far stronger than the differential accelerations shown in Fig. 12-1, so mass does not fly off the surface at $x_1$ and $x_2$. If the material in $m_p$ were incompressible, the $\Delta m$'s at $x_1$ and $x_2$ would weigh less than they would if $m_s$ were not there but the $\Delta m$'s would not move relative to the center. But $m_p$ will still distort into the egg shape unless it is perfectly rigid as well as being incompressible. (When acted on by forces, an incompressible substance preserves its volume but may change shape, whereas a rigid substance does not change shape.) The reason is that points in $m_p$ that do not lie on the line joining the centers of $m_p$ and $m_s$ also experience a differential acceleration relative to $x_{cm}$. But now the differential accelerations are not perpendicular to the surface and are therefore not compensated by the self-gravity that accelerates mass elements more or less toward the center of $m_p$. This is shown in Fig. 12-2, where the differential acceleration at one point is resolved into its equivalent components perpendicular and tangential to the surface. These components are shown as dashed arrows and are equivalent to the solid arrow, in that together they produce the identical acceleration. The self-gravity, shown as an arrow directed toward the center of $m_p$ in the top part of Fig. 12-2, compensates the vertical component of the differential acceleration but not the tangential component. If $m_p$ were entirely fluid, the uncompensated tangential components of the differential accelerations would cause fluid to flow toward the points on $m_p$ closest to $m_s$ *and* furthest away from $m_s$, until $m_p$ would resemble the bottom part of Fig. 12-2. In this ellipsoidal shape, the self-gravity is now no longer perpendicular to the surface but has a tangential component opposite that of the differential acceleration due to $m_s$. Only in the distorted shape are all the differential accelerations compensated and the whole planet is accelerated like the center at $x_{cm}$.

If $m_p$ is not fluid but is rigid like rock, part of the compensating acceleration is provided by internal stresses, and the distortion is not as large as it would be for a fluid. Still, no material is perfectly rigid, so there is always some tidal bulge. All materials are at least somewhat compressible as well, and the expansion of $m_p$ in the directions where the

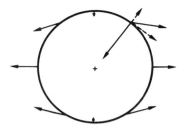

UNCOMPENSATED TANGENTIAL ACCELERATIONS

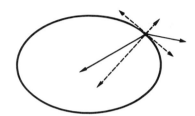

BALANCED BY DISTORTION AND INTERNAL STRESSES

Fig. 12-2 Distortion of fluid, incompressible sphere to bring it into equilibrium in an external gravitational field.

weight of overlying material is reduced by the presence of $m_s$ accentuates the tidal distortion. Note that the tidal distortion is independent of the velocity of $m_p$ relative to $m_s$ and would occur either if $m_p$ and $m_s$ were in a stable orbit or if $m_p$ and $m_s$ were falling directly toward each other. There is no need to invoke centrifugal force to account for the tidal bulge on the side of $m_p$ furthest from $m_s$, although this is often done in erroneous explanations of tidal deformation.

We are now in a position to understand the main features of the Earth tides, given that the Earth tends to be distorted into an egg shape which is aligned toward the Moon and into a lesser distortion aligned toward the Sun. First, from the bottom part of Fig. 12-1, the different magnitudes of the solar and lunar tides follow from the expression for the differential acceleration which is proportional to $m_s/r^3$. The lunar mass is much smaller than the solar mass, but it is so much closer to us that the $r^3$ in the denominator more than compensates for the tiny mass and results in a lunar tide about twice the magnitude of the solar tide. The ocean tide rises and falls at a given point on a coast because the Earth rotates under the Moon. The tidal bulge tends to stay more or less aligned with the Moon, so for our fixed position on the *rotating* Earth, we are carried through alternate high and low points of the tidal deformation. The high tides occur when we pass under the Moon and when we are on the opposite side of the Earth from the Moon, and low tides occur when we see the Moon rise or set. Hence, there are two high and two low tides each day. The ocean tide is obvious to us on a coast because the solid Earth is rigid and distorts less than the fluid ocean.

Close observation reveals that the ocean tide is more complex than the simple picture we have just presented. The high tide can occur at times considerably removed from the time the Moon passes overhead. This is because the continents interrupt the free passage of the tidal wave. The ocean basins are like giant bathtubs in which the water likes to slosh with a period of oscillation given by $w/\sqrt{gd}$, where $w$ is the width of a given ocean, $d$ is the average depth, and $g$ is the acceleration of gravity at the Earth's surface. This period is comparable to the half-day tidal period, which means the reflection of the tidal wave off a continental boundary and the resulting sloshing of the ocean water can result in high tides occurring at odd times relative to the overhead passage of the Moon. As an example, high tide occurs about two hours before the Moon reaches its highest point in the sky at Santa Barbara, California. Only the Earth has such a substantial ocean interrupted by continents, so all other solar-system bodies are more likely to conform to the situation where their tidal bulges are nearly aligned with the direction to the tide-raising body. This simplicity is extremely useful when we consider the dissipation of tidal energy in a more or less solid satellite or planet with no complicating ocean.

We pointed out earlier that the lunar tide on Earth is about twice as large as the solar tide, because this is the ratio of the corresponding differential accelerations or forces causing the tide. In other words, for the small tidal amplitudes observed, the amplitude is proportional to the disturbing force. This is exactly the way a spring behaves—the stretch is proportional to the stretching force. So we can represent a solid satellite or planet now as a collection of springs such as that shown in Fig. 12-3, where the restoring force of the springs is a combination of gravity and elasticity of the material. As the planet or satellite rotates relative to the tide-raising body, the springs are alternately stretched and relaxed. No material is perfectly elastic, so part of the energy stored in the stretched spring is converted to heat. This conversion of energy is represented by the dashpots attached to each spring in Fig. 12-3. (A *dashpot* is a device for turning the kinetic energy of motion of the mass at the end of the spring into heat. It can be thought of as a piston in a pot of oil where the oil is "dashed about" as the piston moves. An everyday example of a dashpot is a shock absorber on an automobile.) The springs follow the forcing term, that is, reach their maximum stretch when the differential acceleration is greatest, if the period of variation of the disturbing force at a given point on the body is sufficiently long. On the Earth, the period of variation of the disturbing force is half the time between moonrises.

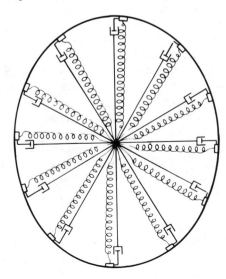

Fig. 12-3 Idealized representation of a solid elastic planet by a collection of springs. The planet is shown distorted by the field of an exterior body.

This condition of slow variation is satisfied on all the solar-system objects subject to deformation.

The dissipation of mechanical energy (conversion into heat) causes each spring to reach its maximum stretch slightly after the maximum force is applied. This means that for a planet or satellite ($m_p$ in Fig. 12-1) uncomplicated by the interaction of oceans and continents, the high tide follows the perturbing body $m_s$, but it occurs at a point after $m_s$ has passed overhead. The amount of the time delay in the response of the spring representing the planet or satellite is proportional to the amount of the dissipation when this dissipation is small. If we define the dissipation function $Q$ by

$$Q = \frac{2\pi E_0}{\Delta E}$$

where $E_0$ is the maximum energy stored in a tidal oscillation and $\Delta E$ is the energy dissipated in one complete cycle, the time delay in the response is proportional to $1/Q$.

This time delay caused by energy dissipation has an interesting consequence shown schematically in Fig. 12-4. The bodies $m_s$ and $m_p$ are reproduced as in Fig. 12-1, except $m_p$ is shown tidally distorted and rotating relative to $m_s$. (There is a similar distortion and perhaps a rotation of $m_s$ that we shall ignore for the time being.) The fact that the maximum tide occurs at a point on $m_p$ after $m_s$ has passed over that point means that the tidal bulge is displaced in the direction of rotation relative to a line to $m_s$. Here we assume the rotation of $m_p$ and the orbital motion of $m_s$ are in the same direction and that the angular velocity of the former is larger than that of the latter. If the orbital motion were faster, the bulge would be displaced in the opposite direction. The angle $\delta$ by which the tidal bulge leads is equal to $1/2Q$. Since low $Q$ implies high dissipation, $\delta$ increases with dissipation. The tidal bulges are about equal in size, but again because the gravitational force decreases as $1/r^2$, the bulge nearest $m_s$ experiences a greater force from $m_s$ than the one further away. These unequal forces $\bar{F}_1$ and $\bar{F}_2$ are represented by the arrows (called *vectors*) of unequal length. (The arrows are drawn in the directions of the forces.) The first consequence of this inequality results from the fact that the forces are not directed through the center of $m_p$. Since $\bar{F}_1 > \bar{F}_2$, a twisting effect or torque is exerted on $m_p$ in a direction which retards the spin angular velocity.

Forces equal to $\bar{F}_1$ and $\bar{F}_2$ but opposite in direction are exerted on $m_s$. These are

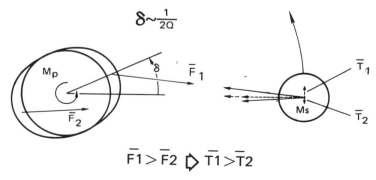

Fig. 12-4  Consequences of dissipation of tidal energy.

represented by the appropriate vectors in Fig. 12-4 drawn from the center of $m_s$. These forces are resolved into two components, one directed toward $m_p$ and one perpendicular to that direction, and they are shown as dashed vectors in Fig. 12-4. The reaction forces on $m_s$ can be replaced by these components to produce the same accelerations. The components in the direction of $m_p$ just contribute to the central attraction responsible for the orbital motion. But the perpendicular components, which are tangent to the orbital motion, result in a net acceleration of $m_s$ in the direction of its orbital motion. That is, since $\bar{F}_1$ is greater than $\bar{F}_2$, $\bar{T}_1$ is greater than $\bar{T}_2$ (Fig. 12-4). This acceleration of $m_s$ causes it to spiral away from $m_p$ as the latter's spin angular velocity is reduced.

In Fig. 12-4 we assume $m_s$ to be orbiting in the equator plane of $m_p$. However, if the spin axis of $m_p$ is not perpendicular to the orbit plane of $m_s$, the tidal bulge is carried out of the plane of the orbit, as illustrated in Fig. 12-5. This results in a twisting action or torque which is not parallel to the spin axis. Part of the torque tends to tilt the spin axis and part tends to change the spin magnitude. By this means, the obliquity (angle between the spin axis and a line perpendicular to the orbit plane) is changed. There is a corresponding but much smaller change in the orbit plane.

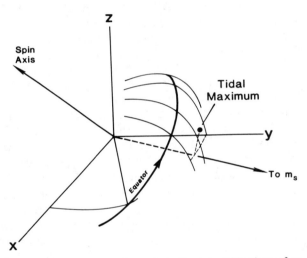

Fig. 12-5  If equator of $m_p$ is inclined to orbit plane of $m_s$, the tidal bulge is carried out of the orbit plane and the resulting torque on $m_p$ is not parallel to the spin axis.

As long as there is any rotation of either $m_p$ or $m_s$ relative to the other or as long as either obliquity is not zero, torques result that cause the orbit and spins to evolve. The ultimate final state, if the evolution is allowed to proceed long enough, is for both $m_p$ and $m_s$ to be rotating synchronously with their orbital motion (no relative rotation) and for both equator planes to be coincident with the orbit planes. Many circumstances, not the least of which is insufficient time, usually prevent this final state from usually being more than partially reached in the solar system.

We can summarize tidal theory as follows:

1. Tides are raised on a solar-system body by differences in gravitational forces across the body.

2. Any relative rotation between $m_p$ and $m_s$ causes tides and hence internal stresses to oscillate.
3. Internal dissipation of tidal energy causes the response of the tidally distorted body to lag the forcing term in phase.
4. The resulting nonalignment of the tidal bulge with the tide-raising body results in an interchange of angular momentum between the spin and orbital motion.
5. Tidal evolution expands the orbits of satellites of rapidly rotating planets while driving spins toward synchronous rotation with zero obliquity.

## 12-3 CONSEQUENCES OF TIDAL EVOLUTION

### 12-3-1 Mercury

Mercury experiences tides raised by the Sun. Reasonable estimates of the rate of dissipation of tidal energy imply that Mercury should be synchronously rotating, and so it was thought for many years. It therefore came as quite a surprise when radar observations revealed Mercury's rotation period to be only two-thirds of its orbital period (Pettengill and Dyce, 1965). Mercury's orbit is quite eccentric ($e = 0.206$). Eccentricity $e$ is a measure of the deviation of the orbit from a circle. As explained by Wasson and Kivelson in Chapter 1, eccentricity $e = 0$ corresponds to a circular orbit, whereas as $e$ approaches 1, the orbit goes to a very narrow ellipse with the Sun at one focus. Because of the high eccentricity, Mercury can rotate stably at several values of rotational angular velocity that are half-integer multiples of its mean orbital angular velocity. These particular rotation states are stabilized by the average torque on the *permanent* axial asymmetry. This asymmetry is like a tidal bulge, except it is not induced by an exterior gravitational field but is supported by internal shear stresses. The stability is such that the long axis of the permanent, egg-shaped asymmetry is oriented toward the Sun as Mercury passes through the perihelion (point in orbit closest to the Sun) of its orbit.

The inferred tidal evolution of Mercury's spin angular velocity is a tidal retardation of the spin through a series of stable states, where the spin angular velocity $\psi$ may be $3n$, $2.5n$, $2n$, and so forth, where $n$ is the mean orbital angular velocity. A state is stable against further tidal evolution if the maximum average torque on the permanent deformation exceeds the tidal torque. Mercury need not be captured in a stable state as it passes through. The probability of capture depends on the detailed tidal model, but all tidal models share the property that the states corresponding to $\psi = An$, where $A$ is a half integer, have a decreasing probability of capture as $A$ increases, ultimately reaching zero probability when $A$ is so large that the state is not stable. Thus we find Mercury in that spin state ($\psi = 1.5n$) which has the highest capture probability of any state Mercury could have passed through as the tides slowed it toward synchronous rotation. Mercury is the only example known where such spin-orbit coupling has led to a stable state other than synchronous rotation (Goldreich and Peale, 1966).

But tidal dissipation is still occurring in Mercury, making the orbit more circular. The stability of the current spin state vanishes when the orbital eccentricity gets sufficiently small. Tides could eventually bring Mercury to synchronous rotation. However, it would take many times the age of the solar system for this to occur.

Mercury's spin axis is very close to being perpendicular to its orbit plane. This is also the result of tidal evolution, which predicts that Mercury's spin should end up in a stable state called a *Cassini state*, where it lies in the plane determined by a line perpendicular to its orbit plane and a line perpendicular to the plane of the solar system. This state is very close to the orbit normal, and Mercury would have been driven there by the tides from virtually any initial obliquity (Peale, 1974).

## 12-3-2 Venus

Venus rotates in a direction which is opposite the sense of its nearly circular orbital motion. This state is described by assigning an obliquity of nearly 180° to the spin axis. This retrograde rotation is not the result of tidal evolution in the ordinary sense of tides raised by the solar gravitational field, since such tides attempt to reduce the obliquity and thereby stand Venus upright (Goldreich and Peale, 1970). If Venus began with an obliquity greater than 90°, say like Uranus whose obliquity $\epsilon$ is now 97°, an interaction between a liquid core and solid mantle could dominate the ordinary gravitational tide and drive Venus to the current 180° obliquity.

If Venus started with $\epsilon$ less than 90°, the core-mantle interaction aids the gravitational tide in eventually driving Venus toward an obliquity of 0°. In this case, the only alternative to account for the current $\epsilon \simeq 180°$ seems to be an atmospheric thermal tide. There is some reason to suspect that such a thermal tide exists on Venus, since we observe one on Earth. The absorption of solar radiation in our upper atmosphere couples with the modes of oscillation of our atmosphere to generate an egg-shaped distribution of atmospheric mass whose maxima occur at 10:00 A.M. and 10:00 P.M. local time. With this phase, the gravitational attraction of this thermal tide by the Sun tends to accelerate the spinning atmosphere and thereby the Earth. This is opposite the effect of the gravitational tide which is decelerating the Earth's spin. The effect of the atmospheric thermal tide on the Earth's spin is only about 10 percent of the effect of the gravitational tide. But Venus's atmosphere has nearly 100 times the mass of our own atmosphere, so if a similar thermal tide is generated on Venus, it may easily dominate the gravitational tide. If we can model that tide by a gravitational tide but simply reverse the torque on the planet, the atmosphere could dominate both the gravitational tide and a core-mantle interaction and drive Venus to its current obliquity from a primordial obliquity less than 90° (Goldreich and Peale, 1970). Observation of the barometric pressure at the surface of Venus for a Venus day would determine the magnitude of the current thermal tide and probably indicate whether or not the atmosphere was capable of turning Venus upside down.

## 12-3-3 Earth

Tidal evolution of the Earth-Moon system has been directly measured. Our day is getting longer at the rate of about 0.0016 sec each century. The Moon's mean orbital angular velocity is decreasing at the rate of about 22 arc · sec/century$^2$. This means that each year, the Moon is moved away from us by about 3 or 4 cm. This rate implies an Earth which is currently more dissipative than it was in the past, for if we extrapolate the Moon's orbit backward in time, assuming the Earth's $Q$ always equaled its current value, the Moon ends up near the Earth's surface only 1.6 Gy ago, about one-third the age of the Earth-Moon system. (A Gy is $10^9$ years.) Most of the tidal dissipation on Earth is in the oceans, and the higher $Q$ in the distant past is usually attributed to a different configuration of the continents, thereby allowing for fewer shallow seas for the dissipation of tidal energy.

The effect of the tide raised on the Moon by the Earth has driven the Moon to the endpoint of its tidal evolution consistent with the current orbital configuration. The Moon's rotation is synchronous with its orbital motion, which we easily ascertain by always viewing the same lunar face from Earth. In addition, its spin axis has been driven to a stable Cassini state, in which it remains coplanar with the perpendiculars to the Moon's and Earth's orbits, with an obliquity of $6°41'$ (Peale, 1969).

Although tidal dissipation in the Moon has been sufficient to slow it to synchronous rotation and drive the spin axis to a Cassini state, the contribution of the heating to the thermal history of the lunar interior has been negligibly small. The Moon was, of course, once much closer to Earth, and the rate of dissipation was much higher than it is now. However, it was pushed out too fast toward distances corresponding to small dissipation for the interior to heat significantly (Peale and Cassen, 1978).

## 12-3-4 Mars

Mars' spin has not been significantly affected by tidal dissipation. It is too far from the Sun, and its satellites, Phobos and Deimos, are very small. The two satellites, on the other hand, are both synchronously rotating. The inner satellite, Phobos, is unusual in the sense that it is so close to Mars (orbit radius $\simeq 6.8$ Martian radii) that its orbit period is less than Mars' rotation period (7.65 h versus 24 h 37 min). Reference to Fig. 12-4 shows that in this case the tidal bulge is behind Phobos and the direction of acceleration of the satellite is opposite its motion. Phobos is therefore spiraling into Mars. The measured acceleration of Phobos' orbital angular velocity leads to a $Q$ of Mars of about 100, which is comparable to that of rocks on the Earth. If $Q$ remains the same as Phobos spirals toward Mars *and* if Phobos is sufficiently strong to remain in one piece, it will strike the surface of Mars in about $10^9$ y.

## 12-3-5 Jupiter

Jupiter is about 300 times more massive than Earth, and its satellites, although three of them are larger than our Moon, total only a little over $10^{-4}$ of Jupiter's mass. The spin of this planet has thus not been significantly altered by tides.

The satellites have experienced significant tidal evolution. All the Galilean satellites (discovered by Galileo)—Io, Europa, Ganymede, and Callisto—are rotating synchronously with their respective orbital motions, as is Amalthea, which is a small satellite inside the innermost Galilean satellite, Io. The orbits have not expanded very far, however, since Io is only about 5 Jupiter radii from the planet's surface. This close proximity after $4.6 \times 10^9$ y of evolution means the $Q$ of Jupiter must exceed $6.4 \times 10^4$. Even larger values of $Q$ might be expected for a planet such as Jupiter, which is essentially completely fluid.

The first three Galilean satellite orbits have mean orbital angular velocities that are commensurate. This means they are nearly in the ratio of small whole numbers. Io's angular velocity is nearly twice Europa's and Europa's is nearly twice that of Ganymede. The satellites are stably locked together in these orbital resonances, and so their orbits must expand together as tides raised on Jupiter push them away. It is possible that the differential tidal expansion of the orbits could have assembled the satellites into the resonances (Yoder, 1979; Yoder and Peale, 1981).

The existence of the orbital resonances has had a profound effect on Io. The condition that conjunctions of Io and Europa (both satellites on the same side of Jupiter on closest approach) always occur when Io is at the pericenter of its orbit, forces the eccentricity of Io's orbit to a value of 0.0041. Even though Io is synchronously rotating with its mean orbital motion, the tide raised on Io by Jupiter increases and decreases in magnitude, and the bulge moves back and forth to follow the motion of Jupiter in Io's sky. Io rotates uniformly, whereas the orbital motion in an eccentric orbit is not uniform. This causes the oscillatory motion of Jupiter relative to Io. The tide raised on Io is enormous, more than 7 km high if Io were fluid and uniformly dense. The variable part of the tide is 100 m in this case, so one expects substantial tidal dissipation. Ordinarily such dissipation would damp the orbital eccentricity to zero, and the dissipation would cease. However, the eccentricity is maintained by the orbital resonance, and the accumulated heat has been sufficient to melt the entire satellite (Peale et al., 1979). Io was observed by *Voyager 1* to be the most volcanically active body in the solar system. This is a case where tidal dissipation has dominated all other energy sources in determining the thermal history of a solar-system body.

### 12-3-6 Saturn

Saturn, like Jupiter, has not evolved significantly through tidal interaction. All the satellites out to and including Iapetus are rotating synchronously with their orbital motion, with the exception of Hyperion. Although Titan's surface has never been seen, it is most likely synchronously rotating, since it is so large and relatively close to Saturn. *Voyager 2* images have shown Hyperion to be slowly rotating and therefore tidally evolved in spite of its small size (see Peale, 1978). However, Hyperion is so far from spherical symmetry and is in such an eccentric orbit that strong gravitational torques on the satellite due to Saturn prevent an evolution to an orderly synchronous rotation. Hyperion will most likely tumble in a chaotic fashion for the remaining lifetime of the solar system (Wisdom et al., 1984).

There are three sets of orbital resonances among the satellites of Saturn that were known before *Voyager 1* and several more among the recently discovered smaller satellites (Peale, 1976). Of the former resonances, the 2:1 resonances between Mimas and Tethys and between Enceladus and Dione could have been assembled by differential tidal expansion of their orbits. However, the 4:3 Titan-Hyperion resonance could have been so assembled only if the $Q$ of Saturn was considerably smaller for the tide raised by Titan on Saturn compared with the $Q$ appropriate to the inner satellites. The proximity of Mimas to Saturn establishes the lower bound $Q \gtrsim 16,000$ for Saturn. A lower $Q$ would have pushed Mimas farther away from the planet than we find it. This value of $Q$ is too high to allow the evolution to the observed state of the Titan-Hyperion resonance from an initial condition appropriate to a presumed tidal assembly. As there are no obvious reasons for imposing a sufficient amplitude or frequency dependence on $Q$ to preserve both the tidal assembly of the inner resonances and that of Titan-Hyperion, it is often assumed that the latter resonance must be primordial, and therefore not due to tidal evolution.

Observations by *Voyager 1* and *2* showed that the surfaces of several of Saturn's satellites had been reworked by some kind of indigenous processes after the major cratering events. Especially puzzling is Enceladus, which is an icy satellite only 250 km in radius but which has areas on its surface which are less than $10^9$ y old. The radioactive content of Enceladus is negligibly small, and application of the tidal-heating hypothesis which was so successful for Jupiter's satellite, Io, also did not provide a sufficient energy source to cause the resurfacing (Squyres et al., 1983). The current orbital eccentricity of Enseladus' orbit, which is forced to 0.0044 by the resonance with Dione, was used in this latter analysis. Still, tidal heating, by a process of elimination, seems to be the only viable energy source to cause the relatively recent resurfacing of Enceladus. If this is so, in the recent past, Enceladus must have participated in one or more other orbital resonances which forced the eccentricity to such a high value that tidal heating became sufficiently large to soften the interior of the icy satellite. The resonance would have to have been destroyed either catastrophically or by evolution into a region of instability in order to allow the satellite to settle to its current low eccentricity forced by Dione alone. Resonances with Tethys and with the small moon 1980S1 have been proposed.

## 12-3-7 Uranus

Uranus has five known satellites which should all be rotating synchronously with their orbital motions. In contrast to the satellites of Jupiter and Saturn, there are no orbital resonances among these satellites, which implies that differential expansion of the orbits has been insufficient over the age of the solar system to assemble them. In fact, the absence of orbital resonances here may be support for the hypothesis of the tidal origin of the resonances among the satellites of Jupiter and Saturn, since there is apparently no constraint that satellites form in orbital resonances.

## 12-3-8 Neptune

Neptune has only two known satellites. The larger one, Triton, is synchronously rotating. This satellite is unusual in the sense that it is approximately the size of Earth's Moon, a large satellite, yet it is in a retrograde orbit, that is, it orbits in a direction opposite to the spin of Neptune. All other retrograde satellites in the solar system are tiny by comparison. Figure 12-4 shows that Triton, like Mars' satellite Phobos, is spiraling into its planet from the tidal interaction. Unlike Phobos, we have no measure of the rate, and have therefore not determined the dissipation function $Q$ of Neptune.

## 12-3-9 Pluto

Pluto has recently been observed to be a binary planet in the sense that the satellite is comparable in size to the primary. The two components of Pluto are very close together, and it is therefore probable that this system is closer to the final state of tidal evolution than any other. Both components would reach synchronous rotation rather quickly. Subsequent tidal evolution may have been sufficient to damp the orbital eccentricity to zero and align both spin axes perpendicular to the orbit plane.

## 12-4 SUMMARY

In summary, we see that tidal dissipation has been effective in altering drastically the spin and orbital configurations in the planet-satellite systems. It is interesting and useful to understand these evolutionary histories in order to deduce the primordial dynamical configurations. These primordial configurations provide constraints on theories of origin of the solar system itself.

## 12-5 REFERENCES

Goldreich, P. and S. J. Peale, 1966, Spin orbit coupling in the solar system, *Astr. J., 71*, 425–438.

Goldreich, P. and S. J. Peale, 1970, The obliquity of Venus, *Astr. J., 75*, 273–283.

Peale, S. J., 1969, Generalized Cassini's Laws, *Astr. J., 74*, 483–489.

Peale, S. J., 1974, Possible histories of the obliquity of Mercury, *Astr. J., 79*, 722–744.

Peale, S. J., 1976, Orbital resonances in the solar system, *Ann. Rev. Astr. Astrophys., 14*, 215–246.

Peale, S. J., 1978, An observational test for the origin of the Titan-Hyperion orbital resonance, *Icarus, 36*, 240–244.

Peale, S. J. and P. Cassen, 1978, Contribution of tidal dissipation to lunar thermal history, *Icarus, 36*, 245–269.

Peale, S. J., P. Cassen and R. J. Reynolds, 1979, Melting of Io by tidal dissipation, *Science, 203*, 892–894.

Pettengill, G. H. and R. B. Dyce, 1965, Rotation of the planet Mercury, *Nature, 206*, 1240.

Squyres, S., R. T. Reynolds, P. M. Cassen and S. J. Peale, 1983, The evolution of Enceladus, *Icarus, 53*, 319–331.

Wisdom, J., S. J. Peale and F. Mignard, 1984, The chaotic rotation of Hyperion, *Icarus, 58*, 137–152.

Yoder, C. F., 1979, How tidal heating in Io drives the Galilean orbital resonance locks, *Nature, 279*, 747–750.

Yoder, C. F. and S. J. Peale, 1981, The tides of Io, *Icarus, 47*, 1–35.

Eugene H. Levy
Department of Planetary Sciences, and
Lunar and Planetary Laboratory
University of Arizona
Tucson, Arizona 85721

# 13

# THE GENERATION OF MAGNETIC FIELDS IN PLANETS

## 13-1 INTRODUCTION

Magnetic fields occur commonly throughout the universe, and in many classes of cosmical objects. These magnetic fields are strong enough to influence or control the dynamical behavior and evolution of many astrophysical systems. Although the details of the fields vary from one kind of object to another, the basic underlying processes of magnetic-field generation seem to be similar in a large variety of natural objects.

The external magnetic fields of Mercury, Earth, Jupiter, and Saturn are predominantly dipolar and nearly aligned with their planets' rotation axes, as illustrated in Fig. 13-1. So far as we know, the magnetic fields of all of these planets, like the geomagnetic field, are quasistationary—with the overall directions and intensities varying only slightly with time, and within relatively narrow bounds. Among the planets, however, we have enough information only for Earth to be sure of the field's long-term behavior. Over long intervals of time Earth's, otherwise relatively quiescent, magnetic field undergoes occasional and randomly occurring polarity reversals. We will return to this point later.

The Sun has a magnetic field. Crudely, the Sun's large-scale magnetic field structure resembles that of the planetary fields. The solar magnetic field has the same general odd spatial symmetry as the dipolar magnetic fields of the planets. However, in detail, the Sun presents a far more tangled magnetic structure than any of the planets. An important

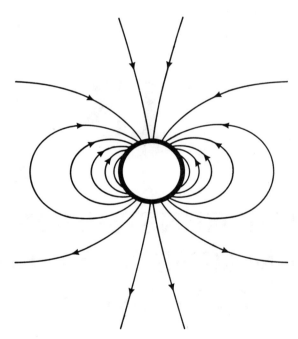

Fig. 13-1 The exterior magnetic lines of force of an idealized, axially aligned dipolar planetary magnetic field. Real planetary magnetic fields are distorted somewhat from this idealized shape. For planets in our solar system these exterior magnetic field lines are severely distorted, at moderate distances from the planet, by their interaction with the solar-wind flow. In the drawing, as in Earth, the field lines emerge from the south and enter at the north.

reason for part of this difference is that planetary magnetic fields are generated in the deep interiors of the planets—in electrically conducting, convecting fluid cores—the more tangled parts of the field are hidden from our view. The Sun's magnetic field is generated in the convecting layer of electrically conducting fluid that occupies the convection zone in the Sun's outer quarter; the top of the field generation region is exposed to our view. The average strength of the magnetic field at the surface of the Sun is about 1 Gauss—not very different from the field strength at the surface of Earth. But the actual field strength on the Sun varies over a large range. The magnetic field seems to have a tendency to clump into strong concentrations, leaving relatively weak fields in the intervening regions.

Of the small-scale structure visible in the Sun's surface magnetic field, perhaps the most prominent is the magnetic structure associated with the sunspots. Sunspots, which appear as small dark regions on the surface of the Sun, contain very strong magnetic fields—some thousand or more times stronger than the general field of the Sun. Sunspots are generally arranged in pairs or pairs of groups, one leading and one following in the direction of the Sun's rotation. Individual sunspot groups appear and disappear over periods of months. At any one time all of the sunspot pairs in the northern solar hemisphere have the same magnetic polarity relation—one polarity in the leading spot and the other polarity in the trailing spot. Sunspots in the southern hemisphere have the opposite relative magnetic polarities.

The temporal behavior of the solar magnetic field differs from that of Earth. Rather than being quasistationary, the Sun's magnetic field undergoes a periodic oscillation, with a full period of approximately 22 y. In this periodic oscillation, the polarity of the polar, dipole-like part of the field reverses every 11 y; in addition, the relative polarities of leading and trailing sunspot pairs reverse in the two hemispheres of the Sun. Although there are other important features and regularities in the solar magnetic behavior, we need not go into these for our present purpose. [For additional information about the exotic dynamical behavior of the solar magnetic field, see Chapter 2 by Zirin; Howard (1977) and Parker (1977).] Interestingly, the Sun's field is drawn far out into space by the solar wind, which continually flows away from the Sun at approximately 300 to 400 km · s$^{-1}$. This streaming tenuous, electrically conducting gas holds the solar magnetic field lines and stretches them to the limits of the heliosphere, probably 50 to 100 AU from the Sun.

Astronomical observations reveal that many stars also possess magnetic fields. Although none of these can be observed with enough resolution to determine their detailed properties, observations do reveal an astonishingly wide range of strengths of stellar magnetic fields. A stellar magnetic field only as strong as the Sun's or Earth's could not be observed readily with present-day astronomical instruments; it is a reasonable presumption that many, and perhaps most stars, have such fields. Indeed, many stars are known to have magnetic fields exceeding 100 Gauss, which is the approximate threshold of detectability for stellar magnetic fields. Some main sequence stars have much stronger average fields, as much as 50,000 times stronger than Earth's. Many compact stars that have evolved off of the main sequence have much stronger magnetic fields yet. White dwarf stars are about the size of Earth; some white dwarfs have magnetic fields of $10^7$ Gauss, 10 million times stronger than Earth's field. Pulsars are believed to be compact neutron stars, some 10 km across, having magnetic fields with intensities exceeding $10^{12}$ Gauss—over a trillion times stronger than the magnetic field of Earth or Sun.

EUGENE H. LEVY

Even the whole galaxy itself has a large-scale magnetic field, which may be generated by a process similar in its essential features to the process that generates the magnetic fields of the planets. Finally many distant extragalactic objects have extensive and relatively strong magnetic fields.

The existence and generation of magnetic fields in cosmical objects are, of course, interesting and important questions. But the fields themselves are not just academic curiosities. The energy stored in these magnetic fields, and the magnetic forces that they produce, are responsible for a large number of dynamical and energetic phenomena. For example, as the solar wind flows past Earth's magnetic field, it severely distorts the field's shape, compressing some parts and stretching other parts. The energy stored in this distorted magnetic field, when it is released, produces hot plasma and energetic particles. Among the many manifestations of such energy release are the auroras clearly visible on Earth at high latitudes. Solar flares seem to be produced by an analogous process.

Similarly exotic behaviors are apparent in more distant and far less well observed objects. The peculiar behaviors of pulsars and the acceleration of energetic particles in extragalactic radio sources seem to result from physical processes in large-scale distorted and stressed magnetic fields. The forces associated with magnetic fields may mediate major evolutionary changes in astrophysical objects. These forces are particularly efficient in transporting angular momentum in rotating objects and thus may play important roles in the evolution of stars and in the formation of stars and planetary systems.

## 13-2 BEHAVIOR OF MAGNETIC FIELDS IN NATURE

In order to understand the generation of magnetic fields in natural bodies, it is useful to first gain a physical picture of the essential features of the environment in which these fields occur and the implications of that environment for the behavior of the fields. For convenience in the following discussion, and to help give intuitive insight into magnetic field behavior, we make use of the geometrical picture of "magnetic lines of force," devised by Faraday, to describe the field. The magnetic force acts on a magnetic dipole, such as a bar magnet, on an electrical current or on a hypothetical unit of magnetic charge. As with all forces, the magnetic force has a strength and a direction—it is a vector quantity. A vector quantity is often represented as an arrow whose direction is aligned with the vector direction and whose length is proportional to the vector magnitude. A line of magnetic force is drawn by starting at some location and drawing a line that remains parallel to the magnetic field as we follow the line through space. The strength of the magnetic field is indicated by the number of lines drawn through a given area perpendicular to the field direction, so the lines are close together where the field is strong and far apart where it is weak. Conceptually, we can follow the evolution of the magnetic field by watching the structure and movement of the lines of force. Figure 13-1 shows the exterior dipole lines of force for an idealized planetary magnetic field.

Correct use of this conceptual picture provides a very powerful tool for gaining insight into the physical behavior of magnetic fields. In accordance with the goal of this book—to aim at a general student audience as well as at students of physical science—I give a completely physical discussion here that does not require great mathematical sophistication to follow. Ultimately, however, a confident understanding of magnetic field behavior requires detailed mathematical analysis; the physical results that I describe for you in this chapter result from such analyses. The physical picture of the magnetic field in terms of a field of lines of force, although it is, in the end, equivalent to the mathematics, does not lend itself as readily to rigorous analysis. The physical picture can, however, be constructed readily from the mathematical description of the field. If $B(x, y, z) = (B_x, B_y, B_z)$ is the functional form of the magnetic field, the lines of force are the solutions to the set of differential equations

$$\frac{B_x}{dx} = \frac{B_y}{dy} = \frac{B_z}{dz} \tag{13-1}$$

The prevalence and importance of magnetic fields in cosmical objects result from the fact that most of the matter in these objects conducts electricity very well; in addition virtually all of space is filled with tenuous gas, which, by virtue of its ionization, contains free electric charges and conducts electricity relatively well. (In this respect, the outer parts of Earth and the other terrestrial planets are anomalies among cosmical bodies.) Magnetic fields result from the motion of electric charges or electric current; the electrical conductivity of the material in which the field is embedded is fundamental to the existence, behavior, and generation of the field.

It is a general fact that when an object moves through a magnetic field, an electric field is induced in that object. Indeed, it is through this phenomenon that electrical current is commonly produced. The wires in the armature of an electrical generator revolve through a magnetic field; the electric field thus induced in the wires drives an electric current that is then distributed to consumers.

Now the large, highly electrically conducting fluids common in cosmical objects cannot support electric fields. Any attempt to set up such a field results in the rapid movement of an electric charge that cancels the electric field. In effect the conducting fluid constitutes an efficient short circuit connecting all points within it and preventing an electric field from being sustained. But we just noted that when an object moves through a magnetic field, it experiences an electric field. If we imagine a large, perfectly electrically conducting fluid, permeated by a magnetic field, and set into some arbitrary motion, we seem to have a contradictory situation. By virtue of its motion past the magnetic field lines, the fluid should have an electric field set up in it; by virtue of the fluids's high electrical conductivity, the electric field cannot persist! The resolution of this dilemma for the magnetic field is that the magnetic lines of force move with the fluid. In a fluid of very high electrical conductivity, we can, to a good approximation, regard the magnetic lines of force as physical entities fixed or frozen into the fluid and moving with it, as illustrated in Fig. 13-2.

The notion that magnetic field lines move exactly with the motion of the fluid in which they are embedded results from our having regarded the fluid as a perfect conductor of electricity—that is, as having no electrical resistivity. No real classical fluid has quite that property; real fluids, while they may have very high electrical conductivities,

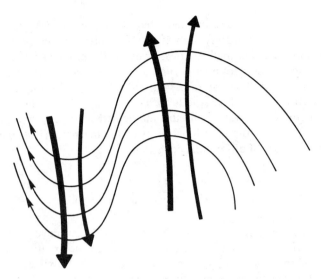

Fig. 13-2 Most cosmical fluids are excellent conductors of electricity. In a fluid that conducts electricity perfectly, magnetic lines of force are constrained to move with the local velocity of the fluid. Here the light lines represent lines of force that are carried with the fluid motion (heavy lines).

always pose some resistance to the flow of electric current. To complete this elementary physical description of magnetic field behavior in electrically conducting fluids, then, we must take into account the effect of the finiteness of the electrical conductivity.

Since the electrical conductivity is not infinite, the electric fields in the fluid are not reduced all the way to zero but only to small values. This allows some relative motion between the fluid and the magnetic field lines. An intimately related consequence of the finiteness of the electrical conductivity is that electric currents decay with time, and the energy of the magnetic field is dissipated as heat. The effect of these processes on the motion of the magnetic field lines is to allow them to *diffuse through* the fluid.

Altogether, then, the physical picture of the behavior of magnetic fields in electrically conducting fluids shows that the magnetic lines of force tend to convect and distort their shape with the local motion of the fluid; at the same time the field lines diffuse through the fluid. It should be remembered that the magnetic field is a vector quantity, having both magnitude and direction. In the course of diffusing through the fluid, magnetic field lines originating from different parts of the fluid may come to overlap; in this case, the resulting field is obtained from the vector sum of the field components. Consequently, by diffusing through the fluid, magnetic field lines from different parts of the fluid can cancel one another, thereby changing the topological structure of the field. This is illustrated in Fig. 13-3.

These results can be derived easily by manipulating Maxwell's equations, which describe the behavior of electric and magnetic fields (Jackson, 1975; Parker, 1979). Depending on the relative values of certain important physical parameters that represent the nature of the fluid and its motion, the evolution of the magnetic field may be dominated by either the convective or the diffusive behavior. These physical parameters may

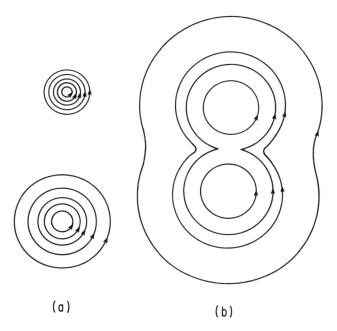

**Fig. 13-3** Because even cosmical fluids resist the flow of electricity to some extent, magnetic field lines move through the fluid in a diffusive manner. (a) Here, if we start out with magnetic field lines confined to a small region of space (top), sometime later (bottom) the field lines expand away in a diffusive manner, moving through the fluid. Magnetic field lines are vector quantities, having both magnitude and direction. When, in the course of diffusing through the fluid, field lines from different regions come to overlap, they must be added together vectorially; the lines may cancel one another locally, thereby causing a change in the topological structure of the magnetic field (b).

be combined into a single dimensionless number, the magnitude of which determines whether convection of the lines of force, or diffusion of the lines of force, dominates the evolution of the field.

Let $v$ be the typical magnitude of the fluid velocity; let $L$ be the typical spatial length scale of the fluid motion; and let $\sigma$ be the electrical conductivity. Then, by defining $\eta = c^2/4\pi\sigma$, the so-called magnetic diffusivity ($c$ is the speed of light), we can write the magnetic Reynolds number as follows

$$R_m = \frac{vL}{\eta} \tag{13-2}$$

When $R_m \gg 1$, the evolution of the magnetic field is dominated by convection of the magnetic field lines with the fluid motion; this occurs in the case of a sufficiently large value of the product of the fluid velocity, the physical scale of the system, and the electrical conductivity. When $R_m \ll 1$, the evolution of the magnetic field is dominated by diffusive motion of the lines of force through the fluid. With $R_m \simeq 1$, the two effects are roughly equally effective. In typical cosmical systems, ranging from the metallic cores of planets to stars and interstellar space, $R_m > 1$, often very much greater than one.[1]

## 13-3 THE PERSISTENCE OF MAGNETIC FIELDS

Before we begin to look specifically at the way magnetic fields are generated in planets and other cosmical bodies, it is instructive to first see why there is a phenomenon here that poses a challenge to our understanding. Magnetic fields and electric fields are closely related manifestations of a single electric force: Electric fields are produced by the presence of electric charge, whereas magnetic fields are produced by the *motion* of electric charge. If free magnetic charges are absent from the universe, or at least very rare, as many studies seem to suggest, then the motion of electric charge—namely, electric current—is the only possible source of most magnetic fields. (The essential relationship between the electric and magnetic fields was clarified by Einstein's theory of special relativity.)

The persistent magnets with which we are familiar in everyday life are the result of a phenomenon called *permanent magnetism*. By virtue of the perpetual motion of electric charges within atoms and molecules, many of these small bits of matter possess tiny permanent magnetic fields. Normally these tiny magnetic fields are randomly oriented in large pieces of matter, so that the magnetic fields tend to cancel one another. Under some circumstances, these individual atomic magnets can largely be aligned in a piece of matter, giving rise to an overall magnetization of the material. In some materials, such as iron, for example, this alignment can be very efficient, owing to the action of interatomic or intermolecular exchange forces and can produce very strong permanent magnetization, a phenomenon known as *ferromagnetism*. The familiar bar and horseshoe magnets are made this way. I noted earlier that the external magnetic fields of planets at least are similar in structure to the magnetic field of a bar magnet; it is a sensible question, then, to ask why Earth or Jupiter could not be just a large permanent magnet.

Permanent magnetism requires that the individual atomic magnetic fields, arising from microscopic electrical currents within the atoms, remain lined up with one another. In real matter, the molecular forces that act to line up the magnetic moments must compete with the jostling and agitation that atoms are subjected to as a result of random thermal motion; this thermal agitation tends to destroy the atomic magnetic alignment. The higher the temperature, the more rapid and thorough the randomization of atomic magnetic moments. At temperatures of a few hundred degrees Celsius, permanent magnetism decays rapidly with time. Above a certain temperature—the so-called Curie point,

---

[1] The behavior of magnetic fields in cosmical situations is dominated by the electrical conductivity of cosmical fluids, that is to say, by the presence of free and mobile electric charges. These free and mobile charges effectively short out any electric fields and thereby enforce the motion of magnetic field lines with the fluid. It is a curious fact that the universe seems to be devoid of analogous magnetic charges. If free magnetic charges were present in abundance, they would tend to move in response to magnetic fields and short out those fields also in cosmical objects.

Interest has recently revived in the possibility of the existence of magnetic charges in the universe. Some theoretical ideas about the unified origin of fundamental forces seem to predict the existence of such magnetic charges and, at the time of this writing, a tentative observation at Stanford University could hint at a surprisingly large abundance of magnetic charges! According to present ideas, the existence and properties of cosmical magnetic fields could be the strongest evidence that free magnetic charges are at least very rare in the universe (Parker, 1970a).

typically above 600° or 700°C—ferromagnetism ceases to exist altogether. The interiors of planets are at temperatures too high to retain permanent magnetization. Furthermore, most of the rest of the matter in the universe is in a fluid state that cannot support permanent magnetization. We will see shortly that there is additional, strong evidence that large-scale natural magnetic fields cannot be the result of permanent magnetization. Even the magnetic field of Earth is not a static phenomenon; the field continually varies in magnitude and direction.

Since, as we have just seen, the large-scale magnetic fields of most natural objects cannot be the result of microscopic electric currents associated with permanent magnetism, the fields must arise from macroscopic electrical currents flowing in the electrically conducting material of which these bodies are composed. Imagine for a moment that the material in which the magnetic field is embedded and in which the electric current is flowing, is static. (The object as a whole may be moving through space or rotating, but the various parts of it are not moving with respect to one another.) In a static electrical conductor, evolution of a magnetic field proceeds through the resistive dissipation of the electric current and the subsequent diffusion of the field out of the material. The characteristic time for the field to decay by about 60 percent is $L^2/\eta$ where, as before, $L$ is the physical size of the conducting body, and $\eta$ is the magnetic diffusivity. Applied to Earth's electrically conducting, molten-iron core, this decay time amounts to about 30,000 y, a very short time in comparison with Earth's 4.5-Gy age. In contrast, paleomagnetic evidence shows that the average intensity of the geomagnetic field has remained nearly unchanged for much of Earth's history. It is apparent that some mechanism must act to regenerate the magnetic field in compensation for its resistive decay.

If we apply a similar consideration to the Sun, we find that the 60 percent decay time is in excess of 1 Gy for a global magnetic field embedded in the Sun's interior; a solar magnetic field, trapped in the Sun at the time of its formation 4.5 Gy ago, would only have decayed by a factor of less than 10. Thus it is conceivable that a stationary magnetic field in the Sun could be remaining from an earlier time. However, we noted earlier that the main solar magnetic field is not stationary, but rather that it oscillates with the very short period of 22 y. The solar field also must result from an active, contemporaneous generation process.

Altogether, using Earth and Sun as examples, it is clear that an explication of magnetic fields in natural bodies must rely on an active regeneration process. Examples could have been drawn from other bodies as well.

## 13-4 MAGNETOHYDRODYNAMIC DYNAMOS

The evolution of a magnetic field in an electrically conducting fluid is the combined result of diffusion of the field lines through the fluid and convection of the field lines with the fluid. We have just discussed how the behavior of measured natural magnetic fields cannot be accounted for on the basis of diffusion alone. Neither the persistence of Earth's field nor the oscillation of Sun's field can be explained on this basis. Both of these phenomena must be a result, then, of the influence of the fluid velocity on the magnetic field. In the remainder of this article we discuss the generation of magnetic fields by fluid motion in natural bodies, through a phenomenon known as the *magnetohydrodynamic dynamo*. This phenomenon seems to be responsible for the presence of persistent magnetic fields in a wide variety of cosmical bodies, including, at least, Earth and the planets, the Sun, and stars.

Now, in order to see how the motion of electrically conducting fluid in a planet or in the Sun generates a magnetic field, we need to have some general idea of how fluid moves in such a body. An understanding of fluid motion in cosmical bodies requires solution of the equations of fluid motion under the influence of natural forces and rotation, and taking into account the constitutive and thermal properties of the fluid as well as the appropriate boundary conditions. This prodigious program so far has not been pushed to completion. Although we have learned a great deal about fluid motion in natural objects, major fundamental issues remain unresolved. For the purpose of this introductory discussion, we do not attempt a strictly realistic treatment of the fluid motion; in any case such an approach is largely beyond our means. Rather, we talk in terms of a simple and plausible class of fluid motions that seems to capture the essential physics of magnetic field generation in real bodies in a particularly clear way.

Instead of working with an idealized solution of the fluid motion equations, we assert a set of fluid motions based on intuitive, heuristic considerations. The particular motions that we assume are chosen because they facilitate a lucid presentation of the physical process of magnetic field generation and capture the essential characteristics of known dynamo systems. In the subsequent discussion, we neglect questions of the forces that produce the fluid motion as well as the fluid's response to those forces. This is called the *kinematical* approach, as opposed to the full *dynamical* approach, in which the equations of fluid motion are solved simultaneously with the electromagnetic equations that describe the generation of magnetic fields. Only a few idealized and limited solutions of this latter sort have been obtained; although the applicability of these idealized dynamical solutions to real objects is not certain, they provide very important insights into some of the dynamical issues.

One of the important features of the magnetic field generation process that is left aside in the kinematical approach is the influence of the magnetic force on the fluid motion. The magnetic field is associated with electric currents; these electric currents, as they flow through the field, produce the so-called Lorentz force. This force can, and indeed must, alter the motion of the fluid in important ways. A most important ultimate alteration is to reduce or change the character of the fluid velocity in such a way as to diminish the fluid's capacity to generate a magnetic field. That this must happen is easy to see; otherwise, once a magnetic field began to be generated in a fluid, the field intensity could grow exponentially without limit in violation of the constraint imposed by conservation of energy.[2]

In an electrical generator this same inhibition of the motion takes place. As electric current flows through the armature, in response to the electric field induced in the wires,

---

[2] Mathematically this result can be readily seen. The kinematic magnetic-field induction equation, which describes the evolution of magnetic fields in conducting fluids, is homogeneous of the first order; it admits normal mode solutions for the magnetic field that vary exponentially in time with a growth rate that depends on the fluid velocity field. Thus once a growing magnetic-field mode, if such exists, is encountered, the field intensity grows exponentially without limit, or until the fluid motion changes. The first major program of kinematical dynamo theory was to demonstrate the existence of such growing or stationary modes. Indeed such modes do commonly exist. The dynamo equations describing them are derived from the induction equation and exhibit the same expected exponential growth.

a Lorentz force acts between the current and the generator's magnetic field. This force acts to slow the motion of the armature, and it is against this force that work must be done to run the generator. The larger the current drawn, the larger the force that must be applied to keep the armature turning.

Because the forces producing fluid motion are not fully treated in the kinematical approach, one of the important questions about natural magnetic-field generation that is beyond reach is that of the intensity of the field that arises in a given physical situation. Other equally significant issues also are bypassed in the kinematical theory; nonetheless, the underlying physical processes are most easily established through this approach.

### 13-4-1 The Fluid Motions

Magnetic fields are efficiently generated by fluid motion in rotating bodies. Rotation, through the action of the Coriolis force, imposes an organization on the flow that is central to the field generation process. Consider an idealized sphere of fluid with an internal energy source. For definiteness, we can imagine that heat is liberated within the sphere by the decay of radioactive nuclides, although there are other possibilities. If heat is liberated at a sufficiently large rate, the resulting steep thermal gradient produces convection of the fluid. As the fluid convects in the radial direction to remove heat from its interior, the tendency of a fluid parcel to conserve angular momentum results in the inner parts rotating with higher angular velocity than the outer parts, producing a state of overall differential rotation (see Fig. 13-4a). This conservation of angular momentum, in motion with respect to a rotating object, is the origin of the Coriolis force.

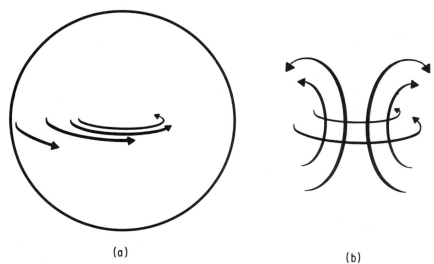

Fig. 13-4 The idealized fluid motion that we consider here consists of two main components—*Overall differential rotation* (a), in which, because of the convecting fluid's tendency to conserve its angular momentum, the inner parts of the convecting core rotate with higher angular velocity than the outer parts of the core, and *smaller-scale cyclonic convection* (b), in which the Coriolis force, acting on the convergence and divergence of the flow toward and away from the axis of a convective cell, induces a cyclonic or helical motion in each small convection cell.

Now we scrutinize the small-scale character of the convection more closely. Picture an idealized convection cell (Fig. 13-4b); there is a radially outward motion of the fluid along the axis of the convective eddy, with a general, subsiding return flow in the surrounding fluid. Associated with this radial motion, then, is convergence of the flow toward the axis of the convection cell at the bottom and a divergence away from the axis at the top of the cell. The Coriolis force associated with this converging part of the flow causes the fluid to turn locally around the axis of the convective eddy, in the manner of a cyclone, and in the same direction as the overall rotation. The divergence of the flow at the top of the convection cell reduces, and may partly reverse, this cyclonic motion. The flow in a convective eddy, then, is helical locally, (**curl V**) $\cdot$ **V** $\neq 0$, with the dominant helicity positive in the north and negative in the south. (Cyclonic motion is discussed in a different context in Chapter 6 by Kivelson and Schubert.)

## 13-4-2 Regeneration of Magnetic Fields

Now we turn to an idealized example to see how these fluid motions regenerate a magnetic field in a planet. For definiteness in the discussion, and because it is the best observed among cosmical bodies, we use Earth as our example. The inner half of Earth, by radius, consists of liquid, molten iron, which is an excellent conductor of electricity. Earth's core also has a small, solid inner part, which is not important to the discussion at this point.

The external geomagnetic field is largely dipolar, with an intensity at the surface of about 0.6 Gauss. The dipole axis is not quite aligned with Earth's rotation axis, deviating by about 11°. Nor is the dipole axis fixed with respect to Earth; over thousands of years it wanders randomly about the geographic pole. The dipolar part of the field is about 80 percent of the total. Subtracting away this dipolar part, the remaining 20 percent consists of patches of magnetic field, roughly of continental size. These patches of field are not stationary. The entire pattern of magnetic patches drifts westward at just under 0.2° of longitude per year; this rate of drift would carry the pattern once around the globe in about 2000 y. In addition, the individual patches grow and decay, apparently randomly, over periods of about 1000 y.

The magnetic field lines at Earth's surface are rooted in the electrically conducting liquid iron core far below. The dynamic variation of the field at the surface reflects the motion of the core fluid, which drags the magnetic field with it. The growing and decaying, continental-scale patches of magnetic field apparently arise as lines of force are dragged about by the convective upwellings of core fluid and as the upwellings expel magnetic flux out through the core's surface. The rate of variation of these magnetic patches as they roil about at Earth's surface, implies a convective-fluid velocity of about $0.1 \text{ mm} \cdot \text{s}^{-1}$.

The westward drift of the entire pattern of magnetic irregularities can be interpreted as a direct reflection of the nonuniform rotation of Earth's core. With the outer part of the core rotating more slowly than the inner part, and with the mantle rotating at the average angular velocity of the core material, the outer layer of the core lags the mantle's rotation. Field lines, forced to rotate with the angular velocity of the core's outer part, then drift westward with respect to Earth's surface. The drift angular velocity of the geomagnetic field—about 0.2° of longitude per year—implies that the differential-rotation velocity in the core is about $0.3 \text{ mm} \cdot \text{s}^{-1}$ in this picture.

For simplicity, neglect the nondipolar part of the geomagnetic field at Earth's surface. The field strength of a magnetic dipole varies inversely with the cube of distance from the dipole. Extrapolating the surface dipole field to the outer boundary of the core at $\frac{1}{2} R_E$ yields a field intensity of about 5 Gauss. The magnetic lines of force of this poloidal field penetrate deeply into the core before again emerging. Since the high electrical conductivity of the fluid traps the magnetic lines of force, the lines move with the local velocity of the fluid. The deeper parts of the core rotate faster than the outer parts, so that each field line is wound like a spring, producing tight coils of toroidal magnetic field (see Fig. 13-5). Using the model of core-fluid motions described above, the field lines wrap around many turns in the time it takes for the field to decay away because of electrical resistance; consequently, the intensity of the toroidal field could be as high as several hundred Gauss—nearly a hundred times greater than the poloidal part of the field. (A *poloidal field*, like a dipole field, has lines of force in meridian planes. A *toroidal field*, in this simple case, has lines of force parallel to the equator.) In the core, then, the dominant magnetic field is toroidal, running in coils around the rotation axis. Such an axisymmetric toroidal field is completely confined to the electrically conducting core and thus is hidden from our view. (In the Sun, the toroidal magnetic field is thought to be observable when loops of it protrude through the solar surface to form sunspot pairs.)

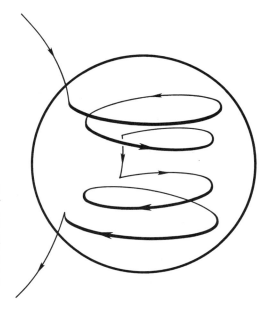

Fig. 13-5  Lines of force of the poloidal magnetic field penetrate deeply into the differentially rotating core before reemerging. Because of the core fluid's high electrical conductivity, the field lines are forced to move with the fluid flow. The inner parts of the core rotate more rapidly than do the outer parts, causing the field to wind into a toroid-like coil. For clarity, only one line of force is drawn.

Although differential rotation can produce a strong toroidal field from an existing poloidal magnetic field, differential rotation by itself cannot hinder the overall decay of both fields as a result of resistive dissipation of associated electric currents. The poloidal field is unaffected by the differential rotation; it decays away resistively. Subsequently, the toroidal field, left without a source, follows.

Maintenance of the general magnetic field requires a mechanism that regenerates the exterior dipole. This mechanism is provided by cyclonic convection. To see how the exterior, poloidal field is regenerated, look at the effect of a single, rising cyclonic convection cell as it moves through the toroidal field lines. In a simple idealized picture, the rising fluid locally carries the lines of force with it, producing a small loop in the toroidal field. Because the convection cell turns in the sense of a cyclone, the loop is likewise turned toward a meridional plane (Fig. 13-6). The overall effect of the convection, then, is to produce many local meridional loops of magnetic field lines.

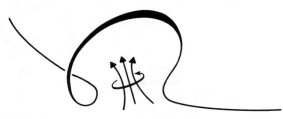

Fig. 13-6   A rising cell of convecting fluid carries a segment of the toroidal field lines with it, raising a localized loop or bubble of field. Because the rising convection cell turns in a cyclonic manner, the loop is twisted so that it lies partially in a meridional plane. Again, for clarity, only one field line is drawn.

Looking at the senses of circulation of the small meridional loops and of the exterior dipole field lines, note that they circulate in the same direction (see Fig. 13-7).

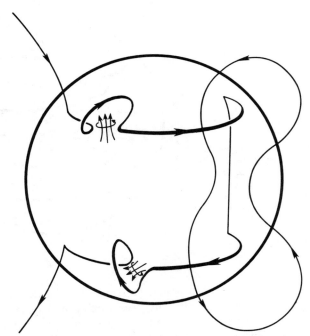

Fig. 13-7   Magnetic loops produced in the core by the cyclonic convection have the same circulatory sense as the exterior dipole field. These loops diffuse outward, expanding and coalescing, and eventually emerge from the core to join and reinforce the exterior dipole magnetic field.

These small loops can reinforce the large-scale, exterior magnetic field. Recall that, in addition to convecting with the local fluid motion, magnetic field lines also diffuse through it. Each of the little magnetic flux loops diffuses outward, expanding, overlapping, and coalescing with its neighbors, and eventually reinforcing and regenerating the exterior dipole field. This completes a regenerative cycle for the magnetic field: The differential rotation produces toroidal magnetic flux from the dipole field lines; the cyclonic convection produces exterior dipole flux from the toroidal field lines. The full regenerative cycle preserves the magnetic field against decay by generating new flux lines to compensate for those lost to resistive dissipation.

We have illustrated the basic cycle of dynamo magnetic-field regeneration through a series of diagrams showing how new magnetic field lines are created as a result of the distortion of the preexisting field by fluid motions and the subsequent topological reconnection of the lines by resistive diffusion. For the reader interested in a detailed mathematical development, accessible discussions can be found elsewhere (Parker, 1970b, 1979; Levy, 1976, 1978; Moffatt, 1978; Krause and Radler, 1980).

### 13-4-3 Discussion

The two competing processes at work in the magnetohydrodynamic dynamo are the dissipation of the magnetic field by electrical resistivity (or, equivalently, diffusion of the magnetic field lines out of the object) and the production of new magnetic field lines by the fluid motions. In order for a nearly stationary field to result, as is the case with Earth, these two processes must effectively be in balance. We stated earlier that magnetic diffusive and convective effects balance when the magnetic Reynolds number, $R_m$, is of the order of unity. In the regenerative cycle we just described, there are two important magnetic Reynolds numbers corresponding to the two essential velocities: the differential rotation velocity and the cyclonic part of the convection velocity. The minimal criterion for the generation of a magnetic field is that the product of these two magnetic Reynolds numbers be about equal to or somewhat greater than unity. This product is often called the *dynamo number*, and it is a measure of the effective vigor of the fluid motions. If the dynamo number is too small, any magnetic field will decay away. If the dynamo number exceeds the minimum needed to generate a magnetic field, there are, in general, many possible states for the magnetic field, including fields that oscillate (like that of the Sun), fields that grow in intensity, and fields that are stationary with increasing structural complexity.

Notice that a requisite of this simple regeneration cycle for a stationary magnetic field is that the small loops of poloidal field have the same circulatory sense as the exterior dipole field loops. Otherwise the new field lines degrade rather than reinforce the preexisting field. The sense of the small poloidal loops is determined by the relative senses of the differential rotation and the helicity of the convection, and by the detailed geometrical structure of the magnetic field (as well as by the angle through which the loops turn). Not all physical situations yield reinforcing loops. In the Sun, because the fluid is highly compressible, the helicity of the convection has its sign opposite to convection in Earth. It is thought that the small poloidal loops generally degrade rather than reinforce the preexisting field. Indeed, this is part of the reason why the Sun's magnetic field runs in an oscillatory mode rather than in a stationary mode like Earth's.

Furthermore, all of the loops generated in a single object need not have the same sense of circulation. Many of the steady-state modes in which the geomagnetic field might operate involve simultaneous production of regenerative and degenerative loops, with the regenerative loops predominating. This could be related to the origin of the phenomenon of the geomagnetic reversal, to which we shortly return.

It should be clear from the foregoing discussion that the dynamo process of magnetic-field generation is one strictly of *regeneration*. A magnetic field cannot be created from a field-free state; some arbitrarily small field must be present from the beginning. The generation of such weak "seed" fields, needed to get the dynamo process started, is not an insurmountable problem. Weak, large-scale magnetic fields are virtually ubiquitous in the universe. Such fields arise in many ways: as a byproduct of the origin of the universe, during the formation of galaxies, or as a result of thermoelectric effects in stars and planets. While these processes are common, the fields that they produce are far too weak to account for the much stronger fields that we actually observe in cosmical bodies.

It is an important characteristic of the dynamo process that the two possible polarities of the magnetic field are equally well generated. If, in the accompanying illustrations, all of the arrows denoting the direction of the magnetic field are reversed at the same time, the regeneration cycle we described continues to work unchanged. The polarity of a stationary field, such as that of Earth, is determined accidentally by whatever stray seed magnetic field is present when the dynamo begins its operation or by any transient event that may subsequently change the polarity.

The specific dynamo cycle with which we illustrated magnetic-field regeneration here is not unique in its details. The feature of the fluid motion that is essential to the dynamo process is the helical, or cyclonic, character of the convection. A large class of fluid motions can be shown to give rise to efficient magnetic-field generation; all of these motions have in common the helical character described above. In fact, dynamo cycles without differential rotation exist. In such cycles, both the toroidal and the poloidal magnetic field components are generated by the cyclonic convection and are of the same general magnitude. Indeed, one important dynamical model suggests the fluid motions, which produce magnetic fields in planets, are of a geostrophic form, in which the Coriolis force is balanced by fluid pressure, with little general differential rotation (Busse, 1975, 1976).

We passed by rapidly the question of the energy source of the fluid motions that generate magnetic fields in planets. Even for Earth this question is not completely resolved. A large quantity of heat is liberated within Earth by the decay of radioactive nuclides—particularly potassium, uranium, and thorium. However, it is uncertain how much of this radioactive material is trapped in Earth's core, where the released energy can contribute to generating the field. Heat sources above the surface of the core contribute to keeping the core material warm, but do not produce or add to convective motions in the core. Therefore, it is only that energy liberated within the core that is significant in this regard.

Additionally, energy can be liberated at the transition between liquid and solid iron at the surface of Earth's solid inner core, if the solid core continues to grow. Energy derived from the continuing growth of the solid core could make a significant contribution to the motions generating Earth's magnetic field (Verhoogen, 1961; Braginskii, 1963).

Similar considerations should apply to the other terrestrial planets—Mercury, Venus, and Mars—but there is considerable uncertainty about the interior states of them all (Stevenson et al., 1983). It is fair to say that the growth in our knowledge about the interior states of planets and in our understanding of magnetic-field generation in planets will proceed together.

In the giant, or Jovian, planets, energy to drive internal convection and generate magnetic fields can come from several sources (Hubbard, 1980). Primordial heat—energy left over from the formation of these planets—may still comprise a significant fraction of the heat emerging from all of these planets. In addition, energy is available from the continuing gravitational contraction of the giant planets as well as from the possible separation and settling of heavy molecular or atomic species, including helium. In these planets, decay of radioactive nuclides probably contributes relatively little to the internal heat.

### 13-4-4 The Long-Term Behavior of the Geomagnetic Field

Because of its proximity, the geomagnetic field is the best studied of all cosmical magnetic fields; it is the only natural magnetic field of which we know the behavior for a significant period of geological history. In Sec. 13-4-2, we briefly reviewed the behavior of the field over periods extending for a few thousand years. Averaging the geomagnetic field over periods of time somewhat longer than that, we can probably regard the external field as closely resembling an axially aligned and axially symmetric dipole. Deviations of the average field from this idealization are interesting, but probably not fundamental to the physical processes we are discussing. When the geomagnetic field is followed for long periods of geological time, interesting behavioral traits are found. We have a fairly complete record of the general behavior of the field for some hundred million years into the past. Isolated measurements give us glimpses of the state of the field for times in the much more distant past.[3]

So far as we are aware, the average intensity of the geomagnetic field has been nearly constant throughout much of Earth's history. The average intensity of the external field being some $\frac{1}{2}$ Gauss at Earth's surface. However, while the long-term average has been steady, the actual intensity varies considerably and in a seemingly random and noisy fashion. Apparently random intensity fluctuations are common, amounting to half or more of the average intensity, and characteristically occurring over intervals of several

---

[3] Our information about the behavior of Earth's magnetic field for long periods of geological history comes to us from the magnetization of old rocks and sediments. For example, lava flows carry small amounts of ferromagnetic material. This ferromagnetic component becomes magnetized in the direction of the local magnetic field at the time that the lava cools after being extruded to the surface of Earth. Similarly, magnetized grains in sediments align along the direction of the local magnetic field as they settle out of solution, leaving a stratified record of the local field.

One of the more fascinating and continuous records of the geomagnetic field's behavior is carried in the crustal rocks of the ocean floor as they spread away from midoceanic ridges. When new crustal material is extruded to the ocean floor and away from the ridge, it cools rapidly through the Curie and magnetic blocking temperatures, thus preserving a continuous record of the local geomagnetic field in bands of magnetized rock. The information is recorded much like the signal on magnetic tape and can be read in a similar fashion by towing a magnetometer just above the ocean floor, transverse to the midoceanic ridges.

thousand years. More provocative still is the behavior that is seen when the field behavior is followed for longer periods of time. After random intervals of time, the magnetic field suddenly, spontaneously, and rapidly reverses its polarity—the north pole and south pole exchange places. The geomagnetic reversal event itself takes place very rapidly, most of the action occurs and is over within less than 10,000 y. The field apparently falls in intensity; the dipole moment shrinks to zero before expanding back to its usual intensity but in the opposite direction. The field does not disappear entirely during a reversal event, it falls to some 20 percent of its usual value, leaving only the higher-order, continental-scale patches of magnetic field that we described earlier.

Much has been made recently in the scientific literature, and in the popular press, of the fact that the intensity of the geomagnetic field has diminished several percent during this century. This fact has produced speculation that we are seeing the onset of a geomagnetic reversal! Perhaps so; but there is no reason to believe it. The geomagnetic field seems to be behaving in its usual fashion.

The intervals between reversal events are distributed randomly. The average interval is about 200,000 y; but actual polarity intervals vary from as short as 10,000 or 20,000 y to 1 million years or more (Cox, 1969, 1975). It is important to distinguish between this random reversal behavior of the geomagnetic field and the regular oscillations of the Sun's magnetic field. The solar cycle arises because the solar dynamo operates in a regular and continuous oscillatory state. Its oscillation period apparently is well determined by characteristic values of several important physical parameters in the Sun's convection zone. Earth's dynamo seems to operate in a stationary (even if noisy) state. Reversal events are isolated and impulsive, occupying much shorter stretches of time than the intervals between them. The intervals between reversal events do not follow any regular pattern.

Before discussing the origin of the geomagnetic reversal phenomenon, it is useful to ask what form of explanation we might expect to obtain. We do not now have a deductive theoretical understanding of the origin of geomagnetic reversals. That is to say, the existence of geomagnetic reversals is not an immediate, predictable, and necessary consequence of our picture of the geomagnetic field's generation. There are two points to be made: First, we have seen that geomagnetic reversal events are distributed randomly and, therefore, may reflect aspects of behavior of Earth's core that are stochastic rather than deterministic. The second and related point is that the actual reversal process itself probably involves details of the physical state of the core fluid and its motions that are beyond our present knowledge.

In the remainder of this chapter, we look more closely at magnetic fields generated by the idealized dynamo mechanism that we pictured above, and see how geomagnetic reversal events with the observed properties might arise in a natural way. Earlier in the chapter we pointed out the necessity that the small loops of poloidal magnetic field reinforce the exterior dipole in order to realize a simple, steady magnetic field of the kind that is present in Earth. These regenerative magnetic loops depend for their polarity on the sign of the toroidal field that is produced by the differential rotation. In the simple idealization with which we illustrated the dynamo process, the average angular velocity of the core fluid increased monotonically inward and all of the poloidal magnetic field lines penetrated unidirectionally into the deep core before turning around and emerging in a unidirectional fashion. As shown in Fig. 13-5, such a magnetic state results in a toroidal magnetic field that has a single polarity in each hemisphere north and south of the equator. The sense of the cyclonic convection was such as to generate, from these toroidal field lines, poloidal loops that reinforced the exterior dipole field (Fig. 13-7). This is an oversimplification and does not, in fact, represent the general case.

Generally the poloidal magnetic field lines follow more complex paths as they wend their way through the core (Parker, 1969; Levy, 1972a). Even the simplest dynamo field modes produce poloidal field lines that first penetrate deeply into the core; turn toward the surface before crossing the equator; and then, upon crossing the equator, turn inward again before finally turning outward and reemerging from the core. From such a poloidal field the nonuniform rotation winds toroidal magnetic field that has both polarities in each of the hemispheres, as shown in Fig. 13-8. As a consequence, the cyclonic convection produces, in some regions of each hemisphere, poloidal loops that reinforce the exterior field, while in other regions of each hemisphere, loops are produced that oppose the exterior dipole magnetic field. In the simplest common modes, loops produced at high latitudes in both hemispheres reinforce the field, whereas loops produced at low latitudes in both hemispheres oppose regeneration of the field. The net magnetic field results from a balance between these regenerative and degenerative processes; in the stationary magnetic dynamo states, high-latitude regeneration prevails in the balance.

The idealized, and probably oversimplified, picture of core-fluid convection that we are using has randomly occurring upwellings in a constantly shifting pattern, with a coherence time of about 1000 y. Following this picture against the background of toroidal magnetic field that we have just described, we expect a continually varying ratio in the rates of production of high-latitude regenerative magnetic loops and low-latitude degenerative loops. While the average balance produces a stationary field, at any instant the dynamo is either overabundant in regenerative loops or overabundant in degenerative loops. Thus the field, at any instant, is either in an actively growing state or is in an actively

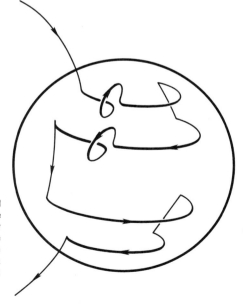

Fig. 13-8 In many dynamo magnetic field states, both regenerative and degenerative magnetic field lines are produced. A steady field is generated because the regenerative processes predominate over the degenerative processes. Fluctuations of the fluid flow in a dynamo magnetic field of this kind can easily precipitate a polarity reversal.

degenerating state. This would account for the generally noisy behavior that is observed to characterize the geomagnetic field.

Now, in this picture, the geomagnetic field's instantaneous behavior—intervals of one to a few thousand years—depends on the actual distribution of convective upwellings, that is to say, on the relative vigor of high-latitude and low-latitude convection in the core. Consider the effect of a large fluctuation in the distribution of convective upwellings. In particular, a large enhancement in the vigor of convection at low latitudes produces a large excess of poloidal magnetic field with polarity opposed to the exterior dipole field. If this pulse of reversed magnetic field is sufficiently large, then it will overwhelm the original field and precipitate a polarity reversal of the overall magnetic field. Remember that, to the dynamo process, the overall polarity of the magnetic field is irrelevant; fields of either sign are equally well regenerated. Thus the field, once reversed by a transient fluctuation in the distribution of convection, is maintained in the reversed state by the usual, average fluid velocity until another similar transient fluctuation precipitates the next reversal event (Parker, 1969; Levy, 1972b, 1972c).

Our understanding of even the average fluid motions in Earth's core is incomplete. Details of the time variations and fluctuations in the motion lie well outside our present knowledge. Nonetheless, it is encouraging that, starting from a simple picture of the core-fluid motions, we can derive an understanding of the geomagnetic field's generation and, on the same basis, account for the phenomenon of the occasional geomagnetic reversals in a natural fashion. This natural susceptibility to reversal by fluctuations in the distribution of fluid motions seems to be a general property of most dynamo stationary states (for another example, see Boyer and Levy, 1981). Estimation of the average interval between reversal events that is implied by this picture would, of course, be a crude and hazardous endeavor. Nonetheless, if one ignores the influence of magnetic forces on the fluid motion and makes the naive presumption that the convection can be represented by an ensemble of randomly occurring upwellings, with at least some degree of statistical independence, then reasonable estimates, based on the picture just described, agree with the observed lengths of polarity intervals (Levy, 1972c).

## 13-5 CONCLUDING REMARKS

The widespread occurrence of strong and persistent magnetic fields among virtually all classes of cosmical objects, in most cases apparently, is the result of an active generation process known as the *magnetohydrodynamic dynamo*. The fields are maintained against dissipation by the regenerative action of fluid motions in these electrically conducting bodies. The regenerative power of dynamo fluid motions derives from the cyclonic or helical character of convection in rotating bodies.

It is nearly 30 y since the regenerative influence of cyclonic convection was demonstrated by Parker (1955) in a seminal paper. Although there are many deep and important aspects of magnetic-field generation and behavior that still elude our understanding, and although many of the effects of magnetic fields on the behavior and evolution of cosmical objects are not yet understood, we can now assert with some confidence that we understand the basic character of magnetic-field generation processes in cosmical bodies.

In the development of our understanding of planetary magnetic fields, as well as other cosmical magnetic fields, Earth has played a central role. Earth's magnetic field, of course, is the best observed. We know details of the geomagnetic field's behavior—its small-scale structure and short-term variations, as well as its general behavior throughout long stretches of geological time—that we scarcely can expect to learn for any other magnetic field. As we have discussed in this chapter, the small-scale, rapid variations in the field are a direct manifestation of the generation process; and the long-term behavior patterns carry important clues about the field's structure and dynamical state. In a similar way, the magnetic fields of solar-system objects are the only cosmical magnetic fields that are close enough to be subjected to detailed scrutiny. By deepening our knowledge of these magnetic fields, we also extend our understanding of objects throughout the cosmos.

## 13-6 REFERENCES

Boyer, D., and E. H. Levy, 1981, Stationary dynamo magnetic fields produced by latitudinally non-uniform rotation, *Astrophys. J., 247*, 282-292.

Braginskii, S. I., 1963, About the structure of the F layer and the cause of convection in the core of the Earth, *Doklady Akademie Nauk SSSR, 149*, 1311 (in Russian).

Busse, F. H., 1975, A model of the geodynamo, *Geophys. J. Roy. Astr. Soc., 42*, 437-459.

Busse, F. H., 1976, Generation of planetary magnetism by convection, *Phys. Earth Planet. Int., 12*, 350-358.

Cox, A., 1969, Geomagnetic reversals, *Science, 163*, 237-245.

Cox, A., 1975, The frequency of geomagnetic reversals and the symmetry of the non-dipole field, *Rev. Geophys. Space Phys., 13*, 185-189, 224-226.

Howard, R., 1977, Large-scale solar magnetic fields, *Ann. Rev. Astr. Astrophys., 15*, 153-177.

Hubbard, W. B., 1980, Intrinsic luminosities of the Jovian planets, *Rev. Geophys. Space Phys., 18*, 1-9.

Jackson, J. D., 1975, *Classical Electrodynamics* (2nd Ed.), John Wiley, New York.

Krause, F., and K.-H. Radler, 1980, *Mean-field Magnetohydrodynamics and Dynamo Theory*, Pergamon Press, Elmsford, N.Y.

Levy, E. H., 1972a, Effectiveness of cyclonic convection for producing the geomagnetic field, *Astrophys. J., 171*, 621-633.

Levy, E. H., 1972b, Kinematic reversal schemes for the geomagnetic dipole, *Astrophys. J., 171*, 635-642.

Levy, E. H., 1972c, On the state of the geomagnetic field and its reversals, *Astrophys. J., 175*, 573-581.

Levy, E. H., 1976, Generation of planetary magnetic fields, *Ann. Rev. Earth Planet. Sci., 4*, 159-185.

Levy, E. H., 1978, Magnetic field generation at high magnetic Reynolds number, *Astrophys. J., 220*, 325-329.

Moffatt, H. K., 1978, *Magnetic Field Generation in Electrically Conducting Fluids*, Cambridge University Press, New York.

Parker, E. N., 1955, Hydromagnetic dynamo models, *Astrophys. J., 122*, 293-314.

Parker, E. N., 1969, The occasional reversal of the geomagnetic field, *Astrophys. J., 158*, 815-827.

Parker, E. N., 1970a, The origin of magnetic fields, *Astrophys. J., 160*, 383-404.

Parker, E. N., 1970b, The generation of magnetic fields in astrophysical bodies I—the dynamo equation, *Astrophys. J., 162*, 665-673.

Parker, E. N., 1977, The origin of solar activity, *Ann. Rev. Astr. Astrophys., 15*, 45-68.

Parker, E. N., 1979, *Cosmical Magnetic Fields*, Clarendon Press, Oxford.

Stevenson, D. J., T. Spohn, and G. Schubert, 1983, Magnetism and thermal evolution of the terrestrial planets, *Icarus, 54*, 466-489.

Verhoogen, J., 1961, Heat balance of the Earth's core, *Geophys. J. Roy. Astr. Soc., 4*, 276-281.

C. T. Russell
Department of Earth and Space Sciences, and
Institute of Geophysics and Planetary Physics
University of California
Los Angeles, California 90024

# 14

# SOLAR AND PLANETARY MAGNETIC FIELDS

## ABSTRACT

Most of the volume of the solar system is filled with plasma, loosely bound electrons and ions, which carries currents and hence contains magnetic fields. Inside many of the planets there are also electrical conductors whose mc.ions generate magnetic fields. It is the interplay of these magnetic fields that makes the exploration of planetary magnetic fields both interesting and difficult. We can classify the observed solar-system magnetic fields as either being mainly localized or global in character. The solar magnetic field, the field in the Venus ionosphere, and the magnetized crust of the Moon are all dominated by localized fields. Mercury, Earth, Jupiter, and Saturn have global fields with strong dipole components. Venus has no intrinsic magnetic field that we have been able to detect. The observations at Mars cannot now be unambiguously interpreted as indicating the presence or absence of an intrinsic magnetic field. Theoretical and some observational evidence suggests that Uranus and Neptune may also have intrinsic magnetic fields.

## 14-1 INTRODUCTION

Magnetic fields extend through all of explored space from the surface of the Earth to the most distant reaches of the solar system. These magnetic fields are either planetary or solar in origin. Although electric currents in interplanetary space contribute to the magnetic field, practically all magnetic field lines intersect either the solar surface or a planetary surface. It is the purpose of this review to examine our observational knowledge of the solar and planetary magnetic fields. However, before examining these observations, we briefly discuss what we understand about the generation of magnetic fields and attempt to provide some framework for the observations to be discussed later.

There are four basic equations governing the interrelationship of charges, currents, and magnetic and electric fields. These are collectively called *Maxwell's laws*. In an electrical resistor, current is usually directly proportional to the applied electric field. In a magnetized flowing conductor, an additional term arises because currents flow due to the motion of the conductor. The relation between the current, the electric and magnetic fields, and the flow velocity is called *Ohm's law*. Ohm's law can be combined with Maxwell's laws to give what is known as the *dynamo equation*. This equation relates the change in magnetic field due to the resistive decay of the currents and the regeneration of the field due to fluid motion. If there is no motion of the conducting fluid, the magnetic field decays with time. The time scale for the free decay of the terrestrial field is a few tens of thousands of years and of the Jovian field a few hundred million years, all much shorter than the age of the solar system. The solar free-decay time, however, is comparable to the age of the solar system. Thus, the solar magnetic field could conceivably be primordial, that is, it could have been present from the time of solar formation, presumably as a result of the compression of the preexisting galactic magnetic field. However, the solar magnetic field near the poles is seen to reverse every 11 y (Babcock, 1959; Howard, 1974; Howard and Labonte, 1981). This reversal has apparently been occurring at least as long as terrestrial magnetic records have been collected (Russell, 1974, 1975). Thus a primordial origin for the external solar magnetic field, as advocated by Piddington (1981), is highly unlikely.

The dynamo equation allows us to generate magnetic fields by specifying velocity fields that tell how the conducting fluid moves. This approach is called the *kinematic method* and is not self-consistent, in that it does not treat the source of the velocity field. Nevertheless, even with this approach it is difficult to find regenerative dynamos in which the seed field, upon which the dynamo acts, is recreated or maintained. Figure 14-1 shows how a planetary dynamo can be regenerative (Parker, 1979). The left-hand panel shows an initially poloidal field, that is, one confined to meridional planes. The middle panel shows the distortion of this poloidal field by differential motions in the core to create a toroidal, or azimuthal, field. This distortion of the poloidal field by differential motion is called the ω-effect. Rising convective cells in the right-hand panel recreate a poloidal field from the toroidal field, and the process continues. This creation of poloidal field is called the α-effect, and the overall dynamo process sketched here is termed an α-ω dynamo. This process is more completely treated in Chapter 13 by Levy.

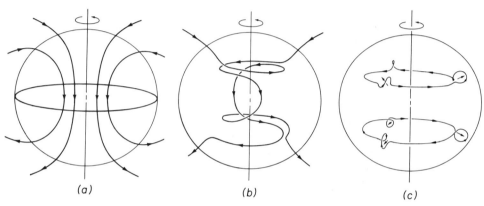

Fig. 14-1 Schematic representation of the α-ω dynamo. (a) The initial poloidal field. (b) The action of differential rotation (the varying rotation rate with depth) in the interior of the conductor, which twists the initially poloidal field around the axis of rotation, thus producing a toroidal field. (c) The rising convective cells twisting the toroidal field into meridian planes and hence reestablishing the poloidal field (Parker, 1979).

In order for the magnetic field to be sustained in the dynamo process, the contributions of the fluid motion that cause the field to grow must balance the rate of decrease from diffusion. The ratio of these two contributions to the dynamo action is of the order of the magnetic Reynolds number, $VL/\lambda$, where $V$ and $L$ are characteristic conductor velocities and magnetic length scales, respectively, and $\lambda$ is the diffusivity which is proportional to the electrical resistance of the conductor. Typically, magnetic Reynolds numbers greater than about 10 are thought to be necessary for a self-sustaining dynamo.

The complete solution to the problem involves not only the dynamo equation but also the generation of the fluid motions responsible for the motion. Despite intense efforts by many capable researchers, this problem is not yet solved, although pieces of the puzzle are beginning to be understood. For example, while we cannot safely predict the amplitude of planetary magnetic fields, we can make some inferences about what energy sources can and cannot drive a dynamo and hence "predict" what planets might have intrinsic magnetic fields. Those readers interested in a more extensive treatment of dy-

> **Magnetic Moments**
>
> Dipolar magnetic fields are those produced by a small loop of wire as observed far from the wire, or, equivalently, the field of a bar magnet observed far from the magnet. Complex arrangements of current loops or combinations of bar magnets produce complex magnetic fields. Mathematically, it is possible to express these complex fields as sums of ordered fields of increasing complexity. The source strengths of these ordered fields are called their *moments*. The strength of the source of the simplest fields is the *dipole moment*. The next most complex, but ordered, source is the *quadrupole moment*, which can be constructed from two closely spaced dipole moments. After that is the *octupole moment*. These so-called higher-order moments result in magnetic fields that decrease in strength with increasing distance from the source faster than the so-called lower-order moments.

namo theory are referred to the recent reviews of Busse (1983) and Stevenson (1983), the monographs of Moffatt (1978) and Parker (1979), and Chapter 13 by Levy.

We may classify the magnetic fields we observe in the solar system as to whether the fields are mainly global or localized. The Sun, the Moon, and the Venusian ionosphere have localized magnetic fields. On the Sun and in the Venus ionosphere, the magnetic field is wrapped up in tight bundles of magnetic field isolated from their neighbors. On the lunar surface, the magnetic field appears to be magnetized in isolated magnetic anomalies. These observations are to be contrasted with the Mercurial, terrestrial, Jovian, and Saturnian fields, which are mainly dipolar at and above their surfaces. This is not to imply that these latter planetary dynamos preferentially generate a dipolar field. Since the higher-order moments have a more rapid radial dependence than the lower-order moments, the dominance of the dipole field far from the core is not necessarily true just outside the core. In fact, if one extrapolates the externally observed magnetic fields of Mercury, Earth, and Jupiter to the expected radius of the core of these planets, one finds almost equal contribution to the field from all orders (Elphic and Russell, 1978). This fact in turn suggests that the planetary dynamo mechanism is chaotic or random. Thus, the difference between, for example, the terrestrial and solar dynamos, may not be as great as would be implied at first glance. Nevertheless, since the basic observations are quite different, we treat them separately. The Sun, the Venus ionosphere, and the Moon are discussed under the heading of shell dynamos. Mercury, Earth, Mars, Jupiter, Saturn, Uranus, and Neptune are discussed under the heading of core dynamos.

## 14-2 SHELL DYNAMOS

In the Sun, convection is thought to be limited to an outer convection zone. Convection in the Venusian ionosphere occurs in a shell a few hundred kilometers thick and about 10,000 km long. Geochemical evidence suggests that as the Moon accreted, the outer 200 km melted (Brett, 1977). The interior of the Moon did not melt until much later, if at all. Each of these systems appears to have produced magnetic lines of force, or magnetic flux, as defined in Chapter 13 by Levy, and to have concentrated this flux into bundles. On the

Sun we see these flux bundles where they break through the solar surface. In the Venusian ionosphere, our spacecreaft can fly right through *flux ropes*, that is, twisted bundles of magnetic flux, and can probe them directly. In the lunar case, we see localized magnetic anomalies. One wonders if they too could be produced as the crust cooled in the presence of a patchy, flux rope-like field. This last interpretation is hard to test, however, due to the small percentage of the lunar surface for which high-resolution lunar magnetic maps exist. In this section, we examine the properties of solar magnetic fields, the magnetic fields in the Venusian ionosphere, and lunar magnetic fields, compare these properties, and investigate whether such intercomparisons lead to further insights.

## 14-2-1 Solar Magnetism

The most obvious manifestation of solar magnetism is the sunspot cycle. (See Chapter 2 by Zirin.) A *sunspot* is a region in the solar photosphere that is cooler than its surroundings. Sunspots have sizeable magnetic fields and occur in bipolar groups of preceding and following polarity. The polarities of the preceding and following groups are opposite in the northern hemisphere and the southern hemispheres and reverse from one 11-y solar cycle to the next. Figure 14-2 shows two ways of observing the magnetic field of the Sun. The top panel is a spectroheliogram of the corona taken from *Skylab* at a wavelength of 285 Å (corresponding to emissions from iron ionized 14 times). The lower panel shows the photospheric line-of-sight component of the magnetic field determined from Zeeman splitting. The white regions indicate fields out of the Sun, and the dark regions, fields into the Sun. It is clear from the top panel, which is an image of the corona well above the photosphere, that arches connect the sunspot groups seen in the bottom panel. The reason why sunspots are cool is not completely understood. Possibly there is greater energy flowing out of the sunspots in the form of "magnetic" waves.

The number of sunspots varies with an approximately 11-y cycle. Figure 14-3 shows this cycle in the data collected since 1610. The initial low number of sunspots is believed to be a real effect and has been called the *Maunder minimum* (Eddy, 1976). It is clear that the amplitude of the 11-y solar cycle is quite irregular.

The latitude of sunspot occurrence also varies throughout the sunspot cycle. Sunspots first appear in a new cycle at high solar latitudes and then proceed to lower latitudes. Figure 14-4 shows this behavior from 1875 to 1977. This same latitudinal behavior can be seen in the average solar magnetic field, as illustrated in Fig. 14-5 (Howard and Labonte, 1981). The top panel shows the total flux, $F(=|F_+| + |F_-|)$ emerging from the solar surface. The middle panel shows the rate of change of this total flux in which all flux changes have been made positive, that is, rectified. The bottom panel shows the rate of change of flux where positive and negative changes have been algebraically added. The magnetic cycle and the sunspot cycle are very similar in latitudinal variation and in amplitude. The region of strong field drifts equatorward during the solar cycle and sunspot minimum corresponds to the period of weakest fields. It should be emphasized that the lifetime of any one magnetic region on the Sun is very short compared with the duration of the solar cycle. New magnetic flux is constantly emerging and old magnetic flux disappearing into the Sun.

Plots such as Fig. 14-5 do little justice to the truly amazing localization of the solar field. This localization into bipolar, that is, radially inward and outward, magnetic regions (BMR) occurs not just on the scale of sunspots as seen in Fig. 14-2, which may contain

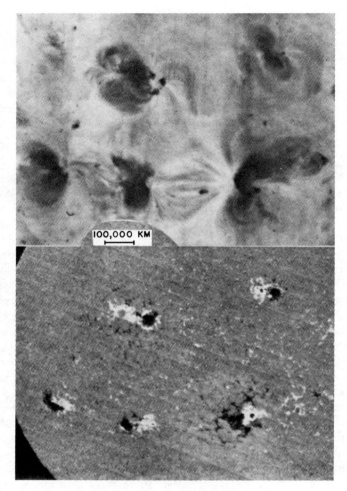

Fig. 14-2 A coronal spectroheliogram from *Skylab* taken at a wavelength (λ 285 Å) of Fe ionized 14 times (top), compared with a photospheric magnetogram of the same region (bottom). The coronal X-rays allow one to observe the extensive magnetic arches connecting the various bipolar magnetic fields. (Courtesy of NASA.)

the order of $10^{14}$ Webers (Wb), but also on much smaller scales. A smaller short-lived BMR that lasts the order of a day contains about $10^{11}$ Wb. Even smaller BMRs containing the order of approximately $10^9$ Wb and having a lifetime of approximately 30 min have been observed. The concentration of the magnetic field into discrete tubes is an unexpected phenomenon and is not simple to explain, but perhaps more surprising is the magnitude of the field in these structures that exceeds 0.1 Teslas (T). (The terrestrial surface field is equal to approximately 50 $\mu$T.) Magnetic fields can exert pressure on gases of charged particles such as those present in the solar photosphere and corona. The pressure in such a flux tube exceeds that of the surrounding photospheric plasma. In order to be in equilibrium, the magnetic field lines in such a structure must be twisted. It is, in fact, pos-

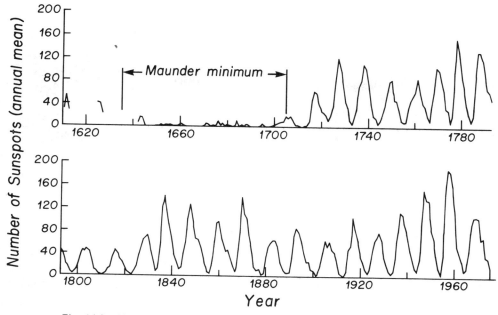

Fig. 14-3  Mean annual sunspot number from the year 1610 to the present. The approximately 70-y period centered about 1670 has been called the *Maunder Minimum* (Eddy, 1976).

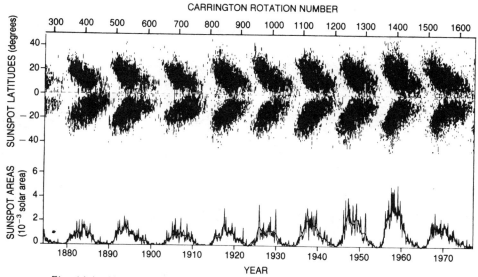

Fig. 14-4  Maunder "butterfly" diagram of sunspot positions and plot of total sunspot area as a function of time. In the upper plot, vertical lines show the range in latitude of sunspots at that particular date. The lower panel gives the total area occupied by sunspots in thousandths of the area of the solar disk. The term *butterfly* refers to the appearance of upper plot. (Data supplied by the Royal Greenwich Observatory. From Newkirk, Jr., G.; and K. Frazier, 1982. The solar cycle, *Physics Today*, 35, 4, 25–36. Used with permission.)

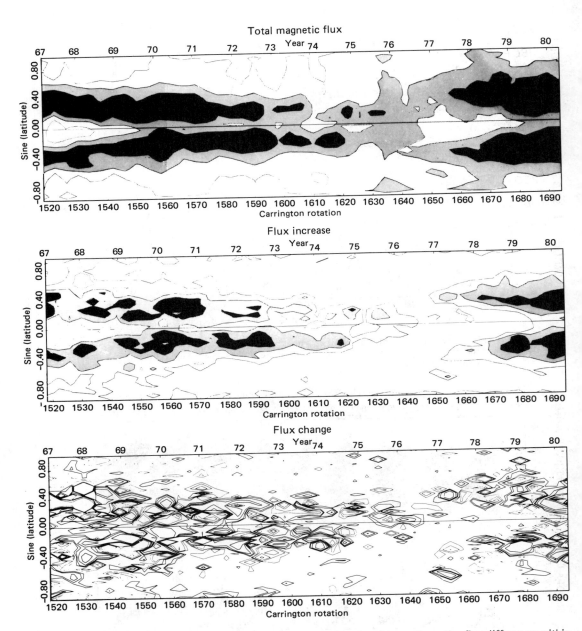

Fig. 14-5 (Top) Latitude distribution of total flux ($|F_+| + |F_-|$) on solar surface from 1967 to 1980. Contour levels are 2, 4, 8, 16, and 32 × $10^{21}$ Maxwells. (Middle) Flux increase as a function of latitude and time. The rate of change of total flux is measured as a function of latitude within 45° of the 0° meridian, the sum of positive differences obtained and corrected to represent the flux increase over 360° of longitude. Contour levels are 4, 8, 16, and 32 × $10^{20}$ Maxwells ∘ day$^{-1}$. (Bottom) Flux change as a function of latitude and time. The sum of positive and negative flux differences within 45° of the 0° meridian are summed and corrected to represent the flux increase or decrease over 360° of longitude. Contour levels are ±1, ±2, ±4, and ±8 × $10^{20}$ Maxwells ∘ day$^{-1}$. (From Howard and Labonte, 1981. Surface magnetic fields during the solar activity cycle, Solar Phys., 74, 131–145. Copyright © 1981 by Reidel Publishing Company, Dordrecht, Holland. Used with permission.)

sible to twist magnetic field tubes into force-free structures that are held together by the twist of the field and not by external plasma pressure, and whose currents are entirely parallel to the magnetic field. Such a structure is more properly referred to as a *flux rope* rather than a flux tube.

Figure 14-6 shows schematically a complex flux rope that is built up of many simpler flux ropes, or fibers (Piddington, 1981). Although it is possible to satisfy pressure balance with a simpler structure, such a complex rope helps explain the complexities of an active region as illustrated in Fig. 14-7. Later evolution of the flux rope might provide the field structure seen outside of the active regions, as shown in Fig. 14-8. Attractive as this model may seem, it depends on the ability of the Sun to create flux ropes in the first place. The mechanism to create these structures is neither trivial nor obvious.

A major feature of the sunspot cycle is the reversal of the polar field that occurs a few years after the peak of the sunspot cycle. At sunspot latitudes, each hemisphere's dominant field has the polarity of the preceding spot in the sunspot pairs in the same hemisphere. It is the motion of this flux to the poles that reverses the polar field (Howard and Labonte, 1981). This motion is episodic, occurring only when following spot polarity is seen near sunspot latitudes. When moving field of like polarity arrives at the pole, the polar field strength increases. Between times of new field arrival, it decays with a decay time of 10 to 20 y. The transport occurs by convection and not diffusion. Convective

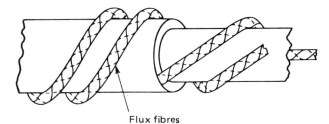

Fig. 14-6 The structure of a complex flux rope consisting of twisted bundles of flux fibers, themselves twisted. (From Piddington, 1981. Used with permission.)

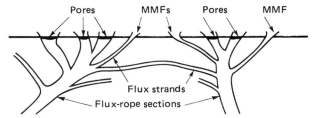

Fig. 14-7 Possible relationship between a solar active region and a flux rope penetrating the photosphere. Sunspots are proposed to develop when sufficient flux is accumulated in one region due to the coalescence of pores and moving magnetic features (MMFs). Pores in general contain less magnetic flux than a sunspot and may or may not develop into one. Moving magnetic features, as their name implies, move across the photosphere and contain even less flux. (From Piddington, 1981. Used with permission.)

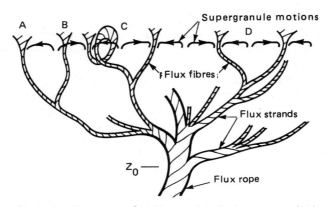

Fig. 14-8 Flux-rope-strand-fiber model of solar magnetic fields outside active regions. The flux rope has unwound from the sunspot configuration and frayed into a number of smaller flux strands and into hundreds or thousands of flux fibers. Below a depth marked $Z_0$ the original rope is intact and retains overall control of all the magnetic fields in the convection and atmosphere. In the photosphere, the projecting flux fibers tend to concentrate in the supergranule boundary regions as for fiber $A$. (Supergranule regions consist of a collection of smaller convection cells with some overall convective order.) The kink instability (see Sec. 14-2-2) creates loops that are observed as low-energy or soft X-ray bright points. (From Piddington, 1981. Used with permission.)

velocities range from 5 to 10 m · s$^{-1}$ during the rising portion of the solar cycle when the poleward moving field and the polar field have opposite polarities, and from 15 to 20 m · s$^{-1}$ in the declining phase when they have the same polarities. It is as if on the Sun, like polarities attract and opposite polarities repel (Bumba and Howard, 1965).

Although the solar polar magnetic field is weak (at most, a few times $10^{-4}$ T), it is important because it is these high-latitude magnetic fields that are expected to be open to the interplanetary space. Figure 14-9 shows the model of Pneumann and Kopp (1971) for the equilibrium of an isothermal corona and a dipole photospheric field. The dashed line shows an undistorted dipolar field. In actuality the global solar field is far from dipolar, and the structure of the interplanetary magnetic field reflects this complexity. The current sheet that separates the distended near-equatorial field is warped. This simple, warped current sheet near the extension of the solar rotational equator into space explains quite well the observed alternating polarities of the interplanetary magnetic field seen near Earth.

In summary, the solar magnetic field is very enigmatic. While the Sun conceivably could have a primordial field, the solar magnetic cycle appears to be driven by dynamo action. Magnetic fields most commonly occur in bipolar magnetic regions and are often of extraordinary strength and of extraordinary complexity. Like fields tend to attract rather than repel. Flux ropes appear to be the answer to some of these enigmatic properties, but if one adopts a flux rope model, then one is left with the equally difficult problem of manufacturing flux ropes. More detailed examinations of solar magnetic fields and solar activity can be found in the reviews by Parker (1977), Stix (1981), Orrall et al., (1981), and Gilman (1983).

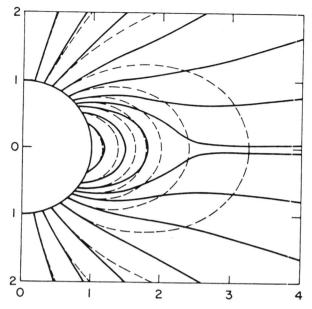

Fig. 14-9 The magnetic field geometry computed by Pneumann and Kopp for the equilibrium of an isothermal corona with a dipole magnetic field imposed on the base of the corona. Expansion of the solar wind into interplanetary space occurs along open magnetic field lines from the "polar" regions. The dashed lines indicate a pure dipole geometry. (From Pneumann, G. W.; and R. A. Kopp, 1971. Gas-magnetic field interactions in the solar corona, *Solar Phys.*, 18, 269. Copyright © 1971 by D. Reidel Publishing Company, Dordrecht, Holland. Used with permission.)

## 14-2-2  The Venusian Ionospheric Magnetic Field

Flux ropes are not solely the province of the Sun; they occur in the Venus ionosphere (Russell and Elphic, 1979a) and in the solar wind-terrestrial magnetosphere interface (Russell and Elphic, 1979b). We will consider only the former structures because they should be nonevolutionary during the measurement interval. However, the availability of better diagnostics in terrestrial flux ropes may eventually help overcome the difficulty imposed by the possible evolution of these features during their period of observation.

Initial observations of the Venus ionosphere with the *Pioneer Venus Orbiter* revealed that the highly conducting ionosphere to first order excluded the solar-wind magnetic field (Russell et al., 1979). A magnetic cushion whose pressure equalled the solar wind dynamic pressure provided a lid for the dayside ionosphere and acted to deflect the solar wind plasma before it intersected the planetary ionosphere. Inside the ionosphere the average magnetic field was quite small, almost two orders of magnitude smaller than the external field. The field that existed in the ionosphere was concentrated in what appeared to be turbulent cells but which later were shown to be quite orderly flux ropes (Russell and Elphic, 1979a).

> **Solar Wind and Dynamic Pressure**
>
> The Sun's upper atmosphere, the *solar corona*, which can be seen during total solar eclipse, is expanding outward at supersonic velocities. The supersonic plasma is called the *solar wind*. The force of this expanding gas against an obstacle is called its *dynamic pressure* and is equal to the density times the square of the velocity.

Figure 14-10 shows an altitude profile of the magnetic field strength and electron density on three phases of *Pioneer Venus* through the subsolar ionosphere. The left-hand panel shows the profiles for conditions of low solar-wind dynamic pressure when the top of the ionosphere, the ionopause, is at high altitudes. The localized bundles of magnetic flux, the flux ropes, are clearly visible. The two right-hand panels show the magnetic profiles for higher solar-wind dynamic pressure. Under these conditions the ionosphere becomes more uniformly magnetized at low altitudes. The magnetic diffusion times in the Venusian ionosphere are of the order of hours (Luhmann et al., 1984) much longer than the convection times at high altitudes. Thus, we interpret the high altitude minimum in these plots as loss due to convection of magnetized plasma to the nightside with replenishment of the dayside with "unmagnetized" plasma due to photoionization. In our conceptual model of the interaction of the solar wind with the ionosphere, the ionosphere becomes magnetized when the solar wind dynamic pressure is high enough that the magnetized magnetosheath (3) is pushed down from its usual high altitudes deep into the photo production region so that the ionosphere becomes magnetized (Russell et al.,

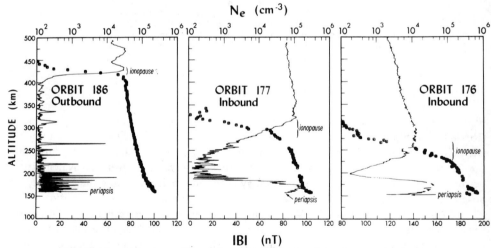

Fig. 14-10 Altitude profiles of the magnetic field and the electron density on three passes through the subsolar ionosphere of Venus. The strength of the magnetic field just above the ionopause and the height of the ionopause are both controlled by the dynamic pressure of the solar wind. On days of low solar-wind dynamic pressure, such as that on orbit 186, the ionospheric magnetic field is weak, consisting of narrow twisted bundles of flux, or flux ropes. On days of high solar-wind dynamic pressure, such as on orbits 176 and 177, a steady field is seen at low altitudes.

1983). At low altitudes, the rate of photoproduction so overwhelms the solar wind that the solar-wind magnetic field is essentially captured by the ionosphere.[1]

The flux ropes in this conceptual model are flux tubes rolled up due to shear in the velocity either at the top of the low-altitude field region (which, of course, is only present occasionally) or at the high-altitude ionopause. These tubes then are convected throughout the Venusian ionosphere as illustrated in the bottom of Fig. 14-11. As the spacecraft passes through one of these flux ropes, it sees a field variation such as that sketched in the top panel of Fig. 14-12. When the spacecraft does not pass exactly through the center of the rope, the variation of the field is three-dimensional. It cannot be contained in one plane. This can be illustrated with the use of *hodograms*, plots of the variation of the magnetic field vectors projected into planes. Hodograms of the field variation, projected in two perpendicular planes, in the principal axis system of a schematic flux rope are shown on the bottom of Fig. 14-12. In the principal axis coordinate system, the $i$-direction is parallel to the direction of the maximum variation; the $k$-direction is parallel to the direction of minimum variation, and the $j$-direction is perpendicular to both $i$- and $k$-

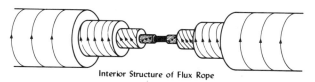

Interior Structure of Flux Rope

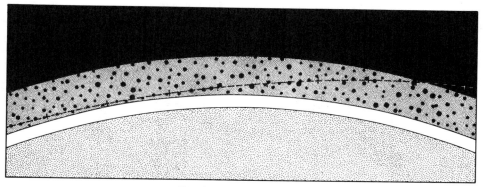

Distribution of Flux Ropes

Fig. 14-11  (Top) Interior structure of a Venus flux rope. In the center of the flux rope the magnetic field is straight and parallel to the axis of the rope. Further from the center the field progressively weakens and is wrapped in a tighter and tighter helix about the axis. (Bottom) Schematic of *Pioneer Venus* passage through the Venus ionosphere containing flux ropes. For the purpose of illustration, the axes of all flux ropes have been taken to be perpendicular to the page (and the orbit), and flux ropes have been depicted as roughly uniformly distributed. In actuality their orientation, size, and occurrence rate are position-dependent.

---

[1] The ionosphere of Venus is composed of electrons and ions, a plasma, created from the neutral atmosphere by photons from the Sun. The pressure of the ionosphere is sufficient to stand off the solar wind at some distance called the *ionopause*. Above the ionopause is the flowing solar wind in the *magnetosheath*. Since the solar wind moves supersonically relative to Venus, a shock forms, and the *magnetosheath* consists of shocked plasma.

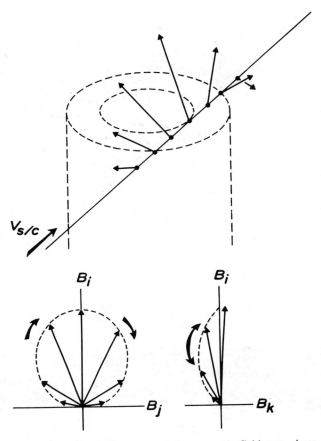

Fig. 14-12 (Top) Variation of the magnetic field seen along spacecraft trajectory for an off-axis pass through a flux rope. (Bottom) Hodograms of the variation of the field in the principal axes of the field variation. For off-axis passes, it is impossible to two-dimensionalize the field variation in a flux rope (Elphic et al., 1980).

directions. The only time-stationary structure that produces such a field variation is shown in the upper panel of Fig. 14-11, that is, a tube of flux with a strong field parallel to the axis of the rope in the center of the rope, weakening and becoming more twisted with increasing distance from the center of the rope.

The magnetic field and helical pitch angle for six flux ropes are shown in Fig. 14-13. The pitch angle is the angle between the magnetic field and the axis of the flux rope. Although they are all qualitatively the same, there are important differences in the radial variation, especially of the pitch angle. The variation on the upper left is narrowly confined to the center of the rope, whereas on the lower right, the twist occurs over a broader region and is not as great (Elphic and Russell, 1983a).

If we attempt to model the variation of the field through these six ropes and to derive the currents parallel and perpendicular to the magnetic field direction, we obtain the results shown in Fig. 14-14. The flux rope on orbit 204 (upper left) is mainly com-

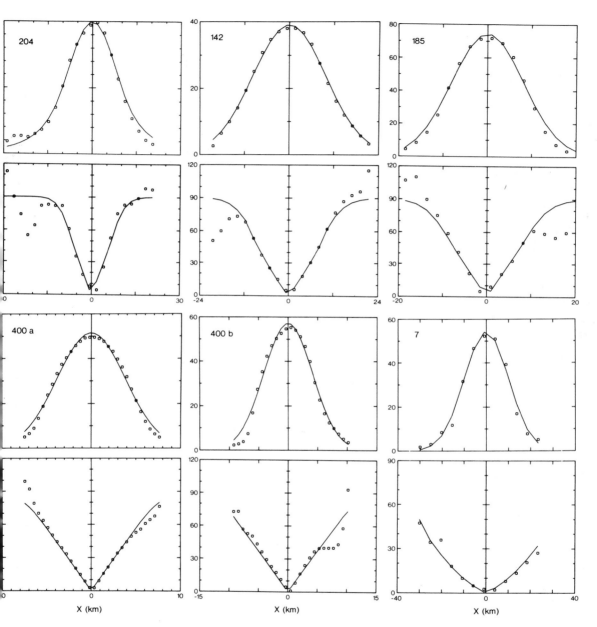

Fig. 14-13 Magnetic field strength and pitch angle of the magnetic field for six flux ropes in order of decreasing helicity (orbits 204, 142, 185, 400a, 400b, and 7). The solid lines show the best fit model used to derive the current structure of these flux ropes (Elphic and Russell, 1983a).

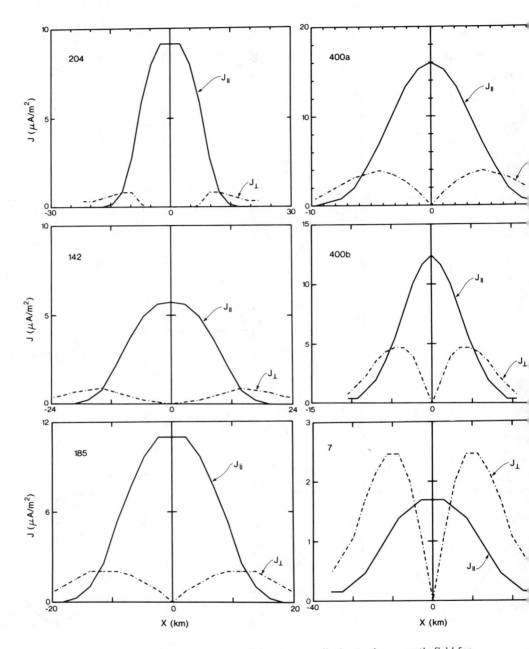

Fig. 14-14  The currents parallel and perpendicular to the magnetic field for the flux ropes of Fig. 14-13. The flux rope on orbit 204 is almost completely composed of parallel currents, that is, it is essentially force-free. At the other extreme, the flux rope on orbit 7 consists largely of perpendicular currents, and the field lines are much less twisted. Hence the thermal plasma pressure inside and outside this rope must be different (Elphic and Russell, 1983a).

posed of parallel currents, whereas the flux rope on orbit 7 (bottom right) has strong currents perpendicular to the field as well. Since there is little perpendicular current for the first rope, and since currents parallel to the field exert no force on the plasma, the flux rope is virtually "force-free," that is, self-balancing. Hence we would expect the plasma pressure inside the rope to be equal to that outside. This should not be true for the flux rope on orbit 7. Here there are large perpendicular currents, and there should be a much lower plasma pressure inside the rope than outside.

Flux ropes decrease in width with decreasing altitude both in the subsolar region and in the terminator region, that is, near the dawn-dusk plane. This is illustrated in Fig. 14-15. They also occur more frequently at lower altitudes. However, as shown in Fig. 14-16, this increase is not as dramatic in the terminator region as it is in the subsolar re-

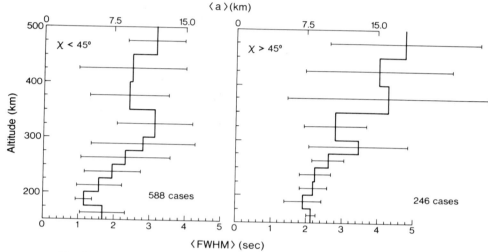

Fig. 14-15 The altitude distribution of the width of flux ropes in the subsolar region at solar zenith angles less than 45° (left) and nearer the terminators at angles from the subsolar point greater than 45° (right) [Elphic and Russell, 1983b].

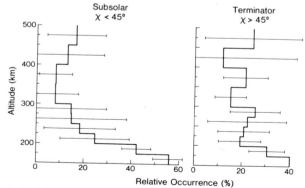

Fig. 14-16 The occurrence rate of flux ropes as a function of altitude in the subsolar region (left) and nearer the terminator (right) (Elphic and Russell, 1983b).

gion. This has been interpreted as a slowing down of the flux ropes as they drift down from the ionopause (Elphic and Russell, 1983b).

The sketch of flux rope distribution given in Fig. 14-11 implies that the axes of the ropes are horizontal. As shown in Fig. 14-17, this is true for the most part in the near terminator region. However, in the subsolar region the flux ropes are more randomly oriented with respect to the horizontal plane. The bottom panels show the distribution separated into low altitudes (dashed lines) and high altitudes (solid lines). In the subsolar region the flux ropes tend to be more vertical than horizontal at low altitudes and more horizontal than vertical at high altitudes. In the terminator regions, the flux ropes are quite horizontal at high altitudes and perhaps randomly oriented at low altitudes.

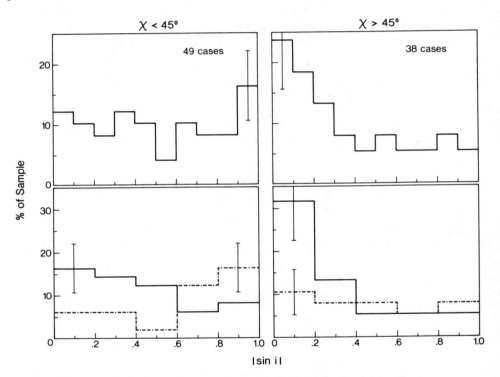

Fig. 14-17 (Top) Flux rope inclination relative to the horizontal for all subsolar cases (left) and all near terminator cases (right) for flux ropes crossed close to the rope axis. Sin $i$ is zero for horizontal ropes (i.e., zero inclination), and sin $i$ is unity for vertical ropes. A distribution with random orientations would be constant in this display. (Bottom) Flux rope inclinations for subsolar (left) and terminator (right) regions subdivided into low-altitude (dashed lines) and high-altitude (solid lines) cases. For the subsolar region, the dividing line between low altitudes and high altitudes is 200 km; for the terminator region it is 300 km.

The explanation for this behavior is given in Fig. 14-18. At lowest altitudes the flux ropes are narrower and more twisted. No longer can they remain straight tubes, but rather they attempt to shed some of their twist by forming a corkscrew. This behavior is known as the *kink instability*. It also helps explain the increase in occurrence rate at low alti-

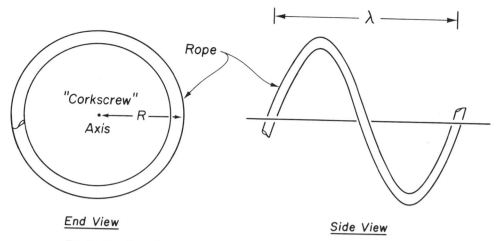

Fig. 14-18   Two views of a kinked flux rope. At low altitudes in the subsolar region, the flux ropes become thinner and more twisted. Eventually the originally straight flux rope assumes a corkscrew shape much like a rubber band that is excessively twisted.

tudes, since the kinking process also increases the total volume occupied by the rope. In fact, if the flux tube is kinked, the spacecraft can pass through it several times on a single orbit.

The flux ropes seen in the Venusian ionosphere are intriguing phenomena, and there is much to learn about them. However, they are all the more important because of the solar analogue. Flux ropes appear to be critically important for solar activity. On Venus we have the opportunity to probe flux ropes directly in a benign environment.

## 14-2-3  The Lunar Magnetic Field

The earliest Russian space probes detected no lunar magnetic field (Dolginov et al., 1961, 1966). *Explorer 35* later confirmed these results but added indirect evidence for patchy surface magnetism (Barnes et al., 1971; Colburn et al., 1971; Mihalov et al., 1971; Sonett and Mihalov, 1972). When the Moon was in the solar wind and the *Explorer* was just about to pass into or had just left the lunar wake region, it often observed an enhancement in the strength of the interplanetary magnetic field, as if some small deflection of the solar wind had occurred in the terminator regions. However, these "limb compressions" were not always present. Their occurrence was correlated with the appearance only of specific regions at the lunar terminator. No "limb compressions" were seen at high latitudes, but *Explorer 35* reached high latitudes only when it was distant from the Moon, so it is possible that the limb compressions were present but had just weakened with altitude. Thus, while the *Explorer 35* measurements gave fairly conclusive, albeit indirect, evidence for patchy lunar magnetism, it did not resolve whether this magnetism is global or just restricted to low latitudes.

Orbital measurements were later obtained at about 100 km altitude from the *Apollo 15* and *16* subsatellites. These measurements confirmed the *Explorer 35* detection of limb compressions. Figure 14-19 shows a limb compression occurring in the top panel

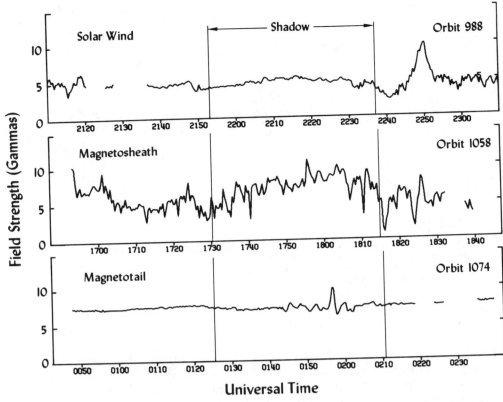

Fig. 14-19 Magnetic field strength versus time for subsatellite orbits while the Moon was in the solar wind (top), the magnetosheath (middle), and geomagnetic tail (bottom). The region labeled "Shadow" corresponds to the period when the satellite was in the optical shadow of the Moon. This is approximately the same as the solar-wind wake when the Moon is in the magnetosheath or the solar wind. The *Apollo* subsatellites orbited the Moon in a roughly 100-km circular orbit. The feature at 2250 UT on orbit 988 is a lunar limb compression associated with the deflection of solar wind by a magnetized region at the terminator. The increase in field strength at about 0155 UT on orbit 1074 is due to the direct observation of the lunar magnetic field in the near-vacuum conditions of the geomagnetic tail. This magnetized feature is near the crater Van de Graaff.

at 2250 universal time (UT) on orbit 988. Figure 14-20 shows where the limb compressions were observed. Many of these source regions coincided with *Explorer 35* source regions (Russell and Lichtenstein, 1975).

The *Apollo* subsatellites flew close enough to the lunar surface to detect the lunar field directly when the Moon was shielded from the solar wind by the geomagnetic tail. The bottom panel of Fig. 14-19 shows an example of the direct measurement of the lunar field. Figure 14-21 shows a mapping of the radial component in the region of low-altitude coverage from the *Apollo 15* and *16* subsatellites (Russell et al., 1977). The patchy nature of the magnetism is clearly visible here. The *Apollo* subsatellite magnetometers were also able to detect the lunar field when the Moon was in the solar wind and when the subsatel-

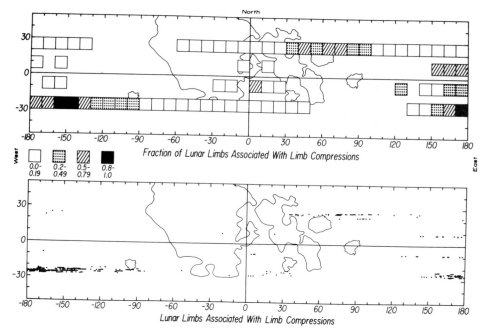

Fig. 14-20 (Top) Limb-compression occurrence rate versus selenographic, or lunar geographic, position for *Apollo 15* and *16*. (Bottom) Lunar positions that were at terminator when limb compressions were observed and that were used in calculating the occurrence rate. Since limb compressions are caused by the deflection of the solar wind by the lunar magnetic field, their occurrence rate is a good indicator of where the lunar surface is magnetized. This plot clearly shows that the lunar surface is far from being uniformly magnetized (Russell and Lichtenstein, 1975).

---

**Flux**

The term *flux* is frequently used in physics to denote the strength of some directed quantity. *Magnetic flux* is synonymous with lines of magnetic force that have strength and direction. *Particle flux* is the quantity that measures the number of particles crossing a unit area per second. For example, *electron flux* is the product of the electron density times the average electron velocity.

---

lite was well within the lunar wake and at low altitudes. A sample of this mapping is shown in Fig. 14-22 (Hood et al., 1979).

The *Apollo* subsatellites were also equipped with energetic electron detectors. These detectors were also able to provide a measure of the lunar surface remanent magnetization through the electron-reflection coefficient, the percent of the electron flux incident on the lunar surface that was scattered back to the satellite. The stronger the surface field, the larger the returned electron flux (Lin et al., 1975). The advantages of this

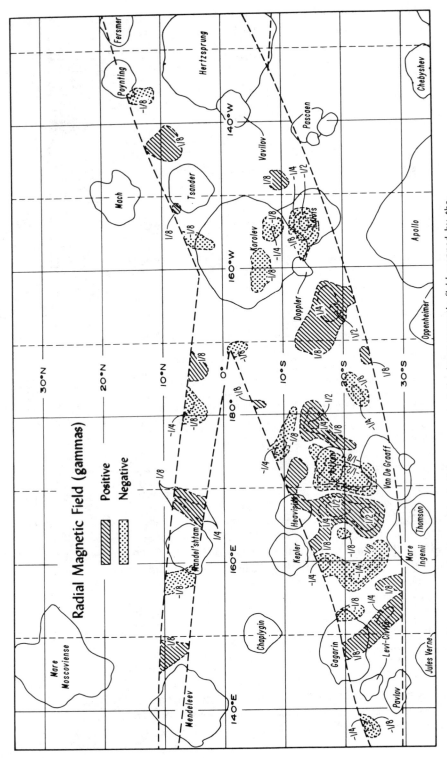

Fig. 14-21 Radial component of the lunar magnetic field mapped by the *Apollo 15* and *16* subsatellites from 135°E to 125°W longitude at altitudes between 65 km and 100 km. Contour levels are ±1/8, 1/4, and 1/2 nT (Russell et al., 1975).

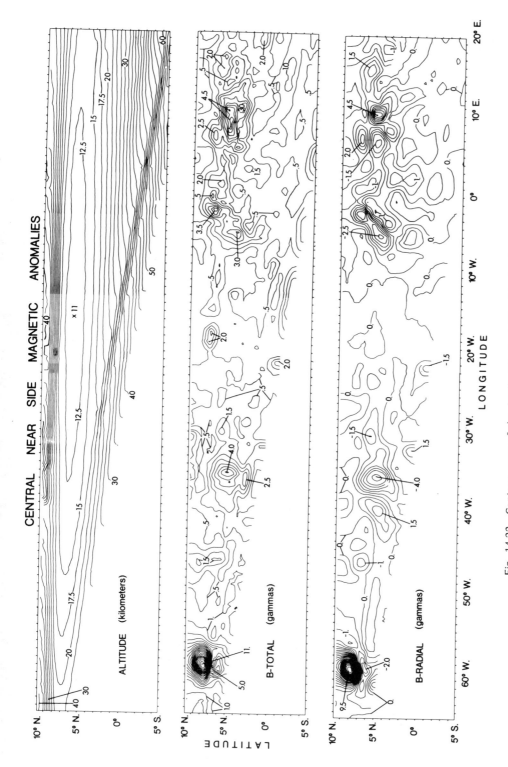

Fig. 14-22 Contour maps of the *Apollo 16* subsatellite altitude, the field magnitude, and radial field component measured by the subsatellite magnetometer across a section of the lunar near side. The contour interval on the altitude plot is 2.5 km and on the field plots 0.5 nT (Hood et al., 1979).

technique are that it is moderately insensitive to the altitude of the subsatellite, it measures the surface field strength, and measurements can be made throughout the orbit in the solar wind, where the Moon spends most of its orbital period. Since this technique is sensitive to field strength and not to direction, maps of the individual north, east, and radial components could not be made. The electron reflectance maps were consistent with the magnetometer data and gave more global, albeit more qualitative, coverage. Still data only with ±35° of the lunar equator were obtained. We are essentially ignorant of lunar magnetism at high latitudes.

Magnetic analysis of the returned lunar samples has considerably aided our understanding of lunar magnetism. Sample analysis provides information on mineral properties, age, and strength of the ancient fields that magnetized the samples. This information cannot be obtained by any other means. The principal finding, and a surprise to many, was that the returned lunar samples possessed stable, natural remanent magnetization (NRM), that is, they were magnetized by some external field in the distant past (Doell et al., 1970; Helsley, 1970; Nagata et al., 1970; Runcorn et al., 1970; Strangway et al., 1970). Lunar NRM is typically carried in fine-grain iron, which is quite different from the terrestrial case, in which the grain size is larger and hence more easily demagnetized. Impact-generated rocks, or *breccias*, contained the strongest magnetization. The strength of the ancient magnetizing field appears to increase with increasing age. Samples formed about 4 gya give paleointensities of less than one-tenth of this value (Stephenson et al., 1975).

Three-component magnetometers (instruments for measuring magnetic fields) were installed at the three landing sites of *Apollo 12, 15*, and *16*, and portable magnetometer readings were taken along traverses on the *Apollo 14* and *16* missions. The surface fields were much greater than one would have expected from the orbital measurements. Since the rate of decay of magnetic fields is faster for smaller regions of magnetization, this observation is consistent with the magnetic fields at the landing sites having small scale-lengths. The inference is consistent with the site-to-site variability and the gradients measured along the surface traverses.

The surface magnetometers together with the *Explorer 35* magnetic measurements permitted electromagnetic sounding for a lunar core (see Chapter 4 by Kaula). The largest core that was allowed by these measurements is about 500 km (Wiskerchen and Sonett, 1978). The principal limitation of the two-satellite lunar-sounding method is that it relies on the precise intercalibration of the two magnetometers. An alternate technique has been used using one magnetometer while the Moon is in the geomagnetic tail. This "deep-sounding" method is consistent with the existence of a core of 400 to 500 km (Russell et al., 1981a). The moment of inertia of the Moon is also consistent with such a core (Gapcynski et al., 1975), and the seismic data are consistent with such a core (Nakamura et al., 1982). Thus, the Moon may have had an ancient lunar dynamo operating deep within its interior.

If the Moon once had a global field, though, there is no trace of it today. *Apollo 15* and *16* subsatellite measurements, again made while the Moon was in the geomagnetic tail, provide an upper limit to the lunar field of $10^9$ T-m$^3$ (Russell et al., 1975). This essentially total absence of a global field led Runcorn (1975a, 1975b) to propose the following scenario. After the formation of a lunar core, an internal dynamo began operating, creating a dipolar field. The dipolar field in turn would magnetize the lunar crust in a dipolar manner as sketched in Fig. 14-23. When the dipolar field of the core ceased, the

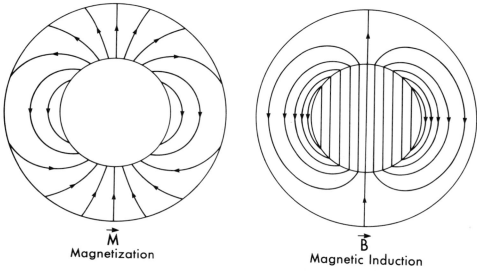

Fig. 14-23 (Left) Dipolar magnetization of spherical shell. (Right) Magnetic field associated with such a pattern of magnetization. The fact that the magnetic field lines do not penetrate the outer surface of such a magnetized shell led Runcorn (1975a, 1975b) to propose that the resolution to the paradox that the lunar surface was magnetized yet the Moon had no global field was that the crust had been magnetized in a dipolar fashion by an ancient internal dynamo.

dipolar magnetism remained. However, the external magnetic field of a uniform spherical shell with dipolar magnetization is zero, as shown in the right-hand panel of Fig. 14-23. In Runcorn's scenario, the local destruction of the uniform magnetization by meteor impact allows the field to leak out of the crust. In support of this picture, Runcorn (1982a) has found three paleopoles, dipole axes that are consistent with the observed lunar magnetization for different epochs. Unfortunately, the available data are sparse. Furthermore, the regions of lunar field do not correspond to impact craters. Thus, we seek alternate means of magnetizing the lunar surface.

Another approach to explaining patchy magnetization is to hypothesize that the material that holds the remanence that is detected at high altitudes is distributed in patches on the lunar surface. Hood et al. (1979) made such an assumption when they hypothesized that lunar "swirl" material had high remanence. However, the lunar swirl material seems to be much younger than the date at which a lunar dynamo could have magnetized the returned lunar samples. Thus, this mechanism seems highly improbable.

Another approach is to propose that the magnetizing field was patchy rather than global. One such proposal was made by Gold and Soter (1976), who proposed that the magnetizing field is the magnetic field provided by comets, which gather up the solar-wind field and press it into the lunar surface upon impact. Again, there is very little evidence for geologic features such as craters being associated with regions of lunar remanent magnetization.

Another means of providing a patchy magnetizing field is the production of flux ropes such as seen on the Sun or in the Venusian ionosphere. The outer 200 km of the

Moon was at one time molten (Brett, 1977). This may or may not have led to core formation as proposed by Stevenson (1980). If a field-generating core were present, the convective shear (that is, nonuniform motion) in this molten shell could have resulted in concentrating the field in flux islands. As the crust cooled, these flux islands would have led to highly irregular magnetization. Such a mechanism would, of course, require flows with a moderately high magnetic Reynolds number, that is, moderately rapid flows. It is not clear that such flows were present. On the other hand, this mechanism is attractive because it concentrates flux from the dynamo, thus relieving the dynamo from directly having to apply such strong field to the crust. Without a flux concentrator, the lunar core would have to experience an ancient field much larger than that in the present-day terrestrial core.

Finally, we note that no method of creating the observed lunar magnetization is completely satisfactory. Lunar magnetism remains an enigma. Those wishing greater detail on the subject are referred to the recent review by Hood and Cisowski (1983) and the highly personal account of Runcorn (1982b).

## 14-2-4 Discussion

The convection zone of the Sun, the Venusian ionosphere, and the molten outer shell of the early Moon are quite different environments. In fact, they are so different that we may not be able to use our understanding of one to discuss the behavior of the others. However, since nature has provided us with only these few examples, we make an attempt to extrapolate. First, let us examine the seed field upon which the dynamo action operates (see also Chapter 13 by Levy). At Venus this seed field is provided by the solar wind either at the ionopause or deep in the ionosphere. In the Sun it is not obvious where the seed field is generated. A primordial field would suffice, but there is little evidence that such a field exists. Furthermore, if the seed field arises from a primordial field, the reversing polarities observed over the 22-y period of the solar cycle seem difficult to explain simply. Perhaps there is a deep-seated dynamo that generates a reversing planetary-like field that is twisted and coiled only in the near-surface layers of the convective zone. On the Moon, the seed field could be provided by an internal dynamo that began well after lunar formation approximately 4 gya but while there was still an actively convecting outer molten shell. However, there is a difficulty with this hypothesis.

The magnetic Reynolds number is simply the magnetic diffusion time divided by the convection time. If the conductivity of the molten shell is that of molten rock ($\sim 1$ mho $\cdot$ m$^{-1}$), then very rapid convection ($\lesssim 1$ day) from the bottom of the shell to the lunar surface is required lest a structure, say, 50 km in radius, diffuse away before it reaches the surface. If the shell has higher conductivities such as that of molten iron ($10^4$ mho $\cdot$ m$^{-1}$) or molten iron sulfide ($10^7$ mho $\cdot$ m$^{-1}$), a much longer time is available. A long-time scale is needed to magnetize the lunar crust if the mechanism is thermoremanent magnetization, that is, cooling of material through its magnetic blocking temperatures, the temperatures at which magnetization becomes frozen into a rock. Since this should take of the order of many years, even the time scales of molten iron seem too short to cause signficant crustal effects. This mechanism, thus, does not appear to solve simply the enigma of lunar magnetism.

We are left with many puzzles for these three bodies. Fortunately, the Venusian ionosphere is accessible to *in situ* probing, and much data exist on Venusian flux ropes.

We are less fortunate with the Moon. We have magnetic coverage over but a limited portion of the lunar surface and have few constraints on ancient lunar conditions. The Sun's photosphere can be observed regularly, but until the solar optical telescope is launched, we may not be able to resolve the finest scale magnetic features wherein most of the solar magnetic flux resides. Furthermore, we will never be able to probe beneath the photosphere. For the properties of this region we must use indirect evidence. Theory too is difficult in this problem. The highly twisted, force-free fields that we see both in the Venusian ionosphere and the solar photosphere are not simple to treat in theoretical developments. It may be many years before significant progress is made in understanding such fields.

## 14-3 CORE DYNAMOS

Inside each of the terrestrial planets there is thought to be a liquid, principally iron, core. In the center of this liquid core there may be a further solid core, formed by the freezing out of material in the liquid core, but this process is not of concern to us except in that it releases energy that could power convection in the core and sustain dynamo action. Jupiter and Saturn are different. While they too may have solid (rocky) cores, their core dynamos contain electrically conducting metallic hydrogen under high pressure. Further out, Uranus and Neptune present us with yet a further possible type of planetary dynamo, an aqueous dynamo. The large moons might also have liquid cores that support dynamo action. Earth's Moon certainly has little, if any, overall global magnetic field (Russell et al., 1975); but still it has signs of having a present-day, highly conducting core (Russell et al., 1981a). The volcanoes of Io show that it is still tectonically alive, and there are indications that Io has its own magnetic field (Kivelson et al., 1979). Titan, however, shows no evidence of having an intrinsic magnetic moment (Ness et al., 1981). In this section we examine first the terrestrial planets, then Jupiter and Saturn, next their large satellites, and finally Uranus and Neptune.

### 14-3-1 The Terrestrial Planets

**The Earth** The Earth has a magnetic field, which, above the planet's surface, has most of its energy in the dipolar component (which has a moment of $8 \times 10^{15}$ T $\cdot$ m$^3$). However, this changes with depth, since higher-order moments fall off more rapidly with distance from the source than lower-order moments. At the depth of the seismically determined core, the contribution to the magnetic field from each of the different "orders" of the magnetic field is quite uniform. As illustrated in Fig. 14-24, one does not have to extrapolate much further to get equal contribution from each order (Elphic and Russell, 1978). This exercise illustrates two basic points about planetary magnetism. First, the mechanism that generates the field need not have much order, despite the fact that the external magnetic field seems quite ordered. Second, the relative power in the various harmonics, as we detect them above the planetary surface, do not in themselves indicate any differences in the dynamo mechanisms at the different planets; rather, it probably reflects only differences in source depth.

We can learn something about the dynamo process by observing the temporal variation of a planetary magnetic field, but this has only been performed successfully for the Earth. The motion of features in the field gives us a measure of the rates of motion in the

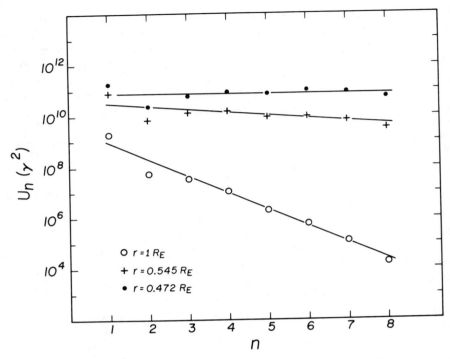

Fig. 14-24 Spatial power spectrum of the geomagnetic field at the surface, at the core-mantle interface, and at the depth at which the extrapolated spectrum is flat (Elphic and Russell, 1978).

core relative to the crust. The secular variation can be used together with the idea of frozen-in flux in the highly conducting core to give the radius of the core (Hide, 1978), and the speed of reversals of the dipole field has also been used to put limits on the conductivity of the Earth's core. (See also Chapter 13 by Levy.) However, again, we cannot apply these techniques to any planet but the Earth.

**Mercury** Of the three terrestrial planets, Mercury has the magnetic field that most resembles the terrestrial field despite the fact that it looks most like our Moon and has a radius of only 2440 km, which is not much larger than that of the terrestrial Moon (1738 km). However, it is much more dense than the Moon (5.4 g · cm$^{-3}$), compared with the lunar 3.3 g · cm$^{-3}$. Mercury has been visited by only one spacecraft, *Mariner 10*, which made two close encounters on the dark side of Mercury in March of 1974 and 1975. A fully developed magnetosphere was observed with a bow shock, a magnetopause, and a magnetotail. The magnetic field strength increased rapidly as the planet was approached, reaching a maximum of nearly 100 nanoTeslas = 10$^{-9}$ Teslas at an altitude of 723 km on the first encounter and over 400 nanoTeslas = 10$^{-9}$ Teslas at an altitude of 330 km on the last pass (Ness et al., 1974, 1975a).

Despite the fact that Mercury has an almost classic magnetosphere, it has been very difficult to determine an accurate planetary dipole moment and even harder to derive the higher-order moments. Figure 14-25 shows the magnetic field lines in Mercury's noon-midnight meridian (i.e. a plane through the planet containing both the noon and midnight

> ### Magnetospheric Terminology
>
> The highly electrically conducting solar-wind plasma confines planetary magnetic fields to a cavity called a *magnetosphere*. The solar wind is deflected around the obstacle with the help of a shock wave, which also heats the plasma. The region behind the shock is called the *magnetosheath*. The boundary between the flowing magnetosheath and the magnetic cavity is called the *magnetopause*. The solar wind also pulls on the planetary magnetic field and stretches that field into a long tail behind the planet, known as the *magnetotail*.

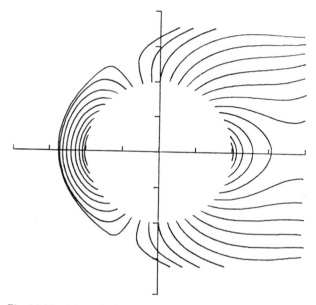

Fig. 14-25 Magnetic field lines in the noon-midnight meridian of the magnetosphere of Mercury. (From Jackson and Beard, 1977, AGU/*J. of Geophys. Res., 82,* 2828. Used with permission.)

meridians) from a model by Jackson and Beard (1977). This illustrates part of the problem. The magnetosphere of Mercury is not much larger than the planet itself. Thus, the various magnetospheric current systems distort the field significantly even at low altitudes. One would hope to get a zeroth-order estimate from the location of the bow shock and the magnetopause. However, to use this information to determine the dipole moment one needs the solar-wind dynamic pressure, and this was not directly obtainable, since the solar-wind ion experiment failed. Initial values of the moment ranged from about $5 \times 10^{12}$ T · m$^3$ (Ness et al., 1974, 1975a, 1975b, 1976) to about $2.5 \times 10^{12}$ T · m$^3$ (Jackson and Beard, 1977; Ng and Beard, 1979; Whang, 1977). Slavin and Holzer (1979) criticized these analyses because they felt earlier authors had underestimated the dynamic pressure of the solar wind during the encounters and because they did not take into account the tangential stress of Mercury's magnetopause. They deduce that a better estimate of the dipole moment would be $6 \pm 2 \times 10^{12}$ T · m$^3$.

With such uncertainty in the size of the dipole moment, the question of the relative power in the higher-order moments seems moot. Jackson and Beard (1977) and Whang (1977) attempt to calculate higher-order moments; but as Ness (1979) has stated, the quadrupole and octupole moments thus obtained are probably in error because of inadequate spatial coverage to determine these higher orders. Since this planetary encounter was the first in which inversion of magnetic field data along a spacecraft orbit was used to deduce a planetary moment, the accumulated wisdom of the community was meager when this work was performed. That wisdom has become much more sophisticated of late in the inversions of the Jovian and Saturnian fields, and such naiveté would probably not be repeated (Connerney, 1981).

The magnetic field of Mercury is clearly of internal origin. It cannot be caused by an external process such as currents driven by the solar wind (Herbert et al., 1976). Furthermore, it is unlikely to be caused by remanent magnetization. To create such a field would require a large free-iron abundance in an outer cold mantle (Stephenson, 1976). This is inconsistent with plausible thermal-evolution models (Stevenson et al., 1983). The only plausible source is a planetary dynamo.

Clearly, we know little more about the intrinsic field of Mercury other than that it exists. This planet is crying out for further exploration both of its magnetic field as well as its other geophysical and geochemical properties. Such missions are well within the technical and financial capabilities of several nations at present. We hope that this forgotten planet will again be remembered soon (see Chapter 17 by Kivelson).

**Venus** The exploration of Venus has had none of the problems confronting Mercury. At the date of this paper, Venus has been visited by over 25 spacecraft and probes, with more on the drawing boards. Venus is almost the Earth's twin in size; it is 6051 km in radius, compared with the Earth's 6371-km radius. However, Venus rotates much more slowly. Venus rotates on its axis relative to the stars in a retrograde sense once every 243 Earth days. It was clear from the very first probe to Venus—*Mariner 2* in 1962—that the magnetic field of Venus had to be much less than that of the Earth ($\lesssim 4 \times 10^{14}$ T · m$^3$) (Smith et al., 1963). *Mariner 5* flew by Venus on October 19, 1967, coming within 1.4 $R_V$ (Venusian radii) of the optical shadow axis and within 1.7 $R_V$ of the center of the planet. The *Mariner 5* experimenters were able to rule out the possibility of a planetary magnetic field greater than $10^{-3}$ of the Earth's moment (Bridge et al., 1969). *Venera 4* arrived almost simultaneously and made measurements down to 200 km in the Venusian ionosphere. Dolginov et al. (1969) estimated no Venus moment above $10^{-4}$ of the terrestrial moment. Russell (1976) criticized this estimate and revised the upper limit to $6.5 \times 10^{12}$ T · m$^3$, or $8 \times 10^{-4}$ of the Earth's moment. In 1975 *Venera 9* and *10* were placed in orbit about Venus. Dolginov et al. (1978) interpreted the data in terms of a moment of about $2.5 \times 10^{12}$ T · m$^3$, whereas Yeroshenko (1979) interpreted the data as not indicating an intrinsic but rather an induced field and that $2.5 \times 10^{12}$ T · m$^3$ was an upper limit to the field. Finally, in 1978, *Pioneer Venus* was placed in orbit about the planet. The advantages of this spacecraft over previous ones were the greater amount of telemetry coverage, essentially 100 percent of the orbit; the smaller fields attributable to the spacecraft itself; known orientation at all times; and much data at very low (approximately 150-km) altitudes. Data from the first two years that *Pioneer Venus* was in orbit showed no evidence for a planetary magnetic field greater than $3 \times 10^{11}$ T/m$^3$, or $4 \times 10^{-5}$ of the terrestrial moment (Russell et al., 1980). Despite the extremely low limit such measurements place on the moment, the search goes on. Recently Knudsen et al. (1982) have suggested that the steady radial fields on the Venus nightside, observed in narrow regions of depleted ion density called *holes* (Brace et al., 1982), are really planetary in origin and not induced by the solar-wind action. However, Luhmann and Russell (1983) have shown that the polarity and location of these radial fields are, in fact, controlled by the orientations of the interplanetary magnetic field. Thus, there is still no evidence for an intrinsic field at Venus.

Stevenson et al. (1983) show that a stably stratified core with no dynamo action is possible for a Venus thermal model with as much or more sulphur and oxygen in the core as in the Earth's core. Venus, in fact, could lack an inner solid core as shown in Fig. 14-26

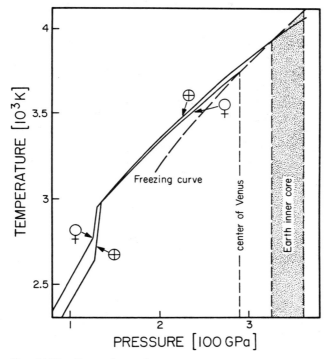

Fig. 14-26 Comparison of the expected interior structure of Venus and the Earth. The temperature-pressure relationships inside the Earth, ⊕, and Venus, ♀, are expected to be very similar (solid lines). The pressure-temperature relationship for the solid-liquid phase change—that is, melting-freezing line (dashed line)—intersects these curves. However, it intersects the Venus curve at a pressure very close to that at the center of the planet. Hence it is not expected that the interior of Venus will freeze. However, much of the terrestrial core is below the freezing line and hence is expected to be solid at the present time. (From Stevenson, D. J., 1983. Planetary magnetic fields, *Reports on Progress in Physics, 46*, 55-620. Copyright 1983 by The Institute of Physics. Used with permission.)

because, since Venus is slightly smaller than the Earth, it has lower internal pressures at its center. A lower pressure decreases the chance of inner-core formation because the adiabat, the temperature pressure relation expected to hold in the convecting interior, is less steep than the freezing curve as illustrated in Fig. 14-26. The mantle temperature of Venus is hotter than that of the Earth because the surface is hotter. We note that, even though Venus has been repeatedly visited and we are beginning to understand its surface and atmosphere very well, we are almost completely ignorant of the properties of its interior. We sorely need long-lived seismic measurements on the Venusian surface. However, the prospect of obtaining such data seems to be off in the very distant future.

**Mars** Of the three terrestrial planets, Mars is the one whose magnetic field is the least understood. Even though Mars has been visited by many spacecraft, only a few of these have carried magnetometers. *Mariner 4*, in 1965, was the only U.S. spacecraft to carry a magnetometer, and the flyby distance was such that only a brief encounter with the planetary bow shock and magnetosheath was observed. The flyby trajectory and magnetic field measurements along this trajectory are shown in Fig. 14-27 (Smith, 1969). Little can be done with observations such as these besides attempting to determine the size of the obstacle responsible for the observed shock. Dryer and Heckman (1967) interpreted these observations as indicating the presence of a small magnetic moment, $2.1 \times 10^{22}$ Gauss · cm$^3$, whereas Spreiter and Rizzi (1972) interpreted the data to be consistent with an obstacle no larger than the planet plus its ionosphere.

*Mars 2* and *3* were injected into Martian orbit on November 27 and December 2, 1971. *Mars 2* had a periapsis of 1300 km and an apoapsis of 28,000 km. (Periapsis and apoapsis are, respectively, the points in the orbit closest to and farthest from the planet.) *Mars 3* had a similar periapsis and an apoapsis of 212,000 km. *Mars 2* was not three-axis stabilized: While it kept one axis pointing to the Sun, it slowly rotated about this direction. Very few *Mars 2* data have been published. *Mars 3* was successfully stabilized about all three axes. Both spacecraft frequently encountered the Martian bow shock and magnetosheath. However, on one pass the *Mars 3* data has been interpreted as showing that the spacecraft entered into a planetary magnetosphere. The data obtained on this day, January 21, 1971, by the magnetometer and the electron and ion probes are shown in Fig. 14-28 (Dolginov et al., 1972, 1973; Gringauz et al., 1974). At point 1, the electron temperature jumps as does the ion current. Dolginov et al. (1973) interpret this to be the bow shock crossing. They then interpret point 2 to be the magnetopause. However, Wallis (1975) questions these identifications. The nonthermal electrons observed beginning at point 1 could be upstream electrons, since the electron detector points back toward the planet away from the Sun. Figure 3 of Vaisberg et al. (1973) indicates thermal ions from the Sunward direction did not appear until point 2. Thus, a plausible interpretation is that point 1 is a change in the direction of the magnetic field in the solar wind, at which time the spacecraft first encounters field lines connected to the bow shock which contain backstreaming particles. Russell (1978a) has interpreted the data in this way and has pointed out that the orientation of the field is consistent with the magnetic field being carried against and draped about a planetary obstacle. We note that there is a sign error in the $Y$ component of the magnetic field in the data displayed by Russell (1978a) due to confusion in the definition of the coordinate system. This was clarified by Dolginov (1978). With the sign correction, the field variation seen in Fig. 14-28 is very similar to *Pioneer Venus* measurements in the Venusian magnetosheath at similar altitudes and also

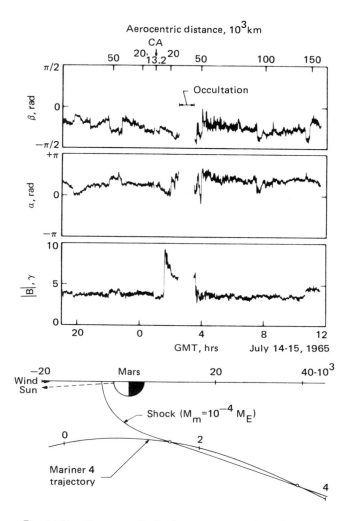

Fig. 14-27 The magnetic field measurements along the trajectory of *Mariner 4* as it passed Mars. The top three panels show the latitude ($\beta$), longitude ($\alpha$), and the magnitude of the magnetic field. The bottom panel shows the distance from the expected solar-wind wake axis plotted versus the distance along it. (Used with permission by American Astronautical Society, from *25 Advances in the Astronautical Sciences*, E. J. Smith, 1969.)

to field profiles from gasdynamic simulations of the interaction. In short, the *Mars 3* data do not provide conclusive proof that there is an intrinsic planetary field.

Because of the smaller orbit of *Mars 2*, its data is potentially valuable for answering the question of a planetary intrinsic field despite its lack of three-axis stabilization. Smirnov et al. (1978) suggest that the Martian magnetic dipole axis lies near the equatorial plane. Figure 14-29 shows the magnetic field measurements from the solar-oriented magnetic sensor and an interpretive diagram showing how the magnetic field in the magneto-

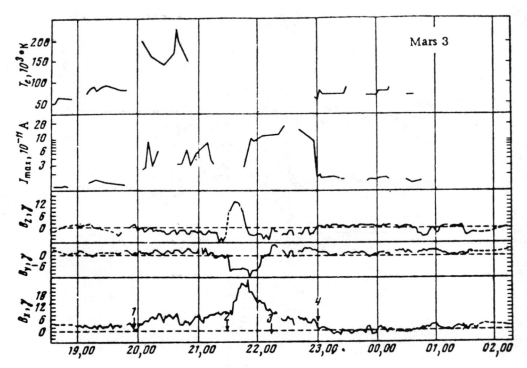

Fig. 14-28 The magnetic field components, the maximum ion current, and the electron temperature measured along the *Mars 3* trajectory on its putative magnetospheric entry. Points 1 and 4 are claimed to be the shock; points 2 and 3 the magnetopause. True zero levels for the baselines are unknown. (From Gringauz, K. I., et al., 1974. Study of solar plasma near Mars and on the Earth–Mars route using charged-particle traps on Soviet spacecraft in 1971–73. II. Characteristics of electrons along orbits of artificial Mars satellites *Mars 2* and *Mars 3*, Cosmic Research, 12, 4, 535–546. Copyright 1974 by Plenum Publishing Corporation. Used with permission.)

tail would change as the planet rotated and the satellite passed through it on two successive passes. While this interpretation is consistent with the data, it certainly is not compelling. An equally plausible interpretation is that the direction of the interplanetary magnetic field reversed about 1300 UT on February 23, causing a reversal in the draped-field pattern. This interpretation is also consistent with the direction of the field at closest approach, at point 2. These measurements do, however, directly contradict Dolginov's (1978) spin-axis-aligned moment. Smirnov et al. estimate a moment of $1.2 \times 10^{22}$ Gauss/cm$^3$ for their equatorial moment.

The last spacecraft to undertake Martian magnetic studies was *Mars 5*, placed into orbit on February 13, 1974. Its periapsis of 1800 km was not as low as that of *Mars 3*, and no apparent dayside entries were seen. A region was detected behind the planet which Dolginov et al. (1976) identified as magnetotail entries. However, these regions are even less like the terrestrial magnetotail than the Venusian magnetotail (Russell et al., 1981b). Depending on whether you interpret these features as Dolginov et al. (1976) did

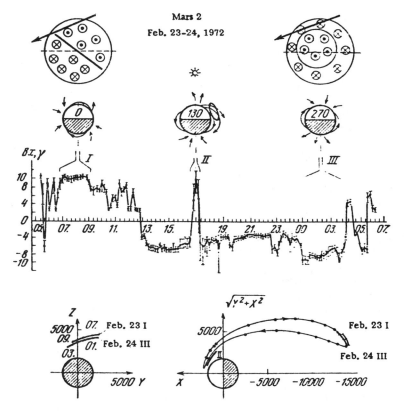

Fig. 14-29 The Sunward-pointing magnetic field component measured by *Mars 2* during its putative Mars magnetotail encounter. The top sketches show the schematic cross sections of the tail during the February 23 and 24 passes. The middle row shows the possible orientation of an equatorial dipole moment in accord with these measurements. The bottom panels show the trajectory. (From Smirnov, W. N., 1978. Possible discovery of cusps near Mars, *Kosmich. Issled.*, 16, 688–692. Copyright 1978 by Plenum Publishing Corporation. Used with permission.)

or as induced by the solar-wind interaction, you obtain a moment varying from $10^{21}$ to $9 \times 10^{21}$ Gauss · cm³ (Russell, 1978b).

In the presence of such ambiguity about the interpretation of the *in situ* observations, several groups have returned to indirect methods. Intriligator and Smith (1979) argue that the observed ionospheric pressure alone is not enough to stand off the average solar-wind dynamic pressure. However, Intriligator and Smith neglect the fact that the Venusian ionospheric electron temperature is twice that of the ions. If the Martian ionosphere is similar to that of Venus, the missing pressure is small. Slavin et al. (1983) have used a gasdynamic model to determine the altitude of the Martian ionopause from the bow shock location. Their best fit has a Mach number of 7.4 (i.e., the ratio of the solar-wind speed to the speed of a compressional wave), a ratio of specific heats of 2, and a subsolar ionopause altitude of $500 \pm 100$ km. This is somewhat higher than the Venusian ionopause, but the solar-wind dynamic pressure is more than a factor of four less at Mars than at Venus, whereas we would expect the Martian ionospheric temperature to be similar to that of Venus, whereas the gravitational field is less. Thus, these data do not demand a nonionospheric obstacle in contradiction to the conclusions of Slavin and Holzer (1983).

The question of the Martian magnetic field is an easy one for any modern Martian mission to solve. All that is needed is magnetic data obtained at low altitudes, especially nightside data. However, recently mission planners have ignored this objective in designing spacecraft. For example, the recent highly successful *Viking* mission carried no magnetic field measuring instruments on either the lander or the orbiters. In light of the evidence, however, it is most probable that Mars has no presently operating internal dynamo. Stevenson (1983) suggests that the Martian core should contain more sulphur than that of the Earth, because it accreted further from the Sun. If the core were to contain more than 15% sulphur by mass, calculations of the thermal evolution indicate that the core should be presently stably stratified (Stevenson et al., 1983). In fact, it cannot be excluded that the core has frozen. If the core had a similar sulphur content to that of the Earth (10% sulphur), the same models would give a large inner core (950 km in radius) with a large toroidal field (approximately 20 Gauss). We emphasize that the absence of a present-day dynamo in no way implies the absence of an ancient Martian dynamo. Future Martian landings should investigate remanent magnetization as well as seismicity and heat flow for clues about the present and past state of the Martian interior.

### 14-3-2 Jupiter and Saturn

The dominant constituents of Jupiter and Saturn are hydrogen and helium, surrounding small rocky cores. Hydrogen forms a metal at pressures of between 2 and 5 Mb (megabars) which are reached inside both planets. Metallic hydrogen is perfectly capable of sustaining dynamo action, as is evidenced by the strong magnetic fields of both planets.

**Jupiter** Jupiter is the largest of the planets—1 $R_J$ (Jovian radius) = 71,400 km—and the most rapidly rotating ($T$ = 09 h 55 min 29.7 sec). It has been known to possess a magnetic field for several decades because of its polarized radio emissions. Much information was inferred about the Jovian field from radio data. The dipole tilt angle was expected to be about 9.5° (Roberts and Komesaroff, 1965; Komesaroff and McCullough, 1967; Morris et al., 1968; Whiteoak et al., 1969; Gardner and Whiteoak, 1977). The field strength was estimated to be between 0.4 and 1.0 Gauss in the radiating region (Komesaroff et al., 1970). Berge (1965) even deduced that the Jovian magnetic moment was northward.

Jupiter thus far has been visited by four spacecraft: *Pioneer 10* flew by Jupiter at a Jovicentric distance of 2.9 $R_J$ in December 1973; *Pioneer 11* at 1.6 $R_J$ in December 1974; *Voyager 1* in March 1979 at 5 $R_J$, and *Voyager 2* in July 1979 at 10 $R_J$.

Figure 14-30 shows the magnetic field strength measured by *Pioneer 10* for the 24 h around closest approach as a function of radial distance (Smith et al., 1974). This display emphasizes the departure of the magnetic field from a dipolar radial dependence beyond about 10 $R_J$, due to the presence of the current sheet. It is this current sheet and the difficulty in modeling it that makes the *Voyager 2* data essentially useful for modeling the internal planetary magnetic field (Connerney, 1981). Although *Pioneer 10* cov-

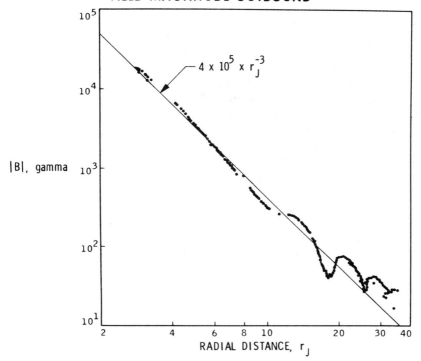

Fig. 14-30 *Pioneer 10* magnetic field strength at Jupiter plotted versus radial distance. (From Smith, E. J., et al., 1974. Magnetic field of Jupiter and its interaction with the solar wind, *Science*, *183*, 305-306. Copyright 1974 by the AAAS. Used with permission.)

ered a wide range of radial distances, it did not cover all longitudes. Although it covered about 315° inside of 10 $R_J$, it covered less than 90° inside of 5 $R_J$.

*Pioneer 11* had the advantages of approaching the planet closer (1.6 $R_J$, Jovicentric) and a retrograde trajectory, so more longitudes were covered at small radial distances. *Pioneer 11* carried two magnetometers, which disagreed in their raw data. This disagreement, together with different analysis techniques, which obscured the differences in the raw data led to much controversy (Acuna and Ness, 1975; Smith et al., 1975; Davis and Smith, 1976). However, tests using the particle data, which should be ordered by the magnetic field orientation, revealed which data were wrong and retrospective recalibration brought the two magnetometers into agreement (Acuna and Ness, 1976a, 1976b; Smith et al., 1976). The final agreed-upon dipole moment was $1.55 \times 10^{30}$ Gauss · cm³, with a ratio of the dipole-to-quadrupole-to-octupole moments of 1.00:0.25:0.20, compared with the Earth's ratio of 1.00:0.14:0.10. This is consistent with the size of the expected conducting core (Elphic and Russell, 1978). Most recently, *Voyager 1* measurements have been carefully analyzed and found to yield identical moments as that determined from the *Pioneer 11* data within statistical error (Connerney et al., 1982). Thus, there is no evidence for temporal variation of the Jovian magnetic field over the short baseline (5 y) for which we have data.

**Saturn** Saturn is the second largest planet, with a radius of 60,330 km and a rotation period of 10 h, 39 min. There were putative observations of radio waves from the magnetosphere of Saturn (Brown, 1975; Kaiser and Stone, 1975). However, these radio waves were not unambiguously detected. The peak frequency of these emissions was about one-eighth of the Jovian waves. Thus, Brown (1975) estimates that the magnetic field of Saturn is about one-eighth of the Saturnian field, or about 1 Gauss at the surface of the planet in the near equatorial regions. On September 1, 1979, *Pioneer 11* reached Saturn, passing within 1.4 $R_S$ (Saturn radii) of the center of the planet. *Pioneer 11* was soon followed by *Voyager 1* on August 22, 1980, passing within 3.1 $R_S$ of the center of the planet and by *Voyager 2* on August 26, 1981, passing within 2.7 $R_S$ of the center of the planet. The results obtained were quite surprising, both in the light of expectations based on the radio wave measurements and based on our experience at other planets.

Figure 14-31 shows the magnetic field strength plotted as a function of distance for the *Pioneer 11* pass (Smith et al., 1980). The first surprise is the magnitude of the field. These data correspond to a dipole moment of $4.8 \times 10^{28}$ Gauss · cm$^3$, or an equatorial surface field of about 0.2 Gauss, much less than the 1 Gauss predicted. Second, as com-

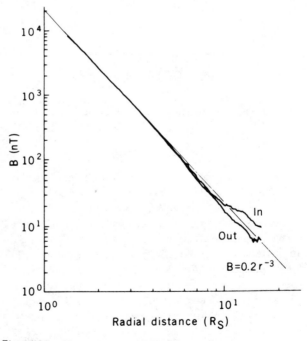

Fig. 14-31 *Pioneer 11* magnetic field strength at Saturn plotted versus radial distance. (From Smith, E. J., et al., 1980. Saturn's magnetic field and magnetosphere, *Science, 207*, 407–410. Copyright 1980 by the AAAS. Used with permission.)

parison with Fig. 14-30 shows, the radial dependence falls much closer to the inverse cube law expected of a dipolar field. In fact, the ratio of the quadrupole moment to the dipole moment for Saturn is 0.07, compared with the terrestrial value of 0.14. The simplest explanation of this harmonic purity is that the conducting core of Saturn is much smaller relative to the radius of the planet than the terrestrial core. The third surprise was the tilt of the dipole moment. The terrestrial dipole is tilted at an angle of 11.5° to the Earth's rotation axis. The Jovian dipole is tilted about 10° to its rotation axis. However, the dipole moment of Saturn is inclined at an angle of 1° or less to the rotation axis (Smith et al., 1980; Ness et al., 1981, 1982a). The only significant asymmetry is an offset of the dipole moment northward along the rotation axis of about $0.05\ R_S$ (Chenette and Davis, 1982). This leaves us with one further puzzle. If the magnetic field is so rotationally symmetric, why are radio emissions seen in synchronism with the rotation of the planet (Desch and Kaiser, 1981)?

Stevenson (1981, 1982) has attempted to explain the low field magnitude and the axisymmetry of the field through helium differentiation and layer structure, as shown in Fig. 14-32 (Stevenson, 1983). In this model the intermediate conducting layer between the almost-insulating, helium-depleted molecular envelope and the helium-enriched, uniformly mixed core is incapable of dynamo action because of the stable stratification provided by a helium gradient. However, it undergoes spin-axisymmetric differential rotation because of the thermal winds due to equator-to-pole temperature differences. This differential rotation effectively filters out any nonspin-axisymmetric components generated in the core below. This process is essentially an electromagnetic skin effect.

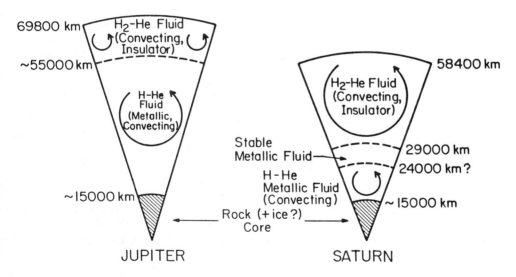

Fig. 14-32 Schematic representations of the possible interior structure of Jupiter and Saturn. (From Stevenson, D. J., 1983. Planetary magnetic fields, *Reports on Progress in Physics*, 46, 555–620. Copyright 1983 by The Institute of Physics. Used with permission.)

### 14-3-3 Large Moons

The planets and the satellites of the solar system form a continuum of which the satellites are not necessarily the end members. The largest satellites, in fact, are in many respects similar to the terrestrial planets. The largest of the moons of Jupiter, Ganymede, and Callisto are larger than Mercury as is Titan, the largest satellite of Saturn. On the other hand, the satellites were probably formed later than the planets. This later formation could affect their chemical composition, as did their smaller size and the location of the satellite relative to its primary and the location of its primary in the solar nebula. One important difference among these large moons and the terrestrial planets is their density. The least-dense terrestrial planet is Mars, which has a density of 3.9 g $\cdot$ cm$^{-3}$. The Earth's Moon has a density of 3.3 g $\cdot$ cm$^{-3}$; Io, 3.4 g $\cdot$ cm$^{-3}$; Europa, 3.1 g $\cdot$ cm$^{-3}$; Ganymede, 1.9 g $\cdot$ cm$^{-3}$; Callisto, 1.8 g $\cdot$ cm$^{-3}$; and Titan, 1.9 g $\cdot$ cm$^{-3}$ (Tyler et al., 1981). With densities this low, there is not much room for an iron core. Io is the best candidate in this respect, being slightly more dense than the terrestrial Moon. The reason for the low density of most of the large moons is that they contain much ice and water. This is true for planets Uranus and Neptune, which, as we see below, are also candidates for dynamo action. However, we do not expect dynamo action in the oceans of these smaller bodies.

If a large satellite of one of the outer planets or Pluto were to have differentiated structure, it should have a stably stratified liquid Fe-S core, because volatile retention was high in the outer solar system, the Fe-S eutectic temperature is low, and the gradual cooling of the core is too slow to sustain convection (Stevenson, 1983). Hence they would not have dynamo-generated fields. This prediction applies, in particular, to Io, where the core should not cool with time because the overlying mantle is kept warm by the tidal heating, first proposed by Peale et al. (1979) (see also Chapter 12 by Peale).

Io, however, is our best candidate for a satellite dynamo. Kivelson et al. (1979) have suggested that the available observations are consistent with an intrinsic field having a moment of 6.5 $\times$ 10$^{22}$ Gauss $\cdot$ cm$^3$. They hypothesize that the moment of Io is antiparallel to that of Jupiter, so that the Jovian and Ionian fields are in a configuration favorable for the process known as *reconnection* (Dungey, 1961). This process connects the Jovian and Ionian magnetic fields and is responsible for generating electric currents and thence radio emissions in the Ionian flux tube. The model is also consistent with the observed ionospheric densities and the behavior of energetic particles in the vicinity of Io.

The interaction of the Jovian magnetosphere with Ganymede is also very puzzling. Transverse perturbations of the magnetic field near the radial distance of Ganymede, but 186 $R_S$ (satellite radii) away, were observed with *Pioneer 11* (Kivelson and Winge, 1976). Plasma cavities were observed by *Voyager 2* within 20 $R_G$ (radii of Ganymede) of Ganymede (Burlaga et al., 1980). It is not clear what the explanation of these features is. The existence of an intrinsic field would increase the cross section of the interaction between Ganymede and the magnetosphere and help explain the surprisingly large width of the disturbed region. However, there is no direct evidence for an intrinsic field.

The only other major satellite that has been probed closely is Titan in the Saturnian system. *Voyager 1* passed within 2.4 $R_S$ of the center of Titan and saw no evidence for an intrinsic magnetic field (Ness et al., 1981, 1982b). In fact, the interaction of Titan with the Saturnian magnetosphere very much resembles the interaction of Venus and the solar wind (Kivelson and Russell, 1983).

In short, then, there is no clear evidence for present-day dynamo action in the small, low-density bodies of the solar system. The only clear dynamo, in a body smaller than the Earth, is in the planet Mercury, which is even denser than the Earth. However, there are intriguing hints at a dynamo in Io and Ganymede. The Galilean satellites deserve further attention, and they are scheduled to get it in the late 1980s, when the *Galileo* will be put into orbit about Jupiter, as discussed in Chapter 17 by Kivelson.

## 14-3-4 Uranus and Neptune

We know very little about Uranus and Neptune. *Voyager 2* is heading out toward them and is scheduled to reach Uranus in 1986. Until then, we have only telescopic observation, modeling, and speculation. Figure 14-33 shows a sketch of a possible model for the interior structure of these two planets (Stevenson, 1983). One of the intriguing facets of this model is the possibility of an ionic water (liquid-ice) dynamo as well as an iron-rich dynamo. If the iron core alone provides the planetary field, we would expect the field to be weak and dominated by the dipolar component. In analogy to Saturn, Stevenson expects the ice layer to filter the nonspin-axisymmetric components. If there is dynamo

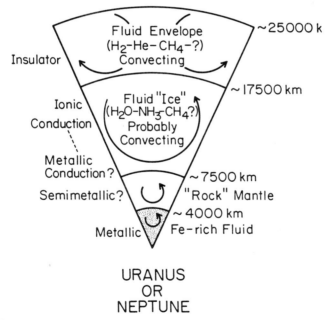

Fig. 14-33 Schematic representations of the possible interior of Uranus and Neptune. (From Stevenson, D. J., 1983. Planetary magnetic fields, *Reports on Progress in Physics*, 46, 555–620. Copyright 1983 by The Institute of Physics. Used with permission.)

action in the ice layer, there should be a substantial field with higher-order harmonics more prominent and a substantial dipole tilt. Stevenson (1983) also speculates on the possible interaction of the iron-core dynamo and the ice-layer dynamo. Intriguing as this may be, we may already see evidence of such types of interactions in the solar magnetic field and the remanent lunar field.

Similar models have been examined by Torbett and Smoluchowski (1980) and Smoluchowski and Torbett (1981). They conclude that a hydrodynamic dynamo can operate in the core of Uranus but probably not on Neptune. On the other hand, they conclude that while the metallic layer of water on Neptune is thick enough for dynamo action, on Uranus it may be too thin. Thus, they would predict that if these two planets are magnetic, their magnetism arises from different sources.

Since the putative observations of radio emissions from Saturn led to an incorrect expectation for the magnetic moment of Saturn, we should not expect to obtain an accurate estimate from the one reported, possible observation of radio emissions from Uranus (Brown, 1976). However, recently there is indirect evidence for a planetary magnetic field from Lyman-$\alpha$ emissions made using the *International Ultraviolet Explorer*. Disk-averaged brightness of $1.6 \pm 0.4$ kilo Rayleighs (Durrance and Moos, 1982) and 430 to 850 Rayleighs (Clarke, 1982) have been reported. The size and variability of this emission suggest the source of the emission to be charged-particle precipitation in the upper atmosphere of Uranus, and by inference these observations suggest that an intrinsic magnetic field is present. These independent observations appear to be very statistically significant. Whether the correct inference from them is the existence of a planetary magnetic field remains for *Voyager 2* to test in 1986.

## 14-4 CONCLUDING REMARKS

The field of planetary and solar magnetism is still in its adolescence. While much has been learned about the present state of the magnetism of the planets, very little is understood about planetary magnetism. We do not have a satisfactory magnetohydrodynamic theory of planetary dynamos, nor do we know for certain the energy source that drives the dynamos whose fields we observe. Part of the answer lies in the continuing efforts of the many excellent mathematicians and geophysicists who are working on the dynamo process; part of the answer lies in further solar and planetary measurements. The measurements that are needed are not simply more magnetic data. Rather, we need information that will lead to further understanding of planetary interiors, moments of inertia, seismic data, heat flow, and so forth. However, we do also need more magnetic data. For understanding solar magnetism, we need finer-resolution magnetographs. For understanding lunar magnetism, we need a complete low-altitude orbital survey of the remanent magnetization of the Moon. For understanding planetary magnetism, we need orbital measurements at both Mercury and Mars. It is indeed ironic that with *Voyager* on its way to Uranus and Neptune and with no terrestrial planetary spacecraft being built, we shall soon understand the magnetism of the outer solar system better than that of the terrestrial planets. With the possible exception of Pluto, the last planet for which we answer the question of whether or not a magnetic dynamo is still operating may well be one of our nearest neighbors, Mars.

## 14-5 REFERENCES

Acuna, M. H., and N. F. Ness, 1975, The main magnetic field of Jupiter: Pioneer 11, *Nature, 253*, 327-328.

Acuna, M. H., and N. F. Ness, 1976a, Results from the GSFC fluxgate magnetometer on Pioneer 11, in *Jupiter* (T. Gehrels, ed.), Univ. Arizona, Tucson, Ariz., pp. 830-847.

Acuna, M. H., and N. F. Ness, 1976b, The main magnetic field of Jupiter, *J. Geophys. Res., 81*, 2917-2922.

Babcock, H. D., 1959, The Sun's polar magnetic field, *Astrophys. J., 130*, 364-365.

Barnes, A. P., P. Cassen, J. D. Mihalov, and A. Eviatar, 1971, Permanent lunar surface magnetism and its deflection of the solar wind, *Science, 172*, 716-718.

Berge, G. L., 1965, Circular polarization of Jupiter's decimetric radiation, *Astrophys. J., 142*, 1688-1692.

Brace, L. H., R. F. Theis, H. G. Mayr, and S. A. Curtis, 1982, Holes in the nightside ionosphere of Venus, *J. Geophys. Res., 87*, 199-211.

Brett, R., 1977, The case against early melting of the bulk of the moon, *Geochim. Cosmochim. Acta, 14*, 443-445.

Bridge, H. S., J. Lazarus, C. W. Snyder, E. J. Smith, L. Davis, Jr., P. J. Coleman, Jr. and D. E. Jones, 1969, Plasma and magnetic fields observed near Venus, in *The Venus Atmosphere* (R. Jastrow and I. Rasool, eds.), Gordon Breach, New York, pp. 533-548.

Brown, L. W., 1975, Saturn radio emission near 2 MHz, *Astrophys. J., 198*, L89-L92.

Brown, L. W., 1976, Possible radio emission from Uranus at 0.5 MHz, *Astrophys. J., 207*, L209-L212.

Bumba, V., and R. Howard, 1965, Large-scale distribution of solar magnetic fields, *Astrophys. J., 141*, 1502-1512.

Burlaga, L. F., J. W. Belcher, and N. F. Ness, 1980, Disturbances observed near Ganymede by Voyager 2, *Geophys. Res. Lett., 7*, 21-24.

Busse, F., 1983, Recent developments in the dynamo theory of planetary magnetism, *Ann. Rev. Earth Planet. Sci., 11*, 241-268.

Chenette, D. L., and L. Davis, Jr., 1982, An analysis of the structure of Saturn's magnetic field using charged particle absorption features, *J. Geophys. Res., 87*, 5267-5274.

Clarke, J. T., 1982, Detection of auroral H Ly $\alpha$ emission from Uranus, *Astrophys. J., 263*, L105-L106.

Colburn, D. S., J. D. Mihalov, and C. P. Sonett, 1971, Magnetic observations of the lunar cavity, *J. Geophys. Res., 76*, 2940-2957.

Connerney, J. E. P., 1981, The magnetic field of Jupiter—A generalized inverse approach. *J. Geophys. Res., 86*, 7679-7693.

Connerney, J. E. P., M. H. Acuna, and N. F. Ness, 1982, *Voyager 1* assessment of Jupiter's planetary magnetic field, *J. Geophys. Res., 87*, 3623-3627.

Davis, Jr., L., and E. J. Smith, 1976, The Jovian magnetosphere and magnetopause, in *Magnetospheric Particles and Fields* (B. M. McCormac, ed.), D. Reidel Publ., Dordrecht, Holland, pp. 301-310.

Desch, M. D., and M. L. Kaiser, 1981, *Voyager* measurement of the rotation period of Saturn's magnetic field, *Geophys. Res. Lett., 8*, 253-256.

Doell, R. R., C. S. Gromme, A. N. Thorpe, and F. E. Senftle, 1970, Magnetic studies of Apollo 11 lunar samples, *Proc. Apollo 11 Lunar Sci. Conf.*, 2097-2120.

Dolginov, Sh. Sh., 1978, The magnetic field of Mars, *Kosmich. Issled., 16*, 257–268.

Dolginov, Sh. Sh., Y. G. Yeroshenko, L. N. Zhuzgov, and N. V. Pushkov, 1961, Investigations of the magnetic field of the Moon, *Geomag. Aeron., 1*, 18–25.

Dolginov, Sh. Sh., Y. G. Yeroshenko, L. N. Zhuzgov, and N. V. Pushkov, 1966, Measurements of the magnetic field in the vicinity of the moon by the artificial satellite Luna 10, *Dokl. Akad. Nauk SSSR, 170*, 574–577.

Dolginov, Sh. Sh., Y. G. Yeroshenko, and L. N. Zhuzgov, 1969, On the nature of the magnetic field near Venus, *Kosmich. Issled, 7*, 747–752.

Dolginov, Sh. Sh., Y. G. Yeroshenko, and L. N. Zhuzgov, 1972, Magnetic field in the very close neighborhood of Mars according to data from the *Mars 2* and *Mars 3* spacecraft, *Dokl. Akad. Nauk SSSR, 207*, 1296.

Dolginov, Sh. Sh., Y. G. Yeroshenko, and L. N. Zhuzgov, 1973, Magnetic field in the very close neighborhood of Mars according to the data from the *Mars 2* and *Mars 3* spacecraft, *J. Geophys. Res., 78*, 4779–4786.

Dolginov, Sh. Sh., Y. G. Yeroshenko, L. N. Zhuzgov, 1976, The magnetic field of Mars according to the data from the *Mars 3* and *Mars 5*, *J. Geophys. Res., 81*, 3353–3362.

Dolginov, Sh. Sh., Y. G. Yeroshenko, and L. N. Zhuzgov, 1978, Magnetic field and magnetosphere of the planet Venus, *Kosmich. Issled., 16*, 827–863.

Dryer, M., and G. R. Heckman, 1967, Application of the hypersonic analog to the standing shock of Mars, *Solar Phys., 2*, 112–116.

Dungey, J. W., 1961, Interplanetary magnetic field and auroral zones, *Phys. Res. Lett., 6*, 47–49.

Durrance, S. T., and H. W. Moos, 1982, Intense Ly $\alpha$ emission from Uranus, *Nature, 299*, 428–429.

Eddy, J. A., 1976, The Maunder Minimum, *Science, 192*, 1189–1202.

Elphic, R. C., and C. T. Russell, 1978, On the apparent source depth of planetary magnetic fields, *Geophys. Res. Lett., 5*, 211–214.

Elphic, R. C., and C. T. Russell, 1983a, Magnetic flux ropes in the Venus ionosphere—observations and models, *J. Geophys. Res., 88*, 58–72.

Elphic, R. C., and C. T. Russell, 1983b, Global characteristics of magnetic flux ropes in the Venus ionosphere, *J. Geophys. Res., 88*, 2993–3004.

Elphic R. C., C. T. Russell, J. A. Slavin, and L. H. Brace, 1980, Observations of the dayside ionopause and ionosphere of Venus, *J. Geophys. Res., 85*, 7679–7696.

Gapcynski, J. P., 1975, A determination of the lunar moment of inertia, *Geophys. Res. Lett., 2*, 353–356.

Gardner, F. F., and J. B. Whiteoak, 1977, Linear polarization observations of Jupiter at 6, 11, and 21 cm wavelengths, *Astron. Astrophys., 60*, 369–372.

Gilman, P. A., 1983, The solar dynamo: Observations and theories of solar convection, global circulation and magnetic fields, in *Physics of the Sun*, Space Science Board, National Academy of Sciences, Washington, D.C.

Gold, T., and S. Soter, 1976, Cometary impact and the magnetization of the Moon, *Planet. Space Sci., 24*, 45–54.

Gringauz, K. I., V. V. Bezrukikh, T. K. Breus, M. I. Verigin, G. I. Volkov, and A. V. Dyachkov, 1974, Study of solar plasma near Mars and on the Earth-Mars route using charged particle traps on Soviet spacecraft in 1971-1973. II—Characteristics of electrons along orbits of artificial satellites *Mars 2* and *Mars 3*, *Kosmich. Issled., 12*, 585–599.

Helsley, C. E., 1970, Magnetic properties of lunar 1002, 10069, 10084, and 10085 samples, *Proc. Apollo 11 Lunar Sci. Conf.*, 2213–2219.

Herbert, F., M. Wiskerchen, C. P. Sonett, and J. K. Chao, 1976, Solar wind induction in Mercury—constraints on the formation of a magnetosphere, *Icarus, 28*, 489-500.

Hide, R. H., 1978, How to locate the electrically conducting fluid core of a planet from external magnetic observations, *Nature, 271*, 640-641.

Hood, L. L., and S. M. Cisowski, 1983, Paleomagnetism of the Moon and meteorites, *Rev. Geophys. Space Phys., 21*, 676-684.

Hood, L. L., P. J. Coleman, Jr., and D. E. Wilhelms, 1979, Lunar nearside magnetic anomalies, *Proc. Lunar Planet. Sci. Conf. 10th*, 2235-2257.

Howard, R., 1974, Studies of solar magnetic fields. I. The average field strengths, *Solar Phys., 38*, 283-299.

Howard, R., and B. J. Labonte, 1981, Surface magnetic fields during the solar activity cycle, *Solar Phys., 74*, 131-145.

Intriligator, D. S., and E. J. Smith, 1979, Mars in the solar wind, *J. Geophys. Res., 84*, 8427-8435.

Jackson, D. J., and D. B. Beard, 1977, The magnetic field of Mercury, *J. Geophys. Res., 82*, 2828-2836.

Kaiser, M. L., and R. G. Stone, 1975, Earth as an intense planetary radio source: similarities to Jupiter and Saturn, *Science, 189*, 285-287.

Kivelson, M. G., and C. T. Russell, 1983, The interaction of flowing plasmas with planetary ionospheres—a Titan-Venus comparison, *J. Geophys. Res., 88*, 49-57.

Kivelson, M. G., and C. R. Winge, 1976, Field-aligned currents in the Jovian magnetosphere—*Pioneer 10* and *11, J. Geophys. Res., 81*, 5853-5858.

Kivelson, M. G., J. A. Slavin, and D. J. Southwood, 1979, Magnetospheres of the Galilean satellites, *Science, 205*, 491-493.

Knudsen, W. C., P. M. Banks, and K. L. Miller, 1982, A new concept of plasma motion and planetary magnetic field for Venus, *Geophys. Res. Lett., 9*, 765-768.

Komesaroff, M. M., and P. H. McCullough, 1967, The radio rotation period of Jupiter, *Astrophys. Lett., 1*, 39.

Komesaroff, M. M., D. Morris, and J. A. Roberts, 1970, Circular polarization of Jupiter's decimetric emission and the Jovian magnetic field strength, *Astrophys. Lett., 7*, 31-36.

Lin, R. P., R. E. McGuire, H. C. Howe, K. A. Anderson, and J. E. McCoy, 1975, Mapping of lunar surface remanent magnetic fields by electron scattering, *Proc. Lunar Sci. Conf. 6th*, 2971-2973.

Luhmann, J. G., and C. T. Russell, 1983, Magnetic fields in the ionospheric holes of Venus: Evidence for an intrinsic field, *Geophys. Res. Lett., 10*, 409-412.

Luhmann, J. G., C. T. Russell, and R. C. Elphic, 1984, Diffusion time scales for large-scale magnetic fields in the Venus ionosphere, *J. Geophys. Res., 89*, 362-368.

Mihalov, J. D., C. P. Sonett, J. H. Binsack, and M. D. Montsoulas, 1971, Possible fossil lunar magnetism inferred from satellite data, *Science, 171*, 892-895.

Moffatt, H. K., 1978, *Magnetic Field Generation in Electrically Conducting Fluids*, Cambridge Univ. Press, New York, 343 pp.

Morris, D., J. B. Whiteoak, and F. Tonking, 1968, The linear polarization of radiation from Jupiter at 6 cm wavelength, *Australian J. Phys., 21*, 337-345.

Nagata, T., Y. Ishikawa, H. Kinoshita, M. Kono, and Y. Syono, 1970, Magnetic properties and natural remanent magnetization of lunar materials, *Proc. Apollo 11 Lunar Sci. Conf.*, 2325-2340.

Nakamura, Y., G. V. Latham, and H. J. Dorman, 1982, Apollo lunar seismic experiment—final summary, *Proc. Lunar Planet. Sci. Conf. 13th*, A117-A123.

Ness, N. F., 1979, The magnetosphere of Mercury, in *Solar System Plasma Physics* (C. F. Kennel, L. J. Lanzerotti, and E. N. Parker, Eds.), North-Holland Publ. Co., New York, 183-198.

Ness, N. F., K. W. Behannon, R. P. Lepping, Y. C. Whang, and K. H. Schatten, 1974, Magnetic field observations near Mercury: preliminary results from *Mariner 10*, *Science, 185*, 151-160.

Ness, N. F., K. W. Behannon, R. P. Lepping, and Y. C. Whang, 1975a, The magnetic field of Mercury confirmed, *Nature, 255*, 204-205.

Ness, N. F., K. W. Behannon, and R. P. Lepping, 1975b, The magnetic field of Mercury: Part One, *J. Geophys. Res., 80*, 2708-2716.

Ness, N. F., K. W. Behannon, and R. P. Lepping, 1976, Observations of Mercury's magnetic field, *Icarus, 28*, 479-488.

Ness, N. F., M. H. Acuna, R. P. Lepping, J. E. P. Connerney, K. W. Behannon, L. F. Burlaga, and F. M. Neubauer, 1981, Magnetic field studies by *Voyager 1*—preliminary results at Saturn, *Science, 212*, 211-217.

Ness, N. F., M. H. Acuna, R. P. Lepping, J. E. P. Connerney, K. W. Behannon, L. F. Burlaga, and F. M. Neubauer, 1982a, Magnetic field studies by *Voyager 2*—preliminary results at Saturn, *Science, 215*, 558-563.

Ness, N. F., M. H. Acuna, K. W. Behannon, and F. M. Neubauer, 1982b, The induced magnetosphere of Titan, *J. Geophys. Res., 87*, 1369-1381.

Newkirk, Jr., G., and K. Frazier, 1982, The solar cycle, *Physics Today, 35*, 25-34.

Ng, K. H., and D. B. Beard, 1979, Possible displacement of Mercury's dipole, *J. Geophys. Res., 84*, 2115-2117.

Orrall, R. Q., 1981, *Solar Corona Explorer: A Mission for the Physical Diagnosis of the Solar Corona*, NASA, Goddard Space Flight Center, Maryland, 80 pp.

Parker, E. N., 1977, The origin of solar activity, *Ann. Res. Astr. Astrophys., 15*, 45-68.

Parker, E. N., 1979, *Cosmical Magnetic Fields—Their Origin and Their Activity*, Oxford, Clarendon, 841 pp.

Peale, S. J., P. Cassen, and R. T. Reynolds, 1979, Melting of Io by tidal dissipation, *Science, 203*, 892-894.

Piddington, J. H., 1981, *Cosmic Electrodynamics* (2nd Ed.), Krieger Publ. Co., Malabar, Fla., 360 pp.

Pneumann, G. W., and R. A. Kopp, 1971, Gas-magnetic field interactions in the solar corona, *Solar Phys., 18*, 258-270.

Roberts, J. A., and M. M. Komesaroff, 1965, Observations of Jupiter's radio spectrum and polarization in the range from 6 to 100 cm, *Icarus, 4*, 127-156.

Runcorn, S. K., 1975a, An ancient lunar magnetic dipole field, *Nature, 253*, 701-703.

Runcorn, S. K., 1975b, On the interpretation of lunar magnetism, *Phys. Earth Planet. Int., 10*, 327-335.

Runcorn, S. K., 1982a, Primeval displacements of the lunar pole, *Phys. Earth Planet. Int., 29*, 135-147.

Runcorn, S. K., 1982b, The Moon's deceptive tranquility, *New Scientist, 96*, 174-180.

Runcorn, S. K., D. W. Collinson, W. O'Reilly, M. H. Battey, A. Stephenson, J. M. Jones, A. J. Manson, and P. W. Readman, 1970, Magnetic properties of *Apollo 11* lunar samples, *Proc. Apollo 11 Lunar Sci. Conf.*, 2369-2387.

Russell, C. T., 1974, On the heliographic latitude dependence of the interplanetary magnetic field as deduced from the 22-year cycle of geomagnetic activity, *Geophys. Res. Lett., 1*, 11-14.

Russell, C. T., 1975, On the possibility of deducing interplanetary and solar parameters from geomagnetic records, *Solar Phys., 42*, 259-269.

Russell, C. T., 1976, The magnetic moment of Venus—*Venera-4* measurements reinterpreted, *Geophys. Res. Lett., 3*, 125-128.

Russell, C. T., 1978a, The magnetic field of Mars—*Mars 3* evidence re-examined, *Geophys. Res. Lett., 5*, 81-84.

Russell, C. T., 1978b, The magnetic field of Mars: *Mars 5* evidence re-examined, *Geophys. Res. Lett., 5*, 85-88.

Russell, C. T., and R. C. Elphic, 1979a, Observations of magnetic flux ropes in the Venus ionosphere, *Nature, 179*, 616-618.

Russell, C. T., and R. C. Elphic, 1979b, ISEE observations of flux transfer events at the dayside magnetopause, *Geophys. Res. Lett., 6*, 33-36.

Russell, C. T., and B. R. Lichtenstein, 1975, On the source of lunar limb compressions, *J. Geophys. Res., 80*, 4701-4711.

Russell, C. T., P. J. Coleman, Jr., and G. Schubert, 1975, The lunar magnetic field, *Space Res. XV*, 621-628.

Russell, C. T., H. Weiss, P. J. Coleman, Jr., L. A. Soderblom, D. E. Stuart-Alexander, and D. E. Wilhelms, 1977, Geologic-magnetic correlations on the Moon—*Apollo* subsatellite results, *Proc. Lunar Sci. Conf. 8th*, 1171-1185.

Russell, C. T., R. C. Elphic, and J. A. Slavin, 1979, Initial *Pioneer Venus* magnetic field results: dayside observations, *Science, 203*, 745-748.

Russell, C. T., R. C. Elphic, J. G. Luhmann, and J. A. Slavin, 1980, On the search for an intrinsic magnetic field at Venus, *Proc. Lunar Planet. Sci. Conf. 11th*, 1897-1900.

Russell, C. T., P. J. Coleman, Jr., and B. E. Goldstein, 1981a, Measurements of the lunar induced magnetic moment in the geomagnetic tail—evidence for a lunar core?, *Proc. Lunar Planet. Sci. Conf., 12B*, 831-836.

Russell, C. T., J. G. Luhmann, R. C. Elphic, and F. L. Scarf, 1981b, The distant bow shock and magnetotail of Venus—magnetic field and plasma wave observations, *Geophys. Res. Lett., 8*, 843-846.

Russell, C. T., J. G. Luhmann, and R. C. Elphic, 1983, The properties of the low altitude magnetic belt in the Venus ionosphere, *Adv. Space Res., 3*, 13-16.

Slavin, J. A., and R. E. Holzer, 1979, The effect of erosion on the solar wind standoff distance of Mercury, *J. Geophys. Res., 84*, 2076-2082.

Slavin, J. A., and R. E. Holzer, 1983, The solar wind interaction with Mars revisited, *J. Geophys. Res., 86*, 11401-11418.

Slavin. J. A., R. E. Holzer, J. R. Spreiter, S. S. Stahara, and D. S. Chaussee, 1983, Solar wind flow about the terrestrial planets. 2. Comparison with gas dynamic theory and implications for solar-planetary interactions, *J. Geophys. Res., 88*, 19-35.

Smirnov, V. N., A. N. Omelchenko, and O. L. Vaisberg, 1978, Possible discovery of cusps near Mars, *Kosmich. Issled., 16*, 688-692.

Smith E. J., 1969, Planetary magnetic field experiments, in *Advanced Space Experiments* (O. L. Tiffany and E. M. Zaitzeff, eds.), American Astronautical Society, Tarzana, CA, pp. 103-130.

Smith, E. J., L. Davis, Jr., P. J. Coleman, Jr., and C. P. Sonett, 1963, *Mariner II*: preliminary reports on measurements of Venus magnetic field, *Science, 139*, 909-910.

Smith, E. J., L. Davis, Jr., D. E. Jones, D. S. Colburn, P. J. Coleman, Jr., and C. P. Sonett, 1974, Magnetic field of Jupiter and its interaction with the solar wind, *Science, 183*, 305-306.

Smith, E. J., L. Davis, Jr., D. E. Jones, D. S. Colburn, P. J. Coleman, Jr., and C. P. Sonett, 1975, Jupiter's magnetic field, magnetosphere and interaction with the solar wind—Pioneer 11, *Science, 188*, 451-455.

Smith, E. J., L. Davis, Jr., and D. E. Jones, 1976, Jupiter's magnetic field and magnetosphere, in *Jupiter* (T. Gehrels, ed.), Univ. Arizona, Tucson, Ariz., pp. 788-829.

Smith, E. J., L. Davis, Jr., D. E. Jones, P. J. Coleman, Jr., D. S. Colburn, P. Dyal, and C. P. Sonett, 1980, Saturn's magnetic field and magnetosphere, *Science, 207*, 407-410.

Smoluchowski, R., and Torbett, M., 1981, Can magnetic fields be generated in the icy mantles of Uranus and Neptune?, *Icarus, 48*, 146-148.

Sonett, C. P., and J. D. Mihalov, 1972, Lunar fossil magnetism and perturbations of the solar wind, *J. Geophys. Res., 77*, 588-603.

Spreiter, J. R., and A. W. Rizzi, 1972, Martian bow wave theory and observation, *Planet. Space Sci., 20,* 205-208.

Stephenson, A., 1976, Crustal remanance and the magnetic moment of Mercury, *Earth Planet. Sci. Lett., 28*, 454-458.

Stephenson, A., S. K. Runcorn, and D. W. Collinson, 1975, On changes in the intensity of the ancient lunar magnetic field, *Proc. Lunar Sci. Conf., 6th*, 3049-3062.

Stevenson, D. J., 1980, Lunar asymmetry and paleomagnetism, *Nature, 287*, 520-521.

Stevenson, D. J., 1981, Models of the Earth's core, *Science, 214*, 611-619.

Stevenson, D. J., 1982, Reducing the non-axisymmetry of a planetary dynamo and an application to Saturn, *Geophys. Astrophys. Fluid Dyn., 21*, 113-127.

Stevenson, D. J., 1983, Planetary magnetic fields, *Reports on Progress in Physics, 46*, 555-620.

Stevenson, D. J., T. Spohn, and G. Schubert, 1983, Magnetism and thermal evolution of the terrestrial planets, *Icarus, 54*, 466-489.

Stix, M., 1981, Theory of the solar cycle, *Solar Phys., 74*, 79-101.

Strangway, D. W., E. E. Larson, and G. W. Pearce, 1970, Magnetic studies of lunar samples—breccia and fines, *Proc. Apollo 11 Lunar Sci. Conf.*, 2435-2451.

Torbett, M., and R. Smoluchowski, 1980, Hydromagnetic dynamo in the cores of Uranus and Neptune, *Nature, 286*, 237-239.

Tyler, G. L., V. R. Eshleman, J. D. Anderson, G. S. Levy, G. F. Lindal, G. E. Wood, and T. A. Croft, 1981, Radio science investigations of the Saturn systems with *Voyager 1*: Preliminary results, *Science, 212*, 201-206.

Vaisberg, O. L., A. V. Bogdanov, N. F. Borodin, A. V. Djachkov, A. A. Zerzalov, F. P. Karpinsky, S. P. Kondakov, Z. N. Mamotko, B. V. Polenov, S. A. Romanov, V. N. Smirnov, and B. N. Khazanav, 1973, Measurements of low-energy particles on *Mars 2* and *Mars 3* spacecraft, *Kosmich. Issled., 11*, 743-755.

Wallis, M. K., 1975, Does Mars have a magnetosphere?, *Geophys. J. Roy. Astr. Soc., 41*, 349-354.

Whang, Y. C., 1977, Magnetospheric magnetic field of Mercury, *J. Geophys. Res., 82*, 1024-1030.

Whiteoak, J. B., F. F., Gardner, and D. Morris, 1969, Jovian linear polarization at 6 cm wavelength, *Astrophys. Lett., 3*, 81-84.

Wiskerchen, M. J., and C. P. Sonett, 1978, On the detectability of a metallized lunar core, *Proc. Lunar Sci. Conf. 9th*, 3113.

Yeroshenko, E. G., 1979, Unipolar induction effects in the magnetic tail of Venus, *Kosmich. Issled., 17*, 93-105.

Marcia Neugebauer
Jet Propulsion Laboratory
California Institute of Technology
Pasadena, California 91109

# 15

# COMETS

## ABSTRACT

Comets are minor members of the solar system. They are thought to be small ($10^2$ to $10^4$ m in diameter), icy conglomerates of frozen gases (largely water–ice) and nonvolatile grains. When their orbits bring them close to the Sun, the surface heats up, and the water vapor and other volatiles boil off, dragging small dust grains with them to form a large, visible coma. The solid grains are forced into a tail by solar radiation pressure. Cometary ions are formed by photoionization and by other, poorly understood processes. The ions are then forced into a highly structured ion tail by the interaction with the solar wind. Most of the details of this interaction can only be surmised, although many similarities to the interaction of the solar wind with Venus are expected.

Comets are thought to originate in a pristine state in the Oort cloud surrounding the Sun at a distance of about $10^5$ AU. Active, long-period comets have probably made only one or a few trips into the inner solar system. Those comets that are caught in short-period orbits decay rapidly.

The importance of comets to solar-system studies lies in the possibilities that (1) they are well-preserved samples of either the interstellar cloud which collapsed to form the solar system or the planetesimals from which the outer planets accumulated, and (2) they provided either the prebiotic complex molecules from which life evolved or some volatiles necessary for the evolution of these molecules.

## 15-1 A VERY BRIEF HISTORY OF COMET RESEARCH

The fact that comets orbit the Sun as do the planets was not discovered until the 18th century. Aristotle believed that comets were an atmospheric or a sublunar phenomenon. It wasn't until 1577 that Tycho Brahe demonstrated that the great comet of that year had to be at least four times more distant than the Moon. Despite his discovery of the laws of planetary motion, Kepler believed that comets moved through the solar system in straight lines. Hevelius correctly stated that comets moved on curved trajectories, but he believed that they originated in the atmospheres of Jupiter and Saturn.

It was Isaac Newton who showed that the trajectories of comets and planets could be accounted for by the same laws of gravitation. In 1705, Edmund Halley applied Newton's methods to a number of comets and concluded that the historical comets of 1531, 1607, and 1682 were successive returns of the same comet on an elliptical orbit with a period of about 75.5 y. He predicted it would return in 1758. This comet was named after Halley when it reappeared nearly on schedule after his death.

Contemporary cometary research focuses on the physical and chemical nature of comets (e.g., why they look and act as they do) and on the role of comets in the history and evolution of the solar system.

## 15-2 NUCLEUS, COMA, AND TAILS

Comets are perhaps at once the most spectacular and the least well understood members of the solar system. The principal reason for our poor understanding of comets is that the cometary bodies from which the visible gas and dust issue have never been resolved by telescope.

The most credible working hypothesis is Whipple's theory (1950, 1951) that a comet is a kind of "dirty snowball." Whipple proposed that cometary nuclei were "icy conglomerates" comprising a variety of frozen gases and small bits of nonvolatile solids. Brightness profiles and spectroscopic data support the idea that the principal constituent of most observable comets is water-ice, with small amounts of other volatiles trapped in the lattice structure to form what is known as a *clathrate*.

Most cometary nuclei are probably a few hundred meters to a few kilometers in diameter. The fact that no dark spot could be seen as Halley's comet crossed the solar disk in 1910 means that its nucleus must be less than 50 km across. Modern models indicate that its diameter is about 5 km (Newburn, 1979).

Most observable comets are in very eccentric orbits, and they spend most of their time very far from the Sun. As a comet absorbs and reradiates sunlight, it remains at an equilibrium temperature well below the freezing point of water. When the snowball gets closer than 2 or 3 AU to the Sun, it heats up to about 200 K, at which temperature the water-ice vaporizes and escapes. The other volatiles in the clathrate are released with the water vapor. As the gas expands into the near-vacuum of interplanetary space, it collides with and exerts a drag on any small solid particles embedded in the ice. Because the comet nucleus is so small, its gravitational field is very weak, and all but the very largest solid grains are expected to be accelerated to speeds greater than the escape velocity. Some ice grains may also be dragged away by the escaping gas. Once free of the gravitational field of the comet, the mixture of gas, dust, and ice freely expands to form the visible atmosphere called the *coma*. The coma of a typical, bright comet may be some $10^5$ km in diameter.

The process is not constant and uniform, however. The coma material often appears to be emitted in spatially and temporally varying bursts, jets, and halos, as illustrated in Fig. 15-1. It has been speculated that after many passages by the Sun, much of the surface is covered by solid, nonvolatile debris and that ices are exposed over only a limited fraction of the surface. If this were the case, and if the cometary body were rotating, the emission would increase every time the active, icy area rotated into the sunlight. The

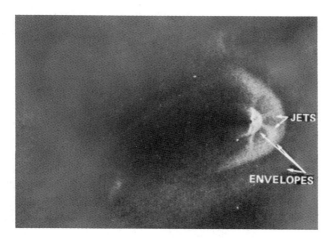

Fig. 15-1  Coma structure of Halley's comet in 1910.

rotation rates of several comets have been determined by timing such outbursts; their rotation periods are usually several hours to about a day in length (Whipple, 1981).

Once the cometary material is in the coma, the gases glow by resonance fluorescence of sunlight. Different gaseous species can be identified by their emission lines. Spectral lines have been found in the ultraviolet, the visible, and radio wavelengths. The chemical species identified to date are listed in Table 15-1. The relative abundances of the different species vary appreciably from comet to comet.

---

TABLE 15-1   Chemical Species in Comets

A. IDENTIFIED SPECIES
   Organic: $C$, $C_2$, $C_3$, $CH$, $CN$, $CO$, $CS$, $\overline{HCN}$, $\overline{CH_3CN}$
   Inorganic: $H$, $NH$, $NH_2$, $O$, $OH$, $\overline{H_2O}$, $S$, $\overline{HN_3}$, $S_2$
   Metals: $Na$, $Ca$, $Cr$, $Co$, $Mn$, $Fe$, $\overline{Ni, Cu, V, Si, K}$
   Ions: $CO^+$, $CO_2^+$, $CH^+$, $Ca^+$, $N_2^+$, $OH^+$, $H_2O^+$, $C^+$, $CN^+$
   Solids: Silicates

B. OTHER SUGGESTED PARENT MOLECULES
   $CO_2$, $CH_4$, $N_2$, $O_2$, $NO$, $CH_3OH$, $H_2CO$, $C_2H_2$, $C_2H_4$, $C_3H_4$, $CS_2$

---

One noteworthy feature of Table 15-1 is that most of the species are molecular fragments or chemically active free radicals. The only stable molecules are those underlined. Most of the original "parent molecules" have evidently been rapidly broken apart by solar ultraviolet radiation, by collisions with energetic particles, or by chemical reactions. The composition of many of the parent molecules can only be surmised, with considerable ambiguity. For example, it is not known whether the observed CN radical comes from hydrogen cyanide (HCN), which has been observed in comets, or from cyanogen ($C_2N_2$), which has not yet been detected. The bottom row of Table 15-1 lists suggested, but so far undetected, parent molecules, many of which have been detected in interstellar clouds. Those chemical models of cometary atmospheres based on the composition of the nucleus typical of interstellar clouds are generally more successful at explaining the observed spectra of comets than are models based on solar-type elemental abundances (Mitchell et al., 1981).

There are forces at work to push the gas and dust out of the coma into the tails. Many comets have two tails—a broad, yellowish tail whose spectrum is consistent with sunlight reflected from dust, and a skinnier, bluish, highly structured tail whose spectral emissions indicate an abundance of ions, especially $CO^+$. The ion tail points nearly radially away from the Sun, whereas the dust tail trails further behind in the comet's orbit.

The size and shape of the dust tail can be explained by the combined effects of drag by the expanding gas and solar radiation pressure (Finson and Probstein, 1968a and b). Theory predicts that the particles are naturally sorted according to their size and the time of emission from the comet, as shown in Fig. 15-2. Comparison of the models with observations has led to calculated size distribution functions for the dust particles. These distributions often peak sharply at one to a few microns. The circles in Fig. 15-2 show the amount of smear of the time-size diagram arising from the initial velocities of the dust

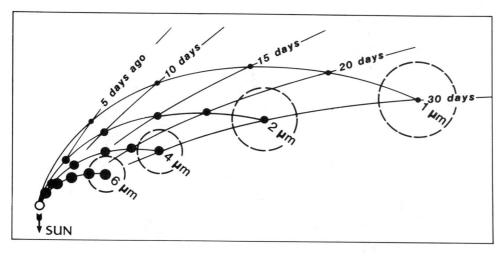

Fig. 15-2 Distribution of dust in a comet's tail as functions of grain size and time elapsed since emission from the nucleus. The dashed circles represent the fuzziness of the trailing edge caused by a finite emission velocity (assumed to be isotropic). (From Delsemme, 1981, figure courtesy of the Jet Propulsion Laboratory.)

grains. For a given grain size, this initial velocity depends on the dust-to-gas ratio of the comet. For most comets, the amount of gas equals or somewhat exceeds the amount of dust.

With time, the dust spreads out along the entire cometary orbit and is observable as a meteor stream. Eventually, orbital perturbations break up the stream, and the dust grains become part of the interplanetary medium, until they are vaporized or eroded to such a small size that they are blown out of the solar system by solar radiation pressure.

Some of the larger dust grains from comets may look like the particle shown in Fig. 15-3. This and other particles like it were collected by Brownlee et al. (1976) on high-flying U-2 aircraft. These particles are known to be of extraterrestrial origin because of their chemical relation to meteorites and because of the presence of large amounts of solar-wind helium in them. Although it has not been proven that these are cometary dust particles, their porous structure suggests that they might have once been filled with volatile material.

It was the study of comet ion tails that led Biermann (1951) to postulate the existence of the solar wind described in Chapter 2 by Zirin. Ion tails point away from the Sun because momentum is transferred from the nearly radially flowing solar wind to the cometary plasma (Biermann et al., 1967). As discussed in Chapter 14 by Russell, there are three general types of interaction between solar-system bodies and the solar wind. Some bodies, like the Earth and Jupiter, have strong magnetic fields, which the solar wind cannot penetrate and so must flow around. It is believed to be very unlikely that bodies as small as comets would have strong enough magnetic fields to stand off the solar wind. At the other end of the scale, there are lunar-type bodies with very little atmosphere or magnetic field; the solar wind can impact the surfaces of these bodies and

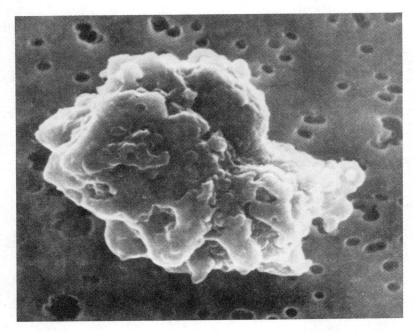

Fig. 15-3  An interplanetary dust grain collected in the stratosphere.

be absorbed, leaving a diamagnetic cavity behind. This type of interaction probably occurs for small or inactive comets and for all comets when they are far from the Sun.

Substantial ion densities have been observed in the inner comas of bright, active comets. In fact, the amount of ionization may be greater than can be accounted for by photoionization alone. Several ionization mechanisms have been suggested, including ionization by energetic electrons accelerated by the interaction of the comet with the solar wind. Whatever its cause, the ion pressure in the ionosphere of a bright, active comet is probably sufficient to balance the pressure of the solar wind at a distance of $10^2$ to $10^4$ km from the nucleus. The interaction with the solar wind may be similar to that at Venus (Russell et al., 1982), except that the comet's ionosphere may be very much larger because of the negligible cometary gravitational attraction.

The interaction of an active comet with the solar wind may be very complex. There may be a contact surface (much like the ionopause of Venus) that separates cometary from solar-wind streamlines. The postulated geometry is shown in Fig. 15-4. Ions can be created by photoionization anywhere in the coma, even very far from the nucleus. And these newly created ions load down the solar wind and force it to become subsonic at a bow shock located very much farther ($10^5$ to $10^6$ km) upstream than would otherwise be the case (Biermann et al., 1967; Wallis, 1973). If the material flowing out from the comet is supersonic, as seems likely, there may be another shock inside the contact surface; its function is to deflect the cometary material tailward. A lot of detailed structure has also been suggested downstream of the nucleus (Wallis and Dryer, 1976). On the other hand,

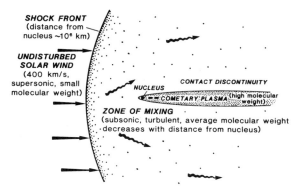

Fig. 15-4 Postulated large-scale geometry of the interaction of an active comet with the solar wind.

collisions between ions and neutral molecules in the coma may play an important role, allowing the solar wind and cometary plasmas to intermix (Wallis and Ong, 1976).

Sometimes an ion tail appears to break off or to disconnect from the coma and then a new ion tail forms. It has been suggested (Niedner and Brandt, 1978) that this phenomenon is caused by the passage of an interplanetary sector boundary (i.e., a boundary across which the direction of the interplanetary magnetic field reverses). If the flow of the solar wind cannot readily penetrate the comet's ionosphere, the interplanetary field lines become draped to form a tail (Alfvén, 1957). If the direction of the interplanetary field reverses, currents may flow in the ionosphere to reconnect the field lines into a different pattern, as shown in Fig. 15-5 (Niedner and Brandt, 1978). This process may allow an island of old tail plasma to flow away downstream and give the appearance of the shedding and regrowth of the ion tail.

There are many fascinating structural forms in comet ion tails, such as kinks, twists, and arches, which may be caused by various types of plasma instabilities and reconnections. The day-to-day variability of ion tails is very great, as shown by the sequence of pictures of the comet Mrkos in Fig. 15-6.

The rather random day-to-day activity is superimposed on the general trend of buildup and decay, as shown in Fig. 15-7 by the approach and recession of Halley's comet in 1910. At its peak, a comet's visible tail can be on the order of 1 AU in length.

Finally, not all the cometary atmosphere flows out through the visible tails. The first ultraviolet studies of comets revealed that comets are surrounded by enormous hydrogen coronas, which often extend to a distance of more than $10^7$ km. Figure 15-8 shows, to the same scale, visible light and Lyman-alpha (ultraviolet) pictures of the comet Kohoutek. The circle shows the size of the Sun relative to the hydrogen cloud visible in the Lyman-alpha spectral line. The hydrogen atoms that cause this emission by the absorption and reradiation of solar Lyman-alpha light probably come from the breakup of water and other parent molecules. These atoms can reach such great distances from the nucleus because of the considerable excess kinetic energy they acquire in the dissociation process (Keller, 1976). Eventually these atoms are ionized by solar ultraviolet radiation, and they too flow away with the solar wind.

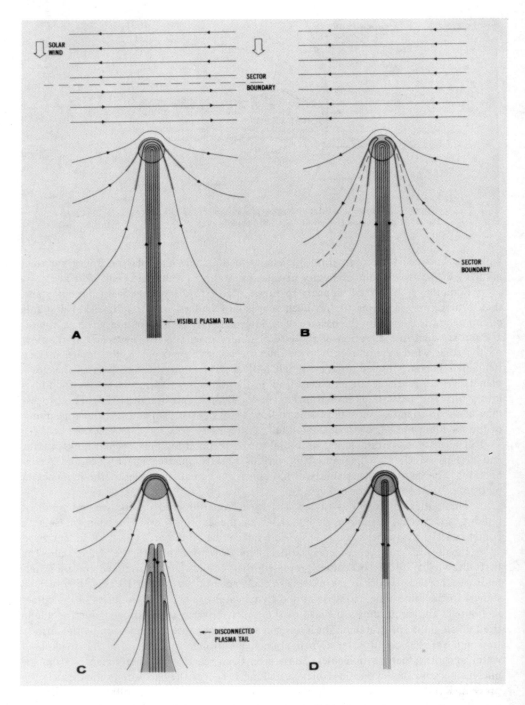

Fig. 15-5  Process of disconnection of a comet tail in response to passage of an interplanetary sector boundary (From Neidner and Brandt, 1978).

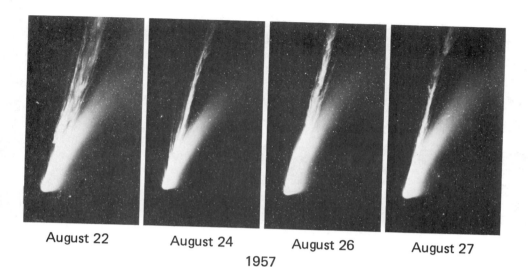

August 22    August 24    August 26    August 27

1957

Fig. 15-6   Comet Mrkos, photographed in 1957 by the 48-in. Schmidt telescope of the Mount Palomar Observatory.

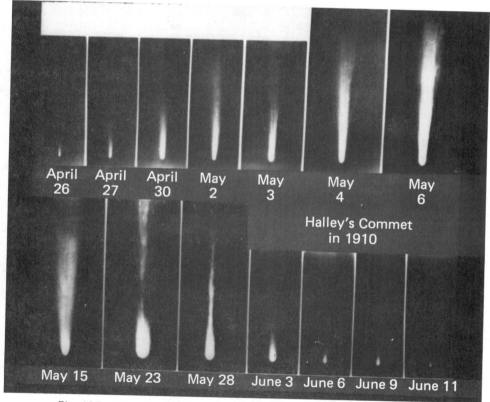

Fig. 15-7   Growth and decay of the tails of Halley's comet in 1910. The comet passed through perihelion on April 20.

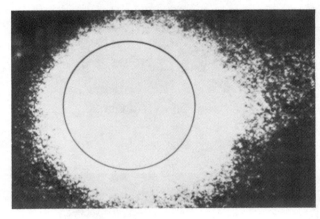

Fig. 15-8 Comet Kohoutek. (Top) Photograph in white light. (Bottom) Photograph in Lyman-alpha radiation (From Whipple, 1979).

## 15-3 ORBITS

The "home" of comets appears to be the very outer fringes of the solar system—at a distance, $R_c$, of about $10^5$ AU, or halfway to the nearest star. The present population of the Oort cloud, as this swarm of comets is called, is at least $1 \times 10^{12}$ comets, with a total mass equal to or greater than that of Earth. There are differences of opinion about whether the comets were created at great distances from the Sun by the sticking together of ice-coated interstellar grains or whether they condensed in the region of the outer planets and were later scattered out to the Oort cloud. The orbital period of a comet in the Oort cloud is about 4 My. During such a period, 20 to 40 stars would be expected to pass within $3\,R_c$ of the Sun. The effect of these passing stars can be looked at as causing diffusion in the orbital elements (see Chapter 1 by Wasson and Kivelson) of the comets. Over the life of the solar system, this diffusion has wiped out any memory of the origin of comets in a disk, and all orbital inclinations are now equally likely. A comet is lost from the cloud by gaining so much energy that it moves out of the solar system or by having its perihelion diffuse down into the planetary region. The latter are the only comets we see. They arrive in the inner solar system from random directions on nearly parabolic orbits. About five such comets are observed per year, one or two of which still have their aphelia in the Oort cloud.

Once or twice a century, Earth dwellers are treated to the sight of a truly spectacular, great comet. We in the twentieth century have been very unfortunate in this respect. The last comet bright enough to be seen in broad daylight appeared in 1910, the same year as the fainter Halley's comet. But even this comet was not very impressive compared with some comets that appeared during the last few hundred years. A great comet is probably a relatively large body on a trajectory which passes within a few tenths of an AU of the Sun. It is probably a "new" comet on its first trip in from the Oort cloud, boiling off all its ices more volatile than the water clathrates; not much of this material can survive such a maiden trip.

If one of these new comets happens to have its perihelion close to a planet, especially Jupiter or Saturn, the resulting trajectory change can be sufficient to temporarily capture the comet on a short-period orbit. About 5 of the approximately 100 known short-period comets reappear each year.

Each time a comet passes close to the Sun, it loses more of its gas and dust. The short-period comets soon become much less active, and even near perihelion their tails become very difficult to detect. There is a rough correlation between the brightness of a comet and its period, as shown in Fig. 15-9. Figure 15-10 shows the comet Encke, a remarkably active short-period comet. Its period of 3.3 y is shorter than that of any other known comet. Its perihelion distance is 0.3 AU.

Halley's comet is one of the brightest of the short-period comets. But it too is fading; Fig. 15-11 shows the brightnesses of comets Halley and Encke as functions of time. Comet Halley's next apparition, in 1985–1986, will be very unfavorable for viewing

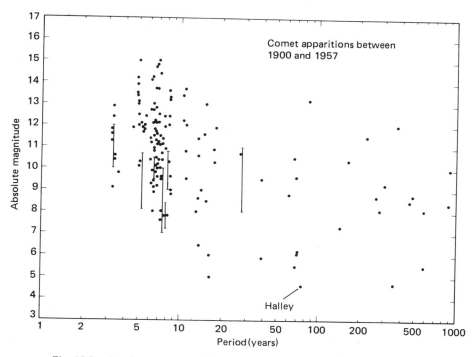

Fig. 15-9 Absolute magnitude (proportional to the logarithm of the inverse of the brightness, so that the brighter comets have lower absolute magnitude) of comets with periods of less than 1000 y. (From Reinhard, 1981. Figure courtesy of the Jet Propulsion Laboratory.)

Fig. 15-10  Comet Encke, photographed by E. Roemer on January 6, 1961.

from Earth. It will be most active just past perihelion when it is on the far side of the Sun from the Earth.

What happens to old comets? Some of them pass too close to the Sun and are ripped apart, presumably by tidal forces. Some comets may be scattered out of the solar system or into the Sun by planetary encounters. But other short-period comets probably just fade away. Some of the Earth-crossing asteroids may be extinct comets. Or perhaps comets continue to get smaller and smaller until there's nothing left.

Four of the short-period comets that have been observed at more than one perihelion passage have been lost (Marsden, 1974). At least two of these showed sudden changes in the nongravitational accelerations caused by the jet action of the escaping gases, so they may be lost because their orbits have changed significantly. Another of the four lost comets split into two parts during its final appearance.

The gas emitted by comets is picked up and carried into interstellar space by the solar wind. The dust grains are lost from the inner solar system (due to vaporization, collisional fragmentation, and radiation pressure) at an average rate of 10 ton · $s^{-1}$. In comparison, the average rate at which comets add dust to the interplanetary medium is estimated to be much less than 1 ton · $s^{-1}$. Thus, unless comets emit much more dust than present models indicate, perhaps in the form of unobservable large particles, the

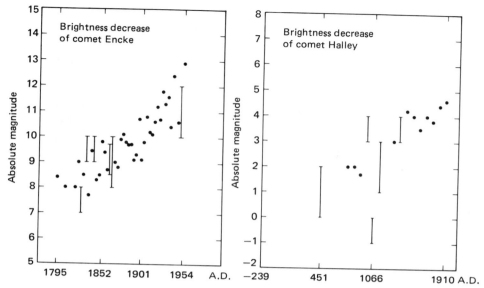

Fig. 15-11  Apparent brightness decreases of comets Encke and Halley (Reinhard, 1981). The *brightness* is expressed in terms of *absolute magnitude*, which is a term astronomers use to measure brightness. It is a logarithmic scale, with a difference of 5 magnitudes corresponding to a factor of 100 in brightness, with a *lower* number representing a *brighter* object. For solar-system bodies, the *absolute magnitude* refers to how bright an object would be if it were situated 1 AU from the Sun and 1 AU from the observer. (Figure courtesy of the Jet Propulsion Laboratory.)

amount of interplanetary dust must be decreasing as a function of time. Much of the present dust cloud may have been ejected from a short-period super comet some twenty thousand years ago. Comet Encke may be the nearly depleted remains of this comet.

## 15-4  SOLAR-SYSTEM EVOLUTION

Several chapters in this book deal with the origin and evolution of the solar system. Here we only point out the possible relevance of comets to those processes. Comets may be nearly unevolved collections of interstellar icy grains. Or their ices may have condensed during the formation of the outer planets. In fact, comets may be the planetesimals from which the outer planets were built. In either case, comets are probably the most pristine samples of early solar-system material available for study, because they have never been incorporated into large bodies and subjected to high pressures, heating, differentiation, and so forth.

Soon after the inner planets formed, they lost their atmospheres during a period of particularly violent solar wind. Then followed a period of intense bombardment of the inner planets by comets as well as by meteorites. If comets were created beyond $10^3$ AU, the early bombardment rate would be only a few times the present rate. But if comets were created in the region of the outer planets and then diffused out to their present location in the Oort cloud, the primordial flux of comets in the inner solar system could have been $10^3$ times greater than the present flux. Comets could have made substantial contributions to the present, second-generation atmospheres of the inner planets. The chemical composition of cometary volatiles may be more hospitable to the evolution of life than were the volcanic gases that contributed much of the Earth's atmosphere. Comets may have been the source of either the prebiotic organic compounds necessary for the evolution of life or some of the volatiles necessary for the synthesis of these compounds (Chang, 1979).

Clearly, comets are important as well as spectacular-appearing members of the solar system. Their study can help us understand the evolution of the solar system as well as plasma-gas interaction processes that occur in many astrophysical settings.

## 15-5 ACKNOWLEDGMENT

This research was conducted at the Jet Propulsion Laboratory of the California Institute of Technology, Pasadena, California, and was supported by the Space Plasma Physics Program of NASA under contract NAS 7-100.

## 15-6 REFERENCES

Alfvén, H., 1957, On the theory of comet tails, *Tellus, 9*, 92–96.

Biermann, L., 1951, Comets' tails and solar corpuscular radiation (in German), *Zeit. Astrophys., 29*, 274–286.

Biermann, L., B. Brosowski, and H. U. Schmidt, 1967, The interaction of the solar wind with a comet, *Solar Phys., 1*, 254–284.

Brownlee, D. E., D. A. Tomandl, and P. W. Hodge, 1976, Extraterrestrial particles in the stratosphere, in *Interplanetary Dust and Zodiacal Light* (H. Elsasser and H. Fechtig, eds.), Springer-Verlag, New York, pp. 279–283.

Chang, S., 1979, Comets—cosmic connections with carbonaceous meteorites, interstellar molecules, and the origin of life, in *Space Missions to Comets* (M. Neugebauer, D. K. Yeomans, J. C. Brandt, and R. W. Hobbs, eds.), NASA Conference Publication 2089, pp. 59–111.

Delsemme, A. H., 1981, Observing chemical abundances in comets, in *Modern Observational Techniques for Comets*, Jet Propulsion Laboratory Report, 81–68, Pasadena, California, pp. 5–13.

Finson, M. L., and R. F. Probstein, 1968a, A theory of dust comets. I. Model and equations, *Astrophys. J., 154*, 327–352.

Finson, M. L., and R. F. Probstein, 1968b, A theory of dust comets. II. Results for comet Arend-Roland, *Astrophys. J., 154*, 353–380.

Keller, H. U., 1976, The interpretations of ultraviolet observations of comets, *Space Sci. Rev., 18*, 641–684.

Marsden, B. G., 1974, Comets, *Ann. Rev. Astr. Astrophys., 12*, 1–21.

Mitchell, G. F., S. S. Prasad, and W. T. Huntress, 1981, Chemical model calculations of

$C_2$, $C_3$, CH, CN, OH and $NH_2$ abundances in cometary comae, *Astrophys. J., 244,* 1087.

Newburn, Jr., R. L., 1979, Physical models of Comet Halley based upon qualitative data from the 1910 apparition, in *Comet Halley Micrometeoroid Hazard Workshop,* European Space Agency Report SP-153, Paris, pp. 35-50.

Niedner, Jr., M. B., and J. C. Brandt, 1978, Interplanetary gas, XXIII. Plasma tail disconnection events in comets—evidence for magnetic field reconnection at interplanetary sector boundaries, *Astrophys. J., 233,* 655-670.

Reinhard, R., 1981, The ESA Mission to Comet Halley, in *Modern Observational Techniques for Comets,* Jet Propulsion Laboratory Report 81-68, Pasadena, California, pp. 284-312.

Russell, C. T., J. G. Luhmann, R. C. Elphic, and M. Neugebauer, 1982, Solar wind interaction with comets—lessons from Venus, in *Comets* (L. L. Wilkening, ed.), Univ. Arizona, Tucson, Ariz., pp. 561-587.

Wallis, M. K., 1973, Weakly shocked flows of the solar wind plasma through atmospheres of comets and planets, *Planet. Space Sci., 21,* 1647-1660.

Wallis, M. K., and M. Dryer, 1976, Sun and comets as sources in an external flow, *Astrophys. J., 205,* 895-899.

Wallis, M. K., and R. S. B. Ong, 1976, Cooling and recombination processes in cometary plasma, in *The Study of Comets* (B. Donn, M. Mumma, W. Jackson, M. A'Hearn, and R. Harrington, eds.), NASA SP-393, pp. 856-879.

Whipple, F. L., 1950, A comet model. I. The acceleration of Comet Encke, *Astrophys. J., 111,* 375-394.

Whipple, F. L., 1951, A comet model. II. Physical relations for comets and meteors, *Astrophys. J., 113,* 464-474.

Whipple, F. L., 1979, Scientific need for a cometary mission, in *Space Missions to Comets* (M. Neugebauer, D. K. Yeomans, J. C. Brandt, and R. W. Hobbs, eds.), NASA Conference Publication 2089, pp. 1-31.

Whipple, F. L., 1981, On observing comets for nuclear rotation, in *Modern Observational Techniques for Comets,* Jet Propulsion Laboratory Report 81-68, Pasadena, California, pp. 191-201.

Typhoon Lee*
Department of Terrestrial Magnetism
Carnegie Institution of Washington
5241 Broad Branch Road., N.W.
Washington, D.C. 20015

# 16

# FROM BIG BANG TO BING BANG

## (From the Origin of the Universe to the Origin of the Solar System)

*Now at: Institute of Earth Sciences, Academia Sinica, P.O. Box 23-59, Taipei, Taiwan 107, R.O.C.

# ABSTRACT

The Sun was born 4.55 Gy ago which corresponds to about 5 Gy to 10 Gy after the universe originated in the "big bang." During the long interval prior to the birth of the Sun, our Milky Way Galaxy evolved considerably by converting gas to stars and by enriching the gas with heavy elements synthesized in the stars. The vast majority of elements in our solar system came from many preceding generations of stars. A number of short-lived radioactive nuclides seem to have been present in primitive meteorites when they formed in the early solar system. The planets are rich in deuterium, whereas the present Sun has none. Both of these observations indicate that our planetary system was formed around a young Sun, most likely in a primeval nebula consisting of gas and dust. Some of the radioactive nuclides suggest that the interval between their production in stellar sources and the accumulation of planets of at least a few kilometers in size was only a few million years. Therefore, the initial steps of forming our solar system took place at a relatively fast pace. The proximity of those stellar sources, presumably supernovae, to the protosolar cloud in space and time means that the formation of our Sun may have taken place in a stellar association and may have been preceded immediately by spectacular celestial fireworks that some scientists have dubbed "bing bang." Such explosions may even have been the triggering mechanism for the formation process itself.

## 16-1 INTRODUCTION

An explosion is usually thought of as a destructive, rather than constructive, process. However, in the evolution of the universe, great events of creation are often associated with explosions, not unlike the rise of a new phoenix from the ashes of a great fire, according to Egyptian mythology. The universe seems to have come into existence during one gigantic explosion that is referred to as the "big bang." A large number of chemical elements are produced in the violent explosive death of the massive stars known as *supernovae*. There are some hints that supernova explosions can initiate the formation of another generation of stars. Finally, in the past few years, some exciting evidence has been found indicating that our own solar system may also have been created this way. Models in which the explosions of the last or last few supernovae led to the formation of our solar system are called the *"bing bang" theory*. In this paper, we briefly outline the evolution of our Milky Way Galaxy between big bang and bing bang, pin down its formation time, present the evidence for the relics of the last supernova explosions, and discuss the possible connection between the last fireworks and the formation process itself.

The history of the solar system is a part of the history of the universe. The major events, from the provincial point of view of human beings, can be divided into two parts using the time of solar-system formation as the reference (zero) point (see Fig. 16-1). The universe is believed to have started with the "big bang" about 5 to 10 Gy before zero. The relic radiation from this gigantic primeval fireball still permeates the entire universe today (Weinberg, 1977). The galaxies, including our own Milky Way, formed early, probably within the first billion years. During the subsequent several billions of years, many generations of stars went through their evolution. Chemical elements heavier than

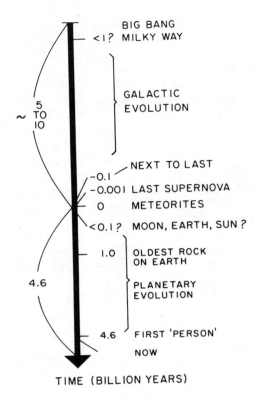

Fig. 16-1 Important events in the cosmic heritage of human beings. Reference zero time is the formation time of our solar system.

helium were gradually synthesized, primarily by those massive stars that ended their lives explosively as supernovae. Many supernovae have contributed to the solar-system material. There is evidence that the explosions of the last two such supernovae took place about 100 My and 1 My, respectively, before the solar-system formation. Most of our solar system was in place within the first 100 My. However, on some of the more active planetary bodies, physical and chemical evolution continued for a long time, generally with intensities decreasing with time. On Earth, these early activities were so intense that no rocks from the first 800 My survived. Human beings appeared only a few million years ago, practically yesterday, when compared with the 4.6-Gy age of our solar system. Such is the overview of our cosmic heritage.

## 16-2 PRESOLAR EVOLUTION

The evolution of galactic matter involves two primary processes: (1) the synthesis of heavy elements from light elements and (2) the conversion of gas to stars. Initially cosmic matter consisted mainly of protons (i.e., hydrogen $^1H$) and neutrons. Nucleosynthesis (i.e., element production) during the big bang made a substantial amount of $^4He$. However, nuclear species at 5 and 8 atomic mass units (Fig. 16-2) are so unstable that the reactions $^1H + {}^4He$ and $^4He + {}^4He$ could not produce any heavier nuclei. Furthermore, the density was too low to permit many three-body reactions; hence, the nucleosynthesis

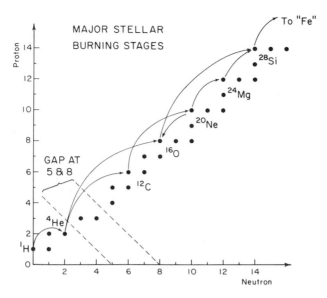

Fig. 16-2  Chart of nuclides from hydrogen to silicon. The number of protons of each nuclide is plotted against its number of neutrons. The big-bang nucleosynthesis could not bridge the gaps at masses 5 and 8, and hence it only produced $^4$He, $^2$D, and small amounts of $^3$He and $^7$Li. The nucleosynthesis associated with the successive thermonuclear burning stages inside massive stars is responsible for the production of progressively heavier species along the arrows in the diagonal direction.

could not proceed further. Thus, at the end of the big bang, the universe was made of mostly $^1$H and $^4$He with some $^2$D, $^3$He, and $^7$Li. The production of heavier elements awaited the high density in stellar interiors.

Stars form out of gas. As the density increases, the $^1$H ions combine to produce $^4$He, releasing energy in the process. As the density further increases, three $^4$He can combine to make $^{12}$C. $^{16}$O is made by capturing another $^4$He. More massive stars evolve further into burning stages in which $^{12}$C is converted to heavier nuclei between $^{20}$Ne and $^{31}$P, $^{16}$O is converted to nuclei between $^{28}$Si to $^{40}$Ca, and finally $^{28}$Si is converted to nuclei near $^{56}$Fe (see Fig. 16-2). These burning processes all release thermonuclear energy to sustain the radiation from the stars. The rare elements beyond the iron group elements are produced by neutron capture, which proceeds concurrently with some of the major burning stages. In this way individual stars are able to build up a rich variety of nuclear species, many of which are vital constituents of our solar system as well as ourselves. At their deaths and sometimes before, stars return some of their mass back to the interstellar gas (see Fig. 16-3). The net effect of stellar evolution is the gradual enrichment of the galactic matter by the heavy elements synthesized in stars. Also, some of the matter is permanently buried in stellar remnants such as white dwarfs, neutron stars, and black holes, causing a gradual decrease of the amount of galactic gas with time.

The Sun is a latecomer in the universe. The age of the universe has been estimated

## GALACTIC LIFE CYCLE

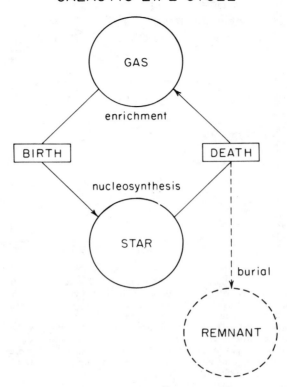

Fig. 16-3  Galactic life cycle. Stars form out of gas, make heavy elements, return some matter back to gas at their death, and bury some matter in the remnants, such as white dwarfs, neutron stars, and black holes. The net effect of cycling mass between stars and gas is the progressive enrichment of the gas with heavy elements and the gradual depletion of galactic gas.

by using three methods. One can measure the present rate of expansion of the universe and extrapolate back to determine when the expansion started. A lower limit can be obtained by dating the oldest stars using the stellar evolution theory. Yet another lower limit is the age of the elements based on the abundance of long-lived radio nuclides at the solar-system formation time and nucleosynthesis theory. All three methods give an age between 10 to 15 Gy (Schramm, 1974). Since the age of our solar system is 4.6 Gy (see Section 16-3), it was born 5 to 10 Gy after the beginning of the universe. Before the birth of the solar system, many generations of stars had gone through their galactic life cycles, and the heavy-element (everything heavier than $^4$He) concentration in the galaxy has built up to about 2% of the total mass. The composition of the matter that was to make the solar system is believed to have been close to that illustrated in Fig. 16-4, the so-called normal or cosmic abundance. This represents the grand average of the products of many supernovae.

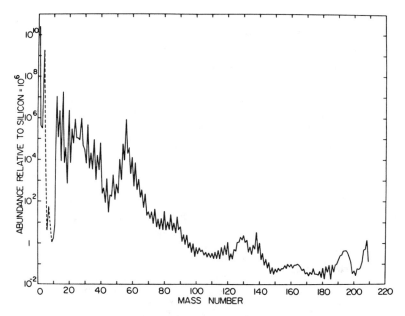

Fig. 16-4 "Cosmic" (or normal) abundance of nuclides relative to that of silicon whose abundance is defined as $10^6$. This is believed to be the initial composition of our solar system. It represents the grand average of the distinct production patterns of all individual supernovae that have contributed to the solar system. The general features of this pattern can be explained using the theoretical production patterns calculated for those processes depicted in Fig. 16-2. (Compiled by A. G. W. Cameron. From Barnes, C. A.; D. N. Schramm; and D. D. Clayton, 1982. *Essays in Nuclear Astrophysics*, p. 32. Copyright © 1982 by Cambridge University Press. Used with permission.)

Some interstellar matter that was to become the solar system was probably in the form of dust. In terms of mass, this component is insignificant (approximately one percent). However, since a significant fraction of the rock-forming heavier elements may have been in the dust, this component is important for the planetary system. This conclusion is inferred from the observation of the present-day interstellar medium. The abundances of elements in the interstellar gas between us and the hot star $\zeta$ Oph. is summarized in Fig. 16-5. The depletion of an element relative to the *cosmic abundance* is plotted against its condensation temperature. It is clear that most refractory elements (i.e., those condensed at high temperatures) are highly depleted in the gas. Since it may be reasonably assumed that the interstellar medium should have a composition not too different from the normal cosmic pattern, the factors of 100 to 1,000 depletion for refractory elements (e.g., Si, Fe, etc.) in the gas are most likely because they are locked up in the dust. Similar observations have been made for the interstellar medium in front of many other stars.

Thus, we can envisage that 5 to 10 Gy after the beginning of the universe there existed an interstellar gas cloud in the disk of the Milky Way Galaxy about 25,000 light years away from the center. The cloud was composed of mostly hydrogen and helium, but also had a total heavy-element concentration of about 2% by mass owing to the

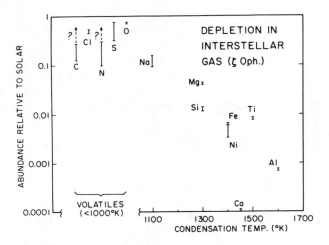

Fig. 16-5 Abundance pattern for the gas between the hot star ζ Oph and Earth (as compiled by Ted Snow, 1981). The abundance of elements relative to their solar value is plotted. The horizontal axis is the temperature at which the elements would condense into solid minerals from a hot gas of cosmic composition. Note the extreme depletion for the refractory elements (i.e., they condense at high temperatures). Since the total interstellar medium is expected to have an abundance approximately cosmic, this depletion in the gas implies that most of the refractory elements are in the dust.

nucleosynthesis in many previous generations of stars. Most rock-forming elements in the cloud resided in the dust. Parts of the cloud were poised to collapse, and a new cluster of stars was about to be born. This was probably the site for the formation of our solar system.

## 16-3 TIME OF FORMATION

Before describing the formation process itself, it is interesting to pin down the time of formation more precisely. There are large uncertainties involved in estimating this time from the beginning of the universe. It is much better to measure backward from the present, that is, the "age" of the solar system can be determined more precisely. The best method is to use radioactive decay to determine the age of crystallization of the oldest rocks in the solar system. The principle of radioactive dating is illustrated in Fig. 16-6 (Chapter 5 by DePaolo also discusses this subject.) $^{87}$Rb atoms decay into $^{87}$Sr atoms with a half-life of 49 Gy. The constituent mineral phases (P1, P2, etc.) of a rock formed $t$ years ago from a homogeneous parcel (e.g., liquid or gas) of material must start from a uniform initial $^{87}$Sr:$^{86}$Sr isotopic ratio but have different $^{87}$Rb:$^{86}$Sr due to chemical fractionation; thus different mineral phases appear in the plot on a horizontal line. Because one $^{87}$Rb atom decays into one $^{87}$Sr atom, the composition for each phase evolves along the 45° dotted lines in Fig. 16-6. The present $^{87}$Sr:$^{86}$Sr ratios for these phases plot

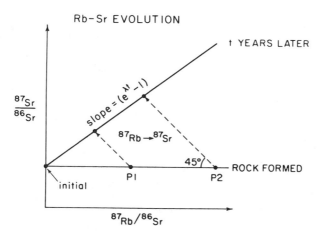

Fig. 16-6 Principle of dating using radioactive nuclides. In a rock formed from a homogenous material, different constituent mineral phases initially all have the identical $^{87}Sr:^{86}Sr$ ratio, although quite different $^{87}Rb:^{86}Sr$ ratios, depending on their chemical properties. The decay of $^{87}Rb$ atoms into $^{87}Sr$ changes the composition along the 45° dotted lines, causing the initially horizontal line to attain steeper slopes. If they remain undisturbed, the compositions for different phases always plot on a straight line, whose slope gives the elapsed time.

on straight lines with a slope that is a function only of the elapsed time $t$. Mathematically, this linear relation can be expressed as

$$\left(\frac{^{87}Sr}{^{86}Sr}\right)_p = \left(\frac{^{87}Sr}{^{86}Sr}\right)_0 + (e^{\lambda t} - 1)\left(\frac{^{87}Rb}{^{86}Sr}\right)_p \qquad (16\text{-}1)$$

where $p$ denotes present values for phase $p$, 0 denotes the initial ratio, and $\lambda = 0.693/$ (half-life) is the decay constant. Therefore, by measuring $(^{87}Sr:^{86}Sr)_p$ and $(^{87}Rb:^{86}Sr)_p$ of mineral phases in a well-preserved rock and fitting a straight line to the data, the elapsed time $t$ can be determined. In this way the time interval between the last crystallization and the present for old rocks may be measured.

The oldest rocks in the solar system are meteorites. The Rb-Sr data for the Guarena meteorite are shown in Fig. 16-7. Indeed, a straight line fits the data for all phases and its slope corresponds to an age of 4.56 ± 0.08 Gy. Many meteorites have been dated using $^{87}Rb$ as well as other radioactive nuclides. The vast majority of them have ages around 4.55 Gy. The oldest lunar rocks also have the same age. Furthermore, in spite of the lack of any preserved terrestrial rocks older than 3.8 Gy, it is possible to obtain an age for the Earth by using simultaneously the two decays $^{238}U \to ^{206}Pb$ and $^{235}U \to ^{207}Pb$. Again, it turns out to be close to 4.5 Gy. Therefore, sizable solid objects have been in existence in the solar system since 4.55 Gy ago, and terrestrial planets formed soon afterward. Since most of the meteorites seem to be solar-system objects, the solar system must have formed earlier than 4.55 Gy ago. But was it born *much* earlier? As we shall see later, there were still some short-lived radionuclides in the solar system when meteorites first

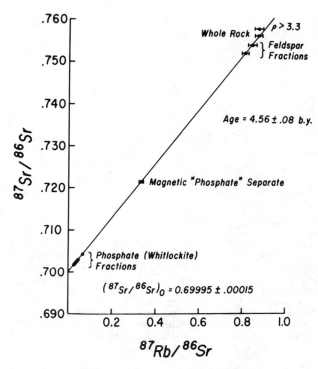

Fig. 16-7  The application of the Rb-Sr dating method to the age of meteorites. Data for the different phases of meteorite Guarena (reported by G. J. Wasserburg, D. A. Papanastassiou, and H. G. Sanz and reviewed by G. W. Wetherill, 1975, Radiometric chronology of the early solar system, *Ann. Rev. Nucl. Sci.*, 25, 283–328) fall on a straight line, and the slope of the line corresponds to an age of 4.56 Gy.

formed. Since these radio nuclides have half-lives only millions of years, the solar system must have received material freshly synthesized in stellar sources within a few million years of meteorite crystallization. The solar system stopped receiving large amounts of such material from other stars soon after it formed. Therefore, the meteorites must have solidified very early, and the solar system must have formed within a few million years of the time 4.55 Gy ago.

## 16-4 CLUES FOR SOLAR-SYSTEM FORMATION

The formation of our planetary system has been a subject of sustained interest, intense study, and heated debate among scientists ever since the 17th century. It is a difficult problem, because so far there are no other observable planetary sytems for comparison. Furthermore, the subsequent evolution of the system is so complex that it is often hard to determine which of the presently observed features are the real clues for the formation process. A good strategy is to look at the "preserved features," that is, those features that are likely to have been relatively unaffected by subsequent processes. In this manner we may hope to "see through" the complications and get at the initial stages. The first set of such features consists of the dynamical ones believed to have something to do with the conservation of angular momentum. The orbits of all the planets lie nearly in a single plane. Their orbital motions are all in the same direction in this plane. The Sun also rotates in the same sense. The axis of this rotation is nearly perpendicular to the plane of planetary orbits. Clearly, the solar rotation and planetary revolution share the same origin. These remarkable regularities have been known for centuries and are the basis for not only the early, but also the present, theoretical ideas.

More recently, other "preserved features" have received increasing attention. These are the regularities in the isotopic composition. Isotopes share similar chemical properties, but differ from one another in mass and in nuclear composition. Most of the solar-system processes, such as chemical reactions, involve such low energies that they only affect the isotopes through the mass differences. Such mass-dependent isotopic fractionations are mostly minor compared with elemental fractionations in chemical processes. Their effects are generally regular and understood. One exception is cosmic-ray–induced nuclear reactions, but their effects are usually feeble even when integrated over billions of years. The other exception is the decay of radioactive nuclides, whose effects are also regular and well understood. Thus, unlike chemical or mineralogical compositions, the isotopic composition is a preserved feature that reflects the effects of presolar and the earliest solar-system processes.

The first remarkable regularity in isotopic composition is its uniformity. Most samples from the Earth, the Moon, and all types of meteorites have identical isotopic composition to a high degree of accuracy for the vast majority of the elements. This is surprising, because the products from each star should have a distinct isotopic imprint resulting from the particular nucleosynthetic processes that took place. There must have been highly effective mixing mechanisms operating after the nucleosynthesis, but before solar-system solidification. This uniform composition is the normal or the "cosmic" composition mentioned before (see Fig. 16-4). The existence of this normal composition provides all long-lived radioactive chronometers with a common starting isotopic composition. Without this resetting of the clocks, it would have been very difficult to date the age of our solar system using radioactive decays.

The second isotopic feature is the presence of heterogeneities in some elements, notably the major element oxygen (see Schramm and Clayton, 1978). Figure 16-8 is a plot of $^{17}O:^{16}O$ versus $^{18}O:^{16}O$ for various objects. A uniform composition would be a single point on this diagram. Mass-dependent processes would fractionate a single composition into a linear array of compositions having a slope of $\frac{1}{2}$. This is because these processes operate on mass differences, and the mass difference between $^{17}O$ and $^{16}O$ is one-half that between $^{18}O$ and $^{16}O$. Indeed, the samples from the Moon, the Earth, and most basaltic and iron meteorites plot close to a line having a slope of $\frac{1}{2}$. However, other classes of objects do not plot this way, indicating that the solar system is heterogeneous. Apparently the mixing was thorough, but not complete. At least two nucleosynthetic components with distinct isotopic compositions are required. In some other elements, the heterogeneity shows regular and distinct patterns that resemble the production pattern of certain nucleosynthetic processes. Therefore, the material from some

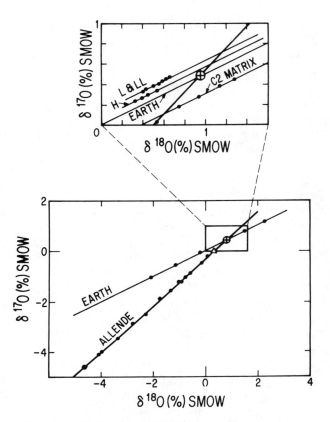

Fig. 16-8 The oxygen isotopic composition of various solar-system objects showing heterogeneity. The variations in $^{17}O:^{16}O$ are plotted against those in $^{18}O:^{16}O$. If the solar system were homogeneous, then its composition would be represented by a single point. At most, the mass-dependent fractionation processes might change the single composition into a line having a slope $\frac{1}{2}$. This is because such processes operate on mass difference between isotopes, and the mass difference between $^{17}O$ and $^{16}O$ is one-half that between $^{18}O$ and $^{16}O$. The Earth, Moon, and most iron and basaltic meteorites do plot on a single line of slope $\frac{1}{2}$ (see line marked "Earth"). However, the refractory phases in carbonaceous meteorites plot on the line marked "Allende." The matrix of type-2 carbonaceous meteorites plot on the line marked "C2 matrix." The upper diagram represents a portion boxed in lower diagram and provides data on other types of meteorites. The L and LL and H, which are different types of ordinary chondrites also do not plot on the line marked "Earth." Therefore, those meteorites and the objects on the "Earth" line could not have been formed from a single homogeneous material. There are thus widespread heterogeneities in the solar system for the major element oxygen.

stellar sources still retains some isotopic memory of its origin in spite of the nearly complete mixing.

The third clue is the important isotope deuterium ($^2$D), the heavy variety of hydrogen (Reeves, 1977). It was produced in the big bang, but is destroyed in stars. The fusion of $^2$D and $^1$H into $^3$He is, in fact, the first nuclear-burning stage in stars. A star like the Sun would exhaust its $^2$D supply in the first few million years. The distribution of $^2$D in the solar system is shown in Table 16-1. Indeed, the Sun has no detectable $^2$D, and the planets have a considerable amount. The Earth and meteorites have high D/H ratio probably because of mass fractionation. The abundance of $^2$D in major planets is similar to that of the interstellar medium. Therefore, the distribution of deuterium implies that the building material of the planets is close to the interstellar medium or the primeval Sun, but not a mature Sun.

TABLE 16-1  Distribution of Deuterium in the Solar System

| Object | $D/H$[a] (ppm)[b] |
|---|---|
| Sun | < 3 |
| Earth | 155 |
| Meteorites | up to 500 |
| Jupiter, Saturn, Uranus | ~20 |
| Interstellar medium | ~10 |

[a] $D/H$ is the ratio between the numbers of deuterium and hydrogen atoms.
[b] ppm = parts per million.

The fourth isotopic clue has to do with those radionuclides whose half-lives are much shorter than the age of the solar system. Of course, they are no longer alive at the present time. Thus, these are termed *extinct nuclides* (see Table 16-2). The extinct nuclides decay quickly, hence the isotopic composition of their daughter elements evolves rapidly with passing time. Therefore, they can be used to constrain the time scales of various events and processes during the formation. Obviously, the shorter the half-life is, the stronger the constraint. Among the extinct nuclides found so far, $^{26}$Al has the shortest half-life, 0.72 My.

TABLE 16-2  Extinct Nuclides

| Nuclide | Half-Life (My) | Found In | Which Source Produced It? |
|---|---|---|---|
| $^{26}$Al | 0.72 | Refractory inclusions | Last |
| $^{107}$Pd | 6.5 | Iron meteorites | Last |
| $^{129}$I | 16 | Variety of meteorites | Both? |
| $^{244}$Pu | 83 | Several meteorites | Next to last |

$^{26}$Al was first discovered in the carbonaceous meteorite Allende, which fell in northern Mexico in February 1969. A fragment of Allende is shown in Fig. 16-9. The nearly spherical white objects are inclusions rich in refractory elements (e.g., Al, Ca, and Ti). A thin section of such an inclusion is shown in Fig. 16-10. The refractory-rich

Fig. 16-9 Fragment of the Allende meteorite. The white, nearly spherical objects are the refractory inclusions, probably the most primitive solids in the solar system. Large oxygen-isotopic heterogeneities and the evidence for $^{26}$Al were discovered in them. The largest inclusion in the center is inclusion WA, whose composition is given in Fig. 16-11. (Photograph courtesy of J. Chen.)

Fig. 16-10 A thin section of a refractory inclusion illuminated from behind. This photo is taken with the section sandwiched between two orthogonally oriented polarizers to show the grains of different mineral phases. Note that these grains are interlocking, well-developed crystals of up to 1 mm in size, reminiscent of minerals formed from a liquid melt. This type of inclusion is not loose aggregates of numerous interstellar dust grains, which are believed to be very tiny, at most 0.001 mm in size. (Photograph courtesy of G. J. Wasserburg.)

Allende inclusions have chemical and mineralogical compositions resembling that of the earliest solids expected to condense from a gas of cosmic composition. Also, they have been shown to be among the earliest objects to solidify using the $^{87}$Rb:$^{87}$Sr chronometer. It is in these seemingly primitive inclusions that the evidence for $^{26}$Al was first discovered.

How does one "prove" in a laboratory today that some radionuclides were alive in a rock 4.6 Gy ago, but have since then become extinct owing to complete decay? One needs to show two things. First, one has to demonstrate that there exists an excess in the abundance of the daughter nuclide above its normal cosmic proportion. Second, one must show that this excess correlates with the abundance of a stable isotope of the parent element. In other words, atoms of the excess daughter nuclide behave as if they have a chemical affinity to the parent element. For the case of $^{26}$Al, which decays into $^{26}$Mg, we thus need to show (1) $^{26}$Mg excess exists and (2) the $^{26}$Mg excess correlates with $^{27}$Al, the stable aluminum isotope, that is, the distribution of excess $^{26}$Mg atoms among different coexisting mineral phases in the rock mimics that of the $^{27}$Al atoms. In this way the logical interpretation would be that these excess Mg atoms were actually Al atoms when these mineral phases first formed, and only the subsequent decay has changed them into Mg atoms.

More rigorously, the total number of $^{26}$Mg atoms measured today in a mineral phase should equal the initial number of $^{26}$Mg atoms plus the initial number of $^{26}$Al atoms that have since decayed into $^{26}$Mg. Dividing these quantities by the number of $^{24}$Mg atoms, we arrive at the basic equation governing the evolution of the $^{26}$Al:$^{26}$Mg chronometer:

$$\left(\frac{^{26}Mg}{^{24}Mg}\right)_p = \left(\frac{^{26}Mg}{^{24}Mg}\right)_0 + \left(\frac{^{26}Al}{^{27}Al}\right)_0 \left(\frac{^{27}Al}{^{24}Mg}\right)_p \qquad (16\text{-}2)$$

where the subscript $p$ denotes values for a mineral phase $p$ measured today, and 0 designates the initial values at the crystallization time of the rock. The implicit assumptions underlying Eq. (16-2) are (a) the reservoir from which the rock formed had a uniform initial isotopic composition for both Mg and Al, and (b) the rock has remained undisturbed since its formation time. In a plot of $(^{26}Mg:^{24}Mg)_p$ versus $(^{27}Al:^{24}Mg)_p$, Eq. (16-2) represents a straight line with the slope and the intercept corresponding to the initial $^{26}$Al:$^{27}$Al and $^{26}$Mg:$^{24}$Mg, respectively. Therefore, by measuring $^{26}$Mg:$^{24}$Mg and the $^{27}$Al:$^{24}$Mg in the mineral phases of a rock, we can hope to obtain a straight-line correlation between the excess $^{26}$Mg and the $^{27}$Al. Such a correlation would be strong evidence that some $^{26}$Al was still alive when this rock solidified.

The method described above has been applied to Allende inclusions. Coexisting mineral phases (with grain size up to 1 mm) are clearly visible in the Allende inclusion shown in Fig. 16-10. Grains of different phases were picked, separated, processed, and analyzed. The data are shown in Fig. 16-11. First, note that there exist large $^{26}$Mg excesses ($\delta$ $^{26}$Mg) above normal of up to 8.5 percent. Second, the excess $^{26}$Mg correlates linearly with $^{27}$Al for phases anorthite, melilite, spinel, and fassaite. A straight line almost passes through the data for all four minerals, implying an initial $(^{26}Al:^{27}Al)_0$ of $5 \times 10^{-5}$ (i.e., initially, there were 50 $^{26}$Al atoms for every 1 million $^{27}$Al atoms). Such relationships have since been observed for at least six inclusions of the Allende meteorite as well as in similar inclusions from the Leoville and Murchison carbonaceous meteorites. Therefore, $^{26}$Al seems to have been present when these primitive objects first formed.

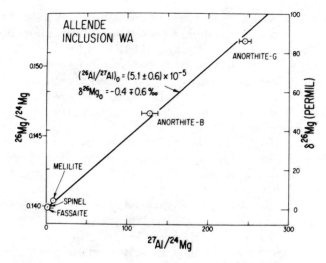

Fig. 16-11 The evidence for the presence of $^{26}$Al in Allende inclusion WA when it crystallized. $^{26}$Mg:$^{24}$Mg is plotted against $^{27}$Al:$^{24}$Mg for four coexisting constituent mineral phases of WA. There are clearly large $^{26}$Mg excesses, displayed as $\delta^{26}$Mg ($\delta^{26}$Mg = [($^{26}$Mg:$^{24}$Mg)$_{sample}$ − ($^{26}$Mg:$^{24}$Mg)$_{normal}$]/ ($^{26}$Mg:$^{24}$Mg)$_{normal}$), where $\delta$ gives the ratio between excess $^{26}$Mg atoms and normal $^{26}$Mg atoms in the sample. There is a strict correlation between $\delta^{26}$Mg and $^{27}$Al:$^{24}$Mg in four minerals, with drastically different $^{27}$Al:$^{24}$Mg ratios that resulted from the chemical fractionation during crystallization; that is, the distribution of excess $^{26}$Mg atoms mimics that of $^{27}$Al as if these Mg atoms had a chemical affinity to Al when these minerals crystallized. The logical interpretation is that these excess $^{26}$Mg atoms were $^{26}$Al at the time of crystallization and became $^{26}$Mg during subsequent decay of $^{26}$Al. (From Lee, T., D. A. Papanastassiou, and G. J. Wasserburg, 1977, $^{26}$Al in the early solar system, fossil or fuel?, Astrophys. J., 201, L107.)

The presence of $^{26}$Al requires the existence of stellar sources to produce $^{26}$Al not more than several million years prior to the crystallization time of Allende inclusions. More quantitatively, we can relate the initial ($^{26}$Al:$^{27}$Al)$_0$ ratio to the production ratio ($^{26}$Al:$^{27}$Al)* in a star:

$$\left(\frac{^{26}\text{Al}}{^{27}\text{Al}}\right)_0 = \frac{^{26}\text{Al}^* e^{-\lambda\Delta}}{^{27}\text{Al}_0} = \left(\frac{^{26}\text{Al}}{^{27}\text{Al}}\right)^* \left(\frac{^{27}\text{Al}^*}{^{27}\text{Al}_0}\right) e^{-\lambda\Delta} \tag{16-3}$$

where $\lambda$ is the decay constant, and $\Delta$ is the interval between the production and the crystallization. Note that $^{27}$Al*/$^{27}$Al$_0$ is the fraction of total $^{27}$Al that came from the source of $^{26}$Al. This term cannot be larger than unity. It is difficult to produce $^{26}$Al abundantly in astrophysical sources. The most extreme estimate for ($^{26}$Al:$^{27}$Al)* in any nucleosynthetic models so far is 1. Thus, the longest possible time interval that is still consistent with the observed $(^{26}Al:^{27}Al)_0 \simeq 5 \times 10^{-5}$ is 9 My. This happens when $(^{26}Al:^{27}Al)^*$ is 1 and $^{27}Al^*:^{27}Al_0$ is 1. That is to say, all the $^{27}Al$ came from the last source, a highly unlikely case. With more reasonable choices, say $(^{26}Al:^{27}Al)^* = 0.01$ and $^{27}Al^*:^{27}Al_0 \simeq 0.01$, then $\Delta$, the interval between production and crystallization, becomes approximately 1 My.

In addition to $^{26}Al$ there is also good evidence for the presence of three other extinct nuclides in the early solar system (Table 16-2). $^{244}Pu$ and $^{129}I$ have been known since the 1960s, and their record has been found in a wide variety of meteorites. Evidence for $^{107}Pd$ has recently been found in several iron meteorites. Note that iron meteorites, in contrast to carbonaceous meteorites, are the result of chemical differentiation that segregated the metallic body from the silicates in a melt during a heating event. Such melt segregation had to take place in bodies of the size of at least a few kilometers across. Thus, the $^{107}Pd$ was still present when chemically differentiated bodies of at least the size of an asteroid already formed.

Did all extinct nuclides come from the last source that made the $^{26}Al$ about 1 My before solidification? The answer is that $^{107}Pd$ probably has to have come from the last source, whereas $^{129}I$ could be from it too but may also contain contributions from earlier events. In the case of $^{244}Pu$, most of it must have come from an earlier event. The evidence for this comes from the negative result of searching for yet another extinct nuclide, $^{247}Cm$ ($t_{1/2} \simeq 17$ My). A well-known conclusion from the nucleosynthetic theory of transuranic elements is that $^{247}Cm$ and $^{244}Pu$ should be coproduced and in comparable amounts, that is, $(^{247}Cm:^{244}Pu)^* \simeq 1$. Therefore, if all $^{244}Pu$ came from the $^{26}Al$ source, then we would expect that a comparable amount of $^{247}Cm$ also came along with it and should measure $(^{247}Cm:^{244}Pu)_0 \simeq 1$ for the early solar system. The observations, however, gave an upper limit 0.06 for $(^{247}Cm:^{244}Pu)_0$. This means that at most 6 percent of the $^{244}Pu$ could have come from the last source and the remaining bulk had to be from an earlier source. This next-to-the-last source was probably separated from the last one by a time interval on the order of 100 My.

## 16-5 NEBULA AND CATASTROPHE THEORIES

The dynamical regularities imply a common origin of solar rotation and the planetary revolution. One way for this to have been accomplished is for the planets and the Sun to have been formed from a rotating interstellar cloud (see Chapter 3 by Lin). The collapse of such a cloud would form a disk and a central core. The core becomes the Sun, and planets grow in the disk. In this way the solar rotation and the planetary revolution are simply different manifestations of the same rotation of the original cloud. This is the "nebula theory," which was advocated and improved by great scholars like Descartes, Kant, and Laplace centuries ago. However, in its simplest form, this theory cannot explain the distribution of angular momentum in the solar system. In a rotating disk, the massive central core (the Sun) is expected to dominate the angular-momentum distribution. However, the present solar rotation is far too slow to dominate the angular-momentum distribution constituting only 2 percent of the solar system total. Because of this problem, an alternative "catastrophe theory" was proposed by Chamberlin and Moulton in the beginning of this century. It soon enlisted the support of and refinements by famous scientists like Jeans and Jeffreys. In the catastrophe theory, the tidal interaction during a near encounter between a passing star and a preexisting Sun caused the latter to rotate and to eject material from which the planets condensed. This theory can account for the present angular-momentum distribution and was quite popular during the first half of this century. However, it does have problems in explaining why all planetary orbits are so circular. Furthermore, the ejected material would be too hot to condense directly into individual objects. Instead, it should form a gaseous nebula, thus blurring the distinction between the catastrophic and the nebula theories.

Do isotopic features help us to distinguish between these theories? The isotopic homogeneities and heterogeneities can be explained in both. In the nebula theory, the collapse of a cloud should have been a highly turbulent process; turbulence could be the mixing mechanism producing homogeneity. The process is described in detail in Chapter 3 by Lin. The heterogeneities are probably due to the interstellar dust. For example, if the dust has a different isotopic composition than the gas and the gas-to-dust ratio is nonuniform, isotopic heterogeneities could arise. In the catastrophe theory, the ejected material is expected to be homogeneous, because it all came from the Sun, whose outer parts are homogeneous. The heterogeneity would then require that some of the matter must come from the other source, the passing star, assuming that it has a different composition than the Sun.

The other isotopic clues—extinct nuclides and the deuterium distribution—tip the balance strongly in favor of the nebula theory. The presence of extinct nuclides clearly requires that the solar system solidified from material containing nuclides made only a few million years earlier. Thus, in the catastrophe theory, the encounter could not happen between two stars of arbitrary ages. One of the two must be, at most, a few million years old, and such stars are rare. Since stellar collisions and near collisions are extremely rare in the first place, the catastrophe theory is seen to require a highly unlikely event.

The deuterium distribution tells essentially the same story. If the catastrophe theory is to remain viable, then one of the parties must be very young. One simply cannot form deuterium-rich planets out of stars that have exhausted their deuterium supplies.

Finally one may emphasize that there is no need to specially manufacture a nebula by the improbable event of a grazing encounter. Observationally young stars are generally surrounded by concentrations of gas and dust, and theoretically they are expected to be cocooned by the material left from their birth. Young stars, then, probably all have their own associated nebulae as a natural consequence of their birth. Therefore, the nebula theory is more plausible than the catastrophe theory and, hence, is widely accepted.

In the nebula theory, the formation of the solar system may be divided into two stages (see Fig. 16-12). Consider a protosolar interstellar cloud consisting of gas and dust with an initial density about $10^{-20}$ g $\cdot$ cm$^{-3}$. Various mechanisms could have compressed this cloud to the point where its self-gravity overcame its internal pressure and the collapse was triggered. The triggering mechanism was probably the compression by a shock wave, which could have been set up by a supernova explosion, galactic spiral density wave, HII region expansion, or cloud-cloud collision. The collapse proceeded rather fast without much opposition from the internal pressure during the first 10 orders of magnitude in-

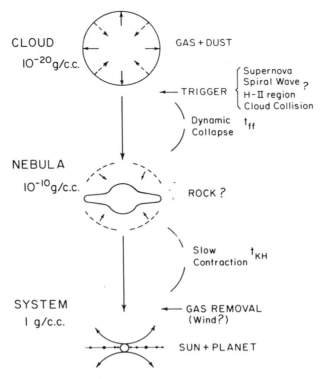

Fig. 16-12 A schematic outlining the nebula theory. A protosolar cloud of interstellar gas and dust with an initial density of $10^{-20}$ g · cm$^{-3}$ was triggered into collapse. After a 10-order-of-magnitude increase in density, it formed a solar nebula on the free-fall time scale of approximately 1 My. Subsequently, solids condensed and accumulated to form planets with a density around 1 g · cm$^{-3}$, and the gas was removed, perhaps by a strong solar wind.

crease in density. The time span in this first dynamical stage would have been roughly the free-fall time scale ($t_{ff}$), given by approximately $10^3/\rho^{1/2}$ seconds[1] where $\rho$ is the initial density. Thus for $\rho \simeq 10^{-20}$ g/cm$^3$, $t_{ff}$ is about 3 My. When the density reached $10^{-10}$ g · cm$^{-3}$, the heat released by the gravitational contraction began to be trapped. The internal pressure built up to counterbalance the gravity and the contraction slowed down. Here we enter the second stage where a solar nebula formed. From then on, further

[1] The free-fall time scale ($t_{ff}$) for a uniform sphere of density $\rho$ may be derived as follows:

$$t_{ff} \simeq \frac{R}{V_{ff}}$$

where $R$ is the radius and $V_{ff}$ is the velocity of the free-fall collapse. $V_{ff}$ may be obtained by equating the total final kinetic energy to the total initial gravitational energy, that is, $\frac{1}{2} m V_{ff}^2 \simeq GmM/R$, where $M$ is the total mass of the sphere and is equal to $\frac{4}{3}\rho\pi R^3$. Thus, we arrive at $t_{ff} = R/(\frac{8}{3}\pi G\rho)^{1/2} = 10^3/\rho^{1/2}$ · s.

contraction was governed by the radiative loss of energy from the surface of the nebula. The time scale for this second, quasi-static phase is the *Kelvin-Helmholtz time scale* ($t_{KH}$) and is about 30 My, using the rate of energy loss of the present Sun. However, since the protosun was probably brighter and larger, $t_{KH}$ was probably much shorter. Sometime during this stage, sizable planetary objects formed, and eventually the gas was removed, possibly by a wind from the Sun, leaving behind our solar system, whose members now have typical density of approximately 1 g · cm$^{-3}$.

Most scientists currently accept the nebula theory, but there is no consensus about the details of the theory. One of the versions is summarized in Cameron (1975) and another by Lin in Chapter 3. Critical questions such as the triggering mechanism of the collapse, the mass of the nebula, the transport mechanism of the angular momentum, the timing of the gas removal, and so forth, remain controversial. In most cases there is no lack of mechanisms; the difficulty lies in choosing the correct one. This problem is compounded by the fact that the process is so complicated that it is not yet possible to follow theoretically the formation of a rotating Sun from the triggering of the collapse to the final ignition of its hydrogen nuclear fuel in a satisfactory manner, let alone the formation of the planets and satellites. Because of the absence of a comprehensive theoretical framework, it is difficult to test the various mechanisms. An example is the old puzzle of angular-momentum distribution. Two major mechanisms have been proposed. Both involve transporting the angular momentum outward until some fast-rotating material at the rim of the nebula is lost. In the theory favored by Lin in Chapter 3, the transport is accomplished by turbulent eddies that pass the angular momentum outward from one eddy to the next through the viscous coupling between them. Another theory does this with magnetic force. One end of the magnetic field line is anchored on the fast-rotating central core; the other, on the slow-moving rim. The differential rotation twists the field, generating a torque to spin down the center and spin up the rim. Both mechanisms seem plausible, but to demonstrate which one would operate in the solar nebula is difficult. Indeed, the correct version may be a combination of the two mechanisms.

Instead of detailing the various possible processes here, it is perhaps more valuable to discuss limitations on these processes imposed by the observed isotopic features, particularly the extinct nuclides. The presence of $^{244}$Pu, but not $^{247}$Cm, implies that the bulk of the solar $^{244}$Pu was produced several hundred million years prior to the solar-system formation. This time scale is close to the time interval between two successive passages of the galactic spiral arm in the Milky Way Galaxy. Since one possible trigger for star formation is the shocking of the local interstellar medium by the passage of the galactic density wave, it is conceivable that a burst of star-forming activities occurred at that time. Perhaps one of the stars in that burst produced the $^{244}$Pu. Then this particular region quieted down, awaiting the passage of the next spiral-arm density wave to trigger the next burst of star formation that included our Sun and the source star for other extinct nuclides.

The presence of $^{26}$Al as well as $^{107}$Pd requires their source to have been active within a few million years before the solidification of refractory inclusions in meteorites. Was this source a supernova that triggered the collapse of the protosolar cloud? Such a profound result would indeed be the case if the following are true: (1) The crystallization of meteoritic inclusions is part of the general solidification process of the solar system; (2) the $^{26}$Al came from a supernova; and (3) the proximity of this supernova was not simply an accident. The first condition is likely to be true, because these inclusions have isotopic compositions identical to the cosmic compositon for the great majority of their constituent elements. Furthermore, their $^{87}Sr:^{86}Sr$ ratios are close to the best estimate of the initial $^{87}Sr:^{86}Sr$ for the solar system, indicating that they could not have been formed much earlier. Condition 2 is probably also true, because there are a number of zones in a supernova or its progenitor star whose physical conditions favor the production of $^{26}Al$. However, it is certainly not impossible to produce $^{26}Al$ in other types of astrophysical sources as novae or red giants. The third condition is more an article of faith. It is true that the occurrence of a supernova in the vicinity of the Sun already starting to form and near the time of the solidification would be quite a spatial and temporal coincidence; but accidents do take place. At the present time we may conclude that the supernova trigger model for the solar-system formation may very well be correct, but not necessarily so. If the supernova trigger model is correct, then what is depicted in Fig. 16-13 may actually have happened. There existed an interstellar cloud complex that contained enough mass to make an entire cluster of hundreds of stars, including the protosolar cloud. A massive star was born in this complex and rapidly evolved and ended its life in a super-

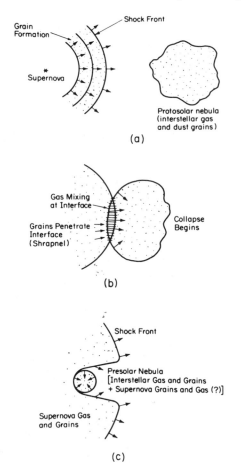

Fig. 16-13 One possible scenario of the supernova trigger hypothesis. A supernova exploded near a protosolar cloud. The compression of its shock wave triggered the collapse of the cloud and injected freshly synthesized nuclides including $^{26}Al$. (From Lattimer, J., D. N. Schramm and L. Grossman, 1978, Condensation in supernova ejecta and isotopic anomalies in meteorites, *Astrophys. J., 219*, 230.)

nova explosion. The shock wave from this supernova hit the protosolar cloud. It compressed the cloud and triggered its collapse and also injected freshly synthesized material including $^{26}$Al into the cloud.

If the formation of our solar system was indeed triggered by a supernova that produced the $^{26}$Al, it becomes interesting to compare the time interval of $\sim 1$ My between the triggering and the solidification of centimeter-sized objects with the free-fall time scale for the cloud collapse. Since these two time intervals are comparable, the accretion of small dust into centimeter-sized aggregate and the crystallization of such aggregates into coarse-grained inclusions observed today must be a rapid process. It must have happened either during the collapse or immediately following the formation of the nebula.

The $^{107}$Pd may provide an additional constraint. This nuclide was apparently still alive when a melted large body (with a radius of several kilometers) cooled down to form the metals and sulfides that now comprise the iron meteorite. The heat source for the melting was likely to have been $^{26}$Al, although other mechanisms are possible. If so, then there are only 4 My available for the large bodies (parent planets of iron meteorites) to accrete from centimeter-sized objects (inclusions); otherwise the concentration of $^{26}$Al in the solar system would have fallen to too low a level to cause melting. Therefore, the accretion of sizable planetary bodies may also have proceeded on the time scale of millions of years.

## 16-6 CONCLUSION

Obviously, there is still a lot about the formation of our solar system that we do not know. But what we do know still represents an impressive monument to the ability of humanity to reflect upon its own cosmic heritage. Let us recount: We know that the Sun was a latecomer in the galaxy by billions of years, but also has been around for exactly 4.55 Gy. We know that many stellar sources contributed to our system, and we even have direct evidence for at least the last two of them. We suspect that the last supernova may have triggered the collapse of the protosolar cloud. We know that the planets must have formed around a very young Sun, perhaps even still being formed, most likely in a nebula consisting of the leftover interstellar gas and dust. We also know that the solidification process in the solar system is rapid. Centimeter-sized objects were able to crystallize during or immediately after the collapse of the protosolar cloud within about 1 My after the initiation of the collapse. The further accumulation to larger bodies of at least a few kilometers in size was probably also a rapid process taking, at most, several million years more. The exact manner in which the nebula evolved, the solid formed, and the planets accumulated are discussed in other chapters.

## 16-7 ACKNOWLEDGMENTS

Incisive comments from Dr. Alan Boss of the Carnegie Institution on a draft of this paper are gratefully appreciated. Dr. Ted Snow kindly provided a compilation of the elemental abundance pattern for the interstellar gas in front of the hot star $\zeta$ Oph. This work is supported in part through the grant AST8115972 from the National Science Foundation.

## 16-8 REFERENCES AND ADDITIONAL READINGS

Cameron, A. G. W., 1975, The origin and evolution of the solar system, *Sci. Am., 233*, 33-41.

Cameron, A. G. W., 1982, Elemental and nuclidic abundances in the solar system, in *Essays in Nuclear Physics*, (C. A. Barnes, D. D. Clayton, and D. N. Schramm, eds.), Cambridge Univ. Press, New York, 23-43.

Reeves, H., 1977, The origin of the solar system, *Mercury, 6*, 7-14.

Schramm, D. N., 1974, The age of the elements, *Sci. Am., 230*, 69-77.

Schramm, D. N., and R. N. Clayton, 1978, Did a supernova trigger the formation of the solar system?, *Sci. Am., 239*, 124-139.

Trimble, V., 1977, Man's place in the universe, *Am., Scientist, 65*, 76-86.

Weinberg, S., 1977, *The First Three Minutes*, Basic Books, New York.

Margaret G. Kivelson
Department of Earth and Space Sciences, and
Institute of Geophysics and Planetary Physics
University of California
Los Angeles, California 90024

# 17

# FUTURE PLANETARY EXPLORATION

# ABSTRACT

Other chapters in this volume have reviewed the evidence from which our current models of solar-system origin and evolution have emerged. Data to test and improve the models will be collected, some in laboratories or astronomical observatories, some through computer calculations, some through further measurements on spacecraft. This chapter describes some relevant missions either in development or under consideration, and comments on their relevance to scientific issues. The chapter concludes with some remarks on long-range planning.

## 17-1 INTRODUCTION

Models of the origin and evolution of the solar system have become increasingly credible in the last decade, largely because they are increasingly constrained by data (see Chapters 1, 5, and 16 by Wasson and Kivelson; DePaolo; and Lee, respectively). Much, though not all, of the new information was obtained from planetary spacecraft. We learned about the complex structure of Saturn's rings. We discovered the volcanoes of Io, saw images of some parts of the surface of Venus, found unanticipated geological structures on Mars. We learned that the rare-gas isotopic ratios did not vary as predicted from one to another of the terrestrial planets. The list of recent discoveries is long, and we will not try to complete it. Rather, we remark that as we learn more, new questions arise or old ones become more pressing. Some important questions can still be answered through further spacecraft exploration of the solar system. What plans exist at the moment for such ongoing explorations?

Fortunately for the further development of solar-system science, the era of planetary exploration will continue. Some data will be obtained from spacecraft already en route to their destinations, some from spacecraft now being constructed and tested. Beyond that, uncertainty increases, but numerous planetary missions have been proposed and studied, and some of them will doubtless be approved and funded. This chapter describes some of the future missions, placing them in the context of a long-range program and summarizing the types of measurements to be made and the expected scientific return.

Before describing the future, let us take a brief look at the past, to help place the currently planned program in perspective. Figure 17-1 illustrates the history of U.S. space launches from 1957 to 1983. The diagram is intended to resemble a histogram and not actually to be read. The lists include both civilian and military launches, and both successful and unsuccessful launches. The peak activity occurred in 1966. The failures are listed on the diagram below the heavy broken line, which steps downward in a satisfying way, reaching the bottom to indicate there have been no failures in the most recent years. Stars on the diagram identify the entries with planetary payloads, and for that subgroup, peak activity also occurred in 1966. Naturally, one cannot gauge the scientific value of a program of planetary exploration only on the basis of the annual number of launches, and the decreasing rate of launches of planetary spacecraft has been partially offset by the increased sophistication of their payloads. Nonetheless, for proponents of an aggressive

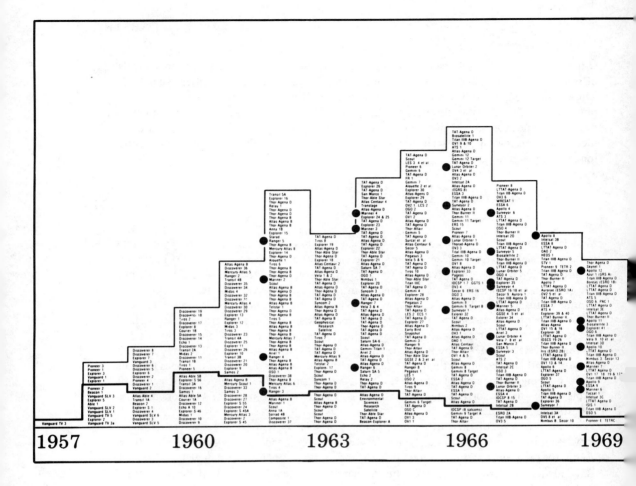

program of planetary exploration, it is disappointing to recognize that there will be no planetary spacecraft launched by the United States before 1986, when the *Galileo* and International Solar Polar missions will set off toward Jupiter.

Only U.S. launches are listed in Fig. 17-1, but the USSR as well has had a program of scientific exploration since 1959. Table 17-1 summarizes the international planetary program through 1980. The USSR has pursued a vigorous program of exploration of the terrestrial planets and the Moon, but only the United States has sent missions to the outer planets.

Planetary exploration proceeds naturally through steps of increasing complexity. The first spacecraft fly past the planet, recording images and sensing the environment for a brief interval of time. This *reconnaissance phase* is followed by an *exploration phase*, in which a spacecraft is placed into a planetary orbit, allowing detailed mapping and prolonged surveillance to take place. Landers may be included in this phase. A natural evolution leads to a phase of *intensive study*, in which specific questions are investigated with specially designed instrumentation, often requiring surface measurements. Table

# U.S. SPACE LAUNCHES
# 1957 - 1983

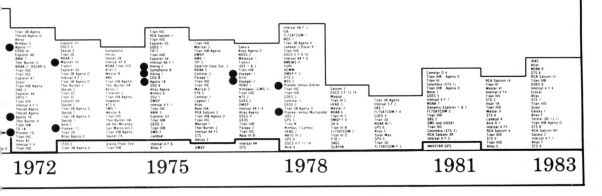

Fig. 17-1 Histogram of U.S. space launches from 1957 to 1983. As noted in the text, the diagram has been reduced in size and is to be examined as a histogram with individual entries too small to be read. (Adapted from *TRW Space Log*, 1982-83.)

17-2 categorizes past U.S. missions and some future missions with the United States or other sponsors according to the level of exploration they represent. Only for the Moon and Earth have investigations appropriately designated "intensive" been carried out. Even for the Moon, lacunae remain. For example, although lunar samples were collected and returned for analysis and near-equatorial regions were mapped with good resolution, the Moon's polar regions remain unexplored. Many aspects of lunar evolution, such as the record of its magnetic history, remain ambiguous partly because of the absence of measurements far from the equator.

Many of the entries in Table 17-2 are speculative, but some of the entries are unambiguous. They refer to past missions. Other entries refer to future plans. Most of them refer to encounters that are reasonably certain to take place. For example, *Voyager 2* is already en route to Uranus and will arrive there in January 1986. Less definite is the indicated encounter with Neptune, and the brackets are used to represent greater uncer-

TABLE 17-1  International Planetary Exploration Program through 1980

| | Moon | | Mars | | Venus | | Outer Planets | |
|---|---|---|---|---|---|---|---|---|
| | USSR | USA | USSR | USA | USSR | USA | USSR | USA |
| 1958 | xxx | | | | | | | |
| 1960 | | | | | | | | |
| 1962 | x | ••• | x | | | x | | • |
| 1964 | | •• | o | • | | o | | |
| 1966 | oxxxx / xxxxx | •• / △△▲▲ / △△△△▲▲▲▲ | | | | xx | | |
| 1968 | oox / ox | ▲□ | | | | x | | • |
| 1970 | oxx / xx | □□□□ / □ | | •• | | xx / x | | |
| 1972 | x / x | □□ / □□ | xxxx | • | | x | | • |
| 1974 | xx | | xxxxxx | | | | | • |
| 1976 | x | | | △△ | xxxx | | | |
| 1978 | | | | | xxxx | △△ | | △ / △ |
| 1980 | | | | | | | | |

| | | | | |
|---|---|---|---|---|
| x Luna (24)[a] | o Zond (1) | o Zond (1) | • Pioneer (2) |
| o Zond (5) | x Mars (11) | x Venera (16) | △ Voyager (2) |
| • Ranger (7) | • Mariner (4) | • Mariner (3) | |
| △ Lunar Orbiter (5 + 1) | △ Viking (2) | △ Pioneer Venus (2) | |
| ▲ Surveyor (7) | | | |
| □ Apollo (10) | | | |

[a]Parenthetical numbers indicate the number of separate spacecraft contributing to the designated program. *Explorer 35* is grouped with *Lunar Orbiters*. The 1974 *Mariner* also went to Mercury.

tainty. The *Galileo* and *ISPM* spacecraft are still in development stages at present. These missions are described in Secs. 17-2 through 17-4 of this chapter. *ISEE 3* is an existing spacecraft which was, for several years, used as a monitor of the solar wind. Originally part of a program of terrestrial magnetospheric investigations at Earth, *ISEE 3*, rechristened *ICE (International Cometary Explorer)*, was redirected to intercept the short-period (6½-y) comet Giacobini-Zinner. The spacecraft has no imaging capability, but it passed through the tail of the comet and obtained data on the plasma and field environment, including some assessment of ion composition. Also it provided evidence on the hazards that await spacecraft in the dusty regions surrounding a comet.

Several entries in Table 17-2 are bracketed to indicate that they are not yet funded. Not all these missions will be undertaken, but they are included as among the most probable of the projects being considered for future funding by the United States or European nations. They are discussed further in Sec. 17-6.

TABLE 17-2  Past U.S. Missions and Some Future Missions (Various Sponsors) Categorized by Phase of Exploration Achieved[a]

|  | Reconnaissance | Exploration | Intensive Study |
|---|---|---|---|
| Mercury | Mariner 10 | | |
| Venus | Mariner 2, 5, 10 | Pioneer Venus  [VRM] | |
| | | Venera 15, 16 | |
| Earth | | | |
| Moon | Ranger (7 s/c) | Surveyor (7 s/c) | Apollo (10 s/c) |
| | | Lunar Orbiter (5 s/c) | |
| Mars | Mariner 4, 6, 7 | Mariner 9, Viking 1, 2 | |
| Jupiter and Satellites | Pioneer 10, 11   Voyager 1, 2 | Galileo | |
| | ISPM | | |
| Saturn and Titan | Pioneer 11   Voyager 2 | | |
| Uranus | Voyager 2 | | |
| Neptune and Pluto | [Voyager 2] | | |
| Asteroids | [Agora], [Galileo], [CRAF] | | |
| Comets | Giotto, Vega, Sakigake, | | |
| | ISEE 3, Ampte, [CRAF] | | |
| Sun | Skylab, SMM | ISPM, [SOHO] | |

[a]Brackets imply uncertainty as for names of projects not yet accepted for funding. It is not yet certain if the *Voyager 2* spacecraft will be sent to Neptune after encountering Uranus.

## 17-2  THE GALILEO MISSION TO JUPITER

The first mission to Jupiter (*Pioneer 10* and *11*) revealed that this giant planet and its satellites are even more complex and interesting than had been anticipated (see Chapters 10 and 11 by Ingersoll and Stevenson, respectively). Therefore, before *Voyager* had reached Jupiter, arguments for a yet more sophisticated mission were being formulated and advanced (e.g., National Aeronautics and Space Administration, 1975). The new mission would be designed to go beyond the phase of pure reconnaissance. It would require an instrumented probe to penetrate beneath the cloud tops and make measurements in regions opaque to radiation. In addition, it would require placing a spacecraft into orbit around Jupiter, so that time variations of the dynamic atmosphere and magnetosphere could be monitored for intervals much longer than the few days of the *Pioneer* or *Voyager* flybys. It would follow a trajectory that would permit close-up views of the major moons, so that these "terrestrial planets" (see Chapter 4 by Kaula) with their unique compositions, histories, and plasma environments could be better understood.

The mission that emerged from the planning process was initially called *Jupiter Orbiter Probe* but is now, fittingly, called *Galileo* after the seventeenth-century Italian scientist who first identified the major moons of Jupiter. Work on the project began in 1977. Originally the spacecraft was to have been launched in 1981 and would have arrived at Jupiter in 1984. Delays, largely associated with delayed development of launch vehicles, have necessitated several postponements. Present plans call for a launch in May 1986. The spacecraft, attached to a *Centaur* rocket, is to be placed in a low-altitude Earth orbit by the space shuttle. Then the *Centaur* engines will be triggered by remote command, and *Galileo* will be shot free of Earth's gravity on a trajectory that will bring it to Jupiter in December of 1988.

Approximately 3 months before closest approach to Jupiter, *Galileo* will separate into two unequal parts. The larger part, the *Orbiter*, will fly very close to the moon Io on its inbound trajectory and then will be placed into orbit around Jupiter. Months later, the *Orbiter* will again approach the planet, this time flying very close to the moon Ganymede. Repeated orbits in a 20-month period will permit 11 close passes of Jupiter's moons, extensive investigation of its magnetosphere, and prolonged monitoring of its atmosphere by spectroscopic means.

The smaller part of the spacecraft, the *Probe*, will follow a very different trajectory. Its path will take it into Jupiter's atmosphere near the equator. The *Probe* will enter at a speed of 48 km $\cdot$ s$^{-1}$, with a heat shield to protect it from burning up despite the hot-gas layer ($T > 8000°K$) that will develop as it decelerates. With the heat shield jettisoned and a parachute to provide a relatively leisurely continued descent, *Probe* instruments (see Table 17-3) will measure atmospheric composition and structure, wind speeds, temperature, and so forth. The relative abundance of He will be measured and compared with solar abundance, so that various accretion models can be tested. Data will be transmitted by radio signals from the *Probe* to the *Orbiter* to be relayed back to Earth. At a pressure level of about 20 Earth atmospheres, high atmospheric density will prevent further transmission, if high temperatures have not caused earlier termination.

TABLE 17-3  Scientific Investigations on *Galileo*

| *Orbiter Investigations* | *Probe Investigations* |
|---|---|
| Radio science | Net flux radiometer |
| Imaging | |
| Infrared | Lightning investigation |
| Plasma probe | |
| Plasma wave detector | Mass spectrometer |
| Ultraviolet spectrometer | |
| Magnetometer | Nephelometer |
| Photopolarimeter radiometer | |
| Energetic particle detector | Atmospheric structure experiment |
| Dust detector | Helium abundance interferometer |

The *Orbiter* investigations (see Table 17-3) include some that make local measurements of fields and particles (plasma detector, magnetometer, etc.) and others that observe either Jupiter or its satellites from great distances. The latter type of observation is best done from a stable (nonspinning) spacecraft, but various advantages of a spinning spacecraft, such as stability, are also desirable. The *Galileo Orbiter* accommodates both requirements through a dual-spin design. One part of the spacecraft spins about an axis pointed toward Earth. The other part is fixed in space and provides a steady platform for imaging and other spectroscopic investigations.

The imaging is done by an instrument that uses CCD technology (charge coupled devices) to enhance low-intensity signals. The close satellite flybys and improved sensitivity are expected to yield improvements in spatial resolution by factors of 20 to 300 relative to *Voyager* photographs.

A recent modification of the mission design will make possible a close flyby of an asteroid as *Galileo* speeds from Earth to Jupiter. Although no definite decision will be made until after launch, the target has been selected, a 200 km diameter asteroid named Amphitrite. Thus, it is possible that in December 1986, another important type of object will have been examined from a range of about 20,000 km for the very first time.

## 17-3 MISSIONS TO HALLEY'S COMET

Comets are thought to consist of primordial matter of the solar nebula, largely unaltered (see Chapter 15 by Neugebauer). Data on their composition and structure are fundamental to solar-system studies, so the scientific value of a cometary mission is indisputable. Yet the poetry of a voyage to a comet also stirs the imagination. Diaphanous tails created by dispersion of volatiles and dust when the heat of the inner solar system warms the comet can be seen as they reflect the light of the Sun. Stargazers have marveled at the displays since ancient times and recorded their appearances. In 1705, the English astronomer Edmund Halley speculated that the comets that had appeared in 1531, 1607, and 1682 were all the same object with an orbital period of about 76 years. He predicted the year of its next return. Fittingly, the comet returned as predicted and was named after Halley. Further examinations of historical records have identified apparitions back to 87 B.C. Giotto, the great Italian painter, painted frescoes for a church in Padua, one of which depicts a comet leading the three kings to Bethlehem. Surely Giotto must have seen the apparition of the first years of the fourteenth century.

Halley's comet will return to the inner solar system in 1986. For the first apparition since the dawn of the space age, three missions are being prepared. They will carry instruments designed to investigate the composition of the coma and to explain the structure and variability of the tail. Imaging instruments will determine the size, shape, and rotation speed of the nucleus, which is obscured by the coma and cannot be observed from Earth even when observing geometry is good.

One of the Halley missions, sponsored by the European Space Agency (ESA), is named *Giotto*. *Giotto* was launched by an *Ariane-2/Sylda* rocket on July 10, 1985, with the Halley's comet encounter on March 13, 1986. Figure 17-2 shows part of the trajectory of the comet and the motion of Earth in its orbit. Notice that the comet's tail is shown pointing away from the Sun in the approximate orientation it will have. The comet moves in orbit in the sense opposite to the Earth's motion, so the relative velocities of the spacecraft and the comet at encounter will be large. Dates on the figure make it clear that at the time Halley's comet is closest to the Sun, it will be behind the Sun as viewed from the Earth. This means that at the time the comet's tail is most fully developed, it will not be visible from Earth. Views from the Earth will be possible in the autumns of 1985 and 1986, before and after closest approach, but only spacecraft instruments will observe the fully developed comet tail.

The primary measurements to be made by *Giotto*, a spinning spacecraft, with a despun antenna platform pointing to Earth, start 3 h, 45 min before closest approach. The distance from the nucleus at closest approach is to be less than 500 km, according to plans, though there are obvious uncertainties in aiming at an object that cannot be directly observed. The spacecraft carries a shield that will partially protect it from the

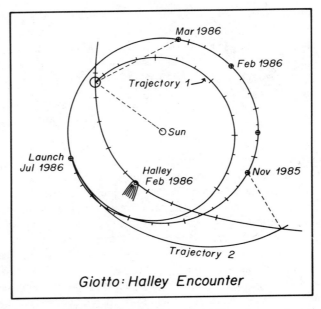

Fig. 17-2  Earth's position in its orbit and key dates and positions of Halley's Comet and the *Giotto* spacecraft.

dusty environment, but it may not survive the close pass anticipated. Color photographs of the surface of the nucleus will have resolution better than 50 m. Ten scientific instruments will measure composition and dynamical properties of the vapor and dust tails and the surface properties of the nucleus.

A two-spacecraft mission called *Vega* (*Venera-Halley*) is also scheduled for an encounter with Halley's comet. The spacecraft was launched by the USSR in December 1984, first targeted at Venus. After flying by Venus, the *Vega* spacecraft continued towards a fly-by of Halley's comet in March 1986. The *Vega* spacecraft are stabilized (not spinning). They carry seven scientific instruments including TV cameras with both narrow- and wide-angle viewing capability. Cameras will be turned on two days before encounter, with intensive study planned for three hours prior to closest approach.

A Japanese mission will also contribute to the studies of Halley's comet. The spacecraft, *Sakigake*, will remain farther away from the comet than *Giotto* and *Vega* and will not carry imaging instruments. Rather, it will monitor plasma conditions in the solar wind to make it possible to study how changes of the large-scale structure of the tail are related to changes of its environment. *Sakigake* will also monitor ultraviolet emissions from Halley's comet. August 1985 was the launch date of this spinning spacecraft.

Additional measurements of the coma of Halley's comet will be obtained by the ultraviolet spectrometer on the *PVO* (*Pioneer Venus Orbiter*), a spacecraft already in orbit around Venus. The spacecraft will come within 0.2 AU of Halley's comet during the time when the comet, not visible from Earth, passes closest to the Sun and is most active.

## 17-4 ISPM—A MISSION TO THE SUN

To understand the solar system, one must, evidently, understand the Sun. Ground-based observations of the Sun have for some time been supplemented by spacecraft observations, but even spacecraft-borne instruments have viewed the Sun only from locations near the ecliptic plane. Thus, the high-latitude features of solar processes of all sorts—electromagnetic emissions, particle emissions, magnetic field structure, and so forth—have been inferred from incomplete evidence. Evidently, then, it would be desirable to send a spacecraft out of the ecliptic plane to view the polar regions of the Sun and to ascertain whether the expected latitudinal variations are present.

The International Solar Polar Mission (ISPM), at present in its developmental phase, was designed to provide out-of-ecliptic measurements and to view the Sun and image it over all heliographic latitudes. Cosponsored by the United States (through NASA) and Europe (through ESA), the mission originally would have involved two spacecraft simultaneously passing over the near-polar regions of the Sun in opposite hemispheres. The United States withdrew support for its spacecraft in the early 1980s, and the reduced, single-spacecraft mission is a European mission on which some American investigators are participants. The European Space Agency renamed the mission *Ulysses* in 1984, so both names are now used.

The large amount of energy per unit mass needed to lift a spacecraft out of the ecliptic plane will be obtained from Jupiter's gravitational field. The *ISPM* spacecraft, like *Galileo*, will be placed into Earth orbit by the space shuttle and thence will be launched in the ecliptic toward Jupiter in May 1986. The spacecraft will reach Jupiter in late 1987, a year earlier than *Galileo*. Near Jupiter, the spacecraft will swing around the planet and emerge heading sunward and toward high latitudes. Polar passes over the Sun will occur in 1989 and 1990, with a Sun-spacecraft separation of approximately 1 AU at closest approach.

The spacecraft carries nine scientific instruments. Some will image the Sun at high heliographic latitudes. Others will measure fluxes of energetic particles and of dust grains, monitor radiofrequency emissions, measure magnetic fields, and so forth. *ISPM* will provide data important for understanding the source of the solar wind, the large-scale structure of the plasma of the solar system, and the way in which plasma and fields act to modify energetic charged particles (cosmic rays) from galactic sources, as well as providing data on the Sun itself. The Jupiter flyby will provide measurements from a trajectory both higher in Jovian latitude and farther toward the dusk meridian than any previous or planned spacecraft trajectory. Thus *ISPM* measurements will in some cases be useful complements to those of *Galileo*.

## 17-5 MISSIONS TO VENUS

Comparative studies of similar planets provide essential evidence on which an understanding of solar-system history rests. Particularly important for such studies are comparisons of Earth and Venus. Near-twins (Venus is at 70% of Earth's distance from the Sun, with 82% of Earth's mass, 95% of its radius, and 88% of its surface gravity), the two planets display many startlingly different properties. For example, the pressure of the atmosphere at the surface of Venus is 90 Earth atmospheres; its surface temperature is 470°C. Surface water and a planetary magnetic field are both absent. Attempts to understand Venus better and thereby to improve our knowledge of planetary evolution are frustrated by the limited data on the geological properties of the surface of Venus, a surface permanently obscured by clouds.

Some mapping of the surface of Venus below the clouds has been obtained from radar imaging. *PVO* gave maps of about 93% of the surface with very low resolution (approximately 500 km). In mid-October 1983, *Venera 15* and *16* were placed into Venus orbit by the USSR. The synthetic aperture radar on these spacecraft have returned images which extend and complement *PVO* coverage. In particular, the *Venera 15* and *16* images are expected to give extensive data on the previously unexamined northern polar regions. Earth-based radar maps cover only a few percent of the surface but allow 50-km resolution. Even better resolution combined with good coverage can be obtained from a spacecraft in orbit using modern radar technology. Such improved resolution is essential for the understanding of the processes important in the evolution of Venus.

The surface can also be investigated by dropping lander probes to make *in situ* measurements. The Soviet spacecraft of the *Venera* series have obtained important data with such probes and have produced remarkable images of the rocky surface terrain near the lander locations.

The exploration of Venus continues to be a prime objective (see Table 17-1) of the planetary program of the USSR. The Soviet *Vega* mission, developed in cooperation with France, was mentioned in Sec. 17-3. Venus flybys, with release of probes to obtain data on the atmosphere and the surface of the planet were important elements of this mission.

The United States has been planning to improve the resolution of the radar maps through an orbital mapping mission. An ambitious mapping spacecraft, *VOIR (Venus Orbiting Imaging Radar)*, was abandoned in 1981 because of its high cost. A less costly mission that can achieve most of the desired objectives for surface mapping has been proposed. Called *VRM (Venus Radar Mapper)*, the proposed spacecraft will be able to map 92% of the surface at better than 1-km resolution and to obtain vertical resolution better than 100 m over much of the planet. Gravity field measurements would supplement those obtained by the previous orbiting mission, *PVO*. Radar images would be valuable for studies of tectonics, volcanism, and cratering. Gravitational data would constrain models of the planetary interior. The VRM mission is proposed for launch in the spring of 1988 and would then arrive at Venus that summer. Approval is anticipated. *VRM* is regarded as the first of a continuing series of U.S. planetary explorers designed to answer quite specific questions at comparatively low cost.

## 17-6 MISSION PLANNING: A LOOK STILL FURTHER AHEAD

*Galileo, ISPM,* and *Giotto* are examples of missions now under way, missions whose objectives are identified and whose observational capabilities have been determined with those objectives in mind. Years of discussion, of scientific salesmanship, and of engineering studies precede this phase of a planetary mission. The solar system is vast, potential objectives many, and the arguments for undertaking one mission rather than another are abundant. How does the process leading from dreams to dollars proceed?

Many steps are involved. Advisory committees, such as the Solar System Exploration Committee of the NASA Advisory Council or the Space Science Committee of the European Science Foundation, identify scientific questions of consequence that can be addressed through new spacecraft observations. The possible missions considered may be ones proposed by the advisory committees themselves or ones that have been proposed by individuals, research teams, or government laboratories. At the initial stages, the proposals specify the targets and the rationale for the choice, the type of orbit required, the type of measurements to be made, and the approximate anticipated costs, but details may be few. Some priority among the interesting possible missions is then suggested by the advisory committees (see, for example, some of the reports listed in Sec. 17-9); the priorities may be accepted or modified by the potential funding agencies such as NASA or ESA.

Missions that survive initial screening are examined in greater depth. Can the objectives be achieved, or are there technological or financial impediments? Requirements on spacecraft performance are also assessed. How stable need the spacecraft be to permit the measurements? How accurately must the instruments be aimed? What data rates are required for scientific telemetry? How will the spacecraft be tracked and the experiments controlled and commanded? What are estimates on the mass and power required for the scientific payload? What trajectories are acceptable for the spacecraft? How will data be provided to science teams for analysis?

At this preliminary assessment stage, the estimates may still be incomplete and may rely more on past experience than on precise analysis, but the mission form begins to be clearly established. Among the missions that have undergone assessment studies, a few are selected for more advanced studies in which engineering feasibility is demonstrated; weight, power, mass, and data rate requirements are determined; and the cost of the mission is calculated more accurately. With this kind of information, the funding agencies (and in the United States, the Congress) select the missions that will be supported and allocate the funds for development and operations.

Some of the planetary missions that are still in early stages of the planning process are listed in Table 17-4. Approximate time frames are indicated, though uncertain. The *Venus Radar Mapper*, described in Sec. 17-5, is proposed for a 1988 launch. This would be followed in 1990 by a *Mars Geoscience/Climatology Orbiter*, which would measure global surface composition and determine the role of water in the Martian climate. This potential NASA mission might usefully be coordinated with a possible ESA mission called Kepler, which would also place a spacecraft (also named *Kepler*) into orbit around Mars. By selecting a near-polar orbit that is fixed in space, one can map the entire surface of the planet as the planet rotates under the orbit. *Kepler* would measure the planetary magnetic field (see Chapter 14 by Russell), seek to understand the structure and energy balance of the upper atmosphere, and characterize the ionosphere.

Both ESA and NASA are considering asteroid missions in the next decade. NASA's *Comet Rendezvous and Asteroid Flyby* mission [CRAF], proposed for launch between 1990 and 1992, would be targeted for asteroid flybys but would also spend some time moving along with a comet, which would allow extended observations, scientifically more informative than those from the fast flybys of Halley's comet. The ESA *AGORA (Asteroid*

TABLE 17-4  Some Candidate Planetary Missions of NASA[a] and ESA[b]

| | Late 1980s | 1990s | Early Next Century |
|---|---|---|---|
| Venus | VENUS RADAR MAPPER | | VENUS ATMOSPHERE PROBE |
| Mars | | MARS GEOSCIENCE/CLIMATOLOGY ORBITER | MARS AERONOMY ORBITER |
| | | Kepler | MARS SURFACE PROBE |
| Moon | | | LUNAR GEOSCIENCE ORBITER |
| Saturn | | TITAN PROBE/RADAR MAPPER | SATURN ORBITER |
| | | Saturn/Titan Orbiter/Probe | SATURN FLYBY/PROBE |
| | | (Cassini) | |
| Uranus | | | URANUS FLYBY/PROBE |
| Comets | | COMET RENDEZVOUS/ASTEROID FLYBY | COMET ATOMIZED SAMPLE RETURN |
| Asteroids | | Agora | MULTIPLE MAINBELT |
| | | | ASTEROID ORBITER/FLYBY |
| | | | EARTH-APPROACHING |
| | | | ASTEROID RENDEZVOUS |

[a] Upper-case letters.
[b] Lower-case letters.

*Gravity, Optical and Radar Analysis*) would rendezvous with as many as three main belt asteroids and fly relatively near to two others. Both missions seek to measure size and density and record surface forms and mineral compositions of representatives of the most primitive bodies in the solar system.

Titan, Saturn's major moon, is also proposed as the primary target for both ESA and NASA missions in the next decade. Titan's dense atmosphere obscures its surface, so the nature of the surface as well as the composition of the atmosphere are still uncertain. An orbiting spacecraft that would investigate Saturn and its rings is also an element in both proposed missions.

Missions for the early years of the next millennium are already being analyzed. In addition to the missions listed in the final column of Table 17-4, Neptune and Pluto orbiters have been suggested, though the long interplanetary cruises needed to reach these planets would place data acquisition in the even more distant future.

One proposed ESA mission not listed in Table 17-4 is in the detailed analysis phase and could be selected in 1986. That is the *SOHO* mission, the *Solar and Heliospheric Observatory*. This mission will observe the Sun, the focus and principal energy source of the solar system. The mission is designed to investigate coronal heating, solar-wind acceleration, and possibly the response of the Sun to gravity waves.

Among the entries in Table 17-4, there are no missions to either Mercury or Jupiter. For Mercury, the absence reflects the majority view (though there are dissenters) that our uncertainties regarding planetary properties are greater for the other terrestrial planets. Jupiter is missing because it is difficult to plan a post-*Galilio* mission until the results of *Galileo* are known. Then remaining areas of ignorance (and there will be many) can be identified, their importance to our understanding of solar-system science evaluated, and a possible post-*Galileo* mission considered. Already, though, scientists have recognized that a Jupiter polar orbiter would provide data unavailable from the *Galileo* orbit and have begun to discuss possible missions.

## 17-7 SUMMARY

> The primary goal continues to be the determination of the origin, evolution and present state of the solar system. . . . Two additional goals include understanding the Earth through planetary studies and understanding the relationship between the chemical and physical evolution of the solar system and the appearance of life, both of which require extensive exploration of other solar system bodies.

Thus the Solar System Exploration Committee of the NASA Advisory Council introduces its report (National Aeronautics and Space Administration, 1983) on future planetary exploration. Earlier chapters of this Rubey volume have described current ideas on the way in which the solar system formed and evolved. On some parts of the story, there is considerable consensus based on evidence from ground-based observations and from the first decades of spacecraft measurements. Gaps remain: Critical speculations must be tested; new ideas must be developed. Some of the missions described in this chapter, and others yet to be proposed, will contribute in essential ways to the development of a version of solar-system history that is complete, consistent, and convincing.

## 17-8 ACKNOWLEDGMENTS

This work was partially supported by NASA under Grant NGL 05-007-004 and by the Jet Propulsion Laboratory under contract NAS 9-55232. It is a pleasure to acknowledge the hospitality of the Section d'Astrophysique, Observatoire de Paris, Meudon, France, while this paper was completed.

## 17-9 REFERENCES AND ADDITIONAL READINGS

Much of the information on future planetary missions is ephemeral, for the planning process is necessarily responsive to changes whether political, economic, or scientific. The best sources, therefore, are recent issues of scientific news magazines such as *Science* (weekly), *Aviation Week and Space Technology* (weekly), and *The Planetary Report* (bimonthly), which are available in most libraries. Another useful source of information is the *TRW Space Log*, published annually by TRW Corporation, One Space Park, Redondo Beach, California 90278. Relatively recent major reports on U.S. and European planetary programs are:

European Science Foundation, 1982, *Planetary Science in Europe, Present State and Outlook for the Future,* ESF, 1, quai Lezay-Marnesia, F-67000 Strasbourg, France.

National Academy of Sciences, 1975, *Report on Space Science 1975,* NAS, Washington, D.C.

National Academy of Sciences, 1976, *Strategy for Exploration of the Inner Planets—1977-1987,* NAS, Washington, D.C.

National Academy of Sciences, 1979, *Strategy for Exploration of Primitive Solar System Bodies—Asteroids, Comets, and Meteoroids—1980-1990,* NAS, Washington, D.C.

National Aeronautics and Space Administration, 1975, *Entry Probes in the Outer Solar System,* NASA, Washington, D.C.

National Aeronautics and Space Administration, 1983, *Planetary Exploration Through the Year 2000* (2 parts), NASA, Washington, D.C.

TRW, 1984, *TRW Space Log 1982-83,* Redondo Beach, California.

# APPENDIX

TABLE A-1  Orbital Elements and Physical Properties of the Planets[a]

| Planet | Mean Distance From Sun (AU)[b] | Sidereal Orbital Period (y) | Orbital Eccentricity | Orbital Inclination (degrees) | Mass[e] (kg) | Equatorial Radius (km) | Mean Density (g/cm³) | Sidereal Rotational Period (h) | Inclination of Equator to Orbit (degrees) |
|---|---|---|---|---|---|---|---|---|---|
| Mercury | 0.3871 | 0.24085 | 0.2056 | 7.00 | $3.302 \times 10^{23}$ | 2,439 | 5.42 | 1408 | 0° |
| Venus | 0.7233 | 0.61521 | 0.0068 | 3.39 | $4.871 \times 10^{24}$ | 6,052 | 5.25 | 5832[c] | 2° |
| Earth | 1.0000 | 1.0000 | 0.0167 | | $5.975 \times 10^{24}$ | 6,378 | 5.52 | 23.93 | 23.44° |
| Mars | 1.5237 | 1.8809 | 0.0934 | 1.85 | $6.421 \times 10^{23}$ | 3,398 | 3.94 | 24.62 | 23.98° |
| Jupiter | 5.2026 | 11.862 | 0.0485 | 1.30 | $1.900 \times 10^{27}$ | 71,398 | 1.31 | 9.841[d] | 3.08° |
| Saturn | 9.5547 | 29.46 | 0.0556 | 2.49 | $5.688 \times 10^{26}$ | 60,330 | 0.69 | 10.23[d] | 29.73° |
| Uranus | 19.2181 | 84.01 | 0.0472 | 0.77 | $8.70 \times 10^{25}$ | 25,400 | (1.19) | 15.5[d] | 97.92° |
| Neptune | 30.1096 | 164.8 | 0.0086 | 1.77 | $1.03 \times 10^{26}$ | 24,300 | 1.66 | (15.8)[d] | 28.8° |
| Pluto | 39.44 | 248.5 | 0.250 | 17.2 | $1.31 \times 10^{22}$ | (1,500) | (0.9) | 153.3 | (>50°) |

[a]Brackets indicate values with uncertainty greater than 10%.
[b]1 AU = 1 astronomical unit = $1.496 \times 10^{11}$ m.
[c]Retrograde.
[d]At equator.
[e]The mass of the sun is $1.99 \cdot 10^{33}$ g.

TABLE A-2  Selected Properties of Major Satellites[a]

| Name | Satellite of | Mean Distance from Planet (10³ km) | Sidereal Period (days) | Orbital Inclination (degrees) | Orbital Eccentricity | Radius (km) | Mass (kg) | Mean Density (g/cm³) |
|---|---|---|---|---|---|---|---|---|
| Moon | Earth | 384 | 27.32 | 6.68 | 0.055 | 1738 | $7.350 \times 10^{22}$ | 3.34 |
| Io | Jupiter | 422 | 1.769 | 0.027 | 0.000 | 1815 | $8.89 \times 10^{22}$ | 3.55 |
| Europa | Jupiter | 671 | 3.551 | 0.468 | 0.000 | 1569 | $4.79 \times 10^{22}$ | 2.96 |
| Ganymede | Jupiter | 1070 | 7.155 | 0.183 | 0.001 | 2631 | $1.490 \times 10^{23}$ | 1.93 |
| Callisto | Jupiter | 1880 | 16.69 | 0.253 | 0.007 | 2400 | $1.075 \times 10^{23}$ | 1.83 |
| Titan | Saturn | 1222 | 15.94 | 0.33 | 0.029 | 2560 | $1.346 \times 10^{23}$ | 1.9 |
| Triton | Neptune | 335 | 5.877 | 160. | 0.000 | (1600) | $3.4 \times 10^{22}$ | (2.0) |

[a]Brackets indicate values with uncertainty greater than 10%.

# INDEX

Abelson, P. H., 166
Abt, H., 33
Abundance ratio, 11
Accretion disks, 29, 36, 43, 57
Accretion heating, 143, 164, 168
Acuna, M. H., 347
Adachi, I., 63
Adams, J. B., 153
Adiabatic state, 256
*AGORA* mission, 409
Ahrens, T. J., 164, 184, 186
Albedo, 120, 134
Alfvén, H., 365
Allegre, C. J., 108
Allende meteorite, 50, 384 *fig.*, 386-88
α-effect, 313
α-ω dynamo, 313
Aluminum (Al) radionuclides, 13, 385-89, 392-94
 Al:Mg isotope ratios, 50, 387
Amalthea, 285
Ammonia ($NH_3$)
 conditions for synthesis, 166-67
 Jupiter and Saturn, 260-61
Amphitrite, 403
Anders, E., 108, 150, 155, 156, 157, 163
Anderson, D. L., 84, 91, 152, 153, 216
Angevine, C. L., 181
Angular momentum
 conservation of, 36
 and Coriolis force, 299
 and "preserved features," 383
 in nebular theory, 29, 30, 31, 389-90, 392
 effect of temperature on, 44
 and formation of planetesimals, 9
 of infall material, 45, 46
 of molecular clouds, 32-33
 and rotation of stars, 16
 transport, 34-41, 43, 44, 47, 53, 392
 of Sun, 16, 30, 31
 transfer from Jupiter to Io, 92
 transport on Venus, 133
Anticyclonic motion, 128
Aphelion, 3 *fig.*, 4
Aphrodite Terra (Venus), 91, 201
Apoapsis, 342
Apocenter, 4
Apogee, 4
*Apollo* missions, 79, 86, 161, 189, 211
 *Apollo 11*, 102
 *Apollo 12*, 334
 *Apollo 14*, 334
 *Apollo 15*, 329, 330-31, 332 *fig.*, 334
 *Apollo 16*, 221, 329, 330-31, 332 *fig.*, 333 *fig.*, 334
Apolune, 4
Archaean Era, 85, 177, 198, 199
Argon (Ar) isotope ratios
 Earth, 107, 108, 112, 165
 Mars, 155-56, 157
 Venus, 155-56, 158, 160
Argyre planita (Mars), 194
Aristotle, 360
Armstrong, R. L., 105
Arrhenius, G., 31
Arsia Mons (Mars), 195
Ascraeus Mons (Mars), 195
Asteroid Belt, 2-3, 11, 12 *fig.*
 composition, 140, 149, 150
*Asteroid Gravity, Optical and Radar Analysis (AGORA)* mission, 409
Asteroids
 composition, 140
 exploratory missions, 401 *tab.*, 403, 407-9
 formation, 9
 and formation of Moon, 14
 mass, 2
 melting and fractionation, 13-14
 and meteorites, 10, 11
 orbital properties, 5 *tab.*, 68
 size, 2-3
 volatile history, 138, 149-50
 weathering, 208, 214 *tab.*, 223-24
Asthenosphere, 182
 Earth, 83
 Moon, 87
Astronomical unit (AU), 4, 81 *tab.*
Atmosphere
 giant planets, 264-66
 Io, 222
 Jupiter and Saturn, 233-49
 outer solar system objects, 147-48
 terrestrial planets, 117-34
  and black-body radiation, 120-21
  circulation, 117-18, 129-34
  energy input, 120
  greenhouse effect, 121-23
  holding on to, 119
  motive forces, 125-29
  optical thickness, 123
  and orbital parameters, 118-19
  planetary properties affecting, 117
  scale height, 119
  vertical thermal structure, 123-24
 Titan, 223
"Atmospheres of the Terrestrial Planets" (Kivelson and Schubert), 117-34
Atmospheric temperatures
 and black-body radiation, 120-21

Atmospheric temperatures *(continued)*
 and distance from Sun, 118
 effective, 117, 121
 greenhouse effect, 121–23
 latitude dependence, 129–30
 thermal tides, 284
 and vertical thermal structure, 123–24, 239
Aurora Borealis, 25, 292
Auroras, Jovian, 232 *fig.*

## B

Babcock, H. D., 312
Balmino, G., 196
Banerdt, W. B., 197
Barnes, A. P., 329
Barnes, C. A., 379 *fig.*
Baroclinic instability, 131
Barsukov, V. L., 219 *fig.*, 220, 225
Basaltic achondritic meteorites, 149–50
Basaltic Volcanism Study Project, 80, 82
Basalts, weathering of, 215, 216, 221, 222
Baumgardner, J. 189
Beard, D. B., 339, 340
Berge, G. L., 346
Berner, R. A., 213
Beta Regio (Venus), 91, 200, 201
Biemann, K., 152
Biermann, L., 363, 364
Big bang theory, 375, 376–77, 385
Binary stars, 33, 34
Binder, A. B., 195
Bing bang theory, 375–94
Bird, Peter, 188, 189, 197
 "Tectonics of the Terrestrial Planets," 83, 86, 87, 90, 146, 149, 177–204
Black, D. C., 31, 72, 156, 157, 158
Black-body radiation, 42, 120–21
Black holes, 377
Blasius, K. R., 154, 197
Bodenheimer, Peter, 32, 33, 45, 47, 49, 72
Bode's law, 4–5, 69
Bogard, D. D., 154
Booth, M. C., 152, 222
Boron side reaction, 18
Boschi, E., 80
Boss, Alan, 394
Bowell, E., 149
Boyer, D., 308
Boynton, W. V., 50
Brace, L. H., 341
Braginskii, S. L., 304
Brahe, Tycho, 360
Brandt, J. C., 365
Brecher, A., 31
Brett, R., 314, 336
Bridge, H. S., 340
Broadfoot, A. L., 161
Brown, G. C., 80
Brown, L. W., 348, 352
Brown, R. A., 147
Brownlee, D. E., 363
Bucher, W. H., 198
Buffon, G. L. L., Comte de, 30
Bumba, V., 320
Bunch, T. E., 224
Burgala, L. F., 350
Burke, K., 86, 167
Burnham, C. W., 107
Burns, B. A., 200, 201
Busse, F. H., 245 *fig.*, 304, 314

## C

Callisto
 density, 350
 ice-rock ratio, 88
 magnetic field, 350
 selected properties, 412 *tab.*
 tidal evolution, 285
 volatile content, 251
 volatile history, 146–47, 148, 169
Caloris Basin (Mercury), 193
Cameron, A. G. W., 31, 32, 33, 34, 37, 38, 39, 45, 50, 51, 53, 54, 55, 57, 58 *fig.*, 65, 70, 146, 272, 379 *fig.*, 392
Campbell, D. B., 200, 201
Canadian Shield, 198
Cannon, W. A., 143, 151, 152, 156
Carbon
 giant planets, 260
 nucleosynthesis, 377
Carbonation weathering, 213
Carbon dioxide ($CO_2$)
 freezing and sublimation on Mars, 131
 weathering, 212–13
 comets, 223
 Earth, 224–25
 Mars, 216, 218
Carr, M. H., 88, 150, 154, 197
Carroll, D., 215
Cassen, P., 45, 146, 148, 285
Cassini state, 284, 285
Catastrophe theory, 30, 389–94
Centrifugal force, 126, 127 *fig.*
 and cyclostrophic balance, 129, 133
 and hydrostatic equilibrium, 34, 54
Ceres, 3, 4, 149
Chaiken, J., 86, 191, 194, 197
Chamberlin, Thomas C., 390
Chang, S., 372
Chemical weathering, 208, 212–14, 215
Chen, J., 386 *fig.*
Cheng, C. H., 192
Chennette, D. L., 349
Chiron, 3
Chondrites, 11–13
 composition, 139–40
 U:Pb isotope ratios, 99
Chondrules, 11
Christ, C. L., 215
Christensen, E. J., 194, 196
Chromosphere, 20
Chromospheric network, 20
Cintala, M. J., 154

Cisowski, S. M., 336
Clark, B. C., 152, 153, 216
Clark, R. N., 146
Clark, S. P., 139
Clarke, J. T., 352
Clathrate, 361
Clayton, R. N., 383
Cloos, M., 199
Clouds, 120. *See also* Interstellar molecular clouds
Cohen, M., 32, 33, 51
Colburn, D. S., 329
*Comet Rendezvous and Asteroid Flyby* mission, 407-9
Comets, 5-6, 359-72
  accretion of matter onto giant planets, 265
  comas, 5-6, 361-62, 364
  composition, 6, 362
  exploratory missions, 400, 401 *tab.*, 403-5, 407-9
  fate of, 370
  history of research, 360
  hydrogen coronas, 365
  mass, 2
  and meteorites, 10
  nuclei, 6, 361
  orbits, 6, 360, 361, 368-71
  role in history and evolution of solar system, 371-72
  source, 6. *See also* Oort cloud
  tails, 362-65, 366 *fig.*, 367 *fig.*
  volatile history, 142, 145
  weathering, 208, 214 *tab.*, 223-24
"Comets" (Neugebauer), 359-72
Comparative volatile studies, 138
Compensation
  and isostatic adjustment, 184
  Mars, 196, 197
Condie, K. C., 106
Condit, C. D., 195
Connerney, J. E. P., 340, 346, 347

"Consequences of Tidal Evolution" (Peale), 276-88
Conservation laws, 36
Convection
  adiabatic constraint, 256
  giant planets, 265, 266
  solar nebula, 39-40, 42, 43-44, 70
  Sun, 18, 24, 314, 319-20
  tectonic, 177, 180, 186-89, 190 *fig.*, 192
Convective instability, 39-40, 44
Copernicus, Nicolaus, 30
Coradini, M., 153
Core dynamos, 337-52
  Earth, 337-38
  Jupiter, 346-48
  large moons, 350-51
  Mars, 342-46
  Mercury, 338-40
  Moon, 334-35
  Saturn, 348-49
  Uranus and Neptune, 351-52
  Venus, 340-42
Coriolis force
  and atmospheric motion, 125-26, 127, 128 *fig.*
  Earth, 130-31
  and magnetic field, 299-300, 304
Coronal holes, 26, 27
Coronas
  cometary, 365
  nebular, 51
  solar, 23, 26-27, 315, 316 *fig.*, 322
Correns, C. W., 215
Cosmic abundance, 378, 379, 383
Cosmogony, modern theories, 29, 31-34
Courtin, R., 265
Covey, C., 117
Cox, A., 103, 306
Cox, J. P., 39
Cox, L. P., 60, 61
Craig, H., 163
Crater count dating, 87, 178
Crater flooding, 184

Craters
  Mars, 90, 194
  Mercury, 89, 193
  Moon, 86, 87, 190-91
  Venus, 177, 201-2
Cuong, P. G., 244 *fig.*
Curie point, 296-97
Curium (Cm) isotope ratios, 389
Cutts, J. A., 154
Cyclonic convection, 299 *fig.*, 300, 301-3, 308
Cyclonic motion, 128, 244
Cyclostrophic balance, 128-29, 133

### D

Dali Chasma (Venus), 201, 202
Daly, R. A., 185
Dashpots, 280
Dating
  by crater counts, 87, 178
  by radioactive isotope rations, 95-99, 378, 380, 381 *fig.*, 383
Daughter nuclides, 96, 98
Davies, G. F., 110
Davis, D. R., 149
Davis, L., 347, 349
DeBergh, C., 158
DeCampli, W. M., 39
Degassing
  Earth, 95, 106-8, 163-68
  fate of volatiles, 143-44
  Io, 148
  Mars, 155-57
  Mercury, 161
  and planetary size, 143
Degenerate electrons, 257-58
Deimos, 285
Delamination, 187 *fig.*, 188-89, 190 *fig.*, 197
Delsemme, A. H., 223, 363 *fig.*
DePaolo, Donald J., 105, 109, 199
  "Isotopic Constraints of Planetary Evolution," 2, 13, 50, 84,

DePaolo *(continued)*
  85, 95-112, 164,
    190, 199, 380, 397
Descartes, René, 29-30,
  389
Desch, M. D., 349
Despinning, 177, 182-83,
  190 *fig.*
  Earth, 197-98
  Mercury, 194
Deuterium (D), 385, 390
Dicke, Robert, 16-17
Differentiation, 11
  chondritic material, 13
  Earth, 85-86, 95, 200
  Mercury, 89
  meteorites, 11
  Moon, 87, 88, 95
Dione
  orbital resonance, 286,
    287
  volatile history, 145
Dip, 180
Dislocation creep, 179,
  181, 185
Dispersion relation, 19
Doell, R. R., 334
Dole, S. H., 63
Dolginov, Sh. Sh., 329,
  340, 342, 344
Donahue, T. M., 65, 160
Donn, B., 145
Downs, G. S., 195
Dreibus, G., 150, 156, 157
Dryer, M., 342
Dry Valleys of Antarctica,
  217
Duke, M. B., 225
Dungey, J. W., 350
Duricrust, 152, 153
Durrance, S. T., 352
Dust storms, 126, 131
Dyce, R. B., 283
Dynamical dynamo theory,
  298
Dynamo equation, 312-14
Dynamo number, 303
Dynamos
  $\alpha$-$\omega$ type, 313
  core type, 337-52
  magnetohydrodynamic,
    266, 297-308
  shell type, 314-37
Dziewonski, A. M., 80
Dzurisin, D., 192, 194

# E

Earth
  age, 381
  albedo, 120
  asthenosphere, 83
  atmosphere, 117, 118,
    119, 120, 121,
    123-24, 126, 127,
    129-31
  circulation, 130-31,
    242-44, 245
  evolution, 65
  formation, 106-8,
    112
  motive forces, 126,
    127, 128
  near surface properties,
    212 *tab.*
  bulk properties, 81 *tab.*
  cataclysmic episodes, 95
  comet apparitions,
    368-69
  composition, 2, 82-83
    D/H ratio, 385
    $H_2O$ abundance, 138,
      142
    main compositional
      parts, 82 *tab.*, *fig.*
    oxygen isotopes, 383,
      384 *fig.*
    rock, 261
    volatiles, 89, 139,
      142-43
  continental crust, 82,
    84, 85, 86, 104-6
    age, 95, 104
    creeping minerals, 181
    differentiation, 200
    formation, 105,
      111-12, 177, 198
    and magmatic de-
      gassing, 107
  cooling of, 85-86
  core, 82-83
    formation, 85, 86, 95,
      98-101, 111
    heat energy, 304
    and magnetic field, 92,
      300-301, 304, 307,
      308
  crust formation, 99, 101,
    103-6
  degassing, 95, 106-8,
    163-68

  density, 12, 88, 140
  differentiation, 85-86,
    95, 200
  escape velocity, 13
  excess volatiles, 137,
    162
  exploratory missions,
    399, 401 *tab.*
  formation, 13, 85
  and formation of Moon,
    14
  heat flow, 84-85, 86
  hydrosphere, 106-8,
    112, 200
  interior
    evolution, 85-86,
      95-112
    structure, 79, 80,
      82-85
  lithosphere, 83, 84, 85,
    197-200
  magnetic field, 300-301,
    304, 307, 308,
    337-38
    decay time, 312
    long-term behavior,
      305-9
    polarity reversals, 290,
      297, 306
    regeneration, 297,
      300-301
    solar wind effects, 292,
      321
    Venus compared, 340,
      341 *fig.*
  mantle, 82, 83, 85
    creeping minerals, 187,
      197
    degassing, 95, 108, 109
    isotope ratios, 98-99
    structure and dy-
      namics, 108-10,
      112
    subduction of crust,
      104, 105
    temperature changes,
      198
    upper, 93
  Mars compared, 90
  meteorites, 10-11
  oceanic crusts, 82, 84,
    85, 103-4
    differentiation, 200
    formation, 177
    subduction, 104, 105

orbital properties, 4, 5 tab., 118, 119 tab., 411 tab.
radioactive isotopes, 95–99
soils, 210
solar constant, 120
solar energy effects, 17–18, 25, 27, 363
tectonics, 177, 181, 184, 185, 188, 190 fig., 197–200, 203
temperature
altitudinal variation, 123–24
effective, 121
latitudinal variation, 129–30, 252
at surface, 19, 117, 118 tab., 121, 123, 198
tidal distortions, 276, 279–80
tidal evolution, 284–85
Venus compared, 88, 91, 405–6
volatile history, 136, 137, 138, 139–40, 142, 153, 155, 156, 158, 162–68
weathering, 208, 211, 214–16, 224–25
Eccentricity, orbital, 4
of asteroids, 11
atmospheric effects, 118
Eclogite, 91, 188
Eddy, J. A., 315, 317 fig.
Einstein, Albert, 16, 17, 296
Elasticity, tectonic, 179, 180
Elastic strain, 180
Electric fields, 293–96, 312
Electron degeneracy, 258
Electron flux, 331
Ellipse, properties, 3 fig., 4
Elmgreen, B. G., 51
Elphic, R. C., 314, 321, 324, 325 fig., 326 fig., 327 fig., 328, 337, 338 fig., 347
Emerging flux region (EFR), 24

Enceladus
orbital resonance, 286, 287
volatile history, 145
Encke comet, 369, 370 fig., 371
Energy, conservation of, 36
Equilibrium condensation models, 138–42, 149, 150, 157, 158, 160, 168, 169
Ernst, W. G., 204
Eshleman, V. R., 223
Eucrites, 163, 164
Europa, 92
bulk properties, 81 tab.
density, 350
exploratory missions, 80, 405
ice-rock ratio, 88
resonant coupling of orbit, 92
selected properties, 412 tab.
terrestrial characteristics, 79
tidal evolution, 285–86
volatile history, 136, 138, 139 fig., 140, 145, 146, 147, 148–49, 169
European Science Foundation, 407
European Space Agency (ESA), 403, 405, 407, 408 tab., 409
Evans, B., 197, 199
Evans, D. L., 153
Excess volatiles, 137, 162
Exfoliation, 212
*Explorer 35*, 329–30, 334
Explosions, 375
Extinct nuclides, 385–89, 390, 392–94
Extragalatic objects, magnetic field, 292
Ezer, D., 32

F

Fanale, Fraser P., 140, 143, 146, 147, 148, 149, 151, 152, 153, 154, 156, 162, 163, 164, 165, 166, 167, 222, 225
"Planetary Volatile History: Principles and Practice," 13, 85, 89, 90, 106, 117, 119, 136–70, 184, 192, 222
Faraday, Michael, 292
Farmer, G. L., 105
Farquhar, R. M., 95
Fast particles, 17, 18
Faulting, 179, 180–81
due to despinning, 182–83
due to nonuniform contraction, 182
due to plate-tectonic convection, 188
and isostatic adjustment, 185
Faure, G., 95, 96 tab.
Fegley, B., 142
Fegley, M. B., 51
Feierberg, M. A., 223
Ferromagnesium, 84
Ferromagnetism, 296–97
Fibrils, 20–21, 22 fig.
Filaments, 21, 22 fig., 26
Finson, M. L., 362
Fishbein, E., 261
Flamini, E., 153
Flasar, F. M., 223
Fleitout, L., 185
Florensky, C. P., 218, 219 fig.
Flux, 331
Flux bundles, 314–15
Flux ropes
Moon, 335–36
Sun, 315, 319, 320, 321
Venus, 315, 321, 322–29
Fox, J., 156
Fractionation, 13–14
France, 406
Frank, E. U., 261
Franklin, F. A., 68
Frazier, K., 317 fig.
Froidevaux, C., 185
"From Big Bang to Bing Bang" (Lee), 375–94
Frost riving, 208, 211, 212, 214 tab.
asteroids, 224
comets, 223

INDEX

421

Frost riving *(continued)*
  Mars, 216
  Titan, 223
"Future Planetary Missions" (Kivelson), 397–410
Fyfe, W. S., 80

## G

Gaffey, M. J., 140, 149
Galactic magnetic field, 32, 312
Galactic matter, evolutionary processes, 376–77
Galaxies, formation, 375
Galilean satellites
  density differences, 251
  discovery, 234
  formation, 62
  ice-rock ratio, 88
  magnetic field, 350, 351
  tidal evolution, 285–86
  volatile history, 136, 145, 146–48, 169
Galileo Galilei, 234, 285, 401
*Galileo* mission, 272, 351, 398, 400, 401–3, 405, 406, 409
Gamma rays, solar, 17, 18
Gancarz, A. J., 99, 100 *fig.*
Ganymede
  density, 350
  *Galileo* mission, 402
  ice-rock ratio, 88
  magnetic field, 350, 351
  resonant coupling of orbit, 92
  selected properties, 413 *tab.*
  tectonics, 146–47
  tidal evolution, 285–86
  volatile content, 251
  volatile history, 146–47, 148, 169
Gapcynski, J. P., 334
Gardner, F. F., 346
Garrels, R. M., 215
Garvin, J. B., 216
Gas pressure gradients. *See* Pressure gradients
Gast, P. W., 108

Gaustad, J. E., 39
Gautier, D., 265
"Generation of Magnetic Fields in Planets" (Levy), 290–309
Geostrophic balance, 127–28
Giacobini-Zinner comet, 400
Giant planets, 255–73
  composition, 255–63
  formation, 54, 55, 61–62, 63, 64, 65, 70, 270–72
  magnetic field, 266, 305
  model construction, 263–70
Gibson, E. K., 152, 154, 161, 217, 222
Gilman, P. A., 320
Giotto di Bondone, 403
*Giotto* mission, 401 *tab.*, 403–4, 406
Goetze, C., 197, 199
Goins, N. R., 190
Gold, T., 335
Goldich, S. S., 99, 215
Goldreich, P., 37, 53, 54, 57, 60, 63, 68, 283, 284
Golombek, M. P., 192
Gooding, James L., 217, 220, 221, 224
"Planetary Surface Weathering," 208–25
Goody, R. M., 117
Grabau, A. W., 211
Granulation, of sun, 18
Gravitational constant, 277
Gravitational energy
  of accretion, Earth, 164, 165
  released from interstellar molecular clouds, 32
  released from nebular material, 41, 43, 44
Gravitational force
  atmospheric effects, 119
  of protosun on nebular material, 34
  tidal distortion due to, 276–79, 282
Gravitational instability, and protoplanet formation, 45, 54–56, 57
Gravitational moments, 264, 266–67
Gravitational scattering, 59–60, 61
Greenberg, R., 59, 60, 61
Greenhouse effect, 121–23, 177
  Earth, 121, 123, 167, 168
  Mars, 123, 152, 168
  Venus, 121, 123, 158
Griggs, D., 211, 215
Gringauz, K. I., 342, 344 *fig.*
Grossman, L., 61
Guarena meteorite, 381, 382 *fig.*
Guest, J. P., 191, 193, 194
Guili, R. T., 39
Gulleys, 150, 167, 168

## H

Hadley cell, 130, 131 *fig.*, 133–34
Haff, P. K., 147
Hafnium (Hf) isotope ratios, 108, 112
Hager, B. H., 110
Haggerty, S. E., 221
Half-life, 96
Halley, Edmund, 360, 403
Halley's comet, 360
  brightness, 369–70, 371 *fig.*
  exploratory missions, 403–4
  tail, 365, 367 *fig.*
Hamann, S. D., 260
Hamilton, P. J., 99, 100 *fig.*, 101
Hanel, R. A., 271
Hanson, G. N., 108–9
Hapke, B., 193
Hart, S. R., 108, 109
Hawaii, 197, 217
Hayashi, C., 38, 54, 62, 64, 65
Head, J. W., 143, 146, 191, 196, 197, 202
Heat flux, in solar nebula, 42, 43, 44

Heavy elements
  efficiency of accumulation, 69-70
  galactic concentrations, 378
  inert gases, 61
  synthesis, 16, 375-77
Heckman, G. R., 342
Hedge, C. E., 95, 99, 109
Heliocentric orbits, 3 *fig.*, 4, 5 *fig.*
Helium
  burned by red giants, 7
  giant planets, 259-60, 265
  Jupiter and Saturn, 234-36, 251
  nucleosynthesis, 17, 376, 377
  protosun, 259
Hellas (Mars), 194, 195
Helsey, C. E., 334
Hennecke, E. W., 108
Herbert, F., 340
Herbig, G., 32, 51
Hertzsprung-Russell (HR) diagram, 6-8
Hess, S. L., 212 *tab.*
Heterogeneous accretion model, 139, 157
Hevelius, Johannes, 360
Hide, R. H., 338
Hodges, R. R., 161
Hodograms, 323, 324 *fig.*
Hoffman, A. W., 109
Hoffman, J. H., 212 *tab.*
Holland, H. D., 165, 166, 167
Holmes, A., 99
Holzer, R. E., 339-40, 346
Homogeneous accretion model, 139, 140-41, 150, 157
Homogeneous convection, 185-86, 190 *fig.*
  Earth, 299
  Mars, 177, 195
Hood, L. L., 331, 333 *fig.*, 335, 336
Horedt, G. P., 51
Houck, J. R., 152
Houtermans, F. G., 99
Howard, H. T., 192
Howard, R., 291, 312, 315, 318 *fig.*, 319, 320

Hoyle, F., 38
Hsui, A. T., 143
Hubbard, W. B., 255, 260, 270, 271, 272, 305
Huebner, W. F., 145
Huguenin, R. L., 152, 217
Hunten, D. M., 65, 143, 147, 148, 166
Hurley, P. M., 105
Hurricanes, 18, 125, 247
Hydration weathering, 213
Hydrogen (H)
  burned by main sequence stars, 6-7, 8
  giant planets, 255, 256, 258, 259, 265
  metallic, 256, 260
  mixing of ices with, 261
  solar conversion, 17
  spheres of, 255-59
Hydromagnetic dynamo, 266
Hydrostatic equilibrium
  Jupiter and Saturn, 239, 258
  solar nebula, 32, 34, 37, 39
Hyperion
  orbital resonance, 286-87
  volatile history, 145

## I

Iapetus, 286
Iben, I., 32
Ice grain region, in solar nebular, 39, 44, 46 *fig.*, 47, 48, 64, 65-66
Iceland, 197
*ICE* mission, 400
Ices
  giant planets, 260-61, 262, 263, 265
  planetary volatile history, 142
Igneous rock
  volatile interaction, 143
  weathering, 213-14, 215
Impact cratering
  and lunar magnetic field, 335
  and weathering, 208, 210, 221, 222

Inclination, of orbits, 2, 5 *fig.*
Infall stage, 38, 45-46, 55
Ingersoll, Andrew P., 244 *fig.*, 248 *fig.*
  "Jupiter and Saturn," 231-52, 401
Inner planets
  bombardment by comets, 371-72
  composition, 2, 10-14
  formation, 9-10
  *See also* Terrestrial planets
*International Cometary Exploration (ICE)* mission, 400
*International Solar Polar Mission (ISPM)*, 398, 400, 405, 406
*International Ultraviolet Explorer*, 352
Interplanetary dust
  from comets, 363, 364 *fig.*, 370-71
  mass, 2
Interplanetary gases, mass, 2
Interplanetary sector boundary, 365
Interstellar dust
  cohesion of, 56-59
  isotopic heterogeneity, 390
  relative mass, 379
  and star formation, 8
Interstellar medium
  D/H ratio, 385
  effect of stellar collapse, 8
Interstellar molecular clouds
  collapse and formation of solar system, 8-9, 29, 32, 45, 379-80, 389, 390-92, 393-94
  cometary samples, 360, 362
  star formation, 8, 31-32
  temperature, 31, 45
Intriligator, D. S., 345
"Introduction: The Sun, the Solar Nebula, and the Planetary

System" (Wasson and Kivelson), 1–14
Io
  atmosphere, 222
  bulk properties, 81 *tab.*
  color, 241
  exploratory missions, 80, 402
  ice-rock ratio, 88
  interior structure and evolution, 92
  magnetic field, 337, 350, 351
  selected properties, 413 *tab.*
  terrestrial characteristics, 79
  thermal relations with Jupiter, 140, 144 *fig.*, 146, 239
  tidal evolution, 79, 92, 285–86
  volatile content, 251
  volatile history, 136, 138, 140, 143, 144 *fig.*, 145, 146, 147–48, 161, 169
  volcanism, 88, 92, 169, 222
  weathering, 208, 214 *tab.*, 222–23
Iodine (I) radionuclides, 96, 98, 385 *tab.*, 389
I:Xe isotope ratios, 107–8
Ionization
  nebular material, 32, 38, 51
  outer planet satellite atmospheres, 147
Ionopause
  Mars, 345–46
  Venus, 322, 323 *fn.*
Ionosphere
  comets, 354, 364
  Venus, 315, 321–29, 336, 337, 345–46
Iron (Fe)
  ferromagnetism, 296–97
  giant planets, 261
  nucleosynthesis, 377
  oxidation weathering, 213
  Mars, 216, 218
  Moon, 221–22

role in planet formation, 9
stability in $CO_2/CO$, 200 *fig.*
terrestrial planets, 79
  Mars, 90
  Mercury, 12, 88
  Moon, 86, 88
*ISEE 3* mission, 400
Ishtar Terra (Venus), 91, 201
Isobars, 127, 128 *fig.*, 129
Isostacy, 184
  Moon, 191
  Venus, 200
Isostatic adjustment, 177, 183 *fig.*, 184–85, 190 *fig.*
Isotherm condensation model, 158–60
Isotopic composition. *See also* Radioactive nuclides
  meteorites, 50
  and planetary evolution, 95–112
  uniformity in solar system, 383
  weathering effects, 215
"Isotopic Constraints of Planetary Evolution" (DePaolo), 95–112
*ISPM* mission, 398, 400, 405, 406

## J

Jackson, D. J., 339, 340
Jackson, J. D., 294
Jacoby, W. R., 188
Japan, 400, 404
Japp, J. M., 217
Jeans, J. H., 30, 390
Jeffreys, H., 30, 390
Jet streams
  Jupiter and Saturn, 242–46, 247, 249
  terrestrial planets, 125, 131
Johansen, L., 154
Johnson, T. V., 140, 149
Johnston, D. H., 186
Johnston, J. H., 212
Judson, S., 154

Jupiter, 231–52, 263 *tab.*
  atmosphere
    dynamics, 238, 242–46, 252
    structure, 237–38, 239–41
    temperature, 266
  bulk physical properties, 231
  colors, 241
  composition, 2, 10, 234–36, 238–41, 260–61, 262–63
    D/H ratio, 385 *tab.*
    helium, 259–60, 265
    hydrogen, 255, 256, 258, 259
    methane, 265
  effect on Mars, 90
  exploratory missions, 398, 401–3, 405, 409
  formation, 54, 65, 67, 70, 88, 270, 271, 272
  gravitational moments, 264
  gravity effects
    on asteroids, 9, 11, 12 *fig.*
    on comets, 6, 369
  Great Red Spot (GRS), 234, 235 *fig.*, 241, 242, 243 *fig.*, 247–49
  heat budgets, 236–38, 249–52
  history of observation, 234
  interior structure, 349 *fig.*
  magnetic field, 266, 290, 314, 337, 340, 346–48
    decay time, 312
    reconnection with Io, 350
  mass, 10, 231, 234
  moment of inertia, 258
  orbital properties, 4, 5 *tab.*, 11, 68, 411 *tab.*
  planetary model, 266–69
  satellites. *See also* Galilean satellites *and main entries for individual satellites*
    exploratory missions, 401, 402

magnetic field, 350, 351
tidal evolution, 285–86, 287
transfer of angular momentum, 92
volatile history, 145, 146
and solar wind, 363
temperature, 236–38
and bulk properties, 256–57
vertical structure, 239
"Jupiter and Saturn" (Ingersoll), 231–52
*Jupiter Orbiter Probe.* See *Galileo* mission

## K

Kaiser, M. L., 348, 349
Kant, Immanuel, 30, 31, 389
Kant-Laplace nebular hypothesis, 30, 31
Kaula, William M., 30, 60, 61, 88, 91, 99, 100, 200, 204
  "Interiors of the Terrestrial Planets: Their Structure and Evolution," 13, 79–92, 95, 107, 181, 190, 193, 264, 334, 401
Keil, K., 217
Keller, H. U., 365
Kellman, S. A., 39
Kelvin, William Thompson, Lord, 179
Kelvin-Helmholtz time scale, 392
Kennedy, G. C., 165
Kepler, Johannes, 360
*Kepler* mission, 407
Kerridge, J. F., 224
Khodakovsky, I. L., 220
Kieffer, H. H., 152
Kinematical dynamo theory, 298
Kinematic method, 313
Kink instability, 328
Kirkwood gaps, 12 *fig.*
Kivelson, Margaret G., 170, 337, 350, 351

"Atmospheres of the Terrestrial Planets," 90, 117–34, 152, 266, 300
"Future Planetary Exploration," 203, 340, 351, 397–410
"Introduction: The Sun, the Solar Nebula, and the Planetary System," 2–14, 32, 61, 69, 84, 88, 164, 283, 368, 397
Klinger, J., 223
Kneading processes, 137–38
Knudsen, W. C., 341
Kobrick, M., 30
Kohoutek comet, 365, 368 *fig.*
Komesaroff, M. M., 346
Kopp, R. A., 320, 321 *fig.*
Kramers, J. D., 99
Krause, F., 303
Kreimendahl, F. A., 220
Kröner, A., 199
Kuhi, L. V., 32, 33, 51
Kumar, S., 147, 148, 161
Kummel, B., 198
Kusaka, T., 34, 54, 57, 63
Kyser, T. K., 108

## L

Labonte, B. J., 312, 315, 318 *fig.*, 319
Lachenbruch, A. H., 188
Lada, C. J., 31
Lambeck, K., 90
Lambert, R. St. J., 105
Lange, M. A., 164
Lanzerotti, L. J., 147
Laplace, P. S. de, 30, 31, 389
Laplace sphere of influence, 63–64
Larimer, J. W., 61
Larson, R., 32
Lattimer, J., 393 *fig.*
Lauer, H. V., 217
Layered convection, 187–88, 190 *fig.*
  Earth, 197, 198, 199
  Venus, 177, 202

Lead (Pb) isotope ratios, 98–101, 108, 112, 381
Lebofsky, L. A., 149
Lee, Typhoon, 50, 143, 149, 163, 386 *fig.*
  "From Big Bang to Bing Bang," 2, 6, 10, 13, 16, 32, 50, 375–94, 397
Legendre polynomial, 264
Leoville meteorite, 387
Leovy, C. B., 117
Le Pichon, X., 197
Levy, Eugene, H., 303, 307, 308
  "Generation of Magnetic Fields in Planets," 14, 38, 266, 290–309, 313, 314, 336, 338
Levy, S., 33
Lewis, J. S., 34, 49, 60, 61, 137, 140, 141 *fig.*, 142, 149, 157, 220, 221
Lichtenstein, B. R., 330, 331 *fig.*
Lin, D. N. C., 39, 42, 45, 47, 49, 53, 54, 62, 63, 64, 65, 71
  "Nebular Origin of the Solar System," 6, 16, 29–72, 85, 138, 389, 390, 392
Lin, R. P., 331
Line of nodes, 5 *fig.*
Linton, M., 260
Lithosphere, 177, 181–82
  convection effects, 186–89
  delamination, 188–89
  despinning effects, 182–83
  Earth, 83, 84, 85, 197–200
  isostatic adjustment, 184–85
  Mars, 90, 177, 194–97
  Mercury, 193–94
  Moon, 86, 190–92
  nonuniform contraction, 182
  thickness of, and tectonic activity, 179

Lithosphere *(continued)*
  Venus, 91, 177, 200–201
Lithostatic state, 182
Lofgren, G. E., 165
Lorentz force, 298–99
Loughnan, F. C., 215
Lugmair, G. W., 96 *tab.*, 102
Luhmann, J. G., 322, 341
Luminosity
  solar nebula, 71
  stars, 6, 7
  Sun, 17
Lüst, R., 36
Lutetium (Lu) isotope ratios, 106, 109, 111
Lynden-Bell, D., 36, 38, 54
Lyttleton, R. A., 63

# M

McCord, T. B., 140, 146, 149, 152
McCulloch, M. T., 105
McCullough, P. H., 346
McDonald, G. J. F., 198
McElhinny, M. W., 80
McElroy, M. B., 156, 157
MacFarlane, J. J., 260, 270, 272
McGill, G. E., 192
McKenzie, D. P., 110
McKinnon, W. B., 191, 194
McLaughlin, R. J. W., 224
Magma oceans, 136, 165, 168
  Earth, 165
  Mercury, 193
  Moon, 190
Magmatism, Earth, 106–7
Magnesium, 9, 261
  Al:Mg ratios, 387
Magnetic fields, 290–309
  behavior in nature, 292–96
  classification, 314
  decay, 312
  Earth. *See* Earth: magnetic field
  equations, 312–14
  and fluid motions, 299–300
  galactic, 32, 312
  giant planets, 266, 305

Jupiter, 266, 290, 312, 314, 337, 340, 346–48, 350
large moons, 337, 350–51
lines of force, 292–96, 297
Mars, 305, 342–46, 352
Mercury, 290, 305, 314, 338–40, 351
Moon, 314, 329–36, 337, 338
Neptune, 266, 337, 350, 352
persistence, 296–97
Pluto, 350
reconnection, 350
regeneration, 300–305
Saturn, 266, 290, 314, 337, 340, 348–49, 352
shell vs. core dynamos, 314
stars, 291
Sun. *See* Solar magnetic field
Uranus, 266, 337, 350, 351–52
Venus, 305, 340–42
  ionosphere, 314, 315, 321–29, 336, 337, 345–46
Magnetic flux, 331
Magnetic moments, 314
  Earth, 337
  Jupiter, 347
  Mars, 342, 344, 345
  Mercury, 340
  Saturn, 348–49
  Venus, 340–41
Magnetic Reynolds number, 295–96, 303, 313, 336
Magnetohydrodynamic dynamos, 297–308
Magnetopause, 339
  Mars, 342
  Mercury, 338
Magnetosheath, 339
  Venus, 322, 323 *fn.*
Magnetosphere, 339
  Mars, 342
  Mercury, 338–39
Magnetotail, 339
  Mars, 343–44

Mercury, 338
Major axis, of ellipse, 4
Malin, M. C., 150, 216
Manuel, O. K., 108
Mare Imbrium (Moon), 102, 191
*Mariner* missions, 152, 178
*Mariner 2*, 340
*Mariner 4*, 342, 343 *fig.*
*Mariner 5*, 340
*Mariner 10*, 192, 338
Marov, M. Ya., 212 *tab.*
Mars
  albedo, 120
  atmosphere, 117, 118, 119, 120, 121, 123, 124, 126, 127, 130, 131, 216
  circulation, 131
  motive forces, 126, 127
  near surface properties, 212 *tab.*
  atmospheric pressure and axial inclination, 150–52
  kneading process, 137–38
  at surface, 117
  bulk properties, 81 *tab.*
  composition, 2
  degassing, 155–57
  density, 12, 88, 140, 350
  exploratory missions, 400 *tab.*, 401 *tab.*, 407, 408 *tab.*
  gravity influence on asteroids, 11, 12 *fig.*
  interior structure and evolution, 79, 90
  lithosphere, 90, 177, 194–97
  magnetic field, 305, 342–46, 350
  orbital properties, 4, 5 *tab.*, 118, 119 *tab.*, 411 *tab.*
  solar constant, 120
  tectonics, 177, 190 *fig.*, 194–97, 201, 203
  temperature
    altitudinal variation, 123, 124
    effective, 121, 123

latitudinal variation, 130
at surface, 117, 118 *tab.*
tidal evolution of satellites, 285
volatile content, 89
volatile history, 136, 137-38, 140, 142, 150-57, 158, 165, 167, 168, 169
weathering, 208, 210, 214 *tab.*, 216-18, 224, 225
Marsden, B. G., 370
*Mars Geoscience/Climatology Orbiter*, 407
*Mars* missions
*Mars 2*, 342, 343-44, 345 *fig.*
*Mars 3*, 342-43, 344
*Mars 5*, 344
Mascons, 191
Mason, B., 99
Mass flux, in solar nebula, 29, 44, 45, 49, 51, 63
Masursky, H., 200, 201
Matson, D. L., 147, 222
Mauna Kea, Hawaii, 217
Maunder butterfly diagram, 317 *fig.*
Maunder Minimum, 24, 315, 317 *fig.*
Maxwell's equations, 294
Maxwell's laws, 312
Melosh, H. J., 182, 191, 194
Melting, 13-14, 183-84
Mendeleev crater (Moon), 190
Mercury
atmosphere, 117, 208
bulk properties, 81 *tab.*
composition, 2
density, 12-13, 88, 140, 160-61
exploratory missions, 401 *tab.*, 409
interior structure and evolution, 79, 89, 90
magnetic field, 290, 305, 314, 338-40, 351
orbital properties, 4, 5 *tab.*, 16, 283, 411 *tab.*

tectonics, 177, 190 *fig.*, 192-94, 203
tidal evolution, 283-84
volatile content, 89
volatile history, 138, 140, 160-61
weathering, 208, 214 *tab.*, 221-22
Meridional circulation, 38
Earth, 130-31
Venus, 133-34
Mestel, L., 31, 32, 38
Metal grain region, 46 *fig.*, 47
Metallic hydrogen, 256, 260
Meteorites, 10-13. *See also* Impact cratering
classification, 11
composition, 11, 50, 383, 385
differentiation, 11
melting due to impact, 183-84
radioactive dating, 381-82
source, 10
Meteoroids, 10
Meteor stream, 363
Methane ($CH_4$)
conditions for synthesis, 166-67
Jupiter and Saturn, 260, 265
surfaces, 145
Titan weathering, 223
Mihalov, J. D., 329
Milky Way Galaxy
age, 2
composition, 2
compression of interstellar cloud by, 32
formation and evolution, 375
magnetic field, 292
Mimas, 286
Mineral riving, 208, 212, 214 *tab.* *See also* Salt riving
Minimum-mass nebula, 33, 37, 47, 51, 57, 62, 64
Mitchell, G. F., 362
Mixing length model, 42
Mizuno, H., 272
Modified Bode's law, 4-5

Modon theory, 249
Moffatt, H. K., 303, 314
Moment-of-inertia ratio, 80, 81 *fig.*, *tab.*
Mars, 90
Moon, 86
Mons Olympus (Mars), 90
Moon
atmosphere, 117, 208
bulk properties, 81 *tab.*
composition, 86
crust formation, 95, 101-3
density, 86, 140, 161, 350
differentiation, 95, 103
exploratory missions, 398, 399, 400 *tab.*, 401 *tab.*, 408 *tab.*
formation, 14
gravity and stress, 87
interior
evolution, 87-88, 95, 99, 101-3, 110, 111, 112
structure, 86-87
Io compared, 92
magnetic field, 314, 329-36, 337, 338
mantle, 110
radius, 3
rock samples
age, 381
magnetism, 334
oxygen isotope ratios, 383, 384 *fig.*
selected properties, 412 *tab.*
surface features, 86-87
tectonics, 177, 189-92, 193, 203
temperature variations, 211
terrestrial characteristics, 79
tidal distortion caused by, 276, 279-80
tidal evolution, 284-85
volatile content, 89
volatile history, 136, 138, 140, 160-61, 163, 164
weathering, 208, 211, 214 *tab.*, 221-22
Moonquakes, 192

Moorbath, S., 99, 104, 106
Moore, C. B., 99
Moore, H. J., 195, 216
Moos, H. W., 352
Moosman, A., 45
Morris, D., 346
Morris, E. C., 217
Morris, R. V., 217
Mouginis-Mark, P. J., 154
Moulton, Forest Ray, 390
Mouschovias, T. Ch., 32
Moving-flame mechanism, 133
Mrkos comet, 365, 367 fig.
Muehlburger, W. R., 105
Mullen, M., 167
Mullen, P. R., 204
Munk, W. H., 198
Murchison meteorites, 387
Murray, B. C., 192, 221
Mussett, A. E., 80

# N

Nagata, T., 334
Nahon, D., 224
Nakagawa, Y., 57, 64
Nakamura, Y., 334
Nash, D. B., 222
National Aeronautics and Space Administration (NASA), 401, 405, 407, 408 tab., 409
Naughton, J. J., 221, 222
Nebular gases
  capture by outer planets, 10
  fractionation, 13
  pressure gradients, 32, 34, 38, 39, 46 fig., 54-55
  removal by solar wind, 392
"Nebular Origin of the Solar System" (Lin), 29-72
Nebular theory, 29-72, 389-94. See also Solar nebula
Nelson, Bruce K., 105
Nelson, R. M., 222
Neodymium (Nd) isotope ratios, 97, 101, 108, 109 fig., 110, 111 fig., 112
Neon, on giant planets, 260
Neptune
  atmospheric temperature, 266
  composition, 2, 3, 10, 255, 256, 261, 263
  density, 10
  exploratory missions, 399, 401 tab., 409
  formation, 67, 272
  gravitational moments, 264
  interior structure, 351 fig.
  magnetic field, 266, 337, 350, 351-52
  orbital properties, 5 tab., 68-69, 413 tab.
  planetary model, 267, 270
  properties, 263 tab.
  tidal evolution, 287
Ness, N. F., 337, 338, 339, 340, 347, 349, 350
Neugebauer, Marcia, "Comets," 5, 6, 265, 360-72, 403
Neutrinos, 17-18
Neutron stars, 377
Newburn, R. L., 361
Newkirk, J. G., 317 fig.
Newton, Isaac, 30, 234, 360
Ng, K. H., 339
Niedner, M. B., 365
Nitrogen, on giant planets, 260
Nonuniform contraction, 177, 182, 183 fig., 190 fig.
  Earth, 198
  Mars, 197
  Mercury, 194
Normal faulting, 181, 182-83
Normal stresses, 180
Nozette, S., 221
Nucleosynthesis, during big bang, 376-77, 383
Nyquist, L. A., 101

# O

Oberon, 145
O'Connell, R. J., 80, 110
O'Connor, J. T., 217
Ohm's law, 312
O'Keefe, J. D., 184
Olivine, 181
Ollier, C. D., 211, 212, 215
Olympus Mons (Mars), 195, 196
$\omega$-effect, 313
Ong, R. S. B., 365
O'Nions, R. K., 109
Oort cloud, 265, 360, 368, 369, 372
Opacity, of solar nebula, 39-40, 51, 62, 70
Optical thickness
  solar nebula, 32, 39, 62, 67
  terrestrial planet atmospheres, 123
Orbital motion, conversion of angular momentum to, 33
Orbits
  of asteroids, 11
  of comets, 6, 360, 361, 368-71
  of planets, 2, 3 fig., 4-5, 411 tab.
    effect on terrestrial atmospheres, 118-19
    inclination, 2, 5 fig.
    and solar system formation, 383
    and tidal evolution, 276-88
Orientale Basin (Moon), 191
"Origin, Evolution, and Structure of the Giant Planets" (Stevenson), 255-73
Orion Nebula, 8
Orrall, R. Q., 320
Outer planets. See also Giant planets
  composition, 2, 10
  exploratory missions, 398, 400 tab.
  formation, 9-10

satellite volatile history, 138, 140, 145–49
Owen, T., 150, 155, 156, 157, 163, 167, 212 *tab.*, 265
Oxidation
 volatile history, 165–66
 weathering effects, 213
 Mars, 217
 Venus, 220
Oxygen
 chemical weathering, 212–13
 giant planets, 260, 261
 isotope ratios, 383–85
 nucleosynthesis, 377
Oyama, V. I., 212 *tab.*

## P

Palagonite, 153
Paleologou, E. V., 32
Palladium (Pd) radionuclides, 385 *tab.*, 389, 392, 394
Papaloizou, John, 39, 42, 53, 54, 62, 63, 64, 65, 71, 72
Papanastassiou, D. A., 101, 102
Parent nuclides, 96, 98
Parker, E. N., 291, 294, 296 *fn.*, 303, 307, 308, 313, 314, 320
Parmentier, E. M., 143, 146
Particle flux, 331
Patchett, P. J., 96 *tab.*, 106, 109
Patterson, C. C., 99
Pavonis Mons (Mars), 195
Peacock, S., 204
Peale, S. J., 88, 148, 283, 284, 285, 286, 350
 "Consequences of Tidal Evolution," 51, 92, 147, 183, 276–88, 350
Pearl, J., 148, 222
Periapsis, 342
Pericenter, 4
Perigee, 4
Perihelion, 3 *fig.*, 4

Perilune, 4
Permanent magnetism, 296–97
Peterman, Z. E., 95
Pettingill, G. H., 283
Phillips, R. J., 88, 90, 91, 159, 160, 200, 201
Phobos, 285, 287
Phoebe, 145
Photons, solar, 18
Photosphere, solar, 18–19, 20 *fig.*, 24, 315–20
Photosynthesis, 166, 168
Physical weathering, 208, 211–12, 214 *tab.*, 215
Piddington, J. H., 312, 319, 320 *fig.*
Pieri, D., 150
Pilcher, C. B., 144 *fig.*, 146
Pine, M. R., 37, 38
*Pioneer* missions, 236, 252
*Pioneer 10*, 234, 346–47, 401
*Pioneer 11*, 234, 264, 346, 347, 348, 350, 401
*Pioneer Venus*, 91, 218, 220 *fig.*, 340–41, 342
*Pioneer Venus Orbiter (PVD)*, 132, 133, 321, 322, 323 *fig.*, 404, 406
*Pioneer Venus Probe*, 156, 158, 160
Plages, 20, 21 *fig.*, 24
Planetary densities, 11–13
 deriving bulk composition from, 11–12
 moment-of-inertia ratio, 80
 reduced, 81 *tfn.*
 of terrestrial planets, 79, 81 *tab.*
Planetary energetics, and volatile history, 136, 142–43
Planetary evolution. *See also* Protoplanet formation
 isotopic constraints, 95–112

weathering effects, 209, 224–25
Planetary exploration, 397–410
 future missions, 397, 399–410
 past missions, 397–98
 phases, 398–99
Planetary models, giant planets, 263–66
"Planetary Surface Weathering" (Gooding), 208–25
Planetary volatile history, 138–70
 asteroids, 149–50
 diversity, 138
 and energy history, 142–43
 fate of degassed volatiles, 143–44
 issues, 136–37
 Mars, 150–57
 outer solar system, 145–49
 at planetary formation, 138–42
 processes, 136, 137
 Venus, 157–60
 and weathering processes, 208, 211, 224–25
"Planetary Volatile History: Principles and Practice" (Fanale), 138–70
Planetesimals, 56
 comets as, 360, 371
 formation, 9, 10, 56–57
 and giant planet formation, 272
 growth of, 59–61, 62–64
 heavy element conversion to, 69
 Laplace sphere of influence, 63–64
 melting due to, 13
Planitae, 194
Plate-tectonic convection, 187 *fig.*, 188, 190 *fig.*
 Earth, 177, 197, 198, 199
Pleiades, 8
Plescia, J. B., 196, 197

Plume convection, 187 *fig.*, 189
  Earth, 197
  Mars, 196
  Venus, 200
Pluto
  capture by Neptune, 68–69
  composition, 3
  exploratory missions, 401 *tab.*, 409
  magnetic field, 350
  mass, 3
  orbital elements and physical properties, 411 *tab.*
  tidal evolution of satellite, 287–88
  volatile history, 145
Plutonium (Pu) radionuclides, 385 *tab.*, 389, 392
Pneumann, G. W., 320, 321 *fig.*
Pollack, J. B., 63, 117, 143, 146, 147, 148, 150, 152, 156, 157, 158, 159, 160, 166, 167
Poloidal fields, 301–2, 303, 306–7, 313
Potassium (K)
  decay product and half-life, 96 *tab.*
  K:Ar isotope ratios, 108, 109
  vapor weathering of Moon, 221
Potential vorticity, 247
Preplanetary nebula, and planetary volatile history, 136, 137, 138, 140, 142, 157, 168. *See also* Solar nebula
Presolar evolution, 376–80
Pressure force, in terrestrial atmospheres, 127–29, 133
Pressure gradients
  solar nebula, 32, 34, 38, 39, 46 *fig.*, 54–55
  terrestrial atmospheres, 117–18, 127, 128, 133

Primordial solar nebula, 29, 272. *See also* Solar nebula
*Principia* (Newton), 30
*Principia Philosophia* (Descartes), 29
Pringle, J. E., 36, 38
Probstein, R. F., 362
Prominences, solar, 21, 26
Proton-proton reaction, 17–18
Protoplanet formation, 54–70
  accumulation in a gas-rich environment scenario, 62–67
  gas-free accumulation scenario, 54, 56–62
  massive gravitational instability scenario, 45, 54–56
  nebular structure at final stages, 67–70
  orbital evolution, 68–69
  phenomenological models, 33–34
  tidal effects on nebular gas, 51–54
Protostar formation, 32, 70
Protosun
  formation, 29, 45, 392
  gravitational force, 34
  helium fraction, 259
  in infall stage, 45, 46, 55
Pulsars, magnetic field, 291, 292
Pyroxine, 181

## Q

Quartz, 181

## R

Radiation. *See also* Solar radiation; Thermal radiation
  from protostars, 32
  transfer through Sun, 18–19
Radioactive nuclides
  dating via isotopic tracers, 95–99, 378, 380, 381 *fig.*, 383
  extinct, 385–89, 390, 392–94
  heat energy from, 13, 143, 168, 304–5
  volatile source, 161, 163–64
Radler, K.-H., 303
Rand, R., 105
Rare elements
  planetary volatile history, 136–37, 156, 159 *fig.*, 160
  production of, 377
Rasool, S. I., 158
Red giants, 7
Ree, F. H., 260
Reeves, H., 385
Refractory elements, 9
  giant planet rock, 261
  meteorite inclusions, 386–87, 392–93
Regolith, 209 *tab.*, 210
  incorporation of volatiles, 143
  Mars, 151, 152, 153–55
Reinhard, R., 369 *fig.*
Retrograde satellites, 287
Reynolds, R. T., 146
Reynolds number, 37, 42, 53, 63, 65–66, 67. *See also* Magnetic Reynolds number
Rhea, 145
Richter, F. M., 110
Rilles
  Mars, 196–97
  Moon, 191
Ringwood, A. E., 84, 88, 106
Rison, W., 108
Rizzi, J. D., 342
Roberts, J. A., 346
Rock
  giant planet composition, 261, 262
  isotopic dating, 97
  mechanics, 179–81
  Moon, 86, 334, 381, 383, 384 *fig.*
  terrestrial planets, 79
  weathering, 208–25
Rosenstock, R. W., 197
Ross, M., 256, 261
Rossbacher, L. A., 154

Rotation
  and Coriolis force, 125–26
  despinning tectonic model, 182–83
  effect on terrestrial atmospheres, 118–19
  and magnetic field, 299
  stars, 16
  sun, 16, 19, 30, 31, 383, 389–90
  and tidal distortion, 276–88
Rubey, W. W., 106, 107, 137, 162, 163, 164, 167, 224
Rubidium (Rb)
  decay product and half-life, 96, 97–98
  Rb:Sr isotope ratios, 101, 102 *fig.*, 109, 380–81, 382 *fig.*, 387
Runaway greenhouse effect, 158
Runcorn, S. K., 334–35, 336
Russell, C. T., 312, 314, 321, 322–23, 324, 325 *fig.*, 326 *fig.*, 327 *fig.*, 328, 330, 331 *fig.*, 332 *fig.*, 334, 337, 338 *fig.*, 340, 341, 342, 344, 345, 347, 351, 364
  "Solar and Planetary Magnetic Fields," 14, 23, 24, 38, 161, 312–52, 363, 407
Russell, H. N., 30

## S

Safronov, V. S., 34, 37, 54, 55, 56, 57, 59, 60, 61, 164
Sagan, C., 167
*Sakigake* mission, 401 *tab.*, 404
Salpeter, E. E., 259, 261, 266
Salt riving
  asteroids, 224
  Io, 222
  Mars, 216

Titan, 223
Samarium (Sm)
  decay product and half-life, 96 *tab.*, 97
  Sm:Nd isotope ratios, 101, 105, 106, 109, 111
Sandford, S. A., 223
Sandwell, D. G., 204
Sass, J. H., 188
Satellites (moons)
  magnetic fields, 337, 350–51
  retrograde, 287
  selected properties, 412 *tab.*
  tidal evolution, 282, 283
Saturn, 231–52, 263 *tab.*
  atmosphere
    dynamics, 242–46, 252
    structure, 237–38, 239–41
    temperature, 266
  bulk physical properties, 231
  colors, 240
  composition, 2, 10, 234–36, 238–41, 263
  D/H ratio, 385 *tab.*
  helium, 259–60, 265
  hydrogen, 255, 256, 258, 259
  methane, 265
  other elements, 260–61, 263
  effect on cometary orbits, 369
  exploratory missions, 401 *tab.*, 408 *tab.*, 409
  formation, 54, 68, 271, 272
  gravitational moments, 264
  heat budgets, 236–38, 249–52
  history of observation, 234
  interior structure, 349 *fig.*
  magnetic field, 266, 290, 314, 337, 340, 348–49, 352
  mass, 231

moment of inertia, 258, 260
  orbital properties, 5 *tab.*, 411 *tab.*
  planetary model, 268 *fig.*, 269–70
  satellites
    tidal evolution, 286–87
    volatile content, 251–52
    volatile history, 142, 145, 146, 147
  temperature, 236–38, 256
Saunders, R. S., 154, 196, 197
Scargle, J. D., 72
Schaber, G. G., 154
Schilling, J.-G., 109
Schmeling, H., 188
Schramm, D. N., 378, 383
Schubert, Gerald, 110, 117, 186, 189, 198
  "Atmospheres of the Terrestrial Planets," 90, 117–34, 152, 266, 300
Schwarzchild, M., 39
Sclater, J. G., 84
Scott, D. H., 195
Sediment, 209 *tab.*, 210
Sekiya, M., 65
Semimajor axis, 3 *fig.*, 4
  of belt asteroids, 11, 12 *fig.*
  Bode's law, 4, 5 *tab.*
Semiminor axis, 3 *fig.*, 4
Seward, T. M., 261
Sharp, R. P., 150
Shaw, D. M., 106
Shear stresses, 180, 181
Shell dynamos, 314–37
  Moon, 314, 329–36, 337
  Sun, 314, 315–20, 336
  Venus ionosphere, 314–15, 321–29, 336
Shemansky, P. E., 161
Shoemaker, E. M., 265
Shu, F. H., 32
Sibson, R. H., 197
Silicon (Si)
  abundance ratios, 11
  giant planets, 261
  role in planet formation, 9

INDEX

431

Singer, R. B., 153, 216
Sjogren, W. L., 200
*Skylab*, 315
Slavin, J. A., 339–40, 345, 346
Smirnov, V. N., 343, 345 *fig.*
Smith, B. A., 146, 148
Smith, E. J., 340, 342, 343 *fig.*, 345, 346, 347, 348, 349
Smith, J. V., 106
Smoluchowski, R., 153, 352
Smythe, W. D., 147, 148
SNC (Shegottites-Nakhlites-Chassigny) meteorites, 156, 157
Snow, Ted, 380 *fig.*, 394
Soderblom, L. A., 153
Sodium weathering, 222
Soil, 209 *tab.*, 210
*Solar and Heliospheric Observatory* (*SOHO*) mission, 409
"Solar and Planetary Magnetic Fields" (Russell), 312–52
Solar constant, 19, 120
Solar corona, 23, 26–27, 315, 316 *fig.*, 322
Solar energy flux, absorption by terrestrial planets, 117, 118, 120, 121–23
Solar flares, 16, 24–26, 292
Solar magnetic field, 20–26, 27, 290–91, 314–20, 336
  age, 312
  flux bundles, 314–15
  generation, 291
  as local (shell) type, 314
  regeneration, 297
  reversal, 291, 297, 303, 312, 319–20, 336, 337
  and rotational speed, 31
Solar mass, 2
Solar nebula, 29–72
  chondrite formation, 11
  convective accretion disk model, 40–41, 43–44
  convective instability, 39–40
  density distribution, 34

detailed models, 42–54
effective transfer of angular momentum, 36–39
evolution, 34–41, 44–54
formation, 375, 391–92
heating of asteroids, 13
heavy element formation, 16
Laplace hypothesis, 30
mass, 33
  distribution, 33–34, 37
  during protoplanet formation, 54
  at final stage, 67
  in infall stage, 45
opacity
  ice grain effects, 44
  and radiative heat flux, 42
  and temperature distribution, 39
phenomenological models, 33–34
planet formation, 8–10, 54–70
temperature
  and bulk composition of terrestrial planets, 88
  distribution, 34, 39
  velocity gradient, 34–35, 36–37, 41
  vertical structure, 39, 42–44
Solar radiation
  absorption factors, 123–24, 129–30
  output, 19
  weathering effects, 217
Solar seismology, 19
Solar system
  age, 2, 378, 380–82
  deuterium distribution, 385
  early conceptual models, 29–31
  evolution and formation, 2, 6, 377–80
  bing bang theory, 375
  clues for, 382–89
  nebular theory, 29–72, 389–90
  role of comets, 371–72

time of, 380–82, 392
mass distribution, 2, 231
Solar System Exploration Committee (NASA), 409
Solar wind
  and comet ion tails, 363–65
  and coronal holes, 26–27
  Earth interface, 321
  effect on planetary magnetospheres, 339
  inner planets, 371
  Mars, 345–46
  Mercury, 339–40
  melting of asteroids by, 14
  Ne:Ar ratio, 158
  removal of angular momentum by, 31
  removal of nebular gas by, 392
  and solar magnetic field, 291
  types of interaction with, 363–64
  Venus interface, 321–23
  volatile source, 161
  weathering by, 222
Solitary-wave theory, 247
Solitons, 247
Solomon, S. C., 86, 99, 100, 191, 194, 196, 197, 201, 202, 204
Solution weathering, 213
Sonett, C. P., 329, 334
Soter, S., 335
Soviet Union (USSR), 329, 398, 400 *tab.*, 404, 406
Spicules, 20
Spitzer, L., 30, 32
Spohn, T., 110
Spreiter, J. R., 331, 342
Sputtering erosion, 147, 169
Squyres, S. W., 146, 149, 287
Stacey, J. S., 99
Star clusters, 8, 30
Stars
  age, 7–8
  classes, 6
  collapse, 8

conversion of gas to, 376, 377
destruction of deuterium, 385
evolution, 6-8, 375-76
formation, 8, 31-33
  phases, 32
  from supernovae, 375
magnetic fields, 291
mass and energy, 7
rotation, 16
Staudacher, T., 108
Stefan-Boltzmann law, 120-21
Stellar evolution theory, 377-78
Stellar wind, 51
Stephenson, A., 334, 340
Stettler, A., 102
Stevenson, David J., 92, 255, 260, 261, 264, 266, 269, 270, 271, 272, 305, 314, 336, 340, 341, 346, 349, 350, 351, 352
  "Origin, Evolution, and Structure of the Giant Planets," 255-73, 401
Stewart, G. R., 61
Stix, M., 320
Stone, R. G., 348
Strain, 179, 180
Strain rate, 179, 182
Strangway, D. W., 80, 334
Stress
  tectonic, 179, 180
  viscous, 35-36, 41
Strickland, E. L., 216
Strike-slip faulting, 181
  due to isostatic adjustment, 185
  due to plate-tectonic convection, 188
Strom, R. G., 192, 193, 194
Strontium (Sr) isotope ratios, 96, 97-98, 101, 102 *fig.*, 108, 112, 380-81, 382 *fig.*, 393
  Rb:Sr ratios, 380-81, 387
Sulfer
  Io weathering, 222
  Jupiter and Saturn, 241

Sun, S. S., 108-9
Sun, 16-27. *See also entries beginning* Solar
  age, 8, 16
  angular momentum, 16, 31
  chromosphere, 20
  composition, 16
  core, 17
  corona, 23, 26-27, 315, 316 *fig.*, 322
  D/H ratio, 385
  diffusion of nebular material toward, 29
  distance from
    and terrestrial bulk composition, 88-89
    and terrestrial surface temperatures, 117, 118
    and volatile content, 136, 140
  effect on comets, 369, 370
  energy production and transport, 17-19
  exploratory missions, 401 *tab.*, 405, 409
  formation, 2, 6, 29, 30. *See also* Protosun
  on HR diagram, 6
  lifetime, 8
  luminosity, 17
  magnetic activity. *See* Solar magnetic field; Sunspots
  mass, 17
    distribution, 18
    loss, 31
    percent of solar system, 2
  oblateness, 16-17
  photosphere, 18-19, 20 *fig.*, 24, 315-20
  and planetary revolution, 2, 3, *fig.*, 383, 389
  relative size, 16
  rotational properties, 16, 19, 30, 31, 383, 389-90
  surface, 19-23
  temperature
    core, 17
    corona, 23

    surface, 17
  tidal effects
    Earth, 276, 279
    Mercury, 283
  U:Pb isotope ratios, 99
"Sun, Inside and Out" (Zinn), 16-27
Sunspots, 19-25
  fibril connection, 20-21
  magnetic cycle, 23-24, 291, 315, 317 *fig.*, 319
  movement, 19
  preceding and following, 23, 24
  and solar flares, 24-25
  toroidal pairs, 301
  umbra and penumbra, 24
Supernovae, 8, 31, 50, 375, 376, 392-94
Superrotation, 131-33
Surface fines, 210
Surface volatile inventory, 142-43
Surkov, Yu. A., 218

T

Tatsumoto, M., 106, 109
Taylor, Geoffrey, 245 *fig.*
Taylor, L. A., 221, 222
Taylor, S. R., 105
Tectonic models, 181-89
Tectonics, 177-203
  data gathering methods, 178 *tab.*
  Earth, 197-200
  limitations of empirical methods, 177-79
  Mars, 194-97
  Mercury, 192-94
  Moon, 189-92
  rock mechanics, 179-81
  Venus, 200-203
"Tectonics of the Terrestrial Planets" (Bird), 177-203
Tera, F., 101, 102, 104 *fig.*
Terrae
  Moon, 86
  Venus, 177
Terraforming, 208

INDEX 433

Terrestrial planets
  atmosphere, 117–34, 212 tab.
  black-body radiation, 120–21
  bulk properties, 81 tab.
  density and size specifications, 79
  inert gases, 61
  interior, 79–92
  magnetic fields, 293, 337–46
  orbital evolution, 69
  protoplanet formation, 54, 55, 56–65, 66
  tectonics, 177–203
  weathering, 208–25
Terrill, Richard, 240 fig.
Tethys
  orbital resonance, 286, 287
  volatile history, 145
Tharsis Montes (Mars), 90, 177, 194, 195–97
Thermal expansion
  tectonics, 179, 180, 182
  and weathering, 211
Thermal instability, 70–71
Thermal-lag time, 143
Thermal radiation
  greenhouse effect, 121–23
  Jupiter and Saturn, 236–38
  latitude dependence, 129–30
  opacity of atmospheres to, 117
Thermal tides, 131, 284
Thermosphere, 123
Thomas-Fermi-Dirac model, 259, 261
Thompson, L., 108
Thorium (Th)
  decay product and half-life 96 tab.
  Th:Pb isotope ratios, 109
Thrust faulting, 180–81
  due to despinning, 182–83
  due to isostatic adjustment, 185
  due to nonuniform contraction, 182
Tidal evolution, 276–88
  consequences, 283–88

theory, 276–83
Tidal heating, 143, 168, 169
  Europa, 148, 149
  Io, 147–48
Tidal truncation process, in solar nebula, 51–54, 55–56, 67, 68, 70, 71–72
Titan
  atmosphere, 223
  density, 350
  exploratory missions, 401 tab., 409
  magnetic field, 337, 350–51
  orbital resonance, 286–87
  selected properties, 412 tab.
  volatile history, 145, 147
  weathering, 208, 214 tab., 223
Titania, 145
Titius-Bode rule, 4
Toksöz, M. N., 143, 186, 192
Tolstikhin, I. N., 163
Toomre, A., 54
Torbett, M., 352
Tornadoes, 126
Toroidal fields, 301, 306–7, 313
Toulmin, P., 152, 153
Townsend, A. A., 37
Trade winds, 130
Tremaine, S., 53, 60, 63, 68
Triton
  selected properties, 412 tab.
  tidal interaction with Neptune, 287
  volatile history, 145
Troposphere, 123
Trubitsyn, V. P., 255
Truran, J. W., 31, 50
T Tauri stars, 7 fig., 29, 32, 33, 65, 71
Tullis, J., 201
Turbulence, in solar nebula, 29, 37–38, 41, 42, 43, 57, 390
Turcotte, D. L., 86, 196, 197

Turekian, K. K., 139, 163
Tyler, G. L., 223, 350

## U

Ugolini, F. C., 215, 217
*Ulysses* mission, 405
United States, 397–98, 399, 400, 401 tab., 405, 406
Universe, age, 375, 376–78
Uranium (U)
  decay product and half-life, 96 tab., 98
  U:Pb isotope ratios, 98, 99–101, 102, 106, 111, 109, 381
Uranus
  composition, 2, 3, 10, 255, 256, 261, 262, 263
  D/H ratio, 385 tab.
  methane, 265
  density, 10
  exploratory missions, 399, 401 tab., 408 tab.
  formation, 272
  gravitational moment, 264
  interior structure, 351 fig.
  magnetic field, 266, 337, 350, 351–52
  orbital properties, 5 tab., 284, 413 tab.
  planetary model, 269 fig., 270
  properties, 263 tab.
  satellites
    volatile history, 142
    tidal evolution, 287
    volatile history, 145

## V

Vaisberg, O. L., 342
Valles Marineris (Mars), 90, 197
Van Hart, D. C., 216
Vastitas Borealis (Mars), 194–95
*Vega* mission, 404, 406

Velocity fields, 313
Velocity gradient, in solar nebula, 34–35, 36–37, 41
*Venera* missions, 200, 210, 218, 220 *fig.*
*Venera 4*, 340
*Venera 9*, 218, 340
*Venera 10*, 218, 220, 340
*Venera 13*, 218
*Venera 14*, 218, 219 *fig.*
*Venera 15*, 406
*Venera 16*, 406
Venus
  albedo, 120
  atmosphere, 117, 118, 119, 120, 121, 123, 124, 126, 128–29
    circulation, 131–34
    near surface properties, 212 *tab.*
  bulk properties, 81 *tab.*
  composition, 2
  density, 12, 88, 91, 140
  exploratory missions, 80, 400 *tab.*, 401 *tab.*, 404, 405–6, 408 *tab.*
  interior structure and evolution, 79, 80, 90–91
  interaction with solar wind, 363
  magnetic field, 305, 340–42
    ionospheric, 314, 315, 321–29, 336, 337, 340–42, 345–46
  orbital properties, 4, 5 *tab.*, 118, 119, 413 *tab.*
  rotation, 119, 284, 340
  solar constant, 120
  tectonics, 177, 200–203
  temperature
    altitudinal variation, 123, 124
    effective, 121
    latitudinal variation, 130
    at surface, 117, 118 *tab.*, 121, 123, 177
  tidal evolution, 284
  volatile content, 89
  volatile history, 136, 138, 140, 144 *fig.*, 155–56, 157–60, 168, 169
  weathering, 208, 210, 211, 214 *tab.*, 218–21, 224, 225
  wind patterns, 117, 118, 126, 132–34
*Venus Orbiting Imaging Radar* (*VOIR*) mission, 406
*Venus Radar Mapper* (*VRM*) mission, 406, 407
Verhoogen, J., 304
Vesta, 79, 157
  bulk properties, 81 *tab.*
  volatile history, 149, 163
Vesta Rupes (Venus), 201, 202
*Viking* missions, 142, 152–53, 156, 178, 194, 210, 346
*Viking 1*, 216
*Viking 2*, 216
Virial theorem, 256
Viscosity, 37, 70
Viscous couple, 35
Viscous frictional force, 35, 45
Viscous stage, 29, 46–51, 70
Viscous stress, 35–36, 41
*VOIR* mission, 406
Volatiles, 2
  and melting point, 184
  and planetary formation, 138–42
  response to freezing, 211
  *See also* Planetary volatile history
Volatile sinks, 136, 137
Volcanism, 177, 183–84, 190 *fig.*, 203
  Earth, 85, 163, 177, 184
  Io, 88, 92, 169, 222
  and isostatic adjustment, 185
  Mars, 90, 177, 194, 195–96, 197
  Mercury, 89, 193
  Moon, 102, 103, 221
  and plume convection, 189
  Venus, 91
  and weathering, 208, 221, 222
Volume addivity, 260
*Voyager* missions, 34, 146–47, 148, 222, 223, 236, 237, 239, 241, 242, 247, 249, 251, 252, 401, 402
*Voyager 1*, 222, 223, 231 *fig.*, 232 *fig.*, 234, 242, 286, 287, 346, 347, 348, 350
*Voyager 2*, 2, 222, 234, 241 *fig.*, 242, 286, 287, 346, 348, 350, 351, 352, 399, 402
VRM mission, 406, 407

## W

Walker, J. C. G., 117, 158
Wall, S. D., 216
Wallace, L., 265
Wallis, M. K., 342, 364, 365
Wänke, H., 150, 156, 157
Ward, W. R., 37, 57, 69, 150
Warner, J. L., 201
Warren, P. H., 86
Wasserberg, G. J., 99, 100 *fig.*, 101, 102, 104 *fig.*, 105, 109, 163, 382 *fig.*, 386 *fig.*
Wasson, John T., 61, 86
  "Introduction: The Sun, the Solar Nebula, and the Planetary System," 2–14, 32, 61, 69, 84, 88, 164, 283, 368, 397
Water ($H_2O$)
  Earth, 106–8, 112, 138, 142, 200
  Jupiter and Saturn, 260
  outer planets, 10
  planetary volatile history, 138
  weathering, 212–13
    asteroids, 223–24
    comets, 223
    Earth, 214–15
    Mars, 216, 218

Weathering, 208–25
  chemical type, 208, 212–14
  comets and asteroids, 223–24
  defined, 208, 209 *tab*., 210
  Earth, 214–16
  Io, 222–23
  Mars, 216–18
  methods of study, 210–11
  Moon and Mercury, 221–22
  physical type, 208, 211–12
  and regolith evolution, 208–9
  significance in planetary evolution, 224–25
  Titan, 223
  Venus, 218–20
Weertman, J., 181
Weertman, J. R., 181
Weidenschilling, S. J., 59, 62
Weinberg, S., 375
Wenner, D. B., 153
Wetherill, G. W., 54, 56, 59, 60, 61, 62, 63, 64, 65, 85, 100, 156, 157, 158, 164, 382 *fig*.
Whalley, W. B., 212

Whang, Y. C., 339, 340
Whipple, F. C., 57, 63, 145, 361, 362
Whipple's theory, 361
White dwarfs, 7 *fig*., 291, 377
Whiteoak, J. B., 346
Willeman, R. J., 196, 197
Williams, G. P., 246 *fig*.
Windley, B. F., 104, 106
Winds. *See also* Solar wind
  and Coriolis force, 126
  Earth, 130–31, 242–44, 245
  Jupiter and Saturn, 233 *fig*., 235 *fig*., 238
  stellar, 51
  Venus, 117, 118, 126, 132–34
Winge, C. R., 350
Wisdom, J., 286
Wiskerchen, M. J., 334
Wolfe, R. F., 265
Wood, J. A., 143
Woolfson, M. M., 30
Wyckoff, S., 145
Wyllie, P. J., 200

# X

Xenon (Xe) isotope ratios, 98, 107–8, 112, 160

# Y

Yang, J., 156
Yellowstone, plumes, 197
Yeroshenko, E. G., 340
Yoder, C. F., 286
York, D., 95
Yuen, D. A., 183, 189
Yund, R. A., 201
Yung, Y. L., 143, 150, 152, 159, 166, 167
Yuty crater (Mars), 154 *fig*.

# Z

Zapolosky, H. E., 259, 261
Zasoski, R. J., 215
Zellner, B., 149
Zharkov, V. N., 255
Ziglina, I. N., 60, 61
Zirin, H., "Sun, Inside and Out," 16–27, 51, 291, 315, 363
Zonal winds
  Earth, 130–31
  Venus, 132–33
Zoo syndrome, 137, 138, 162
Zvygina, Y. V., 61